Two Manuals Combined into One Book

Security Force Assistance

And

Law Enforcement Investigations

Department of Defense

Pentagon
Publishing

Foreword

We remain a nation at war in an era of persistent conflict, but we do not stand alone. Our nation has many multinational partners, equally committed to freedom, rule of law and stability. It is clear that we are stronger when we act with partners in today's operating environment. Therefore, security force assistance is no longer an "additional duty." It is now a core competency of our Army.

As distinctions separating war and peace blur and challenges to security increase, we must seek to help our multinational partners successfully confront their security challenges. Security force assistance builds our multinational partners' capability to defeat regular, irregular, and hybrid threats prevalent in an era of persistent conflict.

The two pillars of security force assistance are the modular brigade and Soldiers acting as advisors. Their tactical efforts build partner capability and improve the security situation. Over time, U.S. forces and partners foreign security forces collectively set the conditions to defeat common threats and ultimately achieve strategic success.

This manual addresses common characteristics and considerations for conducting security force assistance and clarifies what units and individual advisors must understand to work "by, with, and through" their counterparts. Since every situation and foreign security force is unique, units and individuals conducting security force assistance must carefully analyze the operational environment, especially the relationships of foreign security forces and their population. Conducting successful security force assistance requires adaptive units led by well-informed, culturally astute leaders. FM 3-07.1 provides commanders, staffs, and advisors with the common doctrine for achieving this goal.

Conducting foreign security assistance requires great skill in building relationships and "leading from behind." We must all increase our understanding of this important mission.

MARTIN E. DEMPSEY
General, U.S. Army
Commanding General
U.S. Army Training and Doctrine
Command

Field Manual
No. 3-07.1

Headquarters
Department of the Army
Washington, DC, 1 May 2009

Security Force Assistance

Contents

Figures

Preface

This field manual (FM) is the Army's doctrinal publication for security force assistance (SFA). It provides doctrinal guidance and direction for how U.S. forces contribute to SFA. It focuses on the brigade combat team (BCT) conducting SFA and advising foreign security forces. It is based on lessons learned from previous advising efforts and recent combat operations with a view to the future. It supports the Army Education System instruction on the theory and conduct of SFA.

The two primary audiences for this manual are leaders in BCTs conducting SFA and Soldiers assigned as advisors. The BCT is the largest fixed tactical organization and the key formation of the Army's modular organization. Army modularity allows commanders to add selective units to assist the BCT as it conducts SFA. The BCT and higher echelons provide the framework for advisors to function and accomplish the mission—developing foreign security forces and, when appropriate, working by, with, and through foreign security forces to achieve the desired end state of the host nation's internal defense and development. Thus, Soldiers assigned as advisors are the key element of SFA.

This FM provides the conceptual framework for conventional forces to conduct SFA within the construct of full spectrum operations, across the spectrum of conflict. It addresses SFA at operational and tactical levels.

Army doctrine is consistent and compatible with joint doctrine. FM 3-07.1 expands on fundamental principles for SFA expressed in FM 3-07. FM 3-07.1 uses text and concepts developed with North Atlantic Treaty Organization partners and ABCA (America, Britain, Canada, Australia, and New Zealand program armies).

Terms that have joint or Army definitions are italicized and the number of the proponent manual follows the definition. FM 3-07.1 will be the proponent manual for foreign security forces as a term.

FM 3-07.1 applies to the Active Army, Army National Guard/U.S. Army National Guard, and U.S. Army Reserve unless otherwise stated.

Headquarters, US Army Training and Doctrine Command is the proponent for this publication. The preparing agency is the Combined Arms Doctrine Directorate, US Army Combined Arms Center. Send written comments and recommendations on Department of the Army (DA) Form 2028 (Recommended Changes to Publications and Blank Forms) directly to: Commander, US Army Combined Arms Center and Fort Leavenworth, ATTN: ATZL-CD (FM 3-07.1), 201 Reynolds Avenue (Building 285), Fort Leavenworth, KS 66027-1352. Send comments and recommendations by e-mail to leav-cadd-web-cadd@conus.army.mil. Follow the DA Form 2028 format or submit an electronic DA Form 2028.

Introduction

Our strategy emphasizes building the capacities of a broad spectrum of partners as the basis for long-term security. We must also seek to strengthen the resiliency of the international system to deal with conflict when it occurs.

National Defense Strategy, June 2008

In an era of persistent conflict, the United States supports the internal defense and development of international partners, regardless of whether those partners are highly developed and stable or less developed and emerging. While many of these partners are nations, they can also include alliances, coalitions, and regional organizations. U.S. support to these partners ranges from providing humanitarian assistance to major combat operations. U.S. support includes conducting conflict transformation, bolstering partner legitimacy, and building partner capacity. A vital part of these three aspects of U.S. support is assisting partner security forces.

Security force assistance (SFA) is not new for Army forces. In fact, General George Washington's Inspector General of the Army acted as an advisor for Army forces. Baron Friedrick Wilhelm von Steuben instilled discipline and professionalism into an army that previously lacked formalized training. His 1779 *Regulations for the Order and Discipline of the Troops of the United States*, adapted from the Prussian army, formed the doctrinal backbone of the Continental Army throughout the Revolutionary War. Additionally, the lineage of the Army's operations field manual, FM 3-0, can be traced to this document. As a benefactor of advisors such as von Steuben, the Army has since undertaken what is called SFA several times throughout its history.

Army doctrine defines *security force assistance* as the unified action to generate, employ, and sustain local, host-nation, or regional security forces in support of a legitimate authority (FM 3-07). SFA is part of the FM 3-0 construct of full spectrum operations. Similarly, it is conducted across the spectrum of conflict or in any of the operational themes. SFA is normally part of a larger security sector reform effort, while in other instances, SFA is not tied to reform but to building partner capacity.

Three general situations exist in which SFA may occur: an internally focused bilateral relationship, an externally focused bilateral relationship, and a multilateral relationship. Consequently, SFA supports the appropriate partner's plans. When SFA supports a host nation, it also supports that host nation's strategy. If SFA supports a host nation's externally focused efforts, it must support the host nation's national security strategy. SFA may support regional security forces, such as those of the African Union or the Organization of American States. In these cases, SFA supports that organization's plans.

Organization

This FM primarily focuses on the brigade combat team and advisors. It also provides a wider doctrinal framework for SFA. It is organized in the following manner:

- **Chapter 1 focuses above the brigade combat team**. It provides the overall doctrinal context for SFA.
- **Chapters 2 through 6 focus on the brigade combat team level**. These chapters provide a framework for SFA, its considerations for the brigade combat team operations process, considerations for augmenting the modular brigade for security force assistance, its considerations for the unit employment, and its sustainment considerations.
- **Chapters 7 through 10 focus on the individual advisor**. These chapters discuss the advisor, advisor cultural and communication considerations, the advisor working with counterparts, and cross-cultural influencing and communication.
- The appendices discuss Army special operations forces imperatives, legal considerations, and media considerations.

Chapter 1

The Strategic Context

This chapter places security force assistance in strategic context. First, it explains security force assistance and its relationship to full spectrum operations, spectrum of conflict, and operational themes (see FM 3-0). A discussion of U.S. national strategy and defense policies that impact security force assistance follows. This chapter then examines operations and programs related to security force assistance. Finally, it discusses the relationship of security force assistance and related operations and programs.

SECURITY FORCE ASSISTANCE AND THE STRATEGIC CONTEXT

1-1. *Security force assistance* is the unified action to generate, employ, and sustain local, host-nation or regional security forces in support of a legitimate authority (FM 3-07). Security force assistance (SFA) improves the capability and capacity of host-nation or regional security organization's security forces. These forces are collectively referred to as foreign security forces. ***Foreign security forces* are forces—including but not limited to military, paramilitary, police, and intelligence forces; border police, coast guard, and customs officials; and prison guards and correctional personnel—that provide security for a host nation and its relevant population or support a regional security organization's mission.**

1-2. SFA occurs within the framework of full spectrum operations (see FM 3-0). In most situations involving this assistance, there is relatively little weight on offensive and defensive operations from a U.S. perspective. However, when U.S. forces accompany foreign security forces (FSF) in combat, the weight of offensive and defense operations will change to address the situation and align with the foreign security force's efforts. SFA is not just a stability operation, although it is a key contributor to the primary stability tasks of establish civil security and establish civil control.

1-3. SFA can be conducted across the spectrum of conflict, from stable peace to general war. This assistance could focus on improving the security forces of a host nation that is currently under no immediate threat, on paramilitary forces to counter an insurgency, or on advising FSF in major combat operations against an external threat. Figure 1-1 illustrates how SFA spans the spectrum of conflict.

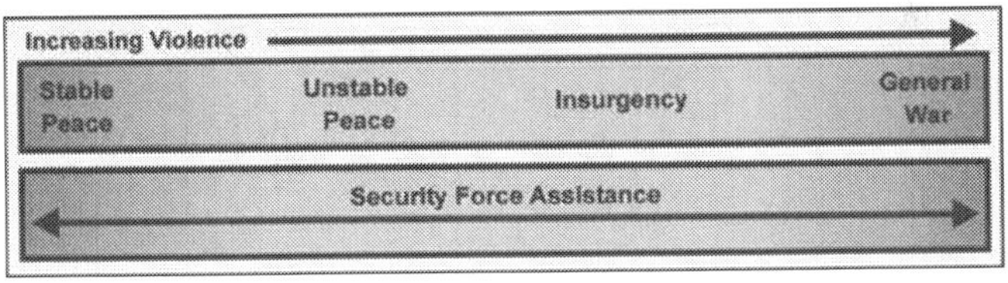

Figure 1-1. The spectrum of conflict and security force assistance

1-4. SFA can occur as part of any operational themes—peacetime limited engagement, limited intervention, peace operations, irregular warfare, and major combat operations. SFA often occurs during peacetime military engagement, peace operations, and irregular warfare. SFA involving paramilitary forces will often be limited to the irregular warfare operational theme, although paramilitary forces should eventually be absorbed, disbanded, or maintained in a long-term relationship with special operations forces.

1-5. While Army units are vital contributors, SFA is part of a comprehensive approach. This comprehensive approach should include close collaboration with military and civilian joint and multinational forces. The host-nation or regional security organization is the key actor within the comprehensive approach. It is vital that SFA is based on an objective, continuous assessment that examines the organization, training, equipment, rebuilding, and advising of the forces involved.

1-6. Law enforcement, military, intelligence, and border forces operate and cooperate within the security sector. U.S. forces understand how these units are intended to operate in the host-nation scheme, not necessarily using a U.S. model. U.S. forces should plan to help develop the respective capabilities so that these units can carry out their security functions.

1-7. Fully developed and institutionalized FSF plan, prepare, and execute at all the appropriate levels. Thus, U.S. forces must integrate SFA into the operations process at all levels. SFA should be designed to work seamlessly with the host-nation government at all levels, from ministries addressing the security sector to initial entry-level FSF training.

NATIONAL STRATEGY AND DEFENSE POLICIES

1-8. U.S. strategy and foreign policy affects all SFA. More importantly, SFA is critical to these policies as it is a primary tool for building partnership capacity with other nations. Key strategy documents for SFA include the National Security Strategy, the National Defense Strategy, the National Military Strategy, the Quadrennial Defense Review, and DODD 3000.05.

1-9. The National Security Strategy outlines the President's vision for providing enduring security for the American people in a volatile, uncertain, complex, and ambiguous strategic environment. The National Security Strategy of 2006 addresses stability operations within the broad engagement strategy for regional conflict. It supports SFA by "tailoring assistance and training of military forces to support civilian control of the military and military respect for human rights in a democratic society." This strategy also states that "[t]he world has found through bitter experience that success often depends on the early establishment of strong local institutions such as effective police forces and a functioning justice and penal system. This governance capacity is critical to establishing the rule of law and a free market economy, which provide long-term stability and prosperity."

1-10. The National Defense Strategy encompasses the Secretary of Defense's vision for the Department of Defense role in protecting the American people and U.S. interests. This strategy stems from the National Security Strategy. The 2008 National Defense Strategy highlights SFA. It states that "[o]ur strategy emphasizes building the capacities of a broad spectrum of partners as the basis for long-term security.... [l]ocal and regional conflicts in particular remain a serious and immediate problem...we will help build the internal capacities of countries at risk...[b]y helping others to police themselves and their regions, we will collectively address threats to the broader international system."

1-11. The National Military Strategy supports the National Security Strategy, National Defense Strategy, and Quadrennial Defense Review. The 2004 National Military Strategy states "[t]he Department will also work to strengthen alliances and partnerships by helping other nations increase their ability to defend themselves and protect common security interests." Working with other nations and strengthening partnerships appears throughout the National Military Strategy, which supports the importance of SFA.

1-12. The Quadrennial Defense Review is a congressionally mandated Department of Defense (DOD) review of strategy, programs, and resources. The 2006 Quadrennial Defense Review directed that joint ground forces "possess the ability to train, mentor and advise FSF and conduct counterinsurgency campaigns." This review also stated that conventional forces need to be capable of training, equipping, and advising local forces; deploying and engaging with partners; and supporting stability operations. Additionally, the review envisioned conventional ground forces continuing to perform more of the tasks traditionally performed by special operations forces. Conventional forces must understand foreign cultures and societies as well as be able to train, mentor, and advise FSF.

1-13. DODD 3000.05 provides guidance on stability operations. DODD 3000.05 establishes that stability operations are a core military mission and equal in importance to combat operations. DODD 3000.05 also reinforces the importance of SFA by stating "[t]he immediate goal often is to provide the local populace with security, restore essential services, and meet humanitarian needs. The long-term goal is to help develop host-nation capacity for securing essential services, a viable market economy, rule of law, democratic institutions, and a robust civil society."

RELATED OPERATIONS AND PROGRAMS

1-14. Several operations or programs directly or potentially relate to SFA. These operations and programs provide the context for SFA. They include security cooperation, security assistance, foreign internal defense, internal defense and development (IDAD), and security sector reform. For more information on operations related to SFA, see DOD 5105.38M, DOD security cooperation guidance, JP 3-07.1, FM 3-07, FM 3-05.137, and AR 11-31.

SECURITY COOPERATION

1-15. *Security cooperation* is all Department of Defense interactions with foreign defense establishments to build defense relationships that promote specific U.S. security interests, develop allied and friendly military capabilities for self-defense and multinational operations, and provide U.S. forces with peacetime and contingency access to a host nation (JP 3-07.1). Finally, security cooperation occurs across the spectrum of conflict and is not exclusively a peacetime activity. Security cooperation includes security assistance programs administered by the DOD as well as activities that enhance interoperability and the collective capability of combined forces (using Title 10 or exercise funding and authority). Security cooperation consists of a focused program of bilateral and multilateral defense activities conducted with foreign countries to serve mutual security interests and build defense partnerships. Security cooperation is governed by various sections of Title 10 and specific public laws addressing DOD interactions with other nations.

1-16. AR 11-31 establishes Department of the Army policy and prescribes responsibilities and procedures for the planning, integration, programming, budgeting, and execution of Army security cooperation activities. It is the policy of the U.S. Army to conduct security cooperation activities in compliance with higher-level guidance and in the execution of Army responsibilities under Title 10 and Title 22 which governs the transfer, exchange, conduct, and development of articles and services via a variety of U.S. Government programs.

1-17. Army security cooperation consists of official, cooperative, and noncombat interactions among any Army elements, Active Army, or Reserve Components. This includes any U.S. Government or nongovernment entity supporting the military and civilian joint and multinational organizations.

1-18. Commanders distinguish security cooperation from SFA based its emphasis in building relationships and capacities by using programmatic activities. Thus, security cooperation may support or be supported by SFA.

SECURITY ASSISTANCE

1-19. *Security assistance* is a group of programs authorized by the Foreign Assistance Act of 1961, as amended, and the Arms Export Control Act of 1976, as amended, or other related statutes by which the United States provides defense articles, military training, and other defense-related services by grant, loan, credit, or cash sales in furtherance of national policies and objectives (JP 3-57). Security assistance is a specific subset of security cooperation and may focus on external or internal threats.

1-20. DOD 5105.38-M describes the scope of security assistance programs in detail. Security assistance allows the transfer of military articles and services to friendly foreign governments. These transfers may be carried out via sales, grants, leases, or loans. If these transfers are essential to the security and economic well-being of allied governments and international organizations, they are equally vital to the security and economic well-being of the United States. U.S. forces can use security assistance to deliver defense weapon systems to foreign governments; to train international students; to advise other governments on improving internal defense capabilities; and to provide guidance and assistance in establishing infrastructures and economic bases for regional stability. Finally, security assistance cannot be conducted using Title 10 or exercise funds.

1-21. The military component of security assistance, implemented by DOD with policies established by the Department of State, has four principal components. They are international military education and training, foreign military sales, foreign military financing program, and peace operations. The foreign military financing program and international military education and training fall inside the military assistance budget process of the Department of State. Security assistance can also include funding peace operations.

International Military Education and Training

1-22. International military education and training contributes to internal and external security of a country by providing training to selected foreign militaries and related civilian personnel on a grant aid basis (see JP 3-07.1). These programs help to strengthen foreign militaries through training for the proper functioning of a civilian-controlled, apolitical, professional military. International military education and training serves as a foreign policy tool where the United States shapes doctrine; promotes self sufficiency in maintaining and operating U.S. acquired defense equipment; encourages the value of rule of law; and occasionally has a marked effect on the policies of the recipient governments. Foreign students, many of who occupy the middle and upper echelons of their country's military and political establishments, are taught U.S. doctrine and weapons systems employment resulting in greater cooperation and interoperability. All international military education and training activities are SFA.

Foreign Military Sales

1-23. Foreign military sales is a nonappropriated program through which foreign governments can purchase defense articles, services, and training from the United States (see JP 3-07.1). Eligible nations use this program to help build national security infrastructures. However, nations that require assistance are often unable to finance their needs. SFA supports foreign military sales when these sales include U.S. forces providing equipment and related training. It does not support foreign military sales when these sales provide equipment without related training.

Foreign Military Financing Program

1-24. The foreign military financing program provides funding to purchase defense articles and services, design and construction services, and training through foreign military sales or commercial channels (see JP 3-07.1). This program can assist nations with weak economies that would otherwise be unable to afford U.S. assistance. The foreign military financing program can fund foreign military sales. SFA and the foreign military financing program may be separate activities, although they can also be complementary.

FOREIGN INTERNAL DEFENSE

1-25. *Foreign internal defense* is the participation by civilian and military agencies of a government in any of the action programs taken by another government or other designated organization to free and protect its society from subversion, lawlessness, and insurgency (JP 3-05). U.S. foreign internal defense efforts involve all instruments of national power to support host-nation IDAD programs. The military instrument of foreign internal defense includes indirect support, direct support (not involving U.S. combat operations), and combat operations. Foreign internal defense can occur across the spectrum of conflict and within any operational theme. Foreign internal defense is often a part of irregular warfare or peacetime military engagement operational theme. SFA may support the military instrument of foreign internal defense. JP 3-07.1 discusses foreign internal defense in detail.

Note: It is a national policy decision for U.S. forces to accompany FSF on combat operations.

Military Instrument of Foreign Internal Defense

1-26. The military instrument of foreign internal defense supports the other instruments of national power through activities ranging from peacetime military engagement to major combat operations. Although foreign internal defense may include capacity building across the host nation, U.S. forces focus on combating internal threats to help the host nation maintain legitimacy and influence over the population. As such, foreign internal defense may be part of irregular warfare.

1-27. SFA and foreign internal defense support the host nation's IDAD. Consequently, forces conducting SFA must have proper mindset—they are part of a developmental and supporting effort. This applies even if SFA is part of working with a potential adversary or an existing enemy. SFA and foreign internal defense focus on working by, with, and through FSF and not conducting unilateral efforts to achieve objectives.

Indirect Support

1-28. Indirect support efforts are foreign internal defense operations that emphasize the principle of host-nation self-sufficiency. Indirect support focuses on building strong national infrastructures through economic and military capabilities that contribute to self-sufficiency. For example, indirect support could include U.S. assistance in improving a host-nation's intelligence infrastructure. Contributions from U.S. forces to indirect support come from security cooperation guidance through security assistance. They are supplemented by multinational exercises, exchange programs, and selected joint exercises. Indirect support that is focused on assisting security forces is SFA.

1-29. Although they indirectly assist those involved, military exchange programs are not SFA. Military exchange programs foster mutual understanding between forces; familiarize each force with the organization, administration, and operations of the other; and enhance cultural awareness. These exchange programs often have long-term implications for strengthening democratic ideals and respect for human rights among supported governments. However, such exchange programs (funded with Title 10 monies) do not themselves become vehicles for security assistance training, as they do not intend to assist security forces but to foster mutual understanding and enhance cultural awareness.

1-30. Joint and multinational exercises strengthen U.S. and host-nation relations and interoperability of forces, but they are not SFA. They complement security assistance and civil-military operations by validating host-nation needs and capabilities and by providing a vehicle for the conduct of humanitarian and civic assistance programs (see JP 3-07.1). Strict legal restrictions clarify the type of support that U.S. forces can provide and on the monetary limits of such support. These efforts are not SFA as they do not intend to directly assist security forces; instead, they intend to improve relations and interoperability. See appendix B.

Direct Support

1-31. Direct support does not involve combat operations but includes the use of U.S. forces providing direct assistance to the host-nation civilians or military. Direct support often focuses on civil-military operations (primarily, providing services to local civilians), psychological operations, intelligence and communications sharing, and logistic support. Sometimes direct support will be SFA, such as psychological operations, military training support, logistic support, and intelligence and communications sharing. SFA is not a civil affairs activity such as foreign humanitarian assistance, humanitarian and civic assistance, or military civic action.

1-32. Direct support differs from security assistance in that direct support is joint- or Service-funded. Direct support does not usually involve transferring arms and equipment and does not often include training local military forces. Direct support operations are normally conducted when the host nation has not attained or regained self-sufficiency and when it still faces social, economic, or military threats beyond its capability.

Combat Operations

1-33. The introduction of U.S. combat forces into foreign internal defense operations requires a Presidential decision. These operations are a temporary solution until host-nation forces can stabilize the situation and provide security for civilians. In all cases, U.S. combat operations support the host-nation IDAD program and remain strategically defensive in nature. When U.S. forces are advising, mentoring, partnering, or augmenting host-nation forces in foreign internal defense combat operations, this is SFA. When Army units conduct independent combat operations alongside FSF, this is not SFA.

1-34. U.S. tactical participation in host-nation internal conflicts requires judicious and prudent rules of engagement and guidelines for applying force. Inappropriate destruction and violence attributed to U.S. forces may easily reduce the legitimacy and sovereignty of the supported government. Adversaries may also use such incidents to fuel anti-American sentiments and assist their cause.

SECURITY RELATIONSHIPS

1-35. A complex relationship exists among security cooperation, security assistance, and the military instrument of foreign internal defense. The left side of figure 1-2 depicts this relationship, including how aspects of foreign internal defense and security assistance overlap. The right side of figure 1-2 illustrates how foreign internal defense focuses on internal threats to a host nation and how security assistance focuses on external threats. The column depicts how security (military, intelligence, and civilian), economic, and governance are considerations common to both foreign internal defense and security assistance.

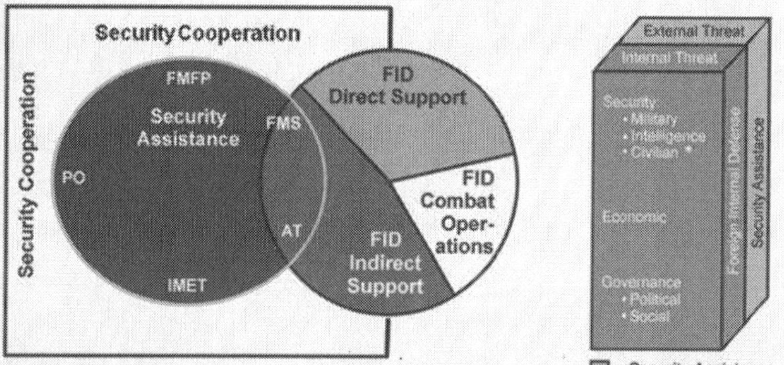

AT antiterrorism
FID foreign internal defense
FMFP foreign military financing program
FMS foreign military sales
IMET international military education and training
PO peace operations

Figure 1-2. Security cooperation, security assistance, and foreign internal defense

INTERNAL DEFENSE AND DEVELOPMENT

1-36. *Internal defense and development* is the full range of measures taken by a nation to promote its growth and to protect itself from subversion, lawlessness, and insurgency. It focuses on building viable institutions (political, economic, social, and military) that respond to the needs of society (JP 3-07.1). IDAD aims to forestall or defeat the threat and to correct core grievances that prompt violence. Ideally it is a preemptive strategy. However, if an insurgency or other threat develops, it becomes an active strategy to combat that threat. To support the host nation effectively, U.S. forces, especially planners, consider the host-nation's IDAD strategy.

1-37. IDAD focuses on building viable institutions that respond to the needs of society. IDAD is a U.S. doctrinal term; it is not used universally by all countries. Internal defense and development blends four interdependent functions to prevent or counter internal threats: balanced development, security, neutralization, and mobilization. Finally, IDAD involves a cyclic interaction of execution, assessment, and adaptation. JP 3-07.1 discusses IDAD and its functions in detail.

Security Sector Reform

1-38. *Security sector reform* is the set of policies, plans, programs, and activities that a government undertakes to improve the way it provides safety, security, and justice (FM 3-07). Security sector reform aims to provide an effective and legitimate public service that is transparent, accountable to civil authority, and responsive to the needs of the public. It may include integrated activities to support defense and armed forces reform; civilian management and oversight; justice, police, corrections, and intelligence reform; national security planning and strategy support; border management; disarmament, demobilization, and reintegration; and concurrent reduction of armed violence. SFA is a vital component of security sector reform when security sector reform includes U.S. assistance to FSF. See chapter 6 of FM 3-07 for an additional discussion of security sector reform.

Security Force Assistance Relationship

1-39. Paragraph 1-35 explained and figure 1-2 supported the complex relationship among security cooperation, security assistance, and the military instrument of foreign internal defense. Figure 1-3 explains the relationship of SFA to security cooperation, security assistance, and the military instrument of foreign internal defense. SFA supports the military instrument of foreign internal defense, much of security assistance efforts, and some security cooperation efforts. The right side of figure 1-3 also shows that SFA addresses security issues, but not economic or governance issues.

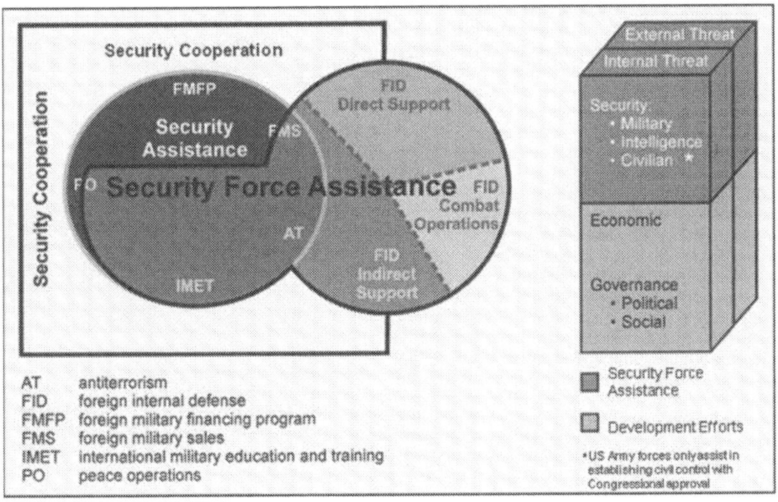

Figure 1-3. Relationship of security force assistance with security cooperation, security assistance, and foreign internal defense

Country Team

1-40. The country team will play a vital role with respect to SFA. The country team is the senior in-country U.S. coordinating and supervising body headed by the U.S. chief of mission. The chief of mission leads the country team and integrates U.S. efforts in support of the host nation. The chief of mission is the highest U.S. civil authority in a foreign country. As the senior U.S. Government official permanently assigned in the host nation, the chief of mission is responsible to the President for policy oversight of all U.S. Government programs. The Foreign Service Act assigns the chief of mission to a foreign country with responsibility for the direction, coordination, and supervision of all government executive branch employees in that country except for Service members and employees under the command of a U.S. area military commander.

1-41. The country team is composed of the senior member of each represented department or agency. As permanently established interagency organizations, country teams represent a priceless resource. They often provide deep reservoirs of local knowledge and interaction with the host-nation government and population. See FM 3-07 and JP 3-07.4 for more information on the country team.

Theater Command Structures

1-42. SFA is inherently multinational. Nations do not relinquish their national interests by participating in multinational operations. As in any multinational operation, commanders conducting SFA prepare to address issues related to legality, mission mandate, and prudence early in the planning process. In multinational operations, consensus often stems from compromise.

1-43. Political considerations heavily influence the ultimate shape of the command structure in which U.S. forces will conduct SFA. However, participating nations should strive to achieve unity of command for the operation to the maximum extent possible. All participants must understand missions, tasks, responsibilities, and authorities. While well defined in U.S. doctrine, command relationships are not necessarily part of the doctrinal lexicon of nations with which the United States may operate in coalition or combined operations.

1-44. Command and control relationships of higher echelons take precedence when conducting SFA. As a result, the brigade combat team (BCT) may be a supported command from a U.S. perspective as well as acting as a supporting command for FSF in the BCT's area of operations. For example, a BCT may support a joint interagency task force conducting international police training. Another example is a joint command supporting a BCT with joint fires; intelligence, surveillance, and reconnaissance; or other capabilities while the BCT supports FSF in training. The basic structures for multinational operations fall into one of three types: lead nation, integrated, or parallel command. Alliance and coalition commands use these structures.

Lead Nation Command Structure

1-45. A lead nation command structure exists when all member nations place their forces under the control of one nation. The lead nation command has a dominant lead nation command and staff arrangement with subordinate elements retaining strict national integrity. This command may also have an integrated staff and multinational subordinate forces. Integrating the staff allows the commander to draw upon the expertise of allied or coalition partners in areas where the lead nation may have less experience. This is the most desirable command structure for SFA.

Integrated Command Structure

1-46. Multinational commands organized under an integrated command structure provide unity of effort in a multinational setting. An integrated, combined command has several distinguishing features. Besides a designated single commander, the staff consists of representatives from all member nations. Subordinate commands and staffs are integrated into the lowest echelon necessary. Although effective for SFA, the integrated command structure works less effectively than the lead nation command structure.

Parallel Command Structure

1-47. Under a parallel command structure, no single force commander is designated. The coalition leadership develops a means for coordination among the participants to attain unity of effort. They can accomplish this by using coordination centers. Nonetheless, because of the absence of a single commander, the use of a parallel command structure should be avoided if at all possible. This is the least desirable command structure for SFA.

Alliance Command Structures

1-48. In combined commands, national political objectives are addressed and generally subsumed within multinational force objectives at the alliance treaty level. Combined command relationships often reflect either an integrated command structure or a lead nation command structure. In combined operations such as SFA, these structures should be used to the maximum practical extent. Combined command and force structures often mirror the degree of allied member participation. Senior military officers from member nations often lead subordinate commands. Effective operations within an alliance, including SFA, require senior political and military authorities to agree on the type of command relationships that will govern the operations of the forces. Potential political sensitivities associated with SFA can impact command relationships and operating procedures.

Coalition Command Structures

1-49. Coalitions often form in response to unforeseen crises that occur outside the area of an established alliance, which is often the case for SFA. Coalition command relationships routinely evolve as a coalition develops. Coalitions often have one of two basic structures: lead nation or parallel. For SFA, all participating coalition members should augment the host-nation headquarters staff with representatives. These representatives might include designated deputies or assistant commanders, planners, and logisticians. Such staffing provides the coalition commander with representative leadership and accessible expertise on the capabilities of the coalition members as well as facilitates the planning process. This integration of the multinational command elements into the coalition planning process should occur as early as practical.

Integration of Conventional and Special Operations Forces

1-50. Many criteria determine when conventional forces, special operations forces, or a combination are appropriate to conduct SFA. Both force levels and force characteristics suggest optimal, acceptable, and undesirable force package options in planning and resourcing SFA. Options for the deployment of a modular brigade augmented for SFA, a select number of conventional military transition teams, or special operations forces depend on conditions of the operational environment, priorities of the IDAD strategy, overall U.S. national policy, and forces available.

1-51. Rarely will U.S. forces conducting SFA be homogenous in terms of conventional or special operations forces, but rather contain both. FM 3-0 and other Army doctrine define command and control relationships and support relationships when both forces operate in the same area. Recent operational experience has shown that, in improving the effectiveness of FSF, the supported commander conducting SFA may be either a conventional force or special operations force commander. Subordinate commanders and staffs of both conventional and special operations forces may act in the supporting role to the U.S. ambassador. Typically this is done through the security assistance officer or the defense attaché at the American Embassy.

1-52. If U.S. forces are to train a large number of FSF in a short time, then the requirement may exceed the capacity of available special operations forces. Large SFA missions occur at the geographic command level or higher headquarters. Planners at these levels determine the number of conventional and special operations forces as part of their mission analysis aimed at training enough FSF in the given time frame. To best prepare for a large SFA, U.S. forces might deploy one or more augmented modular brigades supported by Army special operations forces. These forces conduct complementary foreign internal defense tasks under Title 22 authorities. Another option involves using special operations forces training FSF over several years while using a broad mix of funding and authorities. Another option uses a modular brigade to support a select number of company- or battalion-sized foreign units.

1-53. A force analysis of how best to integrate conventional and special operations forces to accomplish SFA considers host-nation and U.S. national policies. Skill sets among conventional and special operations forces trainers, advisors, and mentors avoid universally excluding or including final, integrated force packages for SFA. Planners leverage skill sets by integrating and coordinating conventional and special operations units with assigned tasks within SFA. Force analysis of the FSF to be trained often suggests a two-phase approach to training. This analysis also suggests a transition to the FSF by either conventional or special operations forces using basic and advanced training. The type of training depends on the nature of the foreign security force being trained. Additionally, integrating the advisor may be a phased process using either conventional or special operations forces as the initial advisor. This method lends itself to the classic crawl-walk-run approach to assisting FSF.

Chapter 2
Framework

This chapter discusses a doctrinal framework for the conduct of security force assistance. This security force assistance framework is composed of the mindset required of units and Soldiers, imperatives for success, inherent tasks and activities, and the three types of security force assistance.

MINDSET

2-1. Conducting successful security force assistance (SFA) requires a specific mindset. This mindset focuses on working by, with, and through foreign security forces (FSF) to support the host nation's internal defense and development (which includes local security requirements) or regional organization's charter. Soldiers conducting SFA must also understand that legitimacy is vital. The relevant population must perceive FSF as legitimate for long-term success. Those conducting SFA must understand that the military instrument of national power is only one part of a comprehensive approach. The imperatives of SFA provide the foundation for proper mindset.

Note: Host nation includes partner organization.

2-2. SFA focuses on how FSF organizations, institutions, systems, capabilities, and limitations—not U.S. organizations, systems, procedures, and equipment—can be enabled to address the host-nation or partner organization problems. Local forces have advantages over outsiders. They inherently understand the local culture and behavior that outsiders simply lack. To tap into those advantages, advisors must resist blatant military solutions. To overcome the temptation to do what they know and do best, whether relevant or not to the situation, advisors must accept that they are bound by unique situations.

IMPERATIVES

2-3. The imperatives of SFA come from the historical record and recent experience. These imperatives do not replace the principles of war or the principles of joint operations. Rather, they provide focus on how to successfully conduct SFA. The six imperatives apply to SFA at every level of war, for any echelon, and for any Soldier.

2-4. Since there is a close relationship between SFA and Army special forces conducting foreign internal defense, SFA planners should also consider the twelve Army special operations forces imperatives, especially at the tactical level (see appendix A).

UNDERSTAND THE OPERATIONAL ENVIRONMENT

2-5. An in-depth understanding of the operational environment—including available FSF, opposing threats, and civil considerations—is critical to planning and conducting effective SFA. Units and Soldiers conducting SFA must clearly understand the theater, population, and FSF with which they are working, especially FSF capabilities. Diplomatic, informational, military, economic, sociological, psychological, and geographic research and understanding are essential prerequisites for successful SFA. Tactically, successful SFA requires identifying the friendly and hostile decisionmakers, their objectives and strategies, and the ways they interact. Further, the conditions of the operational environment can rapidly change, and those conducting SFA must anticipate these changes and exploit potential and possibly fleeting opportunities.

Provide Effective Leadership

2-6. Leadership, a critical aspect of any application of combat power, proves important in the dynamic and complex environments associated with SFA. The operational environment in which SFA occurs places a premium on effective leadership at all levels, from the most junior to the most senior general officer and agency director. Leading FSF or a combined group of U.S. and foreign security forces is inherently more challenging due to differences in culture, language, training, and other aspects. Leaders focus on transitions; their decisions move the foreign security force toward independent operations.

Build Legitimacy

2-7. Ultimately, SFA aims to develop security forces that contribute to the legitimate governance of the local populace. Significant policy and legal considerations may apply to SFA activities. Legitimacy is the most crucial factor in developing and maintaining internal and international support. The United States cannot sustain long-term SFA without legitimacy. Legitimacy is a concept that goes beyond a strict legal definition; it includes the moral and political legitimacy of a host-nation government or partner organization. Local civilians and the international community determine the government's legitimacy based on collective perception of the credibility of its cause and methods.

Manage Information

2-8. Successful SFA disseminates timely and protected relevant information, integrates it during planning, and leverages it appropriately during execution. Effective and efficient information management supports decisionmaking throughout capacity building. Managing information encompasses the collection, analysis, management, application, and preparation of information. Maintaining continuity between rotating SFA elements is critical. For long-term SFA efforts, FSF expect more professional abilities from U.S. advisors. To meet expectations and success in general, units conducting SFA establish and integrate lessons learned. Capturing and incorporating evolving tactics, techniques, and procedures and aspects of the operational environment is vital; with each successive rotation, planners strive to maintain continuity and momentum. Finally, planners synchronize effective information engagement with the entire mission. See appendix C. SFA operations inherently affect the dynamic operational environment.

Ensure Unity of Effort

2-9. SFA often includes many actors, making unity of effort essential for success. SFA will include U.S. and foreign security forces, including conventional forces, special forces, or a combination. Other civilian and military joint and military organizations are often involved in SFA. Planners integrate them into one cohesive effort.

Sustain the Effort

2-10. Sustainability consists of two major components: the ability to sustain SFA effort throughout the operation and the ability of the FSF to sustain their operations independently. While each situation will vary, Army personnel conducting SFA must avoid assisting FSF in techniques and procedures beyond the FSF's capability to sustain. U.S. tactics, techniques, and procedures must be modified to fit the culture, educational level, and technological capability of the FSF. Those involved in SFA must recognize the need for programs that are durable, consistent, and sustainable by both the U.S. and FSF. They must not begin programs that are beyond the economic, technological, or cultural capabilities of the host nation to maintain without U.S. assistance. Such efforts are counterproductive.

TASKS

2-11. SFA aims to establish conditions that support the partner's end state, which includes legitimate, credible, competent, capable, committed, and confident security forces. This requires a force capable of securing borders, protecting the population, holding individuals accountable for criminal activities, regulating the behavior of individuals or groups that pose a security risk, and setting conditions in the operational area that enable the success of other actors.

2-12. This section addresses the six SFA tasks: organize, train, equip, rebuild and build, and advise and assist. When supporting the development of FSF, commanders understand their role with the host-nation or regional security organization. These tasks facilitate SFA planners to assess and allocate resources based on conditions.

ORGANIZE

2-13. Organize is a SFA task that encompasses all measures taken to assist FSF in improving its organizational structure, processes, institutions, and infrastructure. U.S. forces must understand the existing security organizations of FSF to better assist them. Subsequently, SFA personnel may help the host nation organize its security forces to meet the needs of its security environment.

2-14. Organizing a foreign security force depends on the host nation's social and economic conditions, cultural and historical factors, and security threats. SFA aims to create an efficient organization with a command, intelligence, logistic, and operations structure viable for the host nation. Conventional forces with limited special purpose teams (such as explosive ordnance disposal) are preferred. Doctrine or standing operating procedures should apply across the force, as should unit structures. The organization must facilitate the collection, processing, and dissemination of intelligence across and throughout all security forces. As the foreign security force strengthens, U.S. leaders and trainers should expect more independent organizational decisions.

2-15. The host nation determines the structure of its military forces, to include approving all organizational designs. These may include changing the numbers of forces, types of units, and internal organizational designs.

2-16. Organization should address all FSF, from the ministerial level to the patrolling police officer and soldier. Building a competent FSF infrastructure—including civilian information systems—is critical for success. Commanders work with ministries responsible for national and internal security, including the ministry of national defense, the interior ministry, and the justice ministry. Commanders assess strengths and weaknesses of the ministerial organization as well as training requirements of their employees. The joint and multinational advisor team at the ministry level helps the host nation develop a procurement and management system to meet its requirements.

2-17. Organizing a foreign security force requires resolving issues related to—

- Recruiting.
- Promotion screening and selection.
- Pay and benefits.
- Leader recruiting and selection.
- Personnel accountability.
- Demobilization of security force personnel.

Recruiting

2-18. Recruiting is critical when establishing FSF. The recruiting program assimilates local culture and themes that resonate with the populace. The program ensures that FSF have members from all major demographic groups. Joint and multinational partners should encourage and support efforts to recruit from minority populations. Mobile recruiting can target specific areas, ethnic groups, or tribes. Forces should encourage moderate groups and factions within hostile or potentially hostile ethnic groups to join the foreign security force. Often host-nation governments resist recruiting disaffected ethnic groups. However, even moderate success in recruiting from these groups provides enormous payoffs. It builds the security forces' legitimacy and often quiets legitimate fears of such groups regarding their relationship to the host-nation government. Effectively disarming, demobilizing, and reintegrating former enemies or other armed groups should be part of the overall plan. These programs should be mentioned in recruiting efforts.

2-19. A proper recruiting program specifies appropriate behavioral, physical, and moral standards. Ideally, recruits are centrally screened and inducted. Recruiting centers reside in safe areas safe. All recruits undergo a basic security check and are vetted against lists of suspected enemies. Foreign agencies and personnel perform this screening. Membership in illegal organizations needs to be carefully monitored. Past membership need not preclude a person from joining the foreign security force; however, any ongoing relationship with an illegal organization requires constant monitoring. FSF personnel limits the members of an illegal unit, tribal militia, or other militant faction to a single military or police unit.

Promotion Screening and Selection

2-20. Selection for promotion must stem from proven performance and aptitude for increased responsibility. Objective evaluations ensure promotion is by merit and not through influence or family ties. One method of selection identifies the most competent performers, trains them, and recommends them for promotion. The second method identifies those with social or professional status within the training group, then trains and recommends them for promotion. The first method may lead to more competent leaders but could be resisted for cultural reasons. The second method ensures the new leader will be accepted culturally but may sacrifice competence. The most effective solution comes from combining the two methods.

Pay and Benefits

2-21. Appropriate compensation precludes a culture of corruption in FSF. Spending the money needed for adequate wages and producing quality security forces is less costly than ending up with corrupt and abusive forces that alienate the populace. Paying the police adequately is especially important; their duties and contact with civilians often expose them to opportunities for corruption.

2-22. Payroll procedures and systems are also vitally important. They must be transparent and accountable, so FSF members receive their full pay and entitlements. Centralized administration of compensation through secure, automated systems minimizes the risk for corruption and manipulation; however, the supporting infrastructure will have to be available or developed over time.

2-23. Effective FSF can help improve host-nation social and economic development through the benefits each member receives. Every recruit should receive a basic education, job training, and morals and values inculcation.

Leader Recruiting and Selection

2-24. Officer candidate standards should be high. Candidates should have good health and pass an academic test with higher standards than the test for enlisted recruits. Recruiters carefully vet officer candidates to ensure that they do not have close ties to any radical or enemy organization.

2-25. Noncommissioned officers should be selected from the best enlisted FSF members. Objective standards, including proficiency tests, should be established and enforced to ensure that promotion to their ranks comes from merit. Many armies lack a professional noncommissioned officer corps; establishing one for a foreign security force may prove difficult. In the meantime, adjustments will have to be made, placing more responsibility on commissioned officers.

Personnel Accountability

2-26. Host-nation leaders carefully track and account for FSF. Proper personnel accountability reduces corruption, particularly in countries with manual banking systems where soldiers are paid in cash. In addition, many personnel failing to report for duty can indicate possible attacks, low unit morale, or enemy and militia influences on the foreign security force.

Demobilization of Foreign Security Force Personnel

2-27. Host nations develop programs to keep a class of impoverished and disgruntled former officers and soldiers from forming. As the foreign security force mature, officers who perform poorly or fail to meet the standards will need to be removed. Providing some form of government-provided education grants or low-interest business loans enables discharged personnel to earn a living outside the military. All soldiers who serve for several years and are then removed should receive a lump-sum payment or pension to ease their transition to civilian life. These programs should not apply to those guilty of major human rights abuses or corruption. Transition or demobilization planning should start as soon as commanders anticipate the need (it may not be required in all cases). Similar programs may be required when demobilizing nongovernment militias.

TRAIN

2-28. Train is a SFA task to assist FSF by developing programs and institutions to train and educate. These efforts must fit the nature and requirements of their security environment. The imperatives of Army special operations forces for facilitating a comprehensive approach and engaging the threat discriminately apply to the train task (see appendix A).

Training the U.S. Trainers

2-29. Soldiers assigned training missions should receive training on the requirements of developing FSF. The training should emphasize the host nation's cultural background, introduce its language, and provide cultural tips for developing a good rapport. Training should also include protection for troops working with FSF. U.S. trainees must become familiar with FSF organization and equipment, especially weapons not in the U.S. inventory. This training must emphasize the following:

- Sustaining training and reinforcing individual and team skills.
- Using the smallest possible student-to-instructor ratio.
- Developing host-nation trainers.
- Training to standards—not to time.
- Providing immediate feedback; using after action reviews.
- Respecting the culture but learning to distinguish between cultural practices and excuses.
- Learning the language.
- Working with interpreters.

2-30. U.S. forces should show respect for local religions and traditions. They should willingly accept many aspects of the local and national culture, including food (if sanitation standards permit). U.S. forces must stress that they do not intend to undermine or change the local religion or traditions. However, they have to strive to reduce how dysfunctional social practices affect the foreign security force's ability to conduct operations. U.S. trainers and advisors must have enough awareness to identify and stop inappropriate behavior, or at least report it to the appropriate chains of command.

Establishing Training Standards

2-31. Training in SFA involves many of the individual and collective skills performed in conventional military operations. All levels of training for all components should include values training. Metrics for evaluating units should include subjective measures, such as loyalty to the host-nation government, as well as competence in military tasks. Soldiers know how to evaluate military training. However, the acceptance of values, such as ethnic equality or the rejection of corruption, may be a better measure of training effectiveness. Gauging this acceptance is far more difficult than evaluating task performance. While the operational environment varies widely, FSF and trainers can still establish clear measures to evaluate the training of individuals, leaders, and units.

2-32. Effective training programs require clear, detailed individual, leader, and unit performance standards. These standards take into account cultural factors that directly affect the ability of the individual or unit to operate. For example, training a staff or unit to conduct effective operations requires more time in countries where the average soldier is illiterate. Building a security force from the ground up takes far more time than creating one around a trained cadre. Thus, using existing military personnel to form units and cadres for units often proves better than creating novice security forces.

2-33. Often, poorly trained leaders and units commit more human rights violations than well-trained, well-led units. Leaders and units unprepared for the pressure of active operations tend to use indiscriminate force, target civilians, and abuse prisoners. These actions can threaten the popular support and government legitimacy essential for SFA success. Badly disciplined and poorly led FSF can facilitate insurgent recruiting and propaganda efforts.

2-34. Setting realistic measures for FSF and following through on training plans consume time. The pressure is strong to find training shortcuts, employ quick fixes, or train personnel on the job. Trainers should resist such approaches. Such approaches often create more problems than they solve. However, trainers should also avoid the temptation to create long, complex training programs based on unrealistic standards. Effective programs account for the host nation's culture, resources, and short-term security needs. No firm rules exist on how long particular training programs should take. Trainers can use existing and historical training programs to determine how long training should take. To a certain extent, the enemy threat dictates the length of training. As security improves, training programs can expand to facilitate achievement of the long-term end state.

Training Foreign Security Forces

2-35. Members of FSF develop through a systematic training program (individual training and education as well as collective training). The program builds basic skills, then teaches them to work as a team, and finally allows them to function as a unit. Basic military, intelligence, or law enforcement training focuses first on basic skills such as legal considerations, first aid, marksmanship, and fire discipline. Leaders have training in tactics, including patrolling, urban operations, and legal evidence collection. Everyone must master rules of engagement and the law of armed conflict. FSF units should train to standard for conducting the major operations they will face.

Training Leaders

2-36. The effectiveness of FSF directly relates to the quality of their leadership. Building effective leaders requires a comprehensive program of officer, staff, and specialized training. The ultimate success of any SFA effort depends on creating viable FSF leaders able to carry on the fight at all levels and build their nation on their own.

2-37. The standards of leader training reinforce different levels of authority within the foreign security force. Clearly established responsibilities for commissioned and noncommissioned officers specify what is expected of recruits and leaders. To ensure civilian control, subordinate relationships to civilian authorities also are reinforced. In addition, training should establish team dynamics. In some cultures, security forces may need training to understand the vital role of members not in primary leadership positions. For example, noncommissioned officers may be a new or different concept for some FSF.

Employing Newly Trained Forces

2-38. SFA must build the morale and confidence of FSF. Committing poorly trained and badly led forces results in high casualties and invites tactical defeats. While defeat in a small operation may have little strategic consequence in a conventional war, often a small tactical defeat of FSF has serious strategic consequences. If the foreign security force fails, the local populace may begin to lose confidence in the host-nation government's ability to protect them. FSF must prepare for operations so they have every possible advantage. As much as possible, FSF should begin with simpler missions. As their confidence and competence grows, these forces can assume more complex assignments. Collaborating with joint or multinational units can help new forces to adjust to combat stress.

2-39. Newly trained units should enter their first combat operation in support of more experienced foreign units. Operational performance of such inexperienced organizations should be carefully monitored and evaluated to correct weaknesses quickly. The employment plan for FSF should allow enough time for additional training after each operation. Gradually introducing units into combat allows the command to identify poor leaders for retraining or other actions. Competent leaders are also identified and given greater authority and responsibility.

EQUIP

2-40. Equip is a SFA task encompassing all efforts to assess and assist FSF with the procurement, fielding, and sustainment of equipment. All equipment must fit the nature of the operational environment. The SFA principle of ensuring long-term sustainment is a vital consideration for the equip task.

2-41. The strategic plan for security force development should outline equipment requirements. Appropriate equipment is affordable and suitable against the threat. Forces must be able to train on the equipment. Interoperability may be desired in some cases. A central consideration includes the host nation's long-term ability to support and maintain the equipment.

2-42. The initial FSF development plan should use a comprehensive approach with local, regional, national, and other appropriate actors. Fundamentally, it should support the internal defense and development strategy. It should include phases with objectives for FSF to meet over three to four years. Since potential enemies adapt rapidly and situations change, commanders must continually assess the direction and progress of developing FSF.

2-43. The requirement to provide equipment may be as simple as assisting with maintenance of existing formations or as extensive as providing everything from shoes to communications and investigation kits. If the enemy uses heavy machine guns and rocket-propelled grenades, then FSF need comparable or better equipment. This especially applies to police forces, which are often lightly armed and vulnerable to well-armed enemies.

2-44. Primary considerations should include maintainability, ease of operation, and long-term sustainment costs. Few developing nations can support highly complex equipment. In most operations, having many versatile vehicles that require simple maintenance is often better than having a few highly capable armored vehicles or combat systems that require extensive maintenance. Effective maintenance systems for FSF often begin with major maintenance performed by contractors. The program then progresses to arrangements with U.S. forces as they train foreign personnel to conduct the support mission.

2-45. Sources of materials for FSF include U.S. foreign military sales (not technically SFA), multinational or third-nation resale of property, contracts with internal suppliers, or purchases on the international market. The organizations responsible for equipping FSF require flexibility to obtain equipment that meets their needs for quality, timeliness, and cost. As part of their training, FSF also need to learn property accountability to reduce corruption and ensure proper equipment usage. FSF provide equipment the same level of control and protection that U.S. forces provide for similar equipment. (See AR 12-1 and DOD 5105.38M.)

REBUILD AND BUILD

2-46. Rebuild and build is a SFA task to assess, rebuild, and build the existing capabilities and capacities of FSF and their supporting infrastructure. This task requires an in-depth analysis of the capability, capacity, and structures required to meet the desired end state and operational environment. Some FSF may require assistance in building and rebuilding, while other FSF may only need assistance in building.

2-47. FSF need infrastructure support. People need buildings for storage, training, and shelter. Often requirements include barracks, ranges, motor pools, and other military facilities. Construction takes time; the host nation needs to invest early in such facilities if they are to be available when needed. Forces must plan to protect any infrastructures, including headquarters facilities, since these building are attractive targets for insurgents. (See FM 3-34.400 for information on hardening measures to increase infrastructure survivability and improve protection.)

2-48. During some operations, such as counterinsurgency, foreign military and police forces often operate from local bases. Building training centers and unit garrisons requires a long-term force-basing plan. If possible, garrisons should include housing for the host-nation soldiers and their families; government-provided healthcare for the families; and other attractive benefits.

2-49. The host nation may need to make large investments in time and resources to restore or create the nationwide infrastructure necessary to effectively command and control FSF. The host nation will build functional regional and national headquarters and ministries as well as local bases and police stations.

ADVISE AND ASSIST

2-50. Advise and assist is a SFA task in which U.S. personnel work with FSF to improve their capability and capacity. Advising establishes a personal and a professional relationship where trust and confidence define how well the advisor will be able to influence the foreign security force. Assisting is providing the required supporting or sustaining capabilities so FSF can meet objectives and the end state. The level of advice and assistance is based on conditions and should continue until FSF can establish required systems or until conditions no longer require it. Leaders cannot permit FSF to fail critically at a point that would undermine the overall effort. See chapter 6 for a discussion on advisors.

ACTIVITIES

2-51. SFA is inherently a developmental effort. Success is measured by the increase in the foreign security force's capability, capacity, competency, commitment, and confidence in areas in which U.S. forces are assisting them and as a whole. SFA activities include plan and resource, generate, employ, transition, and sustain. Figure 2-1 depicts the SFA activities. These activities blend over time based on the capability and capacity of FSF.

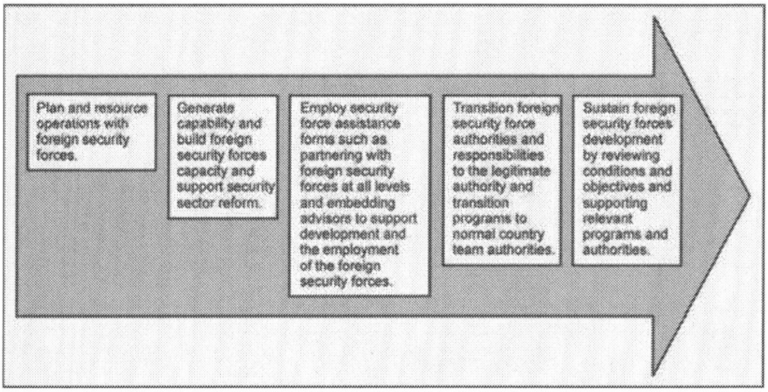

Figure 2-1. Security force assistance activities

PLAN AND RESOURCE

2-52. The plan and resource SFA activity begins as commanders understand the operational environment and determine the requirements of FSF. It also ensures that the United States provides SFA that achieves the objectives and end state of both the host nation and United States. The host nation and United States should then analyze the resource requirements and efforts so that developing FSF have sufficient and appropriate resources. Chapter 4 provides additional detail on planning for facilities and infrastructure.

GENERATE

2-53. The generate SFA activity includes generation of FSF, including the forces, leaders, and warfighting function capabilities (movement and maneuver, intelligence, fires, sustainment, command and control, and protection) based on the desired end state.

EMPLOY

2-54. The employ SFA activity involves FSF transitioning from force generation to mission employment. It does not rely on the maturity of the force or supporting institutions as a whole but is focused on the foreign element in question. Conditions determine when to use an element of FSF. Leaders of U.S. and foreign security forces assess the risk associated with employment and mitigate that risk as much as possible.

TRANSITION

2-55. The transition SFA activity defines the transition between two security forces, when applicable. This could be a transition of authority between U.S. forces to a new foreign security force. Another example could be from a regional foreign security force to a foreign security force with U.S. advisors. It could also be from a host-nation military force to a host-nation police force, with or without U.S. advisors.

SUSTAIN

2-56. The sustain SFA activity occurs when the institutional capacity of the foreign security force has been developed to a point where it is self sustaining. It may continue to have SFA contact through combined exercises, educational opportunity exchange, intelligence sharing, and foreign military sales.

TYPES

2-57. The U.S. or other actors use three types of SFA—augmenting, partnering, and advising—to develop confident, competent, capable, committed, and credible FSF. These types of SFA define the relationship between outside actors and the foreign security force. They may be employed simultaneously, sequentially, or in combination. The progression and types of SFA are determined by the operational environment, the assessment of the FSF, and by resources available. Each of these types requires decidedly different requirements, objectives, and legal considerations.

ADVISING

2-58. Advising is the primary type of security force assistance. Advising is the use of influence to teach, coach, and advise (see chapter 6) while working by, with, and through FSF. This type of SFA relies on the ability of the advisor to provide relevant and timely advice to FSF. Advising helps FSF conduct independent decisionmaking and operations. Advisors may also provide FSF with direct access to joint and multinational capabilities such as air support, artillery, medical evacuation, and intelligence. However, the advisor and advisor team require proper manning and equipment to perform these secondary support functions while staying focused on advising. Multiple sources will pressure advisors as they perform their missions. For example, nearby maneuver units may have other priorities and may focus on accomplishing their missions with or without FSF.

2-59. Advisor teams require a clearly defined and structured chain of command under which to operate. This is not only for logistics and support, but also keeps the advisor focused on developing FSF. Advisor teams will often find themselves answering to their higher military assistance group, the brigade combat team they are attached to, and the host-nation unit with which they are embedded. A well-defined chain of command alleviates confusion concerning who tasks or monitors the team's progress as well as ensures constant and adequate team sustainment.

2-60. Advisors are not partners; U.S. forces act as partners. Advising and partnering are complementary but inherently different activities. Advising requires relationship building and candid discourse to influence development of a professional security force. Partnering incorporates training with combined operations to achieve the same SFA goals. Advisors perform partnership shaping functions, shape discussions with their counterparts, and create opportunities for the partner units.

Partnering

2-61. Partnering attaches units at various levels to leverage the strengths of both U.S. and foreign security forces. As a foreign security force's capabilities mature, the echelon and degree of partnering decrease. As the foreign security force conducts more autonomous operations, U.S. forces still provide quick reaction forces and other assistance as appropriate.

2-62. Partner units should establish combined cells for intelligence, operations, planning, and sustainment. While effective coordination is always required and initial efforts may require completely fused efforts, FSF should eventually build the capability and capacity to conduct all efforts autonomously. These combined cells have several functions. They support transparent operations and a comprehensive approach. They also enhance the relationships among U.S. and foreign security forces by demonstrating trust. Finally, they develop the capacity of FSF in key staff areas. However, combined cells are not without risk. Operations and plans always at risk of compromise and prudent precautions should be taken.

2-63. Another partnering technique establishes relationships among command and staff elements of U.S. and foreign security forces. It requires less reorganization. It also allows counterparts to understand each other since foreign personnel can observe U.S. personnel performing their duties. This construct works well when the foreign culture is extremely sensitive to maintaining honor. In this case, foreign personnel can observe without having to expose their lack of knowledge or appear subordinate.

2-64. Unit partnerships do not replace advisor roles or functions. If partnering and advising are used in combination, it forms a three-part relationship amongst FSF, advisors, and the partner units. Partner units should look to the advisor to identify, shape, and facilitate operational partnering opportunities and training events. Advisors support U.S., coalition, and partner unit objectives, but, depending on the operational phase, the partner unit may support advisors or advisors may support the partner unit. Therefore, some level of advisor skills training should be included in the partner unit training program if those units will be conducting SFA activities.

Augmenting

2-65. Augmenting is an arrangement where FSF provide individuals or elements to combine with U.S. units, or U.S. individuals or elements combine with FSF. Augmentation improves the interdependence and interoperability of U.S. and foreign security forces. Augmentation can occur at many levels and in many different forms. For example, a U.S. squad can be augmented with host-nation individuals, a U.S. company can be augmented with a host-nation platoon, or a U.S. battalion can be augmented with a company from a foreign security force. Similarly, augmentation can be of short duration for a specific operation or of a longer duration for an enduring mission. Augmenting immerses FSF in a U.S. environment to provide language and cultural awareness to the U.S. unit. U.S. forces can also augment FSF.

Combining Tasks, Activities, and Assessment

2-66. Execution of successful SFA links its activities, tasks, and assessment. Figure 2-2 depicts how to arrange SFA activities, tasks, and assessment in time and purpose. The tasks that occur in one SFA activity will normally continue throughout operations. For example, organize, train, equip, rebuild and build, and advise and assist tasks will continue throughout the SFA activities of employment, transition, and sustain.

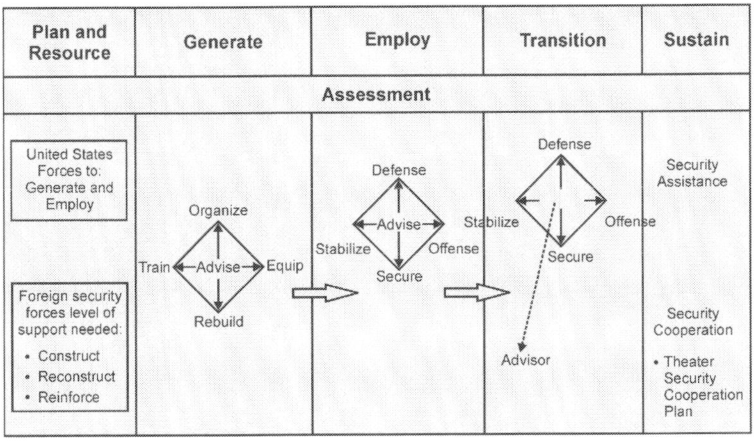

Figure 2-2. Security force assistance tasks, activities, and assessment

Plan and Resource

2-67. The plan and resource SFA activity determines the level of support and number of U.S. forces needed. As all efforts in SFA this activity is part of a comprehensive approach. Planners assess whether they will be constructing a foreign security force from the ground up, reconstructing a foreign security force based on existing capabilities and structure, or merely reinforcing an existing foreign security force. Understanding what capability and capacity the foreign security force should have and what capability and capacity they currently have is vital. Commanders and planners can then determine the size and structure, resources, and funding for the U.S. unit or units that will facilitate or assist the generation of FSF. Chapters 3 and 4 discuss planning and execution considerations for SFA.

Generate

2-68. In the generate SFA activity, U.S. forces focus on organizing, training, equipping, and advising FSF. During generation of FSF, the condition of the existing foreign security force's infrastructure and capabilities determine the assistance needed from U.S. forces. The generate activity adapts to the needs and requirements of the foreign security force. It can be designed to take the individual recruit from initial entry through to an employable unit, or it can augment an existing foreign unit with additional or enhanced capabilities. Like all SFA efforts, generate focuses on bringing a foreign security force's capabilities to a level where it can conduct its mission. With help from advisors in the next phase, FSF can work toward independent and sustained employment. In large SFA efforts, an organization may be required to command and control this activity.

2-69. Once elements of FSF are trained, the focus shifts from organizing, training, equipping, and rebuilding to employing the foreign security force. However, not all FSF units and organizations are ready for employment at the same time. As one unit is employed, others may be part of the generation activity, while still others may be completing the transition activity. FSF may be employed for offensive, defensive, or stability operations while other FSF may be employed against internal or external threats. The primary SFA task for U.S. forces is to advise and assist to facilitate the foreign security force operations.

Transition

2-70. The transition activity focuses on removing the U.S. elements and advisors once FSF are ready to conduct independent operations. All involved actors work in a comprehensive approach to assess when this can be achieved and to coordinate the transition itself. As FSF gradually progress toward the transfer of authority, close relationships forged between the FSF and their partners prove essential to sustainable development and successful transition. Genuine relationships engender trust and confidence, enabling increased responsibility and a well-executed transition process. These relationships also foster a clear understanding of command responsibilities and authorities. Such an understanding ensures FSF approach transition prepared to assume the full weight of their future role in the security sector. Success in developing FSF often depends more on relationships and personalities at the unit level than any other factor.

Chapter 3

Brigade Operations Process

This chapter examines operational considerations for the modular brigade augmented for security force assistance (SFA). It uses battle command and the operations process to examine these considerations.

BATTLE COMMAND

3-1. *Battle command* is the art and science of understanding, visualizing, describing, directing, leading, and assessing forces to impose the commander's will on a hostile, thinking, and adaptive enemy. Battle command applies leadership to translate decisions into actions—by synchronizing forces and warfighting functions in time, space, and purpose—to accomplish missions (FM 3-0). Commanders are central to battle command for brigade combat team (BCT) operations just as they are for any other operation. However, there are key considerations for successful battle command of a BCT conducting SFA. Most importantly, BCT commanders must work with their foreign counterparts, which may require facilitating operations when the foreign security forces (FSF) commander has the lead for the operations process.

UNDERSTAND

3-2. Successful SFA requires that BCT commanders thoroughly understand the operational environment. Conditions of the operational environment affect what SFA commanders provide. Commanders must understand the conditions that make up the current situation, including the relevant population, other actors, and FSF. From this understanding, commanders visualize desired conditions that represent a desired end state. Commanders must understand the capacity and capabilities of foreign security. The host-nation or regional security organization helps to determine desired conditions associated with the foreign security forces. After envisioning a desired end state, commanders then conceptualize how to change the current situation to the desired end state.

3-3. BCT commanders must understand the host-nation's security and political concerns. Security concerns focus on internal and external threats, their capability to affect the operational environment, and the foreign security force's strength and capability to counter those threats. Politically, the considerations associated with the security sector are the legitimacy of the foreign security force, FSF capability and capacity, host-nation or partner organization policy, and the ability to support the host-nation and U.S. Government objectives.

3-4. Commanders use the information obtained from intelligence preparation of the battlefield and intelligence, surveillance, and reconnaissance (ISR) to determine objectives and requirements for assessing and mitigating risk. Risk applies to how well FSF, U.S. Government, host-nation, partner organization, and international organizations can tolerate changes in the operational environment, as well as the challenges and conditions inherent to the operation.

VISUALIZE

3-5. *Commander's visualization* is the mental process of developing situational understanding, determining a desired end state, and envisioning the broad sequence of events by which the force will achieve that end state (FM 3-0). Visualization begins with assignment of the mission, which provides the focus for developing the commander's visualization. However, visualization must be continuous as SFA is dynamic. During planning, commander's visualization provides the basis for developing plans and orders.

3-6. The nature and extent of assistance provided to FSF are determined by this desired end state. SFA is often a slow process that does not lend itself to quick solutions. To ensure long-term success, commanders and host-nation or regional security organizations clarify early what conditions they desire. These conditions include the desired end state for FSF, which often include being—

- **Flexible**. FSF can accomplish broad missions. Flexibility requires an effective command and control organization appropriate for the partner or host-nation. This can include both internal and external threats.
- **Proficient**. The level of proficiency of FSF should include—
  - The capability of working effectively within a comprehensive approach.
  - The capability to integrate their operations with those of multinational partners.
  - Nonmilitary security forces competent in maintaining civil order, enforcing laws, controlling borders, securing key infrastructure, and detaining criminal suspects.
  - Nonmilitary security forces thoroughly trained in modern police ethos and procedures and which understand the basics of investigation, evidence collection, and proper court and legal procedures.
- **Self-sustained**. FSF can manage their own equipment throughout its lifecycle and perform administrative support.
- **Well led**. Leaders at all levels possess sound professional standards and appropriate military values; they are selected and promoted based on competence and merit.
- **Professional**. FSF are—
  - Honest, impartial, and committed to protecting and serving the entire population; they operate under the rule of law and respect human rights.
  - Loyal to the partner organization or host-nation government and serving partner or national interests; they recognize their role as the servants and not masters.
- **Integrated into society**. In the case of a host nation, FSF represent the host nation's major social groups; they are not seen as elements of just one faction.

3-7. These conditions may vary in different countries, but well-trained FSF should—

- Provide reasonable levels of security from external threats without threatening regional security.
- Provide reasonable levels of internal security without infringing on the populace's civil liberties or posing a coup threat to the host-nation government.
- Be founded upon the rule of law, including international principles of the rule of law.
- Be sustainable after joint and multinational forces depart.

3-8. When addressing internal threats, foreign military and police forces often perform unconventional functions. The military may have to fill an internal security role usually reserved for the police, or police may have forces so heavily armed that they would normally be part of the military. In the near term, FSF should—

- Integrate military capabilities with those of local, regional, and national police.
- Maintain the flexibility to transition to more conventional roles of external and internal defense based on long-term requirements.

3-9. To achieve this end state and intermediate objectives, the host nation should develop an internal defense and development (IDAD) strategy (with multinational assistance if required). This strategy should have a section on IDAD that addresses all aspects of SFA.

3-10. U.S. doctrine divides force development into domains: doctrine, organization, training, materiel, leadership and education, personnel, and facilities (DOTMLPF), which may be useful to analyze the IDAD. DOTMLPF elements are tightly linked, simultaneously pursued, and difficult to prioritize. Commanders monitor progress in all domains. Often military forces involved in such programs are tempted to impose their own doctrine and judgment on the host nation. Foreign security force doctrine, like the remaining DOTMLPF domains, must be appropriate to host-nation capabilities and requirements. Paragraphs 3-59 through 3-70 use DOTMLPF as a tool to assess FSF.

DESCRIBE

3-11. After BCT commanders visualize how to conduct SFA, they describe it to their staffs and subordinates to facilitate shared understanding of the mission and commander's intent. Commanders ensure subordinates understand the visualization well enough to begin planning. Commanders express their initial visualization as—

- Initial commander's intent.
- Planning guidance, including an initial concept of operations.
- Commander's critical information requirements.
- Essential elements of friendly information that must be protected.

Commander's Intent

3-12. Like any operation, SFA requires a clear, concise statement of what the force must do and the conditions the force must meet to succeed with respect to the enemy, terrain, and desired end state. In such an operation, the end state should include a description of desired conditions with respect to FSF capability and capacity. The commander's intent links the mission and concept of operations. It describes the key tasks and end state that, along with the mission, form the basis for subordinates' initiative.

3-13. A sample end state could discuss—

- How civilians perceive FSF as credible and legitimate and representing a legitimate authority.
- How FSF have the capability to support stability and security, retain the initiative and freedom of movement, disrupt enemy activities, influence the population, operate with other friendly forces, and sustain themselves.

3-14. Like any operation, how the BCT supports the higher organization's mission is vital. Additionally, the BCT supports FSF. Some sample key tasks are:

- Support the organization of a division of foreign infantry and two battalions of foreign police.
- Train leaders of FSF to create a cadre that can sustain their own training programs.
- Oversee the equipping of FSF and provide new equipment training.
- Support the rebuilding or building of infrastructure required to develop FSF.
- Advise FSF to develop the required capabilities and competencies associated with shared objectives.
- Assist the partner's security forces in meeting their operational requirements.

Planning Guidance

3-15. Commanders develop planning guidance for the staff from the commander's visualization. Planning guidance may be as broad or detailed as circumstances require; however, it must convey to the staff the essence of the commander's visualization. Due to the complex and dynamic nature of FSF, SFA often requires detailed planning guidance. Commanders ensure the staff understands their visualization while still permitting the staff to explore different options. Some considerations are:

- Friendly courses of action (COAs) with differing SFA forms and differing methods on how to augment, partner, or advise.
- Command and control architecture, intelligence architecture, and communications architecture to support dispersed SFA.
- Enemy COAs as they seek to defeat friendly efforts with respect to SFA activities.
- Enemy adaptation given the development of the foreign security force.
- Guidance for security during the conduct of SFA, including the BCT securing its advisor teams. The BCT determines the staging, needs, and ad hoc arrangements of advisor teams.
- ISR guidance to include identifying the focus of effort as it relates to SFA.

Commander's Critical Information Requirements

3-16. Commander's critical information requirements (CCIRs) will be developed for SFA just as for any other operation. These requirements are what commanders need to make decisions or the information they think is important for understanding the operational environment. These CCIRs should be compatible with the FSF commander's information requirements. CCIRs include priority intelligence requirements and friendly force information requirements.

3-17. Priority intelligence requirements are vital to support decisionmaking. During SFA, priority intelligence requirements must be focused on the enemy and the operational environment, not on FSF. U.S. forces do not focus intelligence efforts, such as collection, on the foreign forces they assist.

3-18. Friendly force information requirements provide the information that commanders need to understand the conditions of their own forces, as well as the conditions of the foreign security forces. For example, a BCT commander may direct the staff to better define a foreign security force's critical equipment shortages or maintenance requirements. During execution, commanders may focus on when FSF meet certain objectives or conditions. This information allows the commander and FSF commander to make adjustments, such as changes to the task organization, support requirements, or even the type of SFA.

Essential Elements of Friendly Information

3-19. An *essential element of friendly information* is a critical aspect of a friendly operation that, if known by the enemy, would subsequently compromise, lead to failure, or limit success of the operation, and therefore should be protected from enemy detection (FM 3-0). Essential elements of friendly information should be determined through a comprehensive approach. Information sharing and trust are vital elements of SFA. U.S. commanders must be aware of the inherent risks in this context and employ prudent risk mitigation efforts. Thus, successful SFA requires a balance between sharing information and risk mitigation efforts. Although essential elements of friendly information are not CCIRs, they have the same priority.

Intelligence, Surveillance, and Reconnaissance

3-20. *Intelligence, surveillance, and reconnaissance* is an activity that synchronizes and integrates the planning and operation of sensors, assets, and processing, exploitation, and dissemination systems in direct support of current and future operations. This is an integrated intelligence and operations function. For Army forces, this activity is a combined arms operation that focuses on priority intelligence requirements while answering the commander's critical information requirements (FM 3-0). For the brigade, these efforts will often support FSF. Moreover, the BCT must have means that allow effective intelligence collaboration, such as an intelligence fusion cell with the participation of all appropriate actors.

3-21. ISR efforts focus on the enemy and operational environment within the context of SFA. ISR efforts in SFA are not used on the foreign security force. Rather, U.S. forces assess the foreign security force within a comprehensive approach.

DIRECT

3-22. BCT commanders direct all aspects of operations, although a comprehensive approach, including FSF and other appropriate actors is fundamental to successful SFA. The commander's direction takes different forms during planning, preparation, and execution. Commanders make decisions and direct actions based on their situational understanding, which they maintain by continuous assessment and coordination with FSF. SFA is directed by—

- Preparing and approving plans and orders with FSF.
- Assigning and adjusting missions, tasks, task organization, and control.
- Positioning units to maximize their support to FSF, anticipate threat actions, or create or preserve options.
- Positioning key leaders to ensure observation and supervision of FSF at critical times and places.

- Adjusting SFA and other support priorities and allocating resources based on FSF priorities, as well as opportunities and threats.
- Changing support arrangements, either internally or with respect to FSF.

ASSESS AND LEAD

3-23. Effective battle command requires commanders to continuously assess and lead. Assessment helps commanders to better understand current conditions and determine how the operation is progressing and how best to support FSF. Based on an assessment that is the product of a comprehensive approach, commanders modify plans and orders to better accomplish the mission. This may include changing the type or combination of SFA, support, sustainment, or task organization. If their assessment reveals a significant variance from the original shared commander's visualization, a comprehensive approach must be used to reframe the visualization and develop a new plan.

3-24. Commanders in a BCT conducting SFA lead by example and presence, as Army commanders do in any operation. However, commanders have the additional challenge of leading by, with, and through FSF. Command posts for BCTs conducting SFA have similar challenges in that they should be integrated headquarters as much as practical.

OPERATIONS PROCESS

3-25. As they would in any other mission, BCT commanders conducting SFA balance their time and the staff's time and resources among four major activities in a continuous learning and adaptive cycle called the operations process. The *operations process* consists of the major command and control activities performed during operations: planning, preparing, executing, and continuously assessing the operation. The commander drives the operations process (FM 3-0). While simple in concept, the operations process is complex in execution. The operations process activities may be sequential or simultaneous.

3-26. Figure 3-1 depicts the operations process activities as a cycle that is driven by battle command. This cyclic process applies equally to a BCT conducting major combat operational theme or operations within an irregular warfare operational theme. During the operations process involving SFA, commanders consider the presence and inclusion of other actors, including FSF. Policy, the operational environment, and the forms of SFA determine how U.S. forces, FSF, and other actors interact in the operations process.

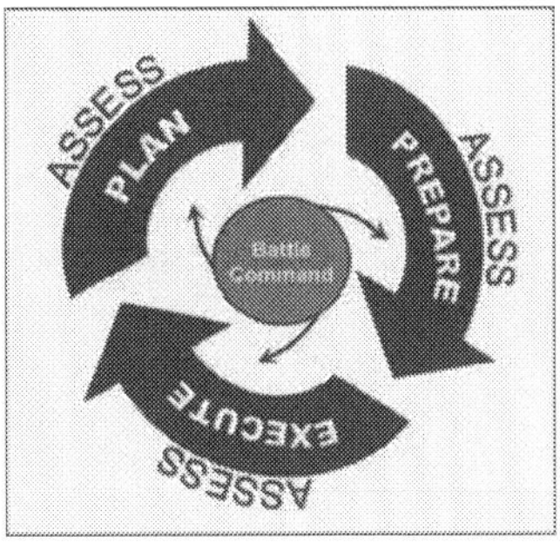

Figure 3-1. The operations process

PLAN

3-27. Planning for SFA, like any other operation, begins either with the anticipation of a new mission or the receipt of mission as part of the military decisionmaking process. *Planning* is the process by which commanders (and the staff, if available) translate the commander's visualization into a specific course of action for preparation and execution, focusing on the expected results (FM 3-0). SFA requires a comprehensive approach to ensure the U.S. commander's visualization synchronizes with policy and the foreign commander's visualization. SFA is the major requirement for a BCT conducting SFA, but the mission may require the brigade to perform other tasks.

3-28. Continuous and open to change, planning for SFA includes identifying how to best assist the FSF and developing a sequence of actions to change the situation. This planning must occur within a comprehensive approach that includes FSF, U.S. forces, and other actors. Similarly, this plan must be nested within policy, IDAD strategy, the campaign plan, and any other higher-echelon plans. It also involves anticipating consequences of actions and developing ways to mitigate them.

3-29. The plan, which includes the commander's intent, provides a common understanding to U.S. and foreign security forces on the actions to take. Plans and orders also provide decision points and branches that anticipate options that enable the force to adapt as the operation unfolds. This is especially important for SFA, as these operations tend to be prolonged efforts. BCTs conducting SFA often rotate before achieving all objectives. As a result, the planning team should establish objectives and milestones that can be achieved during the brigade's mission. These BCT objectives and milestones must support higher-echelon plans, including the campaign plan and IDAD strategy.

Functional Requirements

3-30. The composition of advisor teams is subject to objectives and conditions. For example, security conditions may require three armored vehicles, a crew-served weapon, and nine personnel to conduct independent movement. Another example involves the difference in the number of advisors. Newly generated FSF may require many advisors whereas experienced FSF may only require assistance in a few warfighting functions and consequently need fewer advisors. Understanding the functional requirements provides flexibility to the units assisting FSF. Three examples of determining functional requirements based on the understanding the operational environment include the BCT, a partnering or augmenting unit, and an advisor team.

Brigade Combat Team

3-31. A BCT headquarters has specific functions and requirements. These can include, but are not limited to—

- The functions of—
  - Carrying out SFA tasks and assisting FSF in support of objectives and the end state.
  - Facilitating partnerships between U.S. and foreign security forces.
  - Requesting, coordinating, and synchronizing resources.
  - Providing command and control of subordinate units and assigned advisor teams.
  - Coordinating with other commands and adjacent units.
  - Coordinating with other participants integrated within the comprehensive approach.
  - Developing, implementing, and synchronizing information engagement in assigned areas.
  - Developing relationships and an understanding with appropriate actors involved in SFA.

3-32. The requirements of—

- Having enough staff to handle the communications traffic generated by conditions.
- Being able to analyze information to aid commander and subordinate elements.
- Being able to articulate resource requirements in support of subordinate elements and efforts to conduct SFA tasks.
- Being organized and equipped to conduct operations and support and sustain FSF.
- Including regionally oriented host-nation officers to provide regional expertise.

Partnering or Augmenting Unit

3-33. Functions and requirements depend on the degree and level of partnering or augmenting. No prescriptive way to partner or augment exists; it is based on specific conditions and subject to the needs of policy, higher headquarters guidance, and needs of the FSF. A BCT may designate one battalion to partner or augment while designating other battalions to advise. A partnering or augmenting unit conducting SFA has specific functions and requirements. These can include, but are not limited to—

- The functions of—
  - Partnering with foreign security under the BCT level.
  - Increasing capability, capacity, competency, confidence, and commitment of FSF by conducting combined tactical operations.
  - Providing feedback on performance of FSF and certain shortfalls to embedded advisors and BCT headquarters.
  - Conducting sustainment and medical training with FSF at their home station or on operations.
  - Reporting conditions in the operational environment.
- The requirements of—
  - Having U.S. forces that are capable, competent, committed, and confident.
  - Being interoperable with FSF.
  - Being available for an appropriate period.

Military Advisor Team

3-34. The functions and requirements of a military advisor team are based on conditions—requirements, availability, and footprint. A military advisor team conducting SFA has specific functions and requirements. These can include, but are not limited to—

- The functions of—
  - Under the BCT headquarters, advising, training, and assisting FSF to which it is assigned.
  - Increasing capability, capacity, competency, confidence, and commitment of FSF by providing advice and support during battalion level and higher operations.
  - Assessing partner leaders, staff, and certain shortfalls to BCT headquarters.
  - Conducting sustainment training with FSF at their home or on operations.
  - Reporting on conditions in the operational environment.
- The requirements of—
  - Being manned with trained personnel who are properly equipped so as to be capable of executing their mission in current conditions.
  - Providing personnel with rank, education, and experience corresponding to the echelon and type of unit being advised.

Support and Sustainment Requirements

3-35. Support and sustainment requirements play a vital role in COA development (see chapter 6). The BCT may operate across a broad geographical area when conducting or supporting SFA. Partnering may be on certain levels with FSF and may support numerous embedded advisor teams or other actors. Since FSF often have varied relationships with higher echelons and institutions, the BCT may be involved vertically and spatially. These considerations may require more support and sustainment requirements than an initial analysis might indicate. Support and sustainment considerations include, but are not limited to—

- Distribution of FSF to which U.S. forces are providing SFA.
- The capability of FSF to provide support, sustainment, and medical functions.
- The technical capability and interoperability of U.S. and foreign security forces.
- The infrastructure that supports SFA.
- Overall security conditions where SFA occurs.

- The available contracting support.
- The ability of the environment to accept and absorb the U.S. forces' footprint.

Determining a Course of Action

3-36. COAs for SFA require an understanding of the breadth and width of the security sector. They may have to account for the roles ranging from the individual soldier or policeman to the ministerial level. COA development examines the roles other actors, for example, a gendarme or paramilitary force, in the security sector. Because the United States does not have a gendarme, U.S. forces may not fully appreciate its importance.

3-37. COAs for SFA also need to be flexible and inclusive. Commanders consider the following when examining a partner's security sector:

- The status and challenges of the host-nation security sector or regional security organization.
- The composition and disposition of the partner's security sector.
- Supporting institutions.
- Links among ministries, institutions, and components of the security sector.
- Partner capability and capacity shortfalls and impediments to their development.
- Environmental requirements to enable the building of security capability and capacity.
- Functional requirements to assist building capability and capacity.
- The impact of other actors within the comprehensive approach.

3-38. Once planners analyze the security sector, they can develop COAs that best combine SFA tasks, activities, and types (see chapter 2). While the advising type of SFA can be of any duration, partnering and augmenting tend to be enduring if used.

3-39. All SFA types tend to be more effective if they are long-term efforts. Advising fosters a more personal relationship and infers a stronger commitment to see the partner succeed. Partnering and augmenting include U.S. and foreign security forces conducting operations together. There are levels of partnering. Different advisory relationships use advisor teams embedded within various partner echelons and units, use one advisor team to advise multiple entities, or have a senior leader visit a partner unit to give advice.

3-40. Using any form or combination of forms depends on conditions (such as the level of competency), capability of FSF, and resources available. Policy, guidance, or formal agreements can also help commanders determine which form to use. Historically, a combination of forms appears to be the most effective. However, combining forms often proves more resource intensive and affected by time, geography, and the number of forces. COAs should remain flexible with respect to the form of SFA. Some example changes in conditions are the increased capability of one element in the security sector over another, the desire to increase one element over another based on partner objectives, or the increased targeting of one element over another by the enemy.

3-41. If a desired COA is not feasible given the scope of the security sector or due to other considerations, it may be useful to sequence SFA forms with portions of FSF. For example, a COA that affects smaller portions of the FSF, but in the right area and at the right time, may do more for achieving an objective or end state than a COA that attempts to support all FSF at one time.

3-42. BCTs are task-organized and augmented to accomplish a specific mission. How BCTs are specifically task-organized and augmented is based on the phase and types of SFA that policy and higher headquarters have directed. Chapter 4 describes augmentation in detail.

3-43. Attached units or units under operational control are task-organized under different headquarters as necessary to accomplish assigned and implied SFA missions. Deciding to task-organize specific formations for SFA may reflect previously approved contingency plans or result from current planning efforts. As these missions end, units return to their parent headquarters or are further task-organized by the controlling headquarters.

3-44. As with other operations, gaining commanders task-organize for SFA by designating operational control, attached, or support relationships to another unit or headquarters. Within each BCT, the brigade commander task-organizes all organic, assigned, attached, or operational control units to best suit the operational environment and types of SFA. Commanders may choose to use a support relationship (direct support, reinforcing, or general support reinforcing) among units of the BCT instead of a command relationship. This can also apply to relationships with foreign units.

3-45. SFA is often highly dispersed, as BCTs may have individuals or units assisting FSF over a wide area. This dispersion increases communications and sustainment requirements for the BCT. The nature of SFA also requires BCTs to help FSF meet requests for assets. To do so, the BCT and other units may need to task-organize to provide tactical support or sustainment. Dispersion and the nature of SFA often necessitate a decentralized leadership approach to facilitate effective and rapid support of FSF.

3-46. BCT commanders use the three types of SFA—advising, partnering, or augmenting—to accomplish the mission. The exact nature of a unit's assigned mission, the operational variables, and the mission variables drive modifications to headquarters and maneuver units. For example, a commander may need a unit tasked with advising, a unit tasked with partnering, and a unit tasked with augmenting. In this case, the unit tasked to advise would provide teams for the appropriate FSF while other foreign units continue their traditional missions. The unit that is partnering may provide the host-nation staff to assist the FSF headquarters. The third unit may provide squads to augment foreign platoons. Units may request additional support, sustainment, and medical support. However, BCTs conducting SFA must retain a reserve to conduct full spectrum operations.

3-47. Based on the overall mission analysis, including an analysis of functions and requirements, commanders and staffs begin to determine how to task organize. Due to the level of detail that may be involved, task organization should begin early in the military decisionmaking process, especially during COA development. No standard task organization exists since conditions vary. For example, a squad of infantry and three additional leaders may work as an advisor team for one mission, where another mission may require an advisor team with specialists for different warfighting functions. The art in SFA involves determining where, when, and if to allocate resources; it involves either breaking up or simply moving personnel and units from one foreign unit to another.

3-48. The techniques used by BCT commanders and staffs to allocate other resources and support are just as valid in SFA as in other types of operations. For example, if it is determined that the most important effort is advising, then the BCT commander may weight that effort with additional resources, such as communications equipment or experienced personnel.

3-49. BCTs emphasize unity of effort with FSF. Ideally, interoperability and shared resources extend to the lowest level. This means having one tactical operations center, dining facility, clinic, motor pool, and so on. Not only does this use resources more efficiently, but it also facilitates understanding, trust, and rapport between U.S. and foreign security forces. For Soldiers to effectively conduct SFA work by, with, and through FSF requires a close rapport and high level of trust.

3-50. Part of COA development for SFA includes an analysis of where to focus the brigade's subordinate units. This analysis includes determining which subordinate units support key elements of FSF. The brigade's subordinate units may support provincial reconstruction teams or transition teams that are not organic to the brigade. Another method is for the brigade may provide organic forces to make up transition teams or the military portion of provincial reconstruction teams or transition teams. Transition teams may include, but are not limited to police, border, point of entry, or military transition teams. Figure 3-2 depicts an example for a brigade providing forces as part of a provincial reconstruction team as well as providing the other transition teams. Figure 3-2 is an example of how a brigade may focus its subordinate units. Additionally, the teams are only examples. All of these teams may not be necessary for all brigades, and some brigades may require other types of advisor teams.

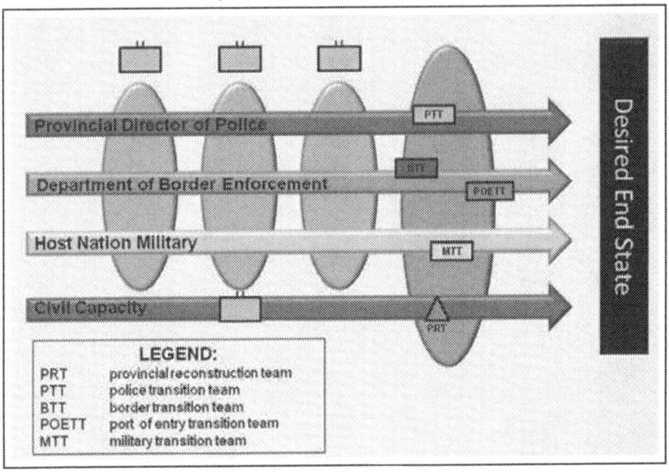

Figure 3-2. Example of brigade focus for subordinate units

PREPARE

3-51. *Preparation* consists of activities performed by units to improve their ability to execute an operation. Preparation includes, but is not limited to, plan refinement; rehearsals; intelligence, surveillance, and reconnaissance; coordination; inspections; and movement (FM 3-0). Like any operation, preparation for SFA creates conditions that improve friendly forces' opportunities for success. It facilitates and sustains transitions, including those to branches and sequels, which are of vital importance for the often dynamic operational environment for SFA. Chapter 5 covers preparation for units conducting SFA, including the BCT.

EXECUTE

3-52. *Execution* is putting a plan into action by applying combat power to accomplish the mission and using situational understanding to assess progress and make execution and adjustment decisions (FM 3-0). Execution of SFA focuses on supporting the needs of the FSF and, like many other operations, is often characterized by phases and transitions between phases. Chapter 4 examines BCT phases for SFA.

Acceptance of Prudent Risk

3-53. SFA, like any other operation, requires prudent, disciplined risk-taking focused on winning rather than preventing defeat. Because uncertainty exists in all operations, every decision involves risk. Among key elements of the art of command are deciding how much risk to accept and minimizing the effects of accepted risk (see FM 6-0). Due to SFA's inherent collaborative nature, determining prudent risk is complex and requires a comprehensive approach.

3-54. Risk reduction measures identified in planning add to the plan's flexibility during execution. A flexible plan can mitigate risk by partially compensating for a lack of intelligence. Again, SFA requires a thorough comprehensive approach to analyzing and agreeing upon risk reduction measures.

Execution Keys to Success

3-55. History and recent operations point to several keys to success for BCTs executing SFA:

- Communication is vital, especially when the BCT shares its operational area with other entities that have cultural differences and lack of communication. It can include FSF or multinational forces, as well as U.S. forces with another mission. For example, an advisor team can develop effective relationships with its foreign security force and local authorities only to have another unit conduct an operation that unhinges those relationships.
- BCTs establish close and continuing relationships with all advisor teams, other actors operating in their operational area, and foreign area officers with local or regional expertise. In addition to advisor teams assigned to military units in direct contact with the BCT, advisors may embed with local police, border, and other forces.
- BCTs establish close and continuing relationships with all foreign units (military, police, and others) operating in their operational areas.
- BCTs establish close and continuing relationships with all political entities and actors in their operational areas.
- Before each operation, as part of the military decisionmaking process, the BCT considers second- and third-order effects of the operation. The team coordinates the commander's intent with advisor teams of FSF and other units in the operational area.
- Unit leaders develop cultural awareness and use this awareness so that operations and relationships achieve the desired end state. In short, the closer a maneuver commander works with advisor teams and the more they interact with local political and cultural leaders, the better the overall chances of mission success.

ASSESS

3-56. The operations process and effective battle command require commanders to continuously assess and lead. Assessment helps commanders to better understand current situations and how SFA is progressing. *Assessment* is the continuous monitoring and evaluation of the current situation, particularly the enemy, and progress of an operation (FM 3-0). Assessment is conducted during planning, preparation, and execution and is evaluated against measures of effectiveness and performances to make decisions and adjustments (see FM 6-0).

3-57. Assessment of SFA requires a comprehensive approach that analyzes the overall operational environment and focuses on FSF and their impact. SFA assessment is performed in close collaboration with the U.S. country team, the host-nation government, and other multinational partners. These partners continually assess the security situation and its influence on other operations. From the assessment, planners develop short-, mid-, and long-range plans. As circumstances change, so do the plans. An insurgency might require employing host-nation forces at earlier stages of development. Some existing security forces may be so dysfunctional or corrupt that the organizations ideally should be disbanded rather than rehabilitated. In some cases, commanders will need to be replaced before their units will become functional. Political realities, however, may make both prospects undesirable. Minimizing and marginalizing particularly corrupt units or leaders while focusing on other units can be effective in this case.

3-58. While every situation differs, BCT leaders assess the following factors throughout planning, preparation, and execution of the operation:

- The operational environment.
- The civil considerations of the operational environment (see appendix B of FM 6-0).
- The core grievances and the threat (possibly including the prerequisites of insurgency).
- FSF elements.
- Methods, successes, and failures of FSF security efforts.
- State of training at all levels, and the specialties and education of leaders.

- Equipment and priority placed on maintenance.
- Logistic and support structure and their ability to meet the force's requirements.
- Extent of acceptance of ethnic and religious minorities.
- Laws and regulations governing the security forces and their relationships to national leaders.

Initial Assessment

3-59. An initial comprehensive assessment offers the BCT a baseline for establishing the scope of effort required. This assessment includes a troop-to-task analysis that determines the type and size of forces needed. FSF may require complete reestablishment or only assistance to increase capacity. They may completely lack a capability (such as internal affairs, corrections, or formal schools for leaders) or they may require temporary reinforcement. As with other military operations, efforts to assist FSF should reinforce success, not failure. For example, instead of building new police stations in every town, improve the good stations and use them as models for other organizations.

3-60. BCT leaders need to make decisions on what shortfalls to address first. The extent of the threats combined with resource limitations inevitably forces commanders to set priorities. However, making efforts to legitimize the host-nation leaders by encouraging them to make the decisions often proves more successful.

3-61. A key effort for SFA is the initial assessment of the DOTMLPF and policy capabilities of a foreign security force. During the early phases of a campaign, a geographic combatant command receives an initial assessment or a unit operating in the area has an assessment of FSF. The incoming BCT updates this assessment as it begins SFA.

3-62. Assessments require objective and subjective reporting, which may require a classification level. At a minimum, a BCT should assess—

- Doctrine.
- Organization.
- Training.
- Materiel and equipment.
- Leadership.
- Personnel.
- Command, control, communications, and intelligence.
- Operational effectiveness.

3-63. The doctrine assessment provides the BCT with an analysis on how the FSF operate. This assessment could focus on formal doctrine or more localized practices.

3-64. The organization portion of the overall assessment provides a perspective on how the FSF are organized. This includes the array of foreign institutions and units.

3-65. The level of training for FSF is a vital consideration. While this assessment is partly subjective, it should objectively examine individual and collective training strategies and mission-essential task lists. In the absence of mission-essential task lists, the BCT should compare FSF to U.S. equivalents.

3-66. The materiel and equipment assessment includes, but is not limited to, weapons, vehicles, equipment, and communications. It should be divided into equipment on hand, equipment readiness, and sustainment. Equipment on hand is quantitative and based on the type of equipment available to security operations. Equipment readiness reflects the operational readiness condition or serviceability of on-hand equipment. Sustainment readiness provides a qualitative and quantitative assessment that reflects how well FSF perform supply, maintenance, transportation, and medical functions. This includes facilities, systems, and structures.

3-67. The leadership assessment analyzes FSF commanders, officers, noncommissioned officers, and other leaders. This includes an assessment of staff members.

3-68. The foreign personnel assessment is quantitative and examines manning, staff manning, and administrative capabilities. It also examines officers, noncommissioned officers, and other personnel.

3-69. The command, control, communications, and intelligence assessment is qualitative and should reflect how well FSF conduct these functions.

3-70. The operational effectiveness assessment examines the foreign security force's ability to execute operations. This portion of the assessment should not be confused with the examination of training, as relatively high levels of training do not necessarily mean that FSF will perform in operations. This assessment is highly subjective.

Assessment During Operations

3-71. Assessment for the BCT conducting SFA should be part of a comprehensive approach. Assessment is directly tied to the commander's decisions throughout planning, preparation, and execution. Critical to effective assessment is developing criteria to evaluate progress toward task accomplishment, achievement of objectives, and attainment of the end state conditions. Assessment criteria are expressed as measures of effectiveness or measures of performance.

3-72. During operations, BCT and FSF commanders and staffs periodically assess the underlying framework of the plan itself. This involves reexamining the logic and assumptions used to develop the original plan, and it requires a deliberately conceived effort to learn as a part of the operation. BCT and FSF collaboration among higher, lower, and adjacent commanders, staffs, and other actors, backed up by qualitative and quantitative assessments, contributes to this learning. Based on this reexamination, commanders update their visualization, direct changes to the order, or develop an entirely new plan to adapt the force to better accomplish the mission.

Chapter 4

The Modular Brigade Augmented for Security Force Assistance

This chapter examines the typical phases for the brigade combat team conducting security force assistance, including a discussion of baseline augmentation by phase. The final section examines some of the subordinate units and elements of the modular brigade augmented for security force assistance.

THE BRIGADE COMBAT TEAM

4-1. The brigade combat team (BCT) is the cornerstone for Army modularity and is designed to operate at the tactical level across the spectrum of conflict. The modular brigade can be augmented, based on the requirements of the operational environment, with enabling assets and capabilities to support distributed security force assistance (SFA). However, the BCT augmented for SFA retains the capability to conduct full spectrum operations—offense, defense, and stability. Any of the three modular BCTs—heavy, infantry, or Stryker—can support SFA. Battalion-sized maneuver, fires, reconnaissance, and sustainment subordinate elements of the BCT support foreign security forces (FSF).

4-2. The gaining theater army force tailors all modular Army forces for use in major combat operations and contingencies. The theater army commander, working for the geographic combatant commander, determines the mix of forces and capabilities required for the BCT. This force tailoring can be done as part of the joint deliberate planning process or as a result of crisis action planning. Force tailoring includes providing BCTs with additional forces, personnel, or capabilities, including the possibility of an embedded provincial reconstruction team. Additional assets and capabilities can include, but are not limited to, command and control, communications, sustainment, engineering, military police, and intelligence.

4-3. In recent operations, the BCT has proven its versatility and capability to learn at the individual and collective levels. It supported other U.S. units conducting SFA and directly increased the effectiveness of FSF. The BCT has operated under varying levels of enemy activity, host-nation capabilities, objectives, and other conditions.

4-4. Because the BCT can operate in nonpermissive and permissive environments, it can conduct SFA across the spectrum of conflict. Thus, it can support civilian and military joint and multinational actors. The support can range from movement security and sustainment to augmentation of reconstruction teams and security for elections.

4-5. BCTs conducting SFA may support FSF development, assist FSF operations, and support and assist the development of host-nation institutions and infrastructure. While providing this support to the host nation, the BCT remains capable of conducting full spectrum operations independently.

BRIGADE COMBAT TEAM PHASES

4-6. If a force lacks the means to accomplish its mission in a single effort, then commanders normally phase the operation. A *phase* is a planning and execution tool used to divide an operation in duration or activity. A change in phase usually involves a change of mission, task organization, or rules of engagement. Phasing helps in planning and controlling and may be indicated by time, distance, terrain, or an event (FM 3-0). Phasing SFA assists in planning and controlling these dispersed and complicated operations. These phases often involve key changes in supporting SFA tasks and types employed.

4-7. BCT phases for SFA are initial response, transformation, and fostering sustainability. These phases mirror the stability operations framework described in FM 3-07 and are based on the operational environment. SFA can start in any phase or may even move to a previous phase due to changes in the conditions of the operational environment.

4-8. Figure 4-1 illustrates an SFA operation that includes all three SFA phases. The horizontal axis depicts an improving security situation, and the vertical axis depicts security capability. The example depicts an ideal situation, although this may not be the case in actual execution. Issues that arise require proper assessment and potentially the use of branches or sequels.

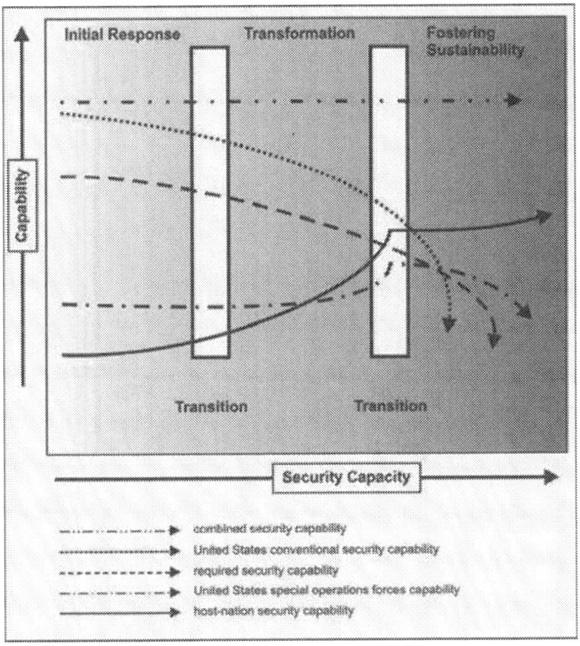

Figure 4-1. Example of phases of security force assistance

4-9. In figure 4-1, the combined security capability represents the overall security capacity of all security forces involved; it is the sum of U.S. conventional security, host-nation security, and the U.S. special operations forces security capabilities. This example assumes that the combined security capability exceeds the required security capability to effectively maintain a secure environment and counter any threat. The U.S. conventional security capability refers to tactical U.S. elements, such as the BCT. It does not refer to higher tactical or operational headquarters or support, sustainment, or medical assets. The U.S. special operations forces capability is depicted as declining in the end of the sustainment phase; as the FSF capability improves, the need for special operations forces declines. The host-nation or regional security organization and U.S. policy determine the end state for special operations.

4-10. For figure 4-1, a major change to enemy resources may require more FSF or additional. Drastic changes to the situation may even require a new operation plan. The figure focuses on capability, not forces, so host-nation security capability may increase without increasing the number or size of host-nation forces.

4-11. Figure 4-2 illustrates how the area of operations for a modular brigade augmented for SFA may change throughout the SFA phases. The differences between the initial response phase and the transformation phase may not change on the surface, but the BCT's relationship with FSF changes drastically. For example, FSF in the north met required conditions so that one of the U.S. brigades conducting SFA was no longer required. The responsibility of the other U.S. brigade thus expanded from providing assistance to one division to two divisions. This brigade's span of control and area of operations similarly expanded. The latter stage of the transformation phase can differ greatly. Areas of operation generally increase in size as they provide SFA to more FSF.

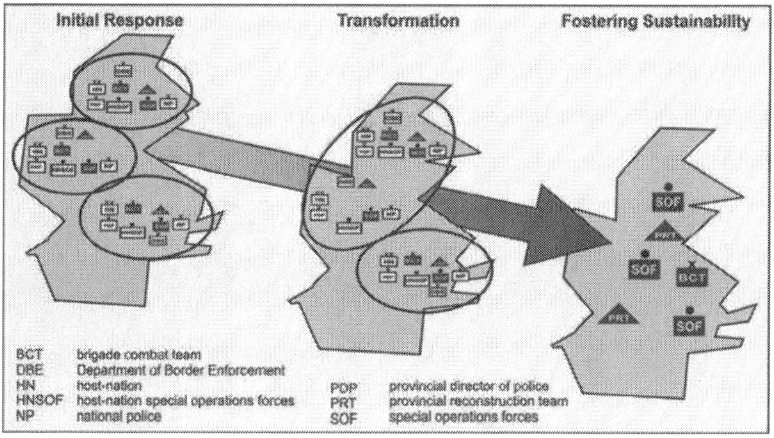

Figure 4-2. Example of changes to a brigade combat team's area of operations

Augmentation By Phase Summary

4-12. BCT augmentation is based on the operational environment and mission. As the three phases are based on the operational environment, they provide a baseline for augmentation. Potential augmentation may require military police, legal, public affairs, civil affairs, psychological operations, engineering, sociocultural experts, sustainment, and military transition team personnel. Table 4-1 illustrates a baseline for augmentation by phase for BCTs.

4-13. An embedded provincial reconstruction team is a key element of a BCT conducting security for assistance. The BCT has the lead on the following tasks: establish civil security, establish civil control (when approved by Congress), and develop and enable foreign security forces. An embedded provincial reconstruction team has the lead for the following tasks: support to economic and infrastructure development, restore essential services, and support to governance. Thus, the BCT and an embedded provincial reconstruction team are able to effectively support FSF and execute all five stability tasks.

Table 4-1. Baseline augmentation for modular brigade

CAPABILITY	INITIAL PHASE (1 HN DIV)	TRANSFORMATION PHASE (2 HN DIV)	TRANSFORMATION PHASE (3 HN DIV)
Police Provincial District	171 - MP Company 2 - MP Team (LTC/SGM) 3 - MP MAJ	171 - MP Company 4 – 2 x MP Team (LTC/SGM) 6 - MP MAJ	171 - MP Company 6 – 3 x MP Team (LTC/SGM) 9 - MP MAJ
Legal HN BDE HN DIV	2 - JAG Team (MAJ/SFC) 2 - JAG Team (CPT/SSG)	4 – 2 x JAG Team (MAJ/SFC) 4 – 2 x JAG Team (CPT/SSG)	6 – 3 x JAG Team (MAJ/SFC) 6 – 3 x JAG Team (CPT/SSG)
Public Affairs	2 - PA (Officer/NCO)	2 - PA (Officer/NCO)	2 - PA (Officer/NCO)
Civil Affairs & Psychological Operations	32 - CA Company – 5 CA Teams 70 – Tactical PSYOP Company – 9 TPTs	32 - CA Company – 5 CA Teams 70 – Tactical PSYOP Company – 9 TPTs	197 - CA Battalion (4 CA Companies) – 20 CA Teams 70 – Tactical PSYOP Company – 9 TPTs
Engineer	8 - FEST-A Team 2 - EOD (CPT/SFC)	8 - FEST-A Team 2 - EOD (CPT/SFC)	8 - FEST-A Team 2 - EOD (CPT/SFC)
Sociocultural Experts	8	8	8
Military Transition Team	56- MTT Augment	4 - MTT Augment	8 - MTT Augment
TOTAL	358	315	493
EPRT	~15-20	~15-20	~15-20
Sustainment Truck Squad Supply Platoon CCT FMST	 21 personnel 32 personnel 4 personnel 4 personnel	 4 personnel 4 personnel	 4 personnel 4 personnel

CA	civil affairs	LTC	lieutenant colonel
CPT	captain	MAJ	major
CCT	contingency contracting team	MP	military police
DIV	division	MTT	military transition team
JAG	Judge Advocate General	NCO	noncommissioned officer
EOD	explosive ordnance disposal	PA	public affairs
EPRT	embedded provincial reconstruction team	SFC	sergeant first class
FEST-A	forward engineering support team – augmenting	PSYOP	psychological operations team
FMST	financial management support team	SGM	sergeant major
HN	host-nation	SSG	staff sergeant
HTT	human terrain team	TPT	tactical psychological operations team

INITIAL RESPONSE PHASE

4-14. The initial response phase seeks to assist FSF stabilize the operational environment in a crisis state. This phase occurs during or immediately after a conflict where the operational environment prevents civilian personnel from operating effectively. The operational environment is typified as nonpermissive.

Thus, the objective of this phase is to improve the security situation, reducing the threat to the populace and creating the conditions that allow civilian personnel to safely operate.

4-15. SFA in the initial response phase is normally required when FSF lack the capability or capacity to provide the required level of security. This phase often requires SFA efforts to help generate and train or assist new and existing FSF. This phase may require a combination of the types of SFA and considerable support, sustainment, and medical resources. BCT activities during the initial response may have to be conducted with multinational major combat operations, to include providing a safe, secure environment for the local populace.

4-16. The BCT's SFA efforts during this phase focus on improving the foreign security force's capability and capacity so all security forces—U.S., other, and foreign security forces—provide a secure environment and reduce the threat. Figure 4-3 illustrates the BCT's span of control. The figure also depicts three BCTs' operational areas in the initial response phase. As security conditions improve, transition to the transformation phase begins.

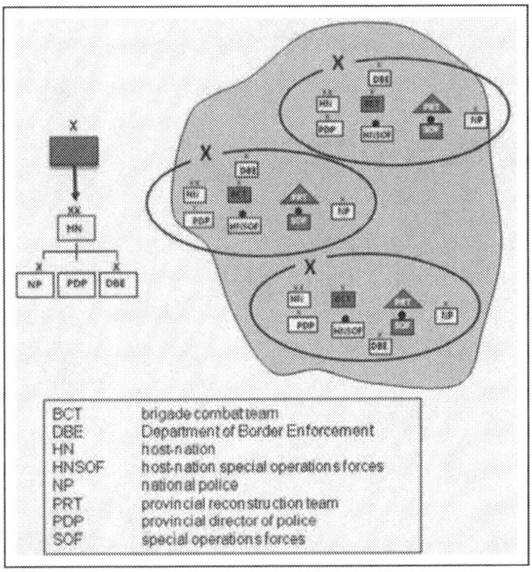

Figure 4-3. Example of initial response span of control and area of operations

4-17. The BCT requires augmentation for the initial response phase. These embedded units provide key enabling capabilities for the brigade, supported FSF, and other supported actors. The embedded provincial reconstruction team is one of the most important of the brigade's augmentation, as it provides interim civilian expertise in developing local institutions to take the lead in national governance, the provision of basic services, fostering economic development, and enforcement of the rule of law. Column two of table 4-1 summarizes augmentation during the initial phase. Although each brigade's mission and conditions in its area of operations are unique, the baseline augmentation for the modular brigade operating in this phase is as follows:

- Embedded provincial reconstruction team.
- Forward engineering support team-augmenting.
- Explosive ordnance disposal team.
- Sociocultural experts.

- Military police company (provincial police transition team).
- Military police team (district police transition team).
- Three military police majors (district police transition team).
- Two judge advocate general teams.
- Civil affairs company.
- Tactical psychological operations company.
- Public affairs team.
- Contingency contracting team.
- Financial management support team.
- Military transition team (specific augmentation personnel include two lieutenant colonels, thirty-one majors, and twenty-three master sergeants).

TRANSFORMATION PHASE

4-18. In the transformation phase, the BCT seeks to assist FSF to stabilize the operational environment in a crisis or vulnerable state. The operational environment in this phase is more permissive than the initial response phase; however, military forces will often be required to provide security to some actors. Activities in this phase normally include a broad range of post-conflict reconstruction, stabilization, and capacity-building efforts, which the embedded provincial reconstruction team is essential for long-term success. Objectives in this phase include continuing efforts to improve the security situation, reducing the threat to the populace, building host-nation capacity across the stability sectors, and facilitating the comprehensive approach to assist FSF.

4-19. The transformation phase represents a broad range of SFA activities that the BCT could perform to support FSF. The initial response phase differs from the transformation phase in the foreign security force's capability to provide for a safe and secure environment. More specifically, FSF have a level of proficiency so they no longer need a permanent U.S. and foreign security forces relationship for tactical operations. However, they may still need full-time advisors and support, sustainment, and medical assistance. The embedded provincial reconstruction team will continue to play a vital role in assisting governance and development efforts throughout this phase.

4-20. Figure 4-4 illustrates the BCT conducting SFA in the transformation phase. In this example, the BCT's span of control expands from assisting one foreign division to assisting two foreign divisions. The figure also depicts how a BCT's operational area has similarly expanded. SFA activities in this phase seek to establish conditions so the host nation's security sector can provide a secure environment with its own security forces.

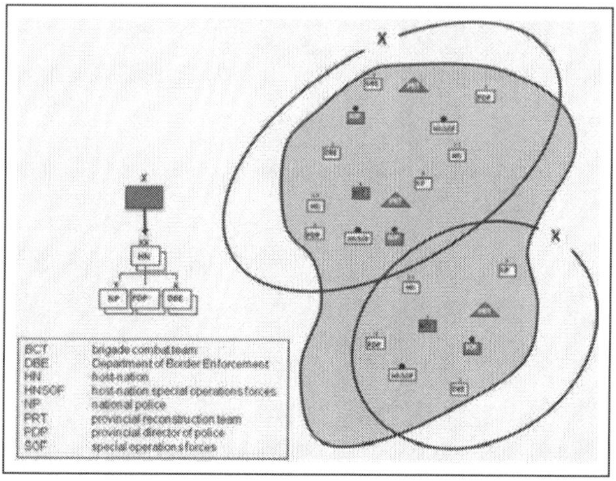

Figure 4-4. Example of transformation phase span of control and area of operations

4-21. The BCT requires augmentation for the transformation phase, like the initial response phase. It often requires similar augmentation due to its greater span of control and area or operations. Column three of table 4-1 summarizes augmentation during the transformation phase with two foreign divisions. The baseline augmentation for the brigade in this phase is as follows:

- Embedded provincial reconstruction team.
- Forward engineering support team-augmenting.
- Explosive ordnance disposal team.
- Sociocultural experts.
- Military police company (provincial police transition team).
- Two military police teams (district police transition team).
- Six military police majors (district police transition team).
- Four judge advocate general teams.
- Civil affairs company.
- Tactical psychological operations company.
- Public affairs team.
- Contingency contracting team.
- Financial management support team.
- Military transition team (specific augmentation personnel include two lieutenant colonels and two master sergeants).

4-22. As conditions of the operational environment improve, the BCT may require a transition to assist three foreign divisions, rather than two divisions as depicted in figure 4-4. This represents a significant expansion of the span of control and area of operations, with a corresponding expansion of required augmentation. Figure 4-5 depicts the BCT supporting three foreign divisions.

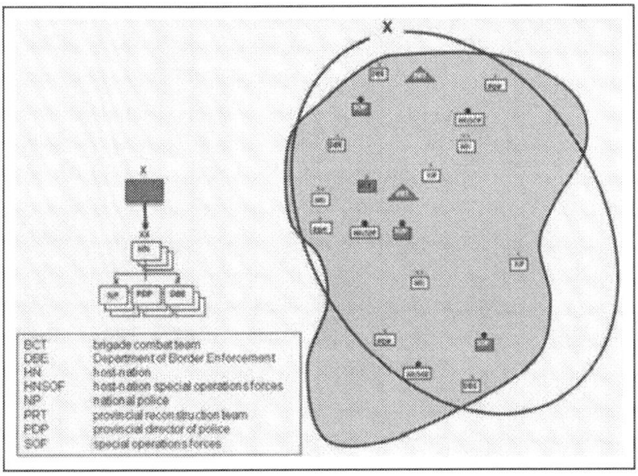

Figure 4-5. Example of a three-division span of control and area of operations

4-23. As in other phases, the BCT requires augmentation to support three divisions. Column four of table 4-1 summarizes augmentation during the transformation phase with three foreign divisions. The baseline augmentation for the brigade in this phase is as follows:

- Embedded provincial reconstruction team.
- Forward engineering support team-augmenting.
- Explosive ordnance disposal team.
- Sociocultural experts.
- Military police company (provincial police transition team).
- Three military police teams (district police transition team).
- Nine military police majors (district police transition team).
- Six judge advocate general teams.
- Civil affairs company.
- Tactical psychological operations company.
- Public affairs team.
- Contingency contracting team.
- Financial management support team.
- Military transition team (specific augmentation personnel include four lieutenant colonels and four master sergeants).

FOSTERING SUSTAINABILITY PHASE

4-24. In this phase, the focus of SFA continues to shift toward assisting institutions required to sustain FSF. This phase encompasses long-term efforts to assist FSF. In this phase, FSF are capable and viable. They conduct independent operations and can provide a safe, secure internal environment. While the brigade may be initially required during this phase, it is normally withdrawn during this phase. The determination to withdraw the brigade is based on the policy and conditions of the operational environment. Special operations forces, provincial reconstruction teams, and other forces may remain to support a theater security cooperation plan. Thus, remaining SFA missions are normally special operations forces missions, not a BCT mission (special operations forces will normally be involved in all phases of SFA).

TRANSITIONS

4-25. SFA can require a vast array of transitions. These transitions are conducted according to the tactical, operational, and strategic conditions identified during planning (which must support the internal defense and development strategy and using an overall comprehensive approach). Transitions can occur simultaneously or sequentially in different levels or war and in separate echelons. This includes potentially having tactical transitions for different units within one brigade's area of operations.

4-26. There are several major transitions. First, the supported unit may be the U.S. BCT in the beginning of an initial response phase but the foreign security forces will transition to be the supported unit. At this point in the transformation phase, the area in which the BCT conducts SFA will expand. This expansion can occur multiple times during the transformation phase, which is based on conditions, especially the capability and capacity of FSF.

4-27. Operations never proceed exactly as planned. To facilitate flexibility, commanders incorporate branches and sequels into the overall plan. Visualizing and planning branches and sequels are important because they involve transitions. Unless planned, prepared for, and executed efficiently, transitions can reduce the tempo of the operation, slow its momentum, and surrender the initiative to the adversary.

SUBORDINATE UNITS

4-28. The modular brigade augmented for SFA has subordinate units whose sole focus is working with FSF. These advisor teams may be formed from brigade organic resources, external augmentation, or a combination. These teams optimally are embedded with the counterpart unit, or they may reside on a U.S. camp and commute to FSF they support. The method depends on U.S. policy, direction from higher headquarters, the conditions of the operational environment, and capacity of the foreign security force's camps to accommodate the U.S. forces.

4-29. The military transition team (MTT) is a key subordinate unit for the brigade conducting SFA. The MTT mission is to assist FSF units. Figure 4-6 illustrates the relationship between a brigade conducting SFA and a MTT assisting a foreign brigade. In this example, the brigade has a battalion and a division MTT as subordinate units, although it would have other subordinate units. The brigade MTT is subordinate to the battalion and has operational control of one of the parent U.S. battalion's company teams.

4-30. When the brigade provides organic forces to form the basis for the brigade MTT, the company team is the foundation of a brigade MTT. The company team will be augmented with additional personnel and assets. The U.S. brigade must facilitate the MTT operations, such as force protection, transportation, communications, and reachback.

4-31. As companies do not have organic intelligence analysts, personnel must be used to form company intelligence support teams (COISTs). A COIST, an ad hoc organization, assists the MTT and its FSF. Ideally, the COIST should consist of six individuals, which will allow for 24 hour operations and the depth in the team to both integrate with the staff and to successfully support the company's mission. The team should consist of an officer, a noncommissioned officer, and four Soldiers, although the cell can be minimally manned with four personnel. The company intelligence cell works most effectively if its work complements the battalion S-2's work.

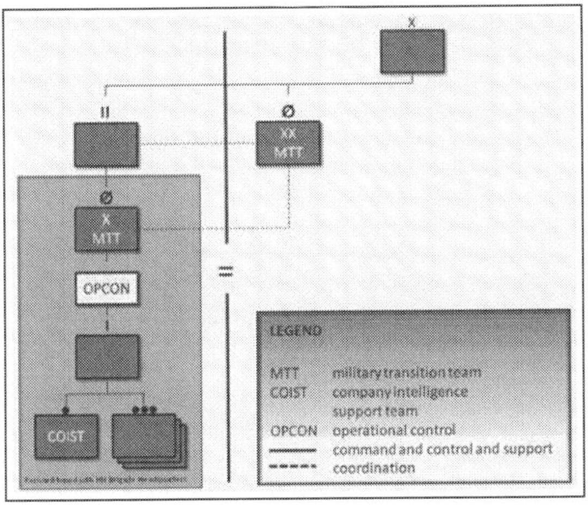

Figure 4-6. Military transition teams

SAMPLE TROOP-TO-TASK MODEL

4-32. The BCT may support assigned to multiple missions, including potentially supporting multiple foreign units in their operational area. Additionally, these FSF organizations may each report through different host-nation government channels and even to different ministries. To synchronize efforts in this case, U.S. forces must achieve unity of command and effort. Similarly, each of the FSF organizational commanders should synchronize their efforts with the host-nation government representatives, as appropriate.

4-33. Figure 4-7 depicts a sample troop-to-task model for a U.S. brigade. It depicts one company team acting as a quick reaction force; thus, the battalion retains the capability to conduct independent full spectrum operations. The figure depicts how U.S. company teams may form the basis of a brigade MTT or a brigade border transition team (and subordinate point of entry transition teams). The figure also depicts how a company team may provide police support. This support is in the form of two platoons supporting police transition teams and a third platoon in a joint security station providing support to a police transition team.

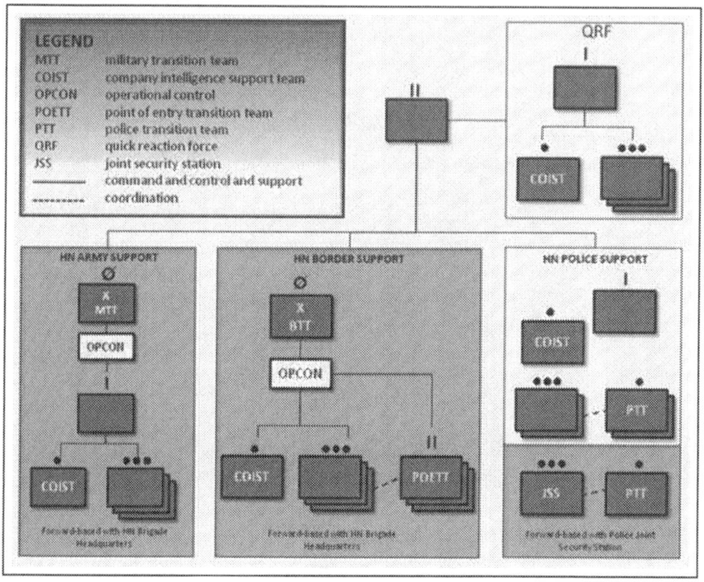

Figure 4-7. Battalions conducting security force assistance

4-34. Subordinate units of BCTs conducting SFA are best located inside a FSF base—not an American base. Co-location facilitates integration with the FSF. It allows the two forces to form mutual understanding and trust. Co-location and the close cooperation it facilitates often improves the population's perception of the legitimacy U.S. and foreign security forces, which can be an essential condition of the overall mission's end state.

4-35. If force protection requires, a U.S. area may be established in the FSF base, although this is not optimal. Key considerations for co-location may include the threat, FSF acceptance, physical space inside the FSF base, sustainment capabilities, medical facilities, and availability of quick reaction forces.

4-36. When U.S. forces are operating out of smaller outposts in an urban environment, the local populace sees the integration and presence of the U.S. and foreign security forces working together. This integration not only enhances overall operational effectiveness and trust, living and working together builds legitimacy of the two forces as well as FSF; it reinforces trust between the FSF and the people they are tasked to protect.

Chapter 5
Unit Operations

Unit operations require continuous activities. Commanders plan for premission, deployment, in-country preparation, employment, redeployment, and postdeployment activities.

PREMISSION

5-1. After receiving a mission, the unit continues detailed preparation activities. The unit obtains supplies, equipment, and training materials and prepares for deployment. Personnel prepare for and rehearse classes to be given in-country. They hone their military skills and conduct extensive area orientation. Paragraphs 5-2 through 5-10 discuss staff actions particular to security force assistance (SFA).

OPERATION PLAN

5-2. The S-3 disseminates the operation plan to unit staff and subordinates. The S-3 ensures predeployment training for Soldiers, to include preparation for training foreign security forces (FSF) and rehearsals for movement. The S-3 reviews the program of instruction for training FSF, to include getting approval from the commander and higher headquarters, if necessary.

CIVIL-MILITARY OPERATIONS

5-3. The civil-military operations section ensures the operation plan minimizes how operations affect the civilian population and addresses ways to mitigate the civilian impact on military operations. Civil-military operations also ensure the operation plan discusses coordination with any augmenting actors. These actors might include civil affairs battalions or brigades, provincial reconstruction teams, or United States Agency for International Development project officers in the operational area. During predeployment training, Soldiers receive training, materials, and briefings on the operational area. This training can cover the history, culture, religion, language, tribal affiliations, local politics, and cultural sensitivities as well as any significant nongovernmental organizations operating in the operational area.

TRAINING

5-4. As discussed in chapter 2, advisors should focus their premission training on the specific requirements of developing FSF. The training emphasizes the host-nation culture and language and provides cultural tips for developing a good rapport with foreign personnel. All premission training for SFA follows the training principles described in FM 7-0.

5-5. Based on the higher commander's training guidance, unit commanders assign missions and approve the draft mission-essential task list that supports SFA. The staff plans, conducts, and evaluates training to support this guidance and the approved mission-essential task list for SFA missions. Commanders prioritize tasks that need training. Since there is never enough time to train in every area, commanders focus on tasks essential for mission accomplishment.

5-6. Once commanders select tasks for training, the staff builds the training schedule and plans on these tasks. The staff provides the training requirements to the commander. After approving the list of tasks to be trained, the commander includes the tasks in the unit training schedule. The staff then coordinates the support and resource requirements with the S-3 and S-4. Finally, the commander ensures standards are enforced during training.

Evaluation

5-7. Evaluations can be either internal or external. Internal evaluations occur at all levels, and they must be inherent in all training. External evaluations are usually more formal and conducted by a headquarters two levels above the unit being evaluated. This subject must be carefully planned and discussed with FSF leaders to account for cultural sensitivities and current capabilities.

5-8. A critical weakness in training is the failure to evaluate each task every time it is executed. Every training exercise provides potential for evaluation feedback. Every evaluation is also a training session. Leaders and trainers must continually evaluate to optimize training. Evaluation must occur as training takes place. Emphasis is on direct, on-the-spot evaluations. However, leaders allow Soldiers to complete the task first. Leaders plan after action reviews at frequent, logical intervals during exercises. This technique allows the correction of shortcomings while they are still fresh in everyone's mind. The after action review eliminates reinforcing bad habits.

Specified Training

5-9. Augmentation elements require area orientation, refresher combat training, field training exercises, and the like. Unit training objectives are for developing capabilities to conduct internal defense and development (IDAD) activities for tactical operations, intelligence operations, psychological operations, populace and resources control operations, and civil affairs and advisory assistance operations in the host-nation language. Units identified for SFA begin intensified training immediately upon deployment notification.

5-10. After deployment to the host nation and before commitment to operations, the unit may receive in-country training at host-nation training centers or at designated training locations. This training helps personnel become psychologically and physically acclimated to the host nation. This training also allows commanders and staff some time to coordinate and plan within their own command and with civilian and military joint and multinational organizations. After commitment, training continues and is stressed between operations, using needed improvements identified in operations as the basis for training.

DEPLOYMENT

5-11. SFA is often conducted in semipermissive environments in which U.S. forces are guests of the host-nation or partner organization. Units move into the operational area by following their deployment standing operating procedures and the operation plan. Paragraphs 5-12 through 5-25 highlight organizations and procedures critical to SFA deployment. This information, however, does not obviate the need for extensive standing operating procedures and an operation plan.

PREDEPLOYMENT SITE SURVEY

5-12. The predeployment site survey unit aims to report accurately to its parent unit the existing FSF mission variables—mission, enemy, terrain and weather, troops and support available, time available, civil considerations (METT-TC). See table 5-1. It also establishes in-country command and control, support, protection, and sustainment relationships for the follow-on unit's mission execution. The site survey coordinates the in-country reception of SFA forces. This coordination wards off most friction during the handoff. This coordination requires all units—outgoing and incoming—to apply a comprehensive approach at all levels of command. Appendix D of FM 3-05.202 discusses site survey procedures in detail.

Table 5-1. Advisor team predeployment site survey checklist

Tasks	*Characteristics*
Area of operations overview	• Map familiarization. • In sector orientation. • Areas of emphasis.
Graphic control measures	• Unit boundaries. • Routes. • Other United States and multinational support locations. • Maintenance support location. • Medical support locations. • Other points of interest.
United States and multinational unit interviews by specialty staff	• Responsibilities. • Battle rhythm. • Required reports. • Standing operating procedures, continuity book. • Stay-behind equipment.
Briefings on advised unit	• Foreign security forces task organization. • Foreign security forces key leadership. • Intelligence assessment. • Unit operational assessment.
Advised unit introductions	• Among commanders. • Among staffs.

Advance Party Deployment

5-13. An advance party deploys to the host-nation or partner organization area of operations after the unit completes its mission planning and the proper commander representative declares it deployable. The advance party conducts predeployment activities and coordinates necessary travel documentation and country clearances. The advance party reviews the rules of engagement and any status-of-forces agreements before deploying. If applicable, the proper staff provides an updated threat briefing to the unit. The S-4 section and support company help with out-loading in the unit area. The unit places all supplies and materials on pallets beforehand and prepares the load manifest.

5-14. Upon arrival, the advance party processes through customs and immediately notifies the higher headquarters of its arrival and status. Often the higher headquarters provides point of contacts to the advance party. These contacts help the advance party obtain the proper identification, documentation, and weapons permits.

5-15. The advance party establishes contact with all U.S. and foreign security forces tactical unit commanders and conducts final premission planning activities. The advance party procures secure working, storage, and living areas for the unit and reconnoiters and prepares the training sites. The advance party coordinates for special support equipment to unload heavy supplies and transport the supplies and personnel to the training site.

Main Body Deployment

5-16. The unit deploys after completing its planning and the appropriate commander declares the unit ready. The unit follows the same procedures performed by the advance party. Members of the advance party meet the main body on its arrival and brief the main body on any required changes to the operation order.

5-17. After arriving in-country, the unit processes through customs and immediately notifies the higher headquarters of its arrival and status. The unit obtains the proper identification, documentation and weapons permits from U.S. or other sources, if applicable. Often the security assistance organization provides to the unit the necessary point of contacts that expedite this process.

5-18. The unit off-loads equipment and personnel onto vehicles for movement to the training site, or it stores the equipment in a secure area until the processing is completed. FSF secure the equipment but the unit members remember the commander is ultimately accountable for the advisor team's physical and personal security. Supply personnel stay with the equipment and help to guard it.

IN-COUNTRY PREPARATION

5-19. Upon arrival, the commander and S-3 brief the higher headquarters on the planned execution of the mission and reconfirm the required command relationship. Local conditions may require the unit to confirm or establish its in-country and external command and control, support, and sustainment relationships from outside its operational area upon arrival. The unit establishes direct working relationships with its next higher in- or out-of-country supporting element to—

- Determine the limits of the available support and expected reaction time between the initiation of the support request and its fulfillment.
- Confirm or establish communications procedures between the supporting element and the unit, to include alternative and emergency procedures for command and control, all available support, and medical evacuation.

5-20. The unit establishes procedures to promote interagency cooperation and synchronization. The unit—

- Identifies the location of the concerned host-nation, U.S., or other agencies.
- Contacts the concerned agency to establish initial coordination.
- Exchanges information or intelligence.
- Confirms or establishes other coordination protocols as necessary.
- Incorporates the newly established or changed procedures into the plans for mission execution.

5-21. The unit immediately establishes operations security procedures to support its mission execution and identifies rally points incorporated into its defensive, evasion, and personnel recovery plans.

5-22. After receiving a detailed briefing and further guidance from the advance party, unit personnel continue to develop effective rapport with the FSF commander and counterparts. They also assess their working, storage, and living areas for security and verify the location of the training site, communications center, dispensary area, and FSF troop area. With the FSF commander, the unit commander—

- Establishes rapport.
- Conducts introductions in a businesslike, congenial manner.
- Briefs on the unit's mission, its capabilities, and the restrictions and limits imposed on the detachment by the higher U.S. commander.
- Ensures all unit personnel fully support FSF and firmly believes a joint U.S.-FSF effort will succeed. Requests counterpart linkup be made under the mutual supervision of the FSF commander and the unit commander.
- Ensures all current unit plans are tentative and that assistance is needed to finalize them.
- Deduces or solicits the actual estimate of unit capabilities and perceived advisory assistance and material requirements.
- Recommends the most desirable courses of action (COAs) while emphasizing how they satisfy present conditions, achieve the desired training, and meet advisory assistance goals.
- Informs the higher in-country U.S. commander of any significant changes in the unit's plan to assist FSF.

5-23. The S-2 supervises the dissemination of intelligence and other operationally pertinent information within the unit and, as applicable, to higher, lower, or adjacent units or agencies. The S-2 also—

- Monitors the implementation of the intelligence collection plans to include updating the commander's priority intelligence requirements (PIRs), conducting area assessment, and coordinating for additional intelligence support.
- Establishes liaison with FSF intelligence and security agencies (within the guidelines provided by applicable higher authority). Assesses the intelligence threat and resulting security requirements, including coordination with the S-3 on specific security and operations security measures.

5-24. Through the S-2, the commander's PIRs are based on the latest information available and requirements for additional PIRs that arise from modified estimates and plans. The S-2 also—

- Analyzes the foreign unit's status to finalize unit plans for advisory assistance. These plans can include task organization of unit with counterparts, staff functions for planning SFA, and advisory assistance for executing SFA.
- Explains analysis to counterparts and encourages them to help with—and participate in—analyzing, preparing, and briefing the analysis to the foreign unit commander.
- Prepares and briefs the plans for training and advisory assistance.
- Helps the foreign unit inspect the available facilities to identify deficiencies. If the unit finds deficiencies, the S-2 prepares estimates of COAs for the FSF commander to correct them.
- Supervises the preparation of the facilities with their counterparts and informs unit and FSF commanders on the status of the facilities.

5-25. The unit ensures its security is based on the present or anticipated threat. Some recommended actions the unit may take include—

- Hardening its positions based on available means and requirements to maintain low visibility.
- Maintaining unit internal guard system with at least one Soldier who is awake and knows the locations of all other unit personnel. The guard reacts to an emergency by following an internal alert plan and starting defensive actions.
- Maintaining communications with all subordinate unit personnel deployed outside the immediate area controlled by the main body.
- Establishing plans for immediate defensive actions in the event of an attack or a loss of rapport with hostile reaction.
- Discussing visible security measures with foreign counterparts to ensure understanding and to maintain effective rapport. Unit personnel do not divulge sensitive information for the sake of possible rapport benefits.
- Encouraging the foreign unit, through counterparts, to adopt additional security measures identified when analyzing the foreign unit's status and inspecting its facilities.
- Coordinating defensive measures with the foreign unit to develop a mutual defensive plan. Unit personnel obtain from the unit's present reaction and defensive plans for attack. They encourage the foreign unit to conduct mutual full-force rehearsals of defensive plans; if unsuccessful, the unit conducts internal rehearsals of the plans.

EMPLOYMENT

5-26. Once the unit has deployed and arrived in-country, it begins employment. Employment is the conduct of SFA to support the FSF commander. Employment occurs within the advisor team itself. It includes foreign counterparts in the mission planning (preparing the FSF for the mission itself) to increase the capability and capacity of their planning processes, as well as increase the probability of success.

PERSONNEL

5-27. During employment, individual unit personnel perform functions described in paragraphs 5-28 through 5-38.

Commander

5-28. Before the mission, the U.S. commander advises and assists the FSF commander. The latter issues planning guidance for planning the execution of the mission and clarifies commander's intent. The U.S. commander advises and assists the FSF unit commander in the command and control process of tactical operations. By accompanying the FSF unit commander when the mission is received from higher headquarters, the U.S. commander assists any subsequent missions. The U.S. commander monitors how FSF understand the commander's intent and all specified or implied tasks.

5-29. During execution of the mission, the U.S. commander helps the foreign unit commander provide command and control during operations. After monitoring the tactical situation, the U.S. commander recommends changes to the COAs to exploit the situation. After monitoring the flow of information, the U.S. commander recommends improvements to use intelligence collection assets and to keep subordinates reporting required information.

Staff

5-30. Before the mission, the advisor staff advises and assists the foreign counterparts in preparing estimated COAs for essential tasks. The staff helps write tentative plans based on the planning guidance and unit's standing operating procedures. These plans include primary, alternate, contingency, and emergency plans.

5-31. During execution, the staff helps foreign counterparts coordinate the execution of the tasks. They disseminate portions of their plans to other personnel, senior and subordinate staff sections, and supporting elements. They help notify higher, lower, or adjacent staff sections of modified estimates and plans. Finally, the staff—with the S-2 and the S-2 counterpart—helps update the commander's critical information requirements with the latest information and requirements that arise from operations.

Executive Officer

5-32. The executive officer performs the organizational analysis of the unit's coordinating staff sections to ensure efficiency during the planning process according to initial planning guidance. With the foreign counterpart, this officer directs foreign staff sections as they develop estimates and plans. The executive officer monitors the liaison and coordination with FSF higher headquarters, recommending changes to improve efficiency.

Adjutant

5-33. The adjutant provides advice, assists, and makes recommendations to the foreign counterpart for conducting human resources. This includes monitoring the maintenance of foreign unit strength, accountability of casualties, unit morale, and postal activities. This may also include concerns with the foreign pay system, leave procedures, and casualty pay procedures.

Intelligence Officer

5-34. The intelligence officer advises and assists the monitoring of FSF operations security to protect classified and sensitive material and operations and recommends improvements. By helping the foreign counterpart update the situation map, the intelligence officer helps to keep the map current. The intelligence officer recommends improvements to the standing operating procedures of the tactical operations center communications so the intelligence section receives situation reports. This officer helps the counterpart monitor the collection, evaluation, interpretation, and dissemination of information. The intelligence officer assists in the examination of captured insurgent documents and material. This officer helps gather and disseminate intelligence reports from available sources to ensure the exploitation of all assets. The intelligence officer helps the counterpart to brief and debrief patrols operating as a part of reconnaissance and surveillance activities. Finally, the intelligence officer assists, with the advisor operations officer, developing reconnaissance and surveillance plans.

Operations Officer

5-35. The operations officer helps the foreign counterpart to prepare tactical plans using estimates, predictions, and information. This officer monitors command and communications nets, assists in preparing all orders and plans, and helps to supervise the training and preparation for operations. Finally, the operations officer monitors the planning process and makes recommendations for consistency with IDAD goals.

Supply Officer

5-36. The supply officer advises and assists the foreign counterpart in maintaining equipment readiness; monitoring the support provided to the foreign unit, its subunits, and attachments; and recommending improvements. Finally, this officer helps to supervise the use of transportation assets.

Civil-Military Team

5-37. Upon deployment, civil-military teams advise the commander and staff on civil-military considerations and coordinate efforts of any civil affairs units supporting the unit. The civil-military team mentors the counterpart team on the supported foreign element staff on civil-military operations, the legal and moral obligations of military commanders to civilian populations under their control, and the importance of respecting human rights. The civil-military team may introduce the counterpart to relevant nongovernmental organizations, United States Agency for International Development project officers, and provincial reconstruction team staff.

Company-Level Advisors

5-38. Advisors sent to elements below battalion level help foreign counterparts to analyze the mission and commander's intent from higher headquarters. They assist FSF leaders restate the mission, conduct an initial risk assessment, identify a tentative decisive point, and define their own intent. They also assist their foreign counterparts to analyze the mission variables. From these variables, advisors help their foreign counterparts to develop a COA that meets the higher headquarters concept of operations and commander's intent. Finally, they advise and assist in the conduct of operations and the flow of information to the FSF higher commander.

TRAINING

5-39. Training FSF during SFA parallels the training of U.S. forces. Training follows the principles of training described in FM 7-0. This training tailors the objectives to account for cultural differences and focus on the FSF mission and problem set.

Training Assessment

5-40. The commander of the advising unit begins with a training assessment of training plans designed during predeployment and passed on from the preceding U.S. units conducted in coordination with the FSF commander. This assessment is important to both evaluate the foreign security force but also as an exercise to establish and enhance the working relationship between U.S. and foreign security forces. The training assessment should cover all aspects of leadership, training, sustainment, and professionalization. To support an assessment, the advising unit should analyze the following specific foreign unit considerations:

- The unit's mission and mission-essential task list and capability to execute them.
- Staff capabilities.
- Personnel and equipment authorization.
- Physical condition.

- Any past or present foreign influence on training and combat operations.
- Operational deficiencies identified during recent operations or exercises with U.S. personnel.
- Sustainment capabilities, to include sustainment training programs.
- Internal training programs and personnel.
- Training facilities.

5-41. The level of professionalism of FSF, both units and individuals, should also be assessed. Adhering to established rules of engagement, ethics that meet the established laws and regulations of the commanding authority, laws for land warfare, and human rights, promotions based on merit and the security forces' support of civilian leaders and political goals all fall within this assessment.

5-42. The advisor unit, working with FSF leaders, also evaluates current members of the security forces for past military skills and positions. Often military reorganizations arbitrarily shift personnel to fill vacancies outside their knowledge and experience.

Training Plan

5-43. After completing the training assessment, advisors analyze the prepared training plan and determine if changes are necessary. The training plan stresses the deficiencies identified in the training assessment. It also identifies those in the host nation able to help train FSF and to strengthen the legitimacy of the process. Finally, the training plan should consider the unit's eventual self-sustainment. As the foreign security force gains sufficient capacity and capabilities to perform independently, advisors transition from a leading role to a mentoring role.

5-44. The plan for training FSF uses a comprehensive approach, supports the IDAD strategy, and has many echelons. Working with other echelons and actors within the area of operations can provide support and expertise that enhance the training and operations process as well as limit redundancy.

Program of Instruction

5-45. In coordination with the FSF commander and training staff, the advising unit develops a program of instruction. This program incorporates all training objectives that satisfy the training requirements identified during the assessment. The training program must support these requirements. The FSF commander approves the program of instruction prior to execution. When executing the program of instruction, advisors adhere to a training schedule consistent with changes in the mission variables. Advisors ensure through their counterparts and the FSF commander that all personnel receive training. Foreign counterpart trainers rehearse all classes approved on the program of instruction.

Presentation of Instruction

5-46. To present the training material properly, trainers follow the lesson outlines approved in the program of instruction. All training clearly states the task, conditions, and standards desired during each lesson, ensuring the foreign students understand them. Trainers state all warning and safety instructions in the host-nation language. The training to reinforce the concepts includes demonstrations of the execution of each task, stressing the execution as a step-by-step process. Trainers monitor students' progress during instruction and practical exercises, correcting mistakes as they are made.

Training Methods

5-47. An effective method of training used by U.S. forces is the crawl-walk-run method of teaching individual tasks, battle drills, collective tasks, and field exercises. (See FM 7-0.) This method is employed to develop well-trained leaders and units. During all phases, the training must include the mission of the unit in the context of the higher unit's mission to assist with the practical application of the training. Identifying the higher commander's mission and intent, as well as the tasks and purposes of other units in the area, would also add context to the training. This method can also be expanded to include the role of other actors.

5-48. At a minimum, units continue individual training to improve and sustain individual task proficiency while training on collective tasks. Collective training requires interaction among individuals or organizations to perform tasks, actions, and activities that contribute to achieving mission-essential task proficiency. Collective training includes performing collective, individual, and leader tasks associated with each training objective, action, or activity.

Collective Training

5-49. Collective training starts at the squad level. Squad battle drills provide key building blocks that support operations. Trainers can link battle drills and collective tasks through a logical, tactical scenario in a situational training exercise. Although this exercise is mission-oriented, it results in more than mission proficiency. Battle drills and collective tasks support situational training exercises, while these exercises support operations. Advisors must understand the operational environment when training FSF; training incorporates how internal and external threats and civilians affect the environment. Flexibility in using U.S. doctrine in training enhances efforts to make training realistic. Trainers modify U.S. doctrine to fit the FSF level of expertise, their command and control, the tactical situation, and sustainment base. Often the structure and capabilities of FSF differ from that required by U.S. doctrine. When FSF counter an insurgency, these exercises emphasize interplay among psychological and tactical, populace and resources control, intelligence, and civil affairs operations. Similarly, scenarios for each tactical exercise must include interplay among the various possible aspects of IDAD.

Individual Training

5-50. Training individuals of FSF emphasizes physical and mental conditioning, tactical training, basic rifle marksmanship, first aid, combatives, and the operational environment. Individual training also includes general tactics and techniques of security operations and the motivation, operations, and objectives of internal and external threats. Tough and realistic training conditions troops to mentally and physically withstand the strain of continuous offensive operations.

5-51. Foreign personnel cross-train on all types of weapons, communications and other equipment, and skills particular to their unit. Personnel losses must never cause weapons, communications equipment, or essential skills to be lost due to a lack of fully trained replacement personnel.

Small-Unit Leader Training

5-52. SFA frequently entail rapidly changing circumstances; thus, FSF small-unit leaders must be able to plan and execute operations with little guidance. Therefore, trainers stress small-unit leadership training which can occur concurrently with individual training. Tools the trainer uses to train leaders are manuals, previously established training, tactical exercises without troops, battle simulations, and unit missions. This training develops aggressiveness, tactical proficiency, and initiative. Small-unit leader training should include combined arms technical training such as procedures for forward observer and close air support. Leadership training must also include land navigation in difficult terrain and under conditions of limited visibility. Mission readiness and the health and welfare of subordinates are also continuous parts of training.

After Action Review

5-53. In any training, the after action review provides the critical link between training and evaluation. It is a professional discussion that includes the training participants and focuses directly on the training goals. An after action review occurs after all collective foreign unit training. Effective after action reviews review training goals with the concerned FSF unit commander. Advisor personnel ask leading questions, surface important tactical lessons, explore alternative COAs, assist the retention teaching points, and keep the after action review positive.

5-54. The advisor encourages the FSF unit commander to review the training with the entire unit or key subordinate leaders. At this review, the FSF commander stresses how to strengthen the chain of command. During the after action review, advisors avoid criticizing or embarrassing the FSF unit commander or subordinates.

5-55. If possible, after action reviews occur during the field portion of the training when the unit assembles at logical stopping points. These reviews provide feedback from instructors and leaders that increases and reinforces learning. This feedback provides a richer database for key points. During the after action review, the senior evaluator draws information from unit leaders to form possible alternative COAs.

Assessment

5-56. In all operations, including SFA, success is generally measured for three periods: short-, mid-, and long-term. Success in SFA is normally defined within the context of these three periods. In the short-term period, FSF make steady progress in fighting threats, meeting political milestones, building democratic institutions, and standing up security forces. In the mid-term period, FSF lead fighting threats and provide security, have a functioning government, and work towards achieving economic potential. In the long-term period, FSF are peaceful, united, stable, and secure; integrated into the international community; and a full partner in international security concerns.

Measures of Effectiveness

5-57. A *measure of effectiveness* is a criterion used to assess changes in system behavior, capability, or operational environment that is tied to measuring the attainment of an end state, achievement of an objective, or creation of an effect (JP 3-0). A measure of effectiveness provides a benchmark against which the commander assesses progress toward accomplishing the mission.

5-58. The following are tools to assess the effects of operations in meeting expectations defined for the short, mid, and long term. Although all SFA is not focused on a host nation, these tools are based on the four IDAD pillars:

- **Balanced development** attempts to achieve national goals through political, social, and economic programs.
- **Security** includes all activities implemented to protect the populace from the threat and to provide a safe environment for national development.
- **Neutralization** is a political concept that encompasses the physical and psychological separation of the threatening elements from the population.
- **Mobilization** provides organized manpower and resources and includes all activities to motivate and organize popular support of the government.

Measures of Performance

5-59. A *measure of performance* is a criterion used to assess friendly actions that is tied to measuring task accomplishment (JP 3-0). This quantitative analysis determines whether the task or action was performed to standard. The standard may be derived from a procedure or may be time-sensitive. Examples of measures of performance are listed in the appropriate Army tasks in FM 7-15.

5-60. The advisor helps the foreign counterpart to monitor the current situation for unanticipated successes, failures, or enemy actions. As the commanders assess the progress of foreign security operations, they look for opportunities, threats, and acceptable progress. Throughout the operation, the commanders visualize, describe, and direct changes to the operation, feeding the assessments of progress or regression back into the planning process.

REDEPLOYMENT

5-61. Redeployment is the return of forces and materiel to the home or mobilization station. Before redeployment, a unit conducting SFA inventories all supplies and equipment (jointly, if possible) to be passed on to the foreign unit or the relieving U.S. or multinational unit. These actions ensure all items are accounted for and the custody chain for property and equipment is unbroken. The outgoing unit also—

- Prepares a final evaluation report showing its estimate of the foreign unit's capabilities and giving an opinion of its future employment.
- Discusses the foreign unit's performance with its commander. The unit submits a copy of the foreign unit's final evaluation to the next higher FSF commander.

- Passes custody of training schedules, lesson plans, foreign operational records, and the foreign unit's final evaluation to the foreign unit and relieving U.S. or coalition unit.
- Presents debriefings, after action reviews, and the foreign unit's final evaluation to the higher in-country U.S. commander.

5-62. The outgoing unit commander supervises the redeployment. This commander approves prepared redeployment plans or, as necessary, directs changes to the redeployment plans. The commander and operations officer assess and select alternatives or changes to the present redeployment COAs based on the intelligence officer's intelligence estimate. They develop and disseminate a fragmentary order for selected alternatives or changes.

5-63. Unit personnel prepare all accompanying supplies and equipment for shipment. They brief ground support personnel on equipment handling procedures, routes, convoy procedures, and actions to take if a terrorist or insurgent incident occurs. They load personnel and equipment for movement to the departure airfield or airport following the load plan in the unit's operation order. They maintain accountability for all their personnel, supplies, and equipment.

Mission Handoff Procedures

5-64. During long-term SFA, conditions determine the rotation of in-theater units. Time is not the only governing factor. Changes in the operational environment may require reshaping force packages as situations change. In addition, internal administrative concerns might prompt or support a commander's decision to rotate units. Regardless, mission handoff is necessary and defined as the process of passing an ongoing mission from one unit to another with no discernible loss of continuity.

5-65. The overall authority for the handoff and the subsequent transfer of authority lies with the commander ordering the change. The authority for determining the handoff process lies with the incoming commander assuming responsibility for the mission. This changeover process may affect conditions under which the mission will continue.

Redeployment Considerations

5-66. Although intended for a direct handoff between U.S. units, commanders must make specific considerations when making a handoff to a multinational force. The considerations include mission, enemy, terrain and weather, troops and support available, time available, and civil considerations. For units relieved of a function by a government agency, procedures typically entail longer handoff times and more complex coordination. However, the other areas of consideration still apply and may in fact be a greater issue for an agency. Outgoing units that have past, present, or future projects planned with agencies prepare to transfer these projects to responsible agents in the incoming unit.

Mission

5-67. The outgoing unit must prepare a detailed study of their unit's mission statement and detail the current and implied mission tasks. The mission may also require an incoming unit with additional skill sets, such as advanced special operations, near real-time connectivity, civil affairs functional specialists, or complex media production ability. Knowing the mission, commander's concept of the mission, commander's critical information requirement, PIRs, and information requirements helps the incoming unit understand the mission. After a complete in-depth study of the operational area, the outgoing unit should help the incoming staff complete the handoff in a manner that allows for continued, uninterrupted mission accomplishment. The changeover must not allow the adversary to gain any operational advantages.

Enemy

5-68. The outgoing unit must provide the latest available intelligence on all threats that affect the mission. This intelligence includes comprehensive data on terrorists and terrorist-related incidents over the previous several months. In addition to the normal intelligence provided to the incoming unit on a regular basis, the situation may call for a liaison from the outgoing unit. Operations security is critical to prevent the enemy from discovering the impending handoff and then exploiting the fluidity of the change and the concentration of U.S. or multinational forces.

5-69. The in-country unit provides continuous information updates to the incoming unit. PIRs and information requirements were established for the original mission along with operational, strategic, and tactical information. The incoming unit must become familiar with the ongoing and upcoming mission PIRs and information requirements along with their links to planned nonlethal effects.

Troops and Support Available

5-70. To the incoming unit, learning about the friendly forces is as important as knowing the enemy situation. The unit must be familiarized with the command and control structure with which it will work daily. The incoming unit must know all friendly units in adjacent operational environments. It also must be aware of joint and multinational units and the capabilities of their mission support base. The incoming unit must also be aware of other operations, units, and their capabilities.

5-71. Mission handoff is the time for the outgoing unit leadership to introduce the incoming unit leadership to leaders of FSF. This practice helps prevent breaks in intelligence networks already developed by the outgoing unit and reduces the time the incoming unit needs to build trust. The outgoing unit supports the incoming unit's plans to prepare for the transition in counterpart relations. However, potential or anticipated friction between a foreign security force and the incoming unit may cause the relief to occur more slowly. Therefore, incoming and outgoing units need time for overlap to allow for in-country, face-to-face contact with their counterparts before the mission handoff. If possible, outgoing unit members should provide biographical data on their counterparts to include photographs. This information allows unit members to become familiar with their counterparts before deployment and to determine which advisor techniques may need more emphasis. Execution of the mission must continue within the capabilities of the incoming unit, the foreign unit, and the available supporting assets.

Time

5-72. The depth and dispersion of units and the number of operations conducted determines the time required to exchange units. Ideally, an overlap occurs to allow the incoming unit to become familiar with the operational environment and to establish rapport with their foreign counterparts. However, the handoff must occur as quickly as possible. The longer it takes, the more vulnerable become personnel in the operational environment. However, both outgoing and incoming units should not sacrifice continued and uninterrupted execution of ongoing operations for speed. The incoming unit needs enough time to observe training techniques and procedures and to conduct debriefing on lessons learned. Outgoing unit personnel need sufficient time to put incoming unit personnel in contact with agency counterparts.

Civil Considerations

5-73. The outgoing unit provides an in-depth area study, giving close attention to local problems. Often sister units can provide general demographic data that can be expanded upon for unit-specific needs. Popular support for U.S. activities occurring in the operational environment may directly influence changes in the mission statement. The outgoing unit provides this critical information and describes all completed and progressing civic action projects. The incoming unit considers how the host-nation government functions and international civilian or government agencies are involved in or influence the situation in its operational environment.

POSTDEPLOYMENT

5-74. The unit commander debriefs to provide an overview of the mission and all relevant information subsets. Topics can range from military geography, political parties, and military forces to insurgents, security forces, and ongoing agency operations. Concerns specific to a foreign security force and relevant in one SFA operation may not be relevant in another operation. The commander should codify and debrief any information subset that may affect mission success.

Postmission Debriefing

5-75. Redeployment is not the end of the mission. Upon arrival at the redeployment location, the unit undergoes an extensive debriefing. The intelligence staff section officer typically organizes and conducts the debriefing. The intelligence officer coordinates with higher-level intelligence organizations to take part in the returning unit's debriefing, particularly if other organizations tasked the unit to obtain information. All deployed personnel, to include attachments, must be available for the debriefing.

Documentation

5-76. After the debriefing, the unit commander and staff prepare two documents—an after action review and a report of lessons learned.

5-77. The after action review states the who, what, when, where, and how of the operation. It is a permanent record of the major activities of the unit from receipt of mission to debriefing. As such, it is an extremely important template on which past missions may be compared and future missions planned. Within 48 hours of being debriefed, the unit normally submits an after action review through command channels to the higher command. The intelligence and operations officers at each echelon keep copies of unit after action reviews.

5-78. Shortly after completion of the after action review, or simultaneously with its submission, the unit submits a report of lessons learned. This report documents the commander and staff's reflection of the operation and recommendations for the future. Units often prepare the lessons according to the six warfighting functions: movement and maneuver, intelligence, fire support, sustainment, command and control, and protection. This method addresses what worked and did not work during the operation, why it did or did not work, and what changes or substitutions are needed for existing tactics, techniques, and procedures in the unit.

5-79. As a minimum, the lessons learned report is forwarded to appropriate Service- and joint-level lessons learned agencies to allow integration into subsequent unit training and leader education. These agencies might include the Center for Army Lessons Learned and Joint Center for Operational Analysis.

Chapter 6
Sustainment

A brigade combat team may be required to execute operations across the spectrum of conflict. Hence, providing security force assistance (SFA) may require the sustainment staff to analyze the mission to determine support requirements and available capabilities. Since brigade combat teams conducting SFA may be responsible for a large operational area, an accurate and thorough assessment of U.S. assets as well as the foreign security force's assets is critical to mission success. The sustainment planner understands that though the mission can change, methods by which logistics are distributed within a theater of operations do not. Army sustainment enables the commander to execute a mission and sustain the force through full spectrum operations. Sustainment assets are allocated based on those requirements; no standard arrangement fits all situations.

As discussed in chapter 2, one of the principles of SFA is to ensure long-term sustainment. This principle dictates that all policies, strategies, and programs must be durable, consistent, and sustainable by foreign security forces (FSF) and their government once U.S. forces have completed assistance. This can be accomplished by developing the foreign security force's capability and capacity for asset production, management, performance, and maintenance.

SUSTAINMENT IN FOREIGN SECURITY FORCES CAPACITY

6-1. Sustainment is the provision of the logistics, personnel services, and medical support necessary to maintain operations until mission accomplishment. The endurance of SFA forces is primarily a function of their sustainment. Sustainment determines the depth to which the forces can conduct decisive operations, allowing the commander to seize, retain, and exploit the initiative. Endurance is the ability to employ combat power anywhere for protracted periods. Endurance stems from the ability to generate, protect, and sustain a force. It involves anticipating requirements and making the most effective, efficient use of available resources. Sustainment also enables strategic and operational reach. SFA forces require strategic and operational reach to deploy and immediately conduct operations anywhere with little or no advanced notice. FM 4-0 discusses sustainment in detail.

CORE LOGISTIC CAPABILITIES

6-2. Similar to any operation, the core logistic capabilities in SFA provide a framework to facilitate integrated decisionmaking, synchronize and allocate resources, and optimize logistic processes. The challenges associated with support cut across all the core logistic capabilities: supply, maintenance, deployment and distribution, engineering, medical support, logistic services, and operational contract support.

Supply

6-3. Supply operations include identifying requirements; selecting supply sources; scheduling deliveries; receiving, verifying, and transferring products; and authorizing supplier payments. To provide responsive supply operations in SFA to both the advising organization and FSF, logistic planning requires a collaborative environment that includes operations planning, maintenance operations, and the distribution system.

Maintenance

6-4. In SFA, maintenance operations aim to reduce repair cycle times by replacing components, modules, and assemblies as far forward as possible as well as maximizing reliance on parts distribution, visibility, and replacement. Advisors and their foreign counterparts require maintenance capacity for their freedom of action. Force movement and maneuver depends on sufficient readiness levels of transportation equipment and weapons systems. In SFA, maintenance encompasses both intermediate and organizational levels and consists of shop maintenance supporting the supply chain, transportation assets, and weapon systems.

Deployment and Distribution

6-5. The deployment and distribution capability moves forces and logistic support on time by meeting the required delivery date and providing time-definite delivery. By sharing critical information, units create unity of effort among diverse distribution organizations. This unity of effort enhances foreign support efforts and provides end-to-end support to satisfy deployment execution and sustainment operations for advisors and host-nation forces.

Engineering

6-6. Engineering provides capabilities to maximize freedom of action. These include capabilities and activities that support the maneuver of land combat forces; modify, maintain, or protect the physical environment; and contribute to a clear understanding of the physical environment by providing geospatial information and services to commanders and staffs.

Medical Support

6-7. A component of Army health systems oversee operational management of the health service support and force health protection missions for training, predeployment, deployment, and postdeployment operations. Army health system includes all mission support services performed, provided, or arranged by Army Medical Department to support health service support and force health protection mission requirements for civilian and military joint and multinational forces. For more information on Army health systems, see FM 4-02.

Health Service Support

6-8. All support and services performed, provided, and arranged by Army Medical Department promote, improve, conserve, or restore the mental and physical well-being of Soldiers and personnel in other Services, agencies, and organizations as directed. This includes casualty care, medical evacuation, and medical logistics. The health service support mission is a function of the sustainment warfighting function.

Force Health Protection

6-9. Force health protection measures promote, improve, or conserve the mental and physical well-being of Soldiers. These measures enable a healthy and fit force, prevent injury and illness, and protect the force from health hazards. They include some prevention aspects of Army Medical Department functions, combat and operational stress control, dental services, and laboratory services. The force health protection mission is a function of the protection warfighting function.

Logistic Services

6-10. Logistic services comprise the support capabilities that collectively enable the United States to rapidly provide global sustainment for U.S. forces. Logistic services include many disparate activities that are highly scalable capabilities. Included in this area are food, water and ice, base camp, and hygiene services.

Contract Support

6-11. Operational contract support provides acquisition functions to obtain materials, services, and construction of facilities for U.S. and host-nation forces in the local operating areas. The Department of Defense increasingly relies on contractors to perform many functions and tasks and then provides technical oversight. The expeditionary nature of SFA creates an even greater need for this capability.

PRINCIPLES OF SUSTAINMENT

6-12. The principles of sustainment are critical to maintaining combat power, maintaining strategic and operational reach, and providing Army forces with endurance. The principles of sustainment are anticipation, responsiveness, simplicity, economy, survivability, continuity, improvisation, and integration. While these principles are independent, they are also interrelated when used in planning and executing operations. For more information on sustainment principles, see FM 4-0.

MISSIONS

6-13. All brigade support battalions have the same general capability to manage distribution and sustainment operations. The brigade support battalion provides support within an assigned operational area. It is force tailored and task-organized and uses subordinate companies, platoons, and teams to perform specific functions. In the sustainment role, the brigade support battalion is primarily concerned with the continuous management and distribution of stocks and allocation of maintenance in the operational area. Such assistance provides operational reach to maneuver commanders.

6-14. The brigade support battalion contains the following sustainment capabilities:

- Provides supplies, field services, field maintenance, recovery, and field feeding for itself and its assigned subordinates.
- Plans and conducts base and base cluster self-defense; defends against level I threats, assists in destruction of level II threats, and escapes or evades against level III threats.
- Has ability to operate as part of an Army or joint force.
- Coordinates host-nation support.
- Has ability to deploy an advance party to support early entry operations.
- Provides opening, distribution, and sustainment management information and advice to commanders and staff within its operational area.
- Exercises technical supervision over operations for all assigned units.
- Provides limited materiel management for internal stocks—classes I, II, III, IV, V, VII, and IX and maintenance management of internal assets.
- Provides a liaison team to augment other headquarters as necessary.
- Manages and maintains the its property records.
- Maintains data in support of Army equipment status reporting database and Army equipment status reporting system.
- Appoints contracting officer representatives to oversee contractor performance, certify receipt of services, and act as liaisons between the requiring activity and the contracting officer.

6-15. A brigade combat team often has attached companies such as an engineering company, a military intelligence company, an antitank company, or a network support company. Similar to the brigade headquarters and headquarters company, these attached companies are supported by the brigade support battalion regardless of where they are located on the battlefield. If one or part of these companies is task-organized to a maneuver battalion, it remains under the support of the brigade support battalion.

Critical Considerations in Building Host-Nation Sustainment Capacity

6-16. Effective SFA planning begins with accurately assessing and understanding U.S. policy objectives, determining the end state, and identifying resources available. Some possible logistic planning considerations—

- Have a holistic perspective.
- Address the causes, not the symptoms.
- Understand the desired end state.
- Understand the level of FSF senior leader support.

6-17. Logistic planning during SFA uses a holistic perspective. Operational environments can range from peacetime cooperation to counterinsurgency, stabilization to reconstruction. SFA forces determine the foreign security force's logistic and reconstruction requirements based on their internal defense and development strategy; they develop a sustainment end state to support the plan. Rebuilding the host nation's infrastructure and industrial capacity not only provides long-term job opportunities, but also builds trust and legitimacy in FSF, the host-nation government, and partner organizations.

6-18. Logistic planning during SFA addresses the causes, not the symptoms. Operational logistic support problems are the symptoms, not the causes. They result from problems that exist in the supporting systems or in the society's industrial capacity and economic strength. Teaching unit-level inventory management alone will not result in any sustainment improvement.

6-19. Logistic planning during SFA understands the desired end state. Depending on the U.S. strategic interest in a host nation and its security forces, levels of engagement can vary significantly to achieve the desired end state. Sustainment capacity-building activities can be force tailored and scaled to assist in any of the sustainment capability areas.

6-20. Logistic planning during SFA understands the level of FSF senior leader support. These leaders develop and set the policies for resource allocation. They support the various development projects and skills enhancement assistance activities. These senior-level advisors must fully understand the problems identified and an assistance strategy developed by SFA planners. From the beginning, SFA planners work with their counterparts to decide the plans and strategy to promote the vision and ensure continued effort.

6-21. SFA depends on capable logistic support for maneuver and sustainment. Without an improvement in sustainment capacity, FSF cannot significantly improve regardless of how much effort is devoted to advising and training. When planning for sustainment capacity building, planners thoroughly understand the logistic systems to identify causes that lead to functional deficiencies. They must have a holistic perspective and understand the various goals and resources required to formulate an operational approach that effectively and efficiently achieves U.S. objectives.

ASSET MANAGEMENT AND PERFORMANCE

6-22. One tool at the disposal of the commander and planners during an evaluation of foreign infrastructure and industrial capacity is called an asset management plan. This plan, commonly employed in the industrial community, manages an organization's infrastructure and assets to deliver a standard of service. An asset management plan provides information to FSF so they can analyze the capacity, condition, and current and future requirements of their facilities and infrastructure. Physical development should include facilities, transportation networks, area development, utilities, and other infrastructure components.

6-23. A description of the perceived problems of current foreign facilities, or problems created by missing assets, is also important. The description includes how the advisor intends to address deficiencies in foreign facilities and infrastructure. Each organization needs to determine each facility's function and value to their organizational mission by analyzing the current asset performance, costs, benefits, and improvements.

Current Asset Performance

6-24. Asset performance is the most difficult portion of the asset management plan process to capture while assessing conditions of facilities and infrastructure. While historical records may exist, combat operations and possible instability of the host-nation government likely render these records irrelevant. Remaining structures, when measured against U.S. construction standards (designed to protect occupants from man-made or natural disasters), often are deficient, even if untouched by hostilities. Most existing facilities require significant modification to be considered for future use. However, FSF may use them in the interim until new facilities are completed.

6-25. Asset performance may also use gap analysis. Gap analysis defines the present state, the desired end state, and the gap between them. Once a gap has been identified, the commander and staff can develop courses of action to rectify or bridge the gap. Gap analysis alone is not adequate for all problem sets, as goals may evolve and emerge during the course of problem solving. Most problems have many alternative solutions.

Costs

6-26. Establishing the short, medium, and long-term costs for facilities and infrastructure, from construction to control, is a critical step in prioritizing which facilities get built or refurbished first, and which get support at all. Units develop a forward-looking cost-profile for operating, maintaining, refurbishing, and replacing assets. This profile explains to civilian and military officials the implications and continued costs to maintain assets. SFA requires a plan for U.S. forces to hand off assets to FSF.

6-27. In constructing these models, the ideal cost-profile extends to cover the life of the longest-lived asset in the system. By doing so, the profile estimates the whole-life cost as well as determines the average annual costs. Furthermore, this cost evaluation may need to be revisited and updated annually, forming the basis for an annual bid for funding.

Benefits

6-28. Assets should provide some measure of benefit to U.S. and foreign security forces that can be measured or explained during and after SFA. This usually involves translating asset costs and projected performance into a monetary figure. Other benefits may be social or environmental, which may be difficult to quantify in monetary terms. Regardless, some attempt to record all the relevant benefits is important; the more quantitative, the better.

6-29. A benefit evaluation is vital to demonstrating that the ongoing cost of an asset system is reasonable given the scale of the benefits. The expenditure incurred by asset management needs to be justified in some way.

Improvements

6-30. How assets of FSF perform may be improved by acquisition, enhancement, or some other means. An improvement may address the potential reduction of performance and its effect on the rest of the asset system. The improvements represent a potential change to the performance and are typically managed as a project. Such management comes with an appraisal of the additional expenditure, comparing different options and selecting a preferred option based on the commander's guidance and host-nation requirements.

OPERATIONAL CONTRACT SUPPORT

6-31. Contracting is a key source of support for deployed forces across the spectrum of conflict. Because of the importance and unique challenges of operational contract support, the commander and staff need to fully understand their role in planning for and managing contracted support in the operational area. This includes understanding the types of contracted support, contract management, and contract close out.

Types of Contract Support

6-32. Current doctrine describes three broad types of contracted support: theater of operations support, external support, and systems support.

Theater of Operations Support

6-33. Contracts for theater of operations support provide support to deployed forces under prearranged contracts or contracts awarded from the mission area. They do so by contracting officers under the command and control of the combat sustainment support battalion. Contractors for theater of operations support acquire goods, services, and minor construction support, usually from local commercial sources, to meet the immediate needs of operational commanders. Theater of operations support contracts are the type of contract typically associated with contingency contracting. Sustainment brigades often are the requiring activity for theater of operations support contract actions.

External Support

6-34. Contracts for external support provide a variety of support to deployed forces. External support contracts may be prearranged contracts or contracts awarded during the contingency. These contracts support the mission and may include a mix of personnel including U.S. citizens, third-country nationals, and host-nation subcontractor employees. The largest and most commonly used external support contract is the logistics civilian augmentation program. This Army program is commonly used to provide life support, transportation support, and other support functions to deployed Army and joint forces.

System Support

6-35. System support contracts are prearranged and include, but are not limited to, newly fielded weapons systems, command and control infrastructure, and communications equipment. System support contractors, made up mostly of U.S. citizens, provide support in garrison and may deploy with the force to both training and real-world operations. The unit does not normally have a significant role to play in planning for or coordinating system support contracts other than coordinating and executing support of system support contract-related personnel.

Contract Management

6-36. Contract management requires nominating and tracking contracting officer representatives for every service contract. It will also nominate a receiving official for all supply contracts. To ensure contractors provide the service or item per the contract, they need a quality contracting officer representative and official support. SFA units must identify and coordinate funding requirements in advance with the supporting resource management staff element to secure the correct amount and type of contract funding associated with the mission.

Contract Close Out

6-37. In long-term operations like SFA, the advising unit ensures direct coordination and transfer of information related to operational contract support before mission handoff. Additionally, when the advance party arrives in the operational area, a designated Soldier actively seeks out current information on local contract support capabilities, policies, and procedures. This individual coordinates the formal handover of existing contract management responsibilities from the redeploying unit. This Soldier must know when recurring service contracts end since it takes 30 to 60 days to obtain funding approval. If the unit waits until the contract is about to expire before requesting additional funds, the sustainment brigade could lose the contracted service until funds are available.

EXTERNAL AGENCIES SUPPORT MANAGEMENT

6-38. Some external agencies help support the management of FSF. External support comes from interagencies, the host nation, and multinational organizations.

Interagency Support

6-39. The Department of Defense performs both supported and supporting roles with other agencies. It often is the lead agency for logistic activities in SFA. When tasked to provide military support to civil agencies, Army forces perform a supporting role. Sustainment forces may be tasked to support the U.S. ambassador that, as the President's representative, serves as senior U.S. Government officer in the country, or may employ resources of other U.S. Government agencies or private firms. Whether supported or supporting, close coordination is the key to efficient and effective interagency operations. In the absence of an American Embassy, or during operations in austere environments or remote locations, the advising unit may be the only organization able to provide supplies, services (including force protection), and life support. More typically, during the early stages of the advising unit deployment, the unit may find the American Embassy to be the primary source of supplies, services, and life support.

6-40. Nongovernmental organizations do not operate in either the military or governmental hierarchy. Their relationship with an advisor organization is neither supported nor supporting. An associate or partnership relationship exists between military forces and engaged nongovernmental organizations, ideally focused on unity of effort. If formed, the focal point where U.S. forces provide coordinated support to nongovernmental organizations would be the civil-military operations center of a joint task force headquarters.

Host-Nation Support

6-41. Host-nation support and local procurement may provide sustainment, operational support, and tactical support. Host-nation support agreements fulfilling the command requirements for support need to be prenegotiated. Such support arrangements must be integrated into the distribution plan and coordinated with joint, allied, and multinational partners to prevent competition for resources and ensure high priority requirements are met. Host-nation support may include functional or area support and may use host-nation facilities, government agencies, civilians, or military units. Preestablished arrangements for host-nation support can reduce the requirement for early deployment of U.S. assets and can offset requirements for early strategic lift by reducing requirements for moving resources to the theater of operations.

Multinational Support

6-42. Multinational support may consist of support provided from one multinational partner to another. One or more of the following organizational and management options facilitates multinational support:

- National support elements provide national support.
- Individual acquisition and cross-servicing agreements provide limited support.
- A lead nation provides specific support to other contributing nation forces.
- A role-specialist nation provides a specific common supply item or service.
- A multinational integrated sustainment unit provides limited common supply and support.
- A multinational joint sustainment center manages common-user logistic support.

6-43. In all cases, the multinational force commander directs specific multinational support within the applicable laws and regulations of the host nation. When operating within a formal alliance, the advising unit executes support following applicable standardization agreements.

FACILITATING THE REDEPLOYMENT PROCESS

6-44. Sustainment assets facilitate the redeployment and the retrograde process through its command and control of the distribution system. Many of the procedures are used to deploy forces, draw pre-positioned stocks, conduct reception staging and onward movement, and distribute supplies within the theater of operations apply to the redeployment process. Two factors in particular complicate redeployment operations: supporting operations and meeting requirements.

6-45. The same elements that operate and manage the theater of operations distribution system during deployment and sustaining operations perform similar roles during redeployment. When redeployment,

deployment, and sustaining operations occur simultaneously, the theater sustainment command may find it necessary to rebalance its forces or change the missions of subordinate units to effectively support ongoing operations and redeployment.

6-46. Requirements vary widely depending on the nature and scale of redeployment operations, infrastructure, and other factors. For example, redeployment operations could range from personnel only to entire units and their equipment. Depending on the political and military strategy, unit rotations may occur while decisive operations continue unabated or during operational transitions. Key considerations include, but are not limited to, several characteristics. These characteristics include the size of the force redeploying and deploying; infrastructure requirements and limitations; security requirements; traffic circulation; staging areas; distribution system capacity; quantity of supplies and materials to be redistributed; the amount and kinds of materials that require disposal agricultural inspections; and establishing accountability of retrograde cargo. The challenge is effective coordination and synchronization, vertically and horizontally, to ensure responsive simultaneous support to not only ongoing distributed operations, but also redeployment.

Chapter 7

The Advisor

Advisors are the most prominent group of U.S. personnel that serve with foreign security forces (FSF). They live, work, and fight with their FSF. The relationship between advisors and FSF is vital. Advisors are not liaison officers, nor do they command FSF units.

ROLES OF THE ADVISOR

7-1. The military advisor has three roles involving different responsibilities. First and foremost, advisors are members of a U.S. military organization with a well-defined chain of command and familiar responsibilities. Second, advisors embed themselves with their counterparts. Third, advisors are interpreters and communicators between U.S. forces and their foreign counterparts.

7-2. As members of military organizations, advisors receive and execute the orders of superiors. These orders may conflict with the orders their counterparts receive. Among other duties, advisors must act unobtrusively, but nonetheless positively, as inspector general—often observing, evaluating, and reporting on the performance of counterparts and their assigned unit.

7-3. Secondly, advisors live, eat, and work with the officers and men of their host units. Often, advisors soon regard themselves as one of them. The sharing of common hardships and dangers forges potent emotional ties. The success and good name of their units become matters of prime and personal importance to the advisor.

7-4. Finally, advisors are interpreters and communicators between U.S. superiors and foreign counterparts. Advisors must introduce and explain one to the other; they help resolve the myriad of problems, misunderstandings, and suspicions which arise in any human organization, particularly when people of starkly different cultures approach difficult tasks together. Advisors with quick and easy access to influential counterparts can sometimes be the best possible means of communicating.

7-5. To be effective, advisors obviously must gain their counterparts' trust and confidence. This relationship, however, is only a prelude to the advisor's major objective: inspiring and influencing a counterpart to effective action. In pursuing this goal—constantly, relentlessly, and forcefully, yet patiently, persuasively, and diplomatically—advisors must recognize conditions which can benefit or handicap their cause.

CONSIDERATIONS OF THE ADVISOR

7-6. An advisor follows ten considerations when serving with FSF.

BY, WITH, AND THROUGH

7-7. All planned operations of a combat advisory mission must be conducted by, with, and through the FSF. Not counting immediate action battle drill responses, the mark of an effective advisory effort is the amount of personal involvement the FSF take in their own operations. Civilians must see that they are secured by their own security forces to promote the legitimacy of the host nation and their capacity and to build trust and confidence. In a word, combat advisors are shadows whose presence is felt at all times.

Empathy Leads to Competency

7-8. Empathy can be defined as identification with and understanding of another's situation, feelings, and motives. This is tough for experienced U.S. leaders and often harder to explain, but it is the key to the success of an advisory mission. Truly understanding other human beings and where they come from allows for honest relationships to develop. In most cultures, the place to begin understanding another person's feelings and experiences is by understanding the other person's narrative. The narrative is a collective group's identity as an interpretation of both ancient history and recent collective experiences. Delving into the narrative, understanding it and how it affects people, is the beginning of empathy.

Success Is Built on Personal Relationships

7-9. Developing a sense of empathy allows advisors to begin nurturing relationships with those they are tasked to advise. No amount of resources and firepower can compensate for the lack of a relationship between advisors and their counterparts. It must be honest, genuine, and heartfelt. Mutual respect, trust, and understanding create success. Both parties rely on each other for mission accomplishment and often for survival. This relationship is likely to be challenged on numerous occasions; only one built on mutual trust can survive and ensure mission success. Furthermore, advisors learn to pick their battles. Camaraderie and rapport with FSF can vanish in an instant if one does not prioritize and define critical lines in the relationship.

Advisors Are Between U.S. and Foreign Security Forces

7-10. Living and fighting with FSF allows for the development of true bonds, empathy, and trust. These desired results have a price to pay. When addressing their own Army, advisors soon realize that they are also not one of them, increasing their level of frustration. Advisors are often alone, navigating between two military systems and two cultures, never quite fitting in with either one.

Advisors Will Never Win, Nor Should They

7-11. Often advisors find it difficult to understand what victory looks like in this environment. Progress tends to occur at a glacial pace and cannot usually be tracked on a day-to-day basis. The advisor attaining a tactical objective does not achieve success; success is achieved by the foreign security force achieving the objective. Advisors find it difficult to satisfy their own units, and they never fully satisfy the demands of their FSF. They are figuratively and literally caught in the middle. Advising is the art of striving to make a win-win situation for all parties. The advisor and foreign security force relationship is one in which the advisor removes the obstacles in the way. Only a plan by FSF will succeed—and it will only succeed if it is, in fact, their plan.

Advisors Are Not Commanders

7-12. Advisors do not command FSF personnel or units. Advisors provide advice, training, and access to coalition resources to FSF. They are not intended to lead FSF in combat; they are responsible for the command and control of their own team of advisors. However, they can and should influence the commander of the foreign unit.

Advisors Are Honest Brokers

7-13. Advisors are advocates for FSF with conventional forces. They transparently assess capabilities and reveal limitations of foreign units to the FSF's and advisors' higher command. They also provide access to U.S. resources—from close air support to medical evacuation—to FSF they advise.

Advisors Live with Shades of Gray

7-14. Black and white is not available to the advisor. Caught between two cultures, systems, and narratives, the advisor works within a gray area. Unlike operations in a conventional unit, many of the checks and balances are not present. Advisors often find themselves isolated with great autonomy and no supervision. These conditions present moral and ethical dilemmas daily. Advisors need to be comfortable in this environment. Often they lose sleep at night and question their actions and those of their FSF. Thus, this mission calls for Soldiers of enormous character, moral courage, and intellect.

Talent Is Everything, But Advisors Must Understand Rank

7-15. Military forces around the world approach rank and prestige in completely different ways. In some cultures, rank equals experience; in others, it is nobility or education. Often many militaries have rewarded loyalty with promotions, which results in rank-heavy armies. Some FSF recognize talent, while others recognize rank. Advisors likely advise counterparts much more senior in rank. They must understand that rank on the uniform is important to many armies.

Make Do

7-16. Advisors will never have everything they believe they need to succeed. However, their creativity often distinguishes between success and failure. Austerity and economy of force can fail to meet the prevalent expectations that FSF have of a U.S. unit in regards to resources. It is up to the advisor team to close the expectation gap by its creativity and imagination. Scrounging, bartering, and negotiating are daily activities of advisors. An enormous amount of energy must be devoted to these activities to achieve mission success and endear them to their counterparts.

PERSONALITY TRAITS OF THE ADVISOR

7-17. Not every Soldier is well suited to perform advisory functions; even those considered to be the best and most experienced have failed at being an advisor. Effective advisors are only the most capable individuals. Advisors are Soldiers known to take the initiative and who set the standards for others; however, they are also patient and personable enough to work effectively with FSF. Recognizing that not all Soldiers are capable of performing as advisors, leaders should immediately remove advisors who do not exhibit these qualities.

7-18. Because advisors operate in very subjective environments, it is difficult to establish objective criteria by which to assess potential advisors. However, research and experience indicate that several personality traits greatly enhance the advisor's ability to adapt and thrive in a foreign culture. These traits include—

- Tolerance for ambiguity.
- Realistic when setting goals and tasks.
- Open-mindedness.
- Ability to withhold judgment.
- Empathy.
- Communicativeness.
- Flexibility.
- Curiosity.
- Warmth in human relations.
- Motivation of self and others.
- Self-reliance.
- Strong sense of self.
- Tolerance for differences.
- Perceptiveness.
- Ability to accept and learn from failure.
- Sense of humor.

7-19. Of the traits listed above, no single trait is paramount. Advisor selection and training programs seek to develop an understanding of the contributions of each of them.

SKILLS OF THE ADVISOR

7-20. Advisor-specific skills fall into two subcategories: enabling skills and developing skills. Enabling skills build on the individual and collective skills of advisors. Developing skills are the main advisor tasks of teaching, coaching, and advising. Advisors use these skills to develop the capabilities of the foreign security force.

ENABLING SKILLS

7-21. Advisors use enabling skills to work individually and in a group. Enabling skills help advisors communicate across cultures, build rapport, influence, and negotiate. Successfully employing these enabling skills sets the conditions for advisors to move forward with their mission.

Individual Skills

7-22. Individual skills help advisors properly perform their mission. Initially advisors should be selected based on subject matter expertise and not just rank. However, tactical proficiency does not necessarily equate to proficiency as an advisor. Modern-day military advisors must be able to teach and advise FSF. To do this, they must be knowledgeable and proficient in procedures and able to impart this knowledge effectively. They may also be able to advise FSF in the areas of intelligence, communications, operations, and logistics. Some advisors may be required to call in U.S. supporting arms, principally artillery and air support, and coordinate both air and ground casualty evacuation. Finally, advisors should be proficient in basic combat skills. These common skills, warrior skills, and survival skills are often required by advisors. Since these skills are critical to an advisor due to the isolated and independent nature of the mission, advisors should refresh them during predeployment training.

Collective Skills

7-23. As a group of advisors form a team, they must practice their warrior tasks, organize to cover the functions associated with any team, and familiarize its members with their duties and responsibilities. As with individual advisor skills, some are refined before deploying and others are learned in country. Figure 7-1 depicts this graphically.

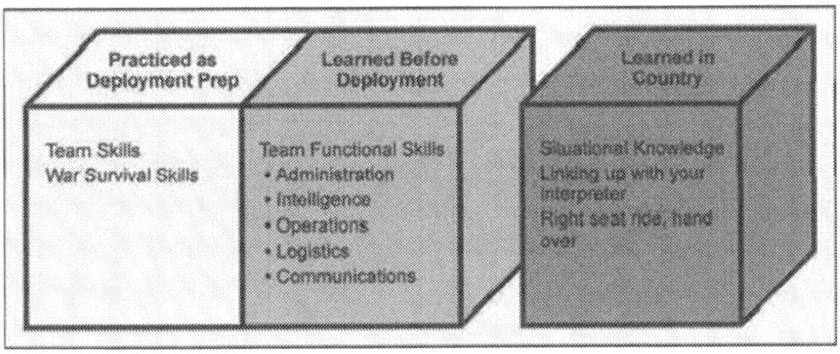

Figure 7-1. Collective advisor skills

DEVELOPING SKILLS

7-24. Developing skills—teaching, coaching, and advising—are required for every advisor. The advisor will make an initial assessment to determine the foreign unit's proficiency to determine the appropriate developing skill with which to start. Figure 7-2 illustrates the appropriate level of advice that corresponds to the foreign unit's capabilities.

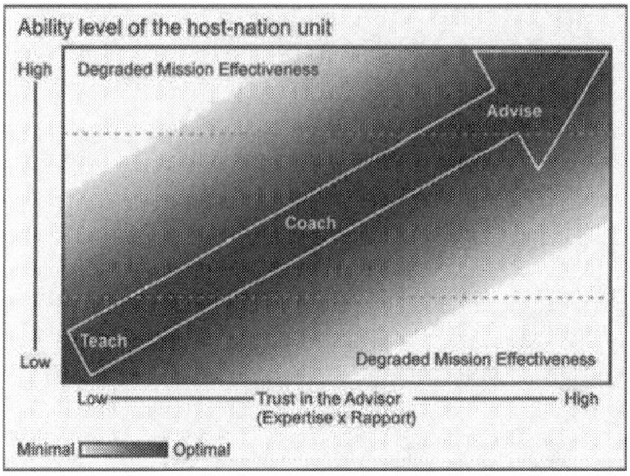

Figure 7-2. Appropriate advising level

7-25. Advisor teams have limited positional power and depend on personal influence to positively affect FSF efforts through teaching, coaching, and advising. Advisor team members often work with people of higher rank or grade than themselves. Advisors should remember the foreign unit's experience and capabilities and carefully choose opportunities to inject or impart knowledge. Foreign units are most receptive to advisor teams that teach unobtrusively. Foreign units most value those who are subtle in their teaching, coaching, and advising. Those advisors who master the ability to create a climate in which FSF personnel feel they are teaching themselves often prove the most effective.

Teaching

7-26. Teaching includes training and education. Methods of teaching can include classroom lectures, seminars, hands-on training, training exercises, and simulations.

Coaching

7-27. Coaching relies on guiding to bring out and enhance capabilities already present. Coaching refers to the function of helping someone through a set of tasks. Those being coached may or may not have appreciated their potential. The coach helps them understand their current level of performance and instructs them how to reach the next level of knowledge and skill. Coaching requires identifying short- and long-term goals and devising a plan to achieve those goals. The coach and the person being coached discuss strengths, weaknesses, and a course of action to sustain or improve the goals.

Advising

7-28. Advising is a combat multiplier that boosts supported unit capability. Advisors have experience in particular areas in which they are advising, but are not required to have similar backgrounds. Advisor relationships are not based on superior to subordinate relationships.

7-29. Advisors provide an expert opinion, advice, and counsel by focusing on both personal development (interpersonal and communication skills) and professional development (technical and tactical knowledge). Advising develops mutual trust and respect.

Advisor Versus Trainer

7-30. Advisors teach, coach, and advise work in a cycle. As FSF master one skill, the advisor can move on to other skills and initiate the process for the new skills. If FSF require additional teaching or coaching, the advisor can take them to that part of the cycle. Eventually, the teaching and coaching should decrease, and most of the advisor's time will be spent as an advisor providing an expert second opinion.

7-31. Ultimately, the question of what is the difference between being a trainer or an advisor will arise. Figure 7-3 shows that every advisor will have to be able to be a trainer, but all trainers are not expected to perform the more sophisticated task of advising. Hence, competent junior officers and enlisted Soldiers can be trainers, but only more experienced individuals can be advisors.

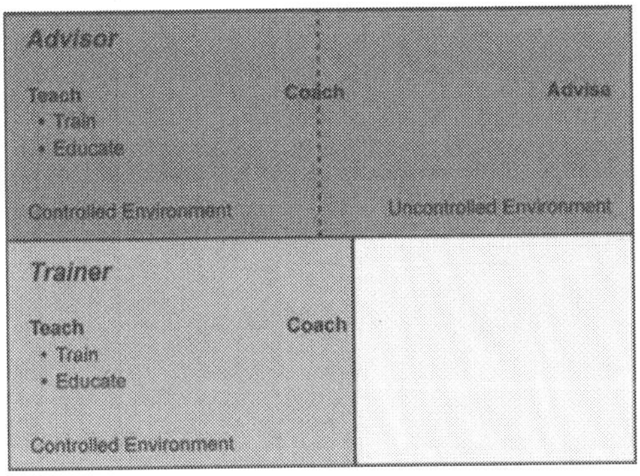

Figure 7-3. Advisor versus trainer

Chapter 8
Culture and Communication

As an advisor, understanding the host-nation population is a crucial element of premission planning and the development of foreign security forces (FSF). Prior knowledge of sociocultural differences aids in building effective relationships and prevents embarrassment, loss of rapport, and compromise of the mission. This chapter discusses three aspects of the sociocultural environment to provide a framework for analysis and comparison of foreign cultures for operational purposes. These aspects are society, culture, and communications.

SOCIETY

8-1. People who have a common culture, share a group identity, occupy a common territory, and are subject to the same political authority can be identified as a society. A society is not easily created or destroyed. Though they share the above factors, societies are not homogeneous nor static. A society usually has a dominant culture but can also have a vast number of secondary cultures, each influencing the other.

8-2. Advisors derive much of their effectiveness from their ability to understand and work with counterparts from a foreign society. Advisors must also consider societies outside their immediate operational environment whose actions, opinions, or political influence can affect the mission.

SOCIAL STRUCTURE

8-3. Society is composed of both social structure and culture. Social structure refers to the relations among groups of persons within a system of groups and is persistent over time. For the purposes of security force assistance, social structure includes groups, institutions, organizations, roles and statuses, and social norms. FM 3-24 contains a more in-depth discussion of social structure.

Groups

8-4. Within societies, groups are composed of two or more people that interact based on mutual interests and interrelated statuses and roles. These groups can be based on race, ethnicity, religion, or kinship. A racial group is defined by its members or by outside groups as different based on physical characteristics or descent. Race is a social category that has no biological basis. An ethnic group is a community whose learned cultural practices, language, history, ancestry, or religion distinguish them from others. Similar to racial groups, ethnic groups can be defined by their members or by outside groups. Religious groups may be subsets of larger ethnic groups and an ethnic group may contain members of different religions. Kinship groups are based on a link by blood relation. It may also be used to indicate a group of persons who live and work together to satisfy basic collective needs and goals. The role of a kinship group is generally seen to care for and educate offspring. This group provides for such needs as food, shelter, and clothing and prepares offspring to be functional members of the society in which they live.

Institutions

8-5. Institutions are the basic building blocks of societies. They are made up of groups that are organized to achieve a common task. Institutions are continuous through many generations and are built to continue to function even as individuals are replaced. An example of an institution from FM 3-24 is a military institution that brings together groups and individuals whose statuses and roles concern defense and security.

Organizations

8-6. Organizations are institutions that have a bounded membership with defined goals and established operations. They include fixed facilities from which they conduct business or operations and a set means of financial or logistic support. The four types of organizations are communicating, religious, economic, and social. Communicating organizations are designed to influence perceptions. Religious organizations are created to regulate societal norms, reaffirm beliefs and values, and provide social support. Economic organizations provide employment and assist in economic development. Social organizations provide support to groups in society, create social networks, and can influence ideologies. Organizations may belong to more than one category and exist to affect the behavior, values, and beliefs of society. As such, advisors must identify influential organizations both inside and outside of the FSF with which they are partnered.

Roles and Statuses

8-7. As members of a group, individuals interact based on assumed or given social positions. These social positions are referred to as statuses. Societies generally associate particular statuses with particular social groups (such as family, lineage, ethnicity, or religion). Each status inherently contains a set of expected behaviors known as a role, which includes how a person of that status should act, feel, and think. A status includes expectations about how others in a society should treat a person of that status.

Social Norms

8-8. Violation of a role derived from a given status results in social disapproval. The standard of conduct for social roles is a social norm. Norms may be either moral (incest prohibition, homicide prohibition) or customary (prayer before a meal, removing shoes before entering a house). Advisors must understand the roles, statuses, and social norms of groups they are advising. This knowledge will clarify the expected behavior of their FSF partners and their role as advisors.

LANGUAGE

8-9. Language is a learned element of culture that is a system of symbols that people use to communicate with one another. Successful communication within a society requires an understanding of the social setting, appropriate behaviors towards different statuses and roles, and nonverbal cues. The languages used in the advisors' operational environment have a major impact on the advising mission. Languages must be identified to facilitate language training during predeployment and employment, as well as in the requisitioning of translators. Translators are critical for communicating with FSF and interacting with local citizens and leaders.

CULTURE

8-10. Social structure comprises the relationships among groups, institutions, and individuals in a society; in contrast, culture provides meaning to individuals within the society. Social structure can be thought of as a skeleton with culture being the muscle on the bones. The two are mutually dependent and reinforcing. A change in one results in a change in the other.

8-11. Culture influences an individual's range of action and ideas, including what to do and not do, how to do or not do it, and whom to do it with or not to do it with. It is neither a random jumble of customs nor a laundry list of accepted practices and taboos that an advisor should know. It influences judgments about what is right and wrong and what is important and unimportant. Cultural rules are flexible in practice; the customs and actions that an individual practices are based largely on that particular environment.

CULTURAL CHARACTERISTICS

8-12. Culture is shaped by history, language, religion, and customs. Members of society use these shared beliefs, values, customs, and behaviors to cope with their world and with one another. It is learned behavior that is shared by members within a society. However, behaviors do not always define a culture.

Frequently, there are more telling cultural signs in the meaning attached to a behavior than in the actual behavior. Culture influences its members' identity, values, belief systems, and cultural forms. As such, it is arbitrary; advisors should make no assumptions regarding what a society considers right and wrong, good and bad. Culture is changeable through the social interactions between people and groups, both inside and outside of that society. TC 31-73 discusses cultural characteristics in great detail.

Identity

8-13. Each individual belongs to multiple groups through birth, assimilation, or achievement. Each group to which individuals belong influences their values and beliefs. Individuals consciously or unconsciously rank their identities into primary and secondary identities. Primary identities are those seen as core to their being; frequently these are national, racial, and religious. In contrast, secondary identities are more peripheral and may include such things as an individual's occupation or hobby.

Values

8-14. Values guide how individuals interact with the world around them. As deeply held beliefs about rightness, values shape how individuals think things should be. Values may never be articulated—in fact, they may not even be consciously held. At times, the beliefs of an individual may even contradict one another. Advisors should have some understanding of a counterpart's value system. Usually values are nonnegotiable. In working with a counterpart with conflicting values, advisors should persuade the individual that the advocated action or behavior really does not conflict with their values.

Belief Systems

8-15. Beliefs primarily take two forms: values (core) and opinions (peripheral). Values, or core beliefs, are not easily changed. Opinions, or peripheral beliefs, stem from thought and decisions. The totality of the identities, values, and beliefs that an individual holds compose that person's belief system.

Forms of Beliefs

8-16. Values are broad moral conclusions about the way to live life. These core beliefs are unstated, taken for granted, resistant to change, and not consciously considered. They reflect strong personal certainties drawn from faith, childhood environments, and life experiences. They lend shape and order to people's lives, helping them to decide if they settled on the right actions, friends, leaders, religions, and careers.

8-17. Opinions, or peripheral beliefs, are immediate conclusions about some aspect of the environment. They are usually quite specific and are open to debate, consciously considered, and easiest to change.

Types of Belief Systems

8-18. Belief systems involve cultural narratives and myths. The interpretation of these narratives and myths provide people with insight into how they should feel, think, and behave. The most prominent systems of beliefs tend to be those associated with religions and ideologies. However, any system of belief in which the interpretation of narratives impacts people's behavior (such as superstitions) can be a contributing component of a society's culture.

8-19. A belief system acts as a filter for new information. It is the lens through which people perceive the world. Beliefs of one group as to what is rational, normal, or true may appear strange, irrational, or illogical to outsiders. Understanding belief systems of various groups in an operational environment allows advisors to work with their foreign counterparts and influence civilians more effectively.

8-20. Belief systems play a fundamental role in the development of a society's values. Careful analysis of a region's belief systems is crucial to understanding the local value system. These systems influence religious doctrine, narratives, and myths and provide enormous insight into the culture of the community. Ultimately, the values which are most prized within a belief system are religious. These beliefs affect that culture's systems of social organization, learned behaviors, and relationship to the physical environment.

Religion as a Belief System

8-21. At its most basic level, religion is both the individual and communal expression of contact with the sacred. Notions of the sacred can vary widely in societies and may include deities and sacred items. In most cases, religions also involve symbols, rituals, and ethics or dogma. Religion can be both an individual practice and a group phenomenon.

8-22. Most societies have some variety of religion—typically a set of sacred beliefs and rituals that provide a common understanding of moral codes and proper conduct. Expressing a common religious understanding helps people to make sense of their world. Often religion provides explanations for human suffering, natural disasters, broken relationships, inequalities, and death. Some societies also have secular equivalents of religion. For instance, America is often remarked to have a civil or civic religion that can be seen in the images, symbols, ideas, and behavior of the political realm.

8-23. Because religion is such an integral part of culture, when advisors carefully prepare and analyze the mission, they must examine the religions and religious groups in the operational environment. Religious beliefs, leaders, and institutions are central to how many societies view the world. When planning any advising operation, planners consider how religion affects FSF and civilians.

Broad Impact of Belief Systems

8-24. In almost every society, followers of dominant religions have a strong societal impact. This influence may be exerted at varying degrees upon the individual citizen, sections of society, the economy, the military, and the political structure. TC 31-73 discusses these in detail.

The Importance of Understanding Culture

8-25. Advisors derive their effectiveness from their ability to understand and work with foreign counterparts from another culture. They must understand enough of their own culture and their counterpart's culture to accurately convey ideas, concepts, and purpose without causing counterproductive consequences.

8-26. Advisors must be aware of aspects of the local culture and history that influence behavior in their operational environment. Advisor team members must understand the reasons and motivations underlying personal interaction and must practice patience when working with their counterparts. Group norms guide individual behavior, and advisors must understand how individuals in a society tend to interact as members of a group, whether a race, ethnic, or kinship group.

8-27. Cultural understanding is not derived from demographic information provided to the military through country briefs prior to deployment. It is gained from studying, interacting, and understanding the people, religion, history, customs, and social and political structures within an area. For true understanding, it is necessary to live among the people, gradually understanding the subtleties and nuances of their culture.

Cultural Education

8-28. Culture influences a person's perception of the world. People from one society tend to view people from other societies through the prism of their own culture. Often, their perceptions are biased, are inaccurate, or involve assumptions of superiority. For example, Americans view their society and its culture very differently from the rest of the world. Likewise, as foreigners, advisors view local cultures through the prism of their own idealized culture. This creates a large gap between the culture the advisor observes and the host nation's real culture.

8-29. To narrow the gap of cultural understanding an advisor will encounter, a program of cultural learning should be undertaken. The end state of this cultural learning should be to gain the trust of foreign partners and civilians. Advisors will do so by demonstrating their understanding and sensitivity towards the host nation's culture and social structures.

8-30. An awareness of the cultural aspects of the operational environment can significantly enhance the effectiveness of an advisor. An extensive study program is best, but even a concise program is worthwhile if it provides advisors with the necessary cultural information. The best programs involve credible teachers who intimately understand the culture and have personal experience in the society. Training is dynamic and moves beyond the standard lecture and handout format. Small groups or a discussion panel work quite well. Cultural behavior and cultural familiarization handbooks can also be used. Since handbooks vary greatly in accuracy and quality, credible subject matter experts should be consulted to determine which documents to incorporate into the training. Table 8-1 discusses the training recommendations and tools.

Table 8-1. Cultural education recommendation

	Training Recommendations
Cultural Education	• Comparison of cultural values and social structures (United States compared to those of the operational area). • Local customs and traditions (for example, greetings do's and don'ts). • Geopolitical history (pre-colonial to contemporary and the orientation of each faction or party). • The role of religion in daily life.
	Training Tools
Resources	• Guest speakers native to the country of interest (for example, nongovernmental organization staff, foreign students, recent immigrants, or selected refugees). • Others who have worked in or studied the mission area (for example, special forces personnel, diplomats, and scholars). • Cultural familiarization handbooks.
Format	• Combination of briefings, small group discussions, and question and answer periods. • Handouts to augment—not replace—speakers. • Visual media, specifically slides and videos of the mission area.

8-31. Advisors with previous operational experience have addressed the importance of cultural education. Many civilian agencies have developed intercultural effectiveness programs designed for their personnel working overseas. Often these programs can be revised and tailored for military and paramilitary groups. If available, these programs are an excellent resource for predeployment preparation.

Cultural Gap

8-32. Each culture has its own societal rules regarding who a person may speak to, how and when the person may speak, and what topics the person may speak about. Many cultures rely on nonverbal signals to communicate. In such cultures, posture, expression, and actions convey more than spoken or written words. Advisors work to bridge the cultural gap by learning the language as well as nonverbal communication cues. To bridge the gap, advisors develop a sense of cultural awareness, recognize differences, learn to adapt, develop tolerance, and maintain personal contact.

Develop a Sense of Cultural Awareness

8-33. Advisors must recognize they are a product of their own culture. They must learn as much as possible about the culture of the people with whom they need to communicate. When communicating with people across cultures, advisors abandon any sentiments of ethnocentrism—the tendency of individuals to judge all other groups according to their own group's standards, behaviors, and customs. Such notions lead an individual to see other groups as inferior by comparison.

Recognize Differences

8-34. Each culture has its own way of accomplishing required daily tasks. Advisors must understand that each particular society may approach things differently; it does not mean that they are inefficient or less valid. Differences should not be seen as negative. Respect for counterparts must be maintained at all times.

Learn to Adapt

8-35. Advisors must be flexible and ready to adapt and adjust their behavior; however, they must be careful not to overdo their adjustment. Individuals who are overly flexible are often perceived as being insincere. The successful advisor must strive to act in a way that is appropriate to the target culture. Above all, advisors should be themselves and show sincerity.

Develop Tolerance

8-36. Advisors must develop a tolerance for deviations from accepted norms. Events or activities that may seem extraordinary to newcomers may be common practice in the culture. Advisors must be aware that members of the foreign culture may be astounded by that which is commonplace in the United States. Careful observation should be made before judgments are rendered about seemingly peculiar behavior.

Maintain Personal Contact

8-37. Personal contact is the most effective way to bridge organizational barriers. Interorganizational stumbling blocks are very real and the prejudices that arise from them are exacerbated by misunderstanding and ignorance. Often it is too easy to attribute negative attitudes and hostile motives to faceless groups. The advisor must employ superior interpersonal skills and work directly and closely with individual members of other organizations. This type of contact can effectively reinforce commonality and diminish the impact of disparity. These personal relationships are the key to effective interorganizational relationships.

CULTURE SHOCK AND ADAPTATION

8-38. The term culture shock describes the anxiety experienced by persons in a completely new environment. It is not the result of a single event. Rather, the condition develops slowly from a series of minor events or conditions. Culture shock expresses the feelings of not having a set direction, not knowing what to do or how to do things in a new environment, and not knowing what is appropriate or inappropriate.

8-39. Culture shock occurs because the mind and body have to go through a period of psychological and physiological adjustment when individuals move from a familiar environment to an unfamiliar one. The cues received by all of the senses suddenly change. During the day the foreigner is bombarded with unfamiliar sights, sounds, smells, tastes, languages, gestures, rules, requirements, interactions, demands, systems, and expectations.

8-40. Some differences that advisors experience between their lives at their home station and their lives when deployed to a foreign location are obvious. These differences include language, climate, religion, food, educational system, and the absence of family and friends. Other differences may not be as obvious. These differences include how people make decisions, spend their leisure time, resolve conflicts, express their emotions, and use their hands, faces, and bodies to express meaning.

Stages of Shock

8-41. Five distinct stages occur in the culture shock process. These include enthusiasm, withdrawal, reemergence, achievement and reentry. Figure 8-1 depicts the stages of shock.

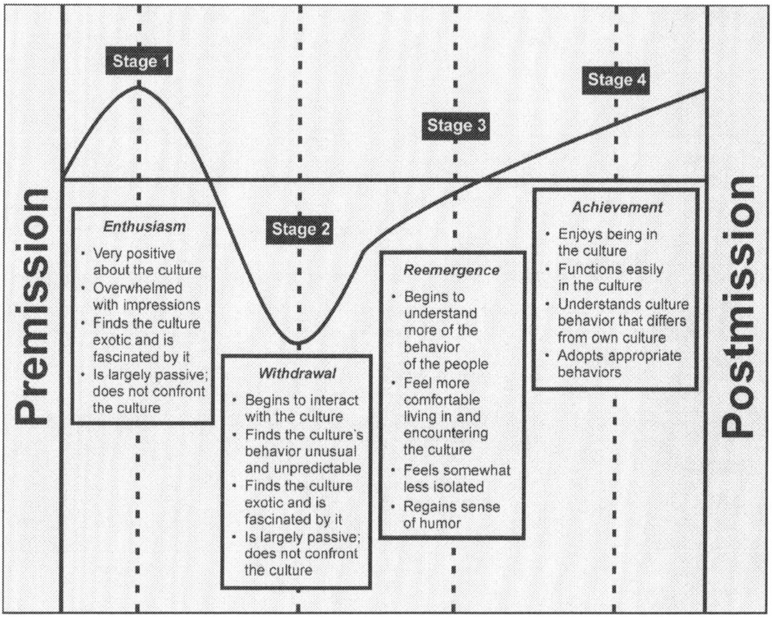

Figure 8-1. Stages of culture shock

Stage 1 – Enthusiasm

8-42. In this stage, a person may feel self-confident and pleasantly challenged. The person may be delighted by all of the new things encountered.

Stage 2 – Withdrawal

8-43. In this stage, a person begins to encounter difficulties and minor crises in daily life. It may be difficult to make oneself understood and foreign counterparts may prove more difficult to work with than anticipated. In this stage, one may have feelings of discontent, impatience, anger, sadness, and even incompetence.

Stage 3 – Reemergence

8-44. This stage is characterized by gaining some understanding of the new culture. A person may not feel as isolated, and a feeling of direction emerges. The individual is more familiar with the environment and is better able to adjust to the circumstances.

Stage 4 – Achievement

8-45. In this stage, the person realizes that the new culture has both good and bad things to offer. This stage is one of integration and the person is increasingly able to function in the new setting. A sense of accomplishment, a reduction of routine annoyances, and a more solid sense of belonging accompany this integration.

Stage 5 – Reentry

8-46. This stage occurs when people return to their own culture. They may find that things are not as they once were. Changes that occurred in their absence—their family, their friends, their communities, and themselves—combine to present a distorted image of home that differs greatly from the one imagined or remembered.

Culture Shock Symptoms

8-47. Being immersed in a society with extreme cultural differences tends to cause feelings of uncertainty and anxiety. The body and mind may react in unusual ways. Some persons may experience more pronounced physical symptoms of stress, such as chronic headaches or upset stomachs. Although uncomfortable, some degree of culture shock is a normal part of the adjustment process. Some common reactions include:

- Feeling irritable with (or even anger toward) one's own group or organization.
- Feeling isolated or alone.
- Tiring easily.
- Changing normal sleep patterns (either too much sleep or not enough).
- Suffering minor (but persistent) body pains, especially in the head, neck, back, and stomach.
- Experiencing feelings of hostility and contempt toward local people.
- Withdrawing from the local population (that is, spending excessive amounts of time alone reading or listening to music).

Overcoming Culture Shock

8-48. An important tool to overcome the obstacles of a new cultural environment is familiarity with the language. An ability to communicate in the new culture, even at the most basic level, reduces the effect and shortens the period needed for adjustment. Other means of overcoming culture shock can include having previous experience in the area (including a regular form of physical activity in personal routines) and acknowledging even slight progress in adjusting to the new culture.

8-49. TC 31-73 discusses culture shock in depth. It identifies the symptoms, cues, and treatment.

CROSS-CULTURAL COMMUNICATIONS

8-50. Communication is the transfer of messages from one person to another. These messages may be passed along verbally, in writing, or by signals (gestures, body language). The sender encodes the message and the receiver decodes it. The type and style of encoding used is based upon the sender's history, beliefs, values, opinions, and belief systems. Communication is determined by culture; as a result it can be either high context or low context. High context communication places an emphasis on clearly stating issues and problems and working towards resolving them in a straightforward manner. Low context communication places a premium on building relationships; as such, problem sets tend to be worked in a less formal matter in societies that utilize this type of communication. Quickly and effectively identifying whether a society uses high or low context communication is a key to effectively conveying material between the advisor and the host-nation population.

8-51. The receiver decodes messages based upon lifestyle, group membership, status and role, worldview, language, and social norms. Communication is a two-way process in which the encoding and decoding methods can affect both sending and receiving. Effective communication occurs when the message is perceived and responded to in the manner the sender intended. Ineffective communication can occur from poorly chosen words, flawed timing, a confused mixture of verbal and nonverbal signals, poor listening skills, and a failure to take culture into account. Communications can be broken into both levels and factors.

LEVELS OF COMMUNICATIONS

8-52. The cross-cultural communications capabilities required by an effective advisor to FSF can be described in three levels. At the lowest level is awareness followed by knowledge. The highest level is reached when these two are combined with well-trained and refined skills. TC 31-73 discusses communications in detail.

Awareness

8-53. Awareness is the basic level of cross-cultural capability. Awareness of cultural differences and their impact is the first prerequisite for successful work with a counterpart. Simply being sensitive to the fact that differences exist and carefully observing actions and reactions can assist the advisor in adjusting behavior and modifying actions to achieve greater influence with the counterpart. Awareness is not region specific and can be instilled in the advisor with relatively little training.

Knowledge

8-54. Knowledge of the details and nuances of a specific culture is the next level of cross-cultural capability. Advisors attain this level through academic study and immersion. Such knowledge is inherently area-specific and rarely transfers from one target area (or culture) to another. Developing in-depth regional knowledge necessary for effective cross-cultural communications requires extensive and time-consuming training. Appropriate personnel assignment policies must support this training.

Skills

8-55. Skills fundamental to effective cross-cultural communications, when combined with awareness and knowledge, form the highest level of cross-cultural capability. These skills include professional competence, language, nonverbal communication, negotiation, interpersonal skills, observation, problem solving, leadership, instruction, and fitness. Although some individuals show greater natural talent for these skills than others, all advisors require continual training in each to achieve and retain their full potential as advisors to foreign counterparts. These skills are discussed in more detail in TC 31-73.

FACTORS AFFECTING COMMUNICATION ACROSS CULTURES

8-56. There are four aspects that can impact communication across cultures. These factors are level of direct communication, perception of time, perception of the individual versus the group, and show of emotion.

Level of Direct Communication

8-57. Cultures fall along a spectrum of formality, with some cultures being direct and explicit in their communication and others less direct and more ambiguous. Cultures that rely on direct communication place more emphasis on verbal communications that are precise and explicit. Context and nonverbal cues are secondary to verbal communication. Cultures that rely on indirect communication rely less on verbal communication and more on context and nonverbal cues. Their verbal communication can seem vague and noncommittal to those from a culture that is more direct.

Perception of Time

8-58. Cultures can differ greatly in their perception of time. This difference divides into two cultural concepts of time: monochronic-time and polychronic-time. In monochronic-time cultures, members emphasize schedules, precise reckoning of time, and promptness. In such cultures, the schedule takes precedence over the interpersonal relation. In polychronic-time cultures, time is viewed as fluid. Members of polychronic-time societies do not observe strict schedules—agendas are subordinate to interpersonal relations.

Perception of the Individual Versus the Group

8-59. Cultures may be individualistic or collectivist in their orientation. An individualist culture is one in which the ties between individuals are loose—where people are expected to take care of themselves and their immediate families. In a collectivist culture, people are raised from birth into strong, cohesive groups. These groups offer a lifetime of protection in exchange for unquestioning loyalty. Advisors must understand from which culture their counterparts come to identify motivations accurately and communicate effectively.

Show of Emotion

8-60. Each culture has its own system of expressing emotion. Some cultures tend to express their emotions and show their feelings openly. Other cultures tend to repress those feelings more. Rather than showing their feelings openly, members of these cultures keep their emotions controlled and subdued. Those from the more expressive culture may view people from the repressive culture as cold or unfeeling. Similarly, those from the more repressive culture may view their more expressive colleagues as eccentric. To communicate effectively, advisors must be able to understand their foreign counterparts' emotional communication and display their own emotions carefully.

Chapter 9
Working with Counterparts

Rapport is a relationship marked by cooperation, conformity, harmony, or affinity. When people discuss good rapport, they describe a relationship founded on mutual trust, understanding, and respect. Relationships characterized by personal dislike, animosity, and other forms of friction often lack rapport. For the advisor, rapport describes the degree of effectiveness.

ESTABLISHMENT OF RAPPORT

9-1. Since advisors are in a unique military position, they establish rapport with their foreign counterparts. That position is one in which the advisor has no positional authority over the actions of their foreign counterparts. This lack of authority means that the doctrinal view of leadership is modified to emphasize interpersonal relationships and deemphasize authoritarian roles. Advisors instead use their people skills to influence the actions and decisions of their counterparts. Advisors use their people skills to building rapport. They have a genuine interest in other people, smile, remember and use people's names, encourage others to talk about themselves, listen to others, discuss what the other person is interested in, and make the other person feel important.

9-2. Advisors obtain certain knowledge before establishing effective rapport. First, they study FM 6-22 to gain basic leadership knowledge for understanding human nature and motivation. Advisors then incorporate information specific to the culture and society of their potential counterparts. This information may take the form of thorough area studies, operational area studies, and other research materials. Knowledge includes the components and techniques of rapport.

COMPONENTS OF RAPPORT

9-3. Three components compose rapport: understanding, respect, and trust. All the components of rapport are two-way streets; to have counterparts share about their culture, language, and experiences, advisors have to be willing and able to share also. Building this rapport may require the advisor to establish a personal level of understanding with their counterparts. Understanding is the first component of rapport. It begins before deployment and may include foreign cultural studies, language training, and equipment and doctrine familiarization. Once in country, advisors continue to broaden their understanding by observing and asking questions.

9-4. Respect is the next component of rapport. It is a reciprocal commodity. The foreign counterpart should grow to respect advisors, what they know, and how they perform. Advisors should look for those things that they respect in their counterparts. This may prove challenging. The counterpart may not fit the mold of U.S. or other coalition officers. Lacking formal training or education, the foreign counterpart may even be illiterate. The uniform standards of the host nation often may differ from the advisors' standards. Start with the basics and expand over time. Additional factors may affect the counterpart's willingness and motivation. For example, by accepting the duties, the counterparts (and their family) may be in mortal danger. The counterpart may have proven to be a fearless warrior, even without formal training. In short, traits deserving respect will exist even before the relationship matures. Mutual respect grows through shared experiences and shared dangers. Advisors should live, eat, and if authorized, fight with their counterparts. By sharing their hardships and dangers, advisors build respect.

9-5. Trust is the final critical component for building rapport. Trust grows gradually from understanding and respect. Building trust needs to begin on day one, but it will not mature until later in the relationship.

Advisors should begin by showing they are reliable and should do everything they say they will do. They should be in the right place at the right time. Two things can enhance this growth. First, the advisor should start out with confidence-building missions. Early success builds self-confidence and trust. Second, advisors should not promise any support they might not be able to deliver.

TECHNIQUES OF RAPPORT

9-6. Foreign security forces (FSF) respect advisors who actually show them the right way to do a task rather than just talk. Experience has shown that foreign units will not hesitate to go into danger if their advisors are right there with them. To further contribute to building rapport, advisors should maintain enthusiasm and a positive attitude; learn the language or how to use an interpreter; respect a counterpart's rank, age, status, and experience; develop negotiation skills; and improve interpersonal skills.

9-7. Other rapport-building techniques can include being confident, competent, and capable. Advisors never expect or demand their counterparts to do something that they are unable or unwilling to do. They demonstrate everything. Advisors learn their counterparts' names and spend time with them at meals and during holidays; they prepare mentally to interact with counterparts at all times. Advisors avoid creating an American-only enclave; they avoid giving the impression of favoring any one group. They frequently assess their counterparts' perceptions. Advisors recognize threats to discipline and enforce the chain of command.

ACCOMPLISHMENT OF RAPPORT

9-8. Advisors measure rapport by how well they can motivate their counterparts to act. The basic techniques of motivation (in the absence of authority) are advising, setting the example, seeking compromise, and coercing. TC 31-73 discusses rapport in detail.

ADVISING

9-9. Advising the counterpart to select a particular course of action is only effective if the counterpart trusts the advisor is professionally competent. If the counterpart does not perceive the proposed solution as realistic, the advisor's competence is questioned. Advisors carefully explain to their counterparts that recommended courses of action are realistic and will be effective.

SETTING THE EXAMPLE

9-10. Setting the example for the counterpart must be an ongoing effort for the advisor to avoid the appearance of a "do as I say, not as I do" attitude. In setting the example, advisors explain that what they are doing is the most effective action for the situation. This is particularly true when the counterpart really does not understand the behavior (or purpose).

SEEKING COMPROMISE

9-11. When seeking compromise with the foreign counterpart, advisors may create a situation in which the counterpart has a personal interest in successful execution. In some cultures, seeking a compromise may allow the counterpart to save face. Furthermore, in certain situations the counterpart—because of practical experience—may have a better solution to the problem at hand.

9-12. When seeking a compromise in certain cultures, the perceived competence of the advisors may suffer. This may be mitigated somewhat by approaching the compromise as two professionals (the advisor and the counterpart) reaching a mutual conclusion. To reach an effective compromise, advisors may have to conduct negotiations.

> ***Note***: Two areas of concern must never be compromised for the sake of maintaining rapport—operations security and human rights.

Coercing

9-13. Coercion is the least desirable method of motivation because it can cause irreparable damage to the relationship. Advisors should use coercion only in extreme circumstances. Advisors must avoid unintentionally forcing their counterparts into action. Instead, they establish and maintain good rapport by conveying they are sincerely interested in their counterparts, their nation, and its cause. Advisors will not belittle their counterparts' efforts. Advisors must demonstrate that they have come to help because they believe the counterparts' goals are just, fair, and deserving of success.

POLICIES AND AGREEMENTS

9-14. Advisors must remember their primary aim is to forward U.S. policy. Therefore, they establish relationships with foreign personnel, other U.S. forces, and other government agencies. Advisors promote U.S. policy through agreements, command relationships, and operations. TC 31-73 discusses policy in greater detail.

United States and Host-Nation Policies

9-15. Advisors not only support U.S. national policy in the areas where they operate, they forward the policies of the host-nation government (unless otherwise directed by higher authority). Advisors maintain communications with higher headquarters to ensure activities align with U.S. objectives.

9-16. Host-nation policies may dictate practices and procedures that appear inefficient or uneconomical. Advisors should avoid criticizing or condemning such practices and procedures until they thoroughly understand them. Advisors should make private recommendations critical or contrary to host-nation policy.

Agreements

9-17. The status of advisors is normally specified by a status-of-forces agreement or other arrangement between the U.S. and host-nation governments. These agreements vary widely; they may offer little political protection or may provide for full diplomatic immunity. If no such agreement exists, advisors are subject to local laws and customs. Even if a status-of-forces agreement or other agreement exists that provides immunity, advisors are still expected to observe local laws.

Command Relationships

9-18. Advisors must use the U.S. chain of command to obtain and disseminate guidance and assistance. They carefully distinguish between the U.S. and foreign security forces' chain of command. In particular, advisors prevent their counterparts from attempting to control subordinates through the U.S. chain of command. Advisors provide recommendations—not orders—to their counterparts. Only the counterpart should issue orders to subordinates.

9-19. Nongovernmental organizations and other government agencies may be co-located in the operational area. Advisors integrate the efforts of all organizations into their mission planning and should seek opportunities for integrated efforts with their counterparts. This type of unified action at the tactical level can be achieved through the planning process to integrate nonlethal actions. Holistically, nonlethal actions include development and diplomatic efforts.

Conduct of Operations

9-20. Advisors can expect to participate in stability operations. These operations use military and paramilitary forces, as part of the interdepartmental team, to support internal defense and development operations. U.S. assistance may include advice on military organization, training, operations, doctrine, and materiel. In addition, the United States may provide and control U.S. logistics and sustainment for FSF. The objective of this assistance is to increase the capability of foreign organizations to perform their missions and operate efficiently in the given operational environment.

LEVELS OF ADVISORS

9-21. Organizations and individuals possessing greater skills and resources assist by imparting their knowledge through assistance efforts. The success of assistance depends largely on effective interaction between advisors and their foreign counterparts. Advisors may operate in the national, operational, and tactical levels. The national level is considered the largest national subdivision. This may be a nation's combined military headquarters, Service components, national-level government officials, or a combination. Operational-level advisors work at a level below the national-level advisor. This may be with military corps or division officers, with regional government officials, or a combination. Depending upon the size of the nation and its military, operational-level advisors may be assigned as low as brigades. Tactical-level advisors normally advise military units at or below the brigade level and provide counsel and assistance across the spectrum of conflict. They advise and assist counterparts in developing unit combat effectiveness and serve as liaisons between FSF and U.S. combat, support, sustainment and protection forces. Chapter 3 of TC 31-73 discusses these levels in detail.

CONSIDERATIONS OF RAPPORT

9-22. When building and maintaining rapport, advisors consider intelligence, human rights, role shock, ambiguity, personal relationships, bureaucracy, responsiveness, and diversity. These considerations are discussed in greater detail in TC 31-73.

INTELLIGENCE

9-23. To promote national policy and attain overall objectives, advisors may need to plan and conduct intelligence training. The advisor should assist the counterpart in developing a local intelligence collection program, training intelligence personnel in their respective specialties, and utilizing trained intelligence personnel properly.

9-24. The advisor should be familiar with area studies and area assessments. Pertinent documents should be compared to detect trends or changes. The advisor also should evaluate the foreign intelligence staff, its standing operating procedures, its chain of command, intelligence projects, and reference material available from other intelligence agencies. Advisors should prepare and maintain a list of essential elements of information.

HUMAN RIGHTS

9-25. Advisors stress the consequences of mistreating suspects, prisoners, or other persons taken into custody. These persons must be treated per Article 3 of the Third Geneva Convention. Article 3 requires care for the sick and wounded, protection of prisoners and detainees of all types from abuse or other harm. Murder, mutilation, and torture are expressly forbidden, as is humiliating or degrading treatment. Advisors must never be active participants in the conduct of such punishment. Furthermore, they must explain to their counterparts that they are obliged to report any atrocities of which they have knowledge.

ROLE SHOCK

9-26. Role shock results from the discrepancy between the roles individuals expect to play and the roles they actually do play. Role shock may also be a product of the tension created by individuals trying to do jobs themselves versus advising others how to do the jobs. Role shock tends to increase until about the anticipated midpoint of the tour and seldom disappears completely until redeployment. Symptoms of role shock resemble those produced by other stressful life events.

9-27. Many advisors become entangled in complexities of bureaucracy—both foreign and American—and become increasingly frustrated when seemingly familiar things fail to respond in expected ways. This continuing frustration can be a key contributor to role shock.

9-28. Although most advisors undergo some degree of role shock, over time they learn to take the personal, living, and social conditions in stride. More challenging are the problems that arise in connection with their jobs—their professional roles, relationships, personal achievement, self-development, self-determination, and self-image. Often advisors find their duties and activities during deployments differ somewhat from those expected. The actual duties and responsibilities are greater in scope, involve technical work outside of their assigned specialties, and require honed administrative (rather than military) skills.

Ambiguity

9-29. Most advisory assignments are ambiguous. What an advisor is expected to achieve is not always made clear. Furthermore, assignments may conflict with what they have trained to do, what others expect of them, or what they want or expect to do.

9-30. U.S. and foreign security forces may spend months—even years—negotiating a project. Still, the overall objectives or definition of the mission to be accomplished might be expressed in general terms. Various interpretations of what is to be accomplished arise. Even more views surface regarding the most appropriate means of reaching those goals.

Relationships

9-31. Ideally, foreign counterparts provide a working context for advisors and help them to understand and adjust to the new culture. Historically, those with good working relationships with their counterparts make better adjustments than those who have no counterparts or bad relations with their counterpart. Good working relationships may be difficult to establish and maintain. Most advisors approach their assignments as teachers and mentors. Foreign counterparts may resist such a relationship, particularly if they were expecting someone to take over and do all or part of the job. In addition, many counterparts want their advisors to share in at least some of the blame if something goes wrong.

9-32. The United States has little control over the availability or capability of counterparts. The assigned individual may be technically incompetent, lack professional commitment, have other interests, or be unavailable when needed. In some cases, numerous counterparts are assigned and rotate in their workings with the advisor. A single counterpart may not be assigned until the advisor has been on the scene for some time. In some scenarios, a specific counterpart is never assigned.

Bureaucracy

9-33. Although some Americans might disagree, the United States maintains a system that is—by world standards—relatively permissive and nonbureaucratic. As such, advisors may become aggravated when confronted with or obstructed by host-nation bureaucracy, red tape, administrative centralization, and a cumbersome decisionmaking process. Cultures that emphasize consensus in the decisionmaking process often prove quite frustrating for advisors. To be successful, advisors strive to understand the bureaucracy in which they operate.

9-34. Advisors should recognize the value of being or knowing the middleman. This role is sometimes one of the key contributions an outsider can make. Middlemen can move in, around, and out of organizations somewhat independently of the local hierarchy, protocol, and customs. They can link persons and offices that might otherwise have difficulty communicating. Many advisors dislike this role and refuse to practice it; however, the middleman can be a significant catalyst in any advisory operation.

Responsiveness

9-35. As frustrating as foreign bureaucracies may be, advisors often get even more irritated with the responsiveness of U.S. forces. Advisors frequently fail to account for distances, lead times, diplomatic issues, and other challenges when conducting security force assistance. Moreover, few Soldiers in the unit conducting security force assistance have had substantive experience in working in an interagency environment. These issues are further complicated by the usual tensions between headquarters and field staffs and between administrators and technical specialists. Expectations, values, and styles of behavior vary from group to group.

9-36. Deploying personnel may not be familiar with the tasks of development and the vast range of problems typical of underdeveloped areas. They come in eager to contribute their knowledge and skill, and anxious to make maximum progress in the limited span of their assignment. In so doing, they may fail to recognize that the factors that frustrate them—such as the lack of professional skills and facilities—are actually their reason for being there.

DIVERSITY

9-37. Organizations attract individuals who differ in abilities, knowledge, skills, work ethics, social behaviors, values, and attitudes. Advisors must be able to recognize personality types and develop techniques to cooperate effectively with each type.

RECOMMENDED PRACTICES IN BUILDING RAPPORT

9-38. Advisors may adopt many practices to help work with counterparts. Practices that help advisors build rapport are discussed in paragraphs 9-39 through 9-51.

ENSURE THAT THE ADVISOR PRESENCE IS UNDERSTOOD

9-39. Advisors entering an operational area work under the sponsorship of a military commander, mayor, village head, or some other recognized local leader. Prior to this, the advisor's arrival may need to be coordinated through the district, sector, or other appropriate local office. Clearances from distant national, state, province, or sector governments cannot compensate for a clear, local explanation of why an American advisor is in the area. This particularly applies to small, isolated communities where it is unusual for a stranger to appear for even an hour without being acknowledged and accepted by local leaders. Without explanations from locally respected persons, civilians often arrive at their own explanations, often to the detriment of the advisor effort.

FIND A BASIS FOR COMMON INTEREST WITH THE LOCAL PEOPLE

9-40. If advisors show appreciation for civilians as individuals, cultural gaps and language barriers can be overcome and common ground can usually be found. Advisors should listen and show interest when the locals speak. Initial conversations usually center on universal matters, such as food, shelter, clothing, health, and education. Over time, discussions may naturally move to the matter the advisor wants them to consider. Advisors are better received if they know something about earlier contributions in such matters as agriculture, folk art, religion, and architecture. Naturally, advisors prove more effective and appreciated if they can speak the language.

TRY TO UNDERSTAND WHY THINGS ARE DONE THE WAY THEY ARE

9-41. Although some local practices may seem strange, they generally have good reasons behind them. Advisors can discover these reasons with careful observation and a creative imagination. Food habits, family traditions, folk cures, and festive celebrations almost always have a great deal of human experience at their root. The advisor also needs to be aware that many villages contain rival subgroups and factions. This tension needs to be accounted for when working with the local people.

START WITH WHERE THE PEOPLE ARE AND WHAT THEY WANT

9-42. Advisors find out what the local people really want and work with them to achieve this aim. The need felt by the local people often is the best starting point. Once advisors address this aim, people become more appreciative and cooperative—they begin to raise their expectations and become interested in working for other improvements. To address the initial desires of the people, the advisor may need to call in personnel with the specialized skills needed for the particular project. Although this can result in a delay, it will help the advisor to achieve better cooperation.

Work Within the Local, Cultural Framework

9-43. Advisors understand basic cultural matters such as ethnic backgrounds of the people, family relationships, leadership patterns, and value systems. They also need some knowledge of local services such as health, education, communications, and transportation.

Help People Believe They Can Improve Their Situation

9-44. Most traditional peoples of South America, Asia, and Africa live in a largely static environment. Through their experience with change, they are more fearful of losing status than they are hopeful of bettering their condition. Therefore, changes suggested by advisors are often viewed with fear. Concrete local projects that yield easily observed benefits help convince villagers that they can improve their situation and make them more willing to cooperate in other projects.

Be Content with Small Beginnings

9-45. Change tends to come slowly in areas where there have been few in recent times. Advisors remember that knowledge—technical or otherwise—is cumulative. Once a small beginning has been made, greater activity and additional changes soon follow. It is easier to achieve momentum than to maintain it. Regardless, advisors start within as promising a framework as possible and with the support needed to sustain the momentum achieved.

Utilize Local Organizations and Recognize Their Leaders

9-46. The surest way for an activity to be continued after the advisor leaves is for it to have been launched and carried forward within the local organizational and leadership framework. People respond best when their local organizations are recognized as important and useful. Advisors consult and encourage recognized local leaders to contribute as they can. A well-conceived technical activity reflects credit on the local leaders associated with it. Advisors give attention not only to officials and family heads of local groups, but also to the quiet, behind-the-scenes leaders.

Help the Government Get Organized To Serve The People

9-47. For advisors to be most effective, they must understand the organization of the local government and how their activities fit into the overall scheme. Usually a set of agreements exists between various local agencies and the national government (usually through some sort of interministerial council). This agreement provides for a coordinated effort in servicing the varied needs of civilians. The advisor may need to work with appropriate agencies to assist in getting such agreements made. If such agreements already exist, advisors should be careful to recognize and strengthen them.

Train and Use Subprofessional, Multipurpose Local Workers

9-48. Selected young people in the villages can be trained and used as subprofessional, multipurpose village workers. Otherwise, advisors' influence is restricted to where they are standing and the immediate vicinity. Furthermore, advisors may spend so much time establishing and maintaining enough rapport with the villagers that they may become incapable of rendering any real service at all. Often the gap between civilians and the advisor is formidable because of their educational and cultural differences. Often the advisor works with villagers who are poor, illiterate, and devoid of outside contacts. Volunteer or paid local workers have proven quite helpful as they serve as a liaison between villagers and advisors.

Expect Slow Progress

9-49. As civilians begin to see successes from their joint efforts and begin to have new hope, they naturally want a larger hand in their own matters. The advisor may sometimes feel they want to assume more responsibility than they can carry. These evidences of growing pains should be appreciated, for they are a necessary part of becoming able to assume responsibility. The advisor needs to adjust to the growing desires of the people to help themselves.

Transfer Controls Constructively

9-50. Advisors need to help civilians see how they can build the new—what they want—upon the foundations of the old—what they already have. From the beginning of a project, the advisor's team needs to talk with local leaders. Together they envision how to train local personnel, gather financial support, and transfer responsibilities before transferring control of activities. If operating responsibility is transferred too early, some breakage—usually of material things—will likely occur. If the team keeps control too long, the civilians who wanted to take over may grow disillusioned with them for not relinquishing control when they thought it should have been. Working with local leaders, the team determines when to phase out each technical activity.

Expect Little Gratitude from Those Helped

9-51. People who benefit from assistance sometimes feel defensive. In accepting assistance they are, in a sense, admitting their own insufficiency. The self-esteem of a person, community, or a nation as a whole is delicate. The advisor's team should not, therefore, expect thanks. Rather, the team should approach the people in a spirit of fraternity and humility, taking satisfaction in their forward progress. The team should do its job the best it can and accept work well done as is its own reward.

DYNAMICS OF ADVISING FOREIGN SECURITY FORCES

9-52. Advisors work in a complex environment that is heavily influenced by the unified effort of multiple organizations, relationships, compartmentalization, and organizational cultures. Paragraphs 9-53 through 9-57 discusses the probable dynamics that an advisor must take into account.

Advisors and Interdependent Operations

9-53. Advisors commonly operate in complex environments with interdependent operations. Interdependent operations require near-seamless integration of agencies representing the instruments of national power—diplomatic, informational, military, and economic. This same degree of integration must be extended to include external organizations and those agencies representing the instruments of power in partner states. The negotiation and cross-cultural skills that enhance their performance as advisors are also well suited to the complex interrelationships that characterize such operations.

Organizational Relationships

9-54. Often organizational relationships can often be misleading and must be clarified. The actual interrelationships among and within organizations seldom follow a line-and-block diagram. Instead, they are heavily influenced by circumstances, personalities, perceptions, and resources. All relationships and lines of authority are subject to negotiation. Advisors understand that clarifying the roles, functions, and responsibilities is a continuous process and that each statement or action sets a precedent for future interrelationships.

9-55. The advisor must become hypersensitive to the most minor organizational disparities. The apparent similarities—the common language, sociological backgrounds, and citizenship—can be disarming and lead the advisor to overlook very real differences. If ignored, these differences can quickly impede synchronization.

Compartmentalization

9-56. Advisors balance security and synchronization. Overcompartmentalizing must be avoided. Often distrust can be prevented simply by not being overly secretive. Although advisors must conduct a careful, continual risk assessment to avoid compromise, the sharing of critical information is necessary for harmonious operations.

Organizational Culture

9-57. Advisors should approach external organizations as they would foreign counterparts—as unique institutions with distinctive cultures. Organizational differences, biases, and approaches must be researched and—whenever possible—accommodated. Each Service in the Department of Defense has unique perspectives that lead to significant differences in objectives and approaches to various problems. Profound differences in perspective may even be found between the different branches of the Army. If not recognized and accommodated, these characteristics may create distrust, disharmony, and disruption during critical operations.

Chapter 10

Cross-Cultural Influencing and Negotiating

Leveraging foreign counterpart cooperation can prove a difficult and critical task. Although coercion may prompt surrogate forces to perform a specific action, it can never lead to established conduct and legitimacy. Lasting advisors can achieve long-term success only by patiently applying their influence.

Advisors analyze their counterparts and determine their motivations and needs. Basic needs, culture, and individual personalities influence people. Advisors determine what these three things are so the appropriate influence techniques can be applied. Then advisors negotiate. Negotiation is fundamental to problem solving using the study, development, and honing of negotiation skills critical to the advisor.

INFLUENCING

10-1. Advisors need to gain influence with their counterparts, leaders, civilians, coalition members, and others beyond their chain of command. Advisors are self-aware, nonjudgmental, and capable. It is not necessary to accept others' beliefs or values, but advisors need to understand others' ways of thinking and operating. Knowing their own culture and understanding why they believe and value what they do helps advisors find a common cause. A common cause unites a diverse set of decisionmakers with whom advisors interact and try to influence.

10-2. Once advisors gain a minimum amount of influence, they can make things happen through others. They can extend their influence beyond their chain of command. Influence is simply getting others to do what you want.

10-3. Advisors identify the types and principles of influencing. They determine the best techniques with which to influence and review vital considerations. Using measures of effectiveness, they check that their influences have the desired results.

TYPES

10-4. Commitment, compliance, and resistance clarify how advisors understand the impact of their attempts to influence their counterparts. These three types of influencing are determined by how committed the foreign security forces (FSF) are to the task.

Commitment

10-5. Commitment is a willing dedication to a requirement or a cause because one wants to. Commitment-focused influence generally produces longer lasting and broader effects. Whereas compliance only changes immediate behavior, commitment reaches deeper—changing attitudes, beliefs, and behavior. For example, when advisors build desire and responsibility for a task among FSF, they often demonstrate more initiative, personal involvement, and creativity.

Compliance

10-6. Compliance is conforming to a requirement or demand because they feel like they have to avoid punishment or to gain a desired reward. Compliance-focused influence is based primarily on the leader's authority. Giving a direct order to a follower is one approach to obtain compliance during a task.

Compliance is appropriate for short-term, immediate requirements and for situations where little risk can be tolerated. Compliance techniques are also appropriate for leaders to use with others who are relatively unfamiliar with their tasks or unwilling or unable to commit fully to the request. If something needs to be done with little time for delay, and there is not a great need for the target of influence to understand why the request is made, then compliance is an acceptable approach. Compliance-focused influence is not particularly effective when a leader's greatest aim is to create initiative and high esteem with a counterpart.

Resistance

10-7. Resistance is a reaction against change. Resistance generally takes one of four forms—reactance, distrust, scrutiny, and inertia. It is manifested behaviorally (I won't do it), affectively (I don't like it), and cognitively (I don't believe it). Recognizing and lowering potential sources of resistance are key components to gaining compliance and commitment.

Motivations

10-8. Different motivations influence behavior. One type of motivation is the desire for something. Another is the fear of something undesirable. As an agent of influence, advisors determine how to increase approach motivations and how to decrease avoidance motivations. For instance, recruiting for a special mission can use special badges as incentive (approach). Advisors can assure Soldiers that the mission is relatively safe by providing extra training and better equipment (lowering risks or avoidance).

Principles

10-9. Six influencing principles are relevant to virtually any culture and any group. Understanding and applying these principles can increase an advisor's effectiveness in persuading a group or individual:

- Principle of scarcity.
- Principle of authority.
- Principle of social proof.
- Principle of liking.
- Principle of reciprocity.
- Principle of consistency.

10-10. With the principle of scarcity, people value more of what they can have less. They typically associate greater value with things that are rare, dwindling in availability, or difficult to acquire.

10-11. In the principle of authority, people are more easily persuaded by individuals perceived to be legitimate authorities or experts. They defer to experts who provide shortcuts to decisions requiring specialized information.

10-12. With the principle of social proof, people often look to the behavior of those around them for direction about what choices to make. This action is heightened when those around them are similar in terms of age, education, social standing, and experience.

10-13. In the principle of liking, people prefer to say yes and to comply with the requests of those they like.

10-14. In the principle of reciprocity, if someone grants favors, invited or uninvited, an overpowering need to repay that favor immediately blooms within the recipient of that favor. This human trait transcends all cultures and races.

10-15. With the principle of consistency, the desire for consistency motivates behavior. The drive to look and be consistent is a highly potent weapon of social influence, often causing people to act in ways clearly contrary to their own best interests. People do not like to appear inconsistent to others. It is human nature that people strive to feel good about themselves, which includes behaving per their important values and beliefs. When the behavior is consistent with whom people are and what they value, they feel good. People align with their clear commitments.

TECHNIQUES

10-16. The art of influencing involves knowing what techniques to use based on the situation. Advisors have to know the limits to which can use their influence; their influence is finite. They need to know when to use their influencing skills to achieve the highest payoffs. Their counterparts will only accept so much influencing if they don't completely trust their advisors. Table 10-1 lists positive and negative techniques delineated in FM 6-22. Table 10-2 lists resistance-based influence techniques. These are used when the counterpart resists the advisor's influences.

Table 10-1. Influencing and rapport techniques

Technique	Description	Characteristics	Outcomes
Pressure	Explicit demands, persistent reminders, sometimes appropriate in high stakes, urgent situations	Frequent checking Nagging	Gains compliance Frequent use may cause resentment, undermine morale, and harm relationship
Legitimate Request	Source of authority is used	Source credibility is important	Gains compliance
Exchange	Quid pro quo Reward offered for task completion; punishment threatened for incomplete task	Offer of reward or punishment not tied directly to getting job done, just something that counterpart wants or doesn't want	Gains compliance
Personal Appeal	Request based on friendship or loyalty	Agent of influence highlights talents of counterpart that make target the right person for the job (this is why you are the right person for the job, will you do it for me?)	Gains commitment Builds relationship
Collaboration	Works with target of influence by providing resources of some type, or by removing barriers to task completion	Offer of assistance is directly tied to request	Gains commitment Builds relationship
Rational Persuasion	Using facts, evidence, or logic in a persuasive message	Relies on agent being viewed as expert	Gains commitment Builds relationship
Apprising	Explaining the consequences of a potential action or decision	Decision is good for unit, career, or mission accomplishment	Gains commitment Builds relationship
Inspiration	Arousing strong emotion to build conviction or commitment	Appealing to values of the counterpart	Gains commitment Builds relationship
Participation	Requests for counterpart to be a part of planning, active participation in development of solution	Promotes commitment Possibly more feasible and culturally appropriate course of action	Gains commitment Builds relationship
Relationship Building	Build positive rapport and trust	Provide the foundation for using influence or inspiring	Gains commitment

Table 10-2. Resistance-based techniques

Technique	Description	Characteristics	Outcomes
Sidestep Resistance	Agent defines own role as consultant or advisor, but not commander Pushes the consequence of choice into future Depersonalize task to gain commitment to general idea Focus on end state instead of expended effort	Does not sell, just advises to prevent immediate resistance Puts less focus on immediate impact, more focus on long-term effect Stating "leaders should" instead of "you should" Stress mission accomplishment	Gains compliance Potentially gains commitment if paired with other influence techniques like rational persuasion
Address Resistance Directly	Address reluctance by lowering costs associated with tasks	Provide some of own resources through collaboration or counterargue concerns	Builds commitment
Address Resistance Indirectly	Build confidence, esteem, and self-efficacy for the task	Acknowledge difficulties while showing support for counterpart's abilities	Builds commitment
Distract Resistance	Interfere with the development of counterarguments	Distracting attention	Gains compliance
Disrupt Resistance	Disrupt complacency to bring attention to message	Shake up to motivate	Gains compliance
Consume Resistance	Resistance is energy	Wear them down prior to opportunities to resist. Persist with requests until they concede	Gains compliance
Use Resistance to Promote Change	Reverse psychology	Issuing a challenge	Gains compliance; risky when you don't know target of influence really well

10-17. To succeed with any influence technique, the influence attempt should be perceived as authentic and sincere. Positive influence comes from leaders who do what is right for Army forces, the mission, the team, and each individual Soldier. Negative influence—real and perceived—emanates from leaders who focus more on personal gain and lack self-awareness. False perception may trigger unintended side effects such as resentment of the leader and the deterioration of unit cohesion. With each and every influence attempt, Army leaders should ensure that the influence objective aligns with Army Values, ethics, the Uniform Code of Military Justice, the Warrior Ethos, and the Civilian Creed.

Considerations when Choosing Techniques for Influencing

10-18. Advisors clearly understand what their counterparts value and fear. This information allows advisors to quickly relate the objective of the influence to something that their counterparts treasure. Also, a strong sense of self supports the advisor's recognition of which techniques they can successfully execute. Someone with an intimidating presence may achieve compliance with pressure more quickly than someone without that presence. Understanding the context of the request is important. For instance, using participation in the midst of flying bullets would not make sense. However, a long-term project that impacts many people might be a worthy situation for participation. Advisors also consider—

- The motivation of their counterparts.
- The relationship they have with their counterparts.
- What they do or not have to offer.
- An understanding of the second- and third-order effects their recommendations will have.
- What has worked in the past.
- Possession of an appropriate power based for use of a technique in particular context.
- Consistency of technique with prevailing social norms and role expectations.
- Level of resistance encountered or anticipated.
- Costs of using a technique in relation to likely benefits.

10-19. The cost of a technique refers to the amount of effort, favor, or resources that accompany the technique. A personal appeal that relies on friendship is quite effective; however, counterparts might expect the favor to be returned. If resources or time are plentiful, then the exchange might not seem expensive, but if they are not plentiful, then advisors must choose a different technique. Advisors balance the costs and benefits of each technique.

Moderating Factors

10-20. Sometimes moderating factors diminish the impact of any influence technique. These factors may include cultural misunderstanding, level of trust, or constraints on the counterpart like threats or corruption. Advisors analyze and learn about these factors. Advisors must be flexible when using influencing techniques. When one technique doesn't work, switching strategies or adding techniques may be appropriate.

CONSIDERATIONS

10-21. Seven considerations need to be addressed to influence behavior. These include determining the goal; who to influence; motives; beliefs, values, and attitudes; cultural beliefs, values, and attitudes; and susceptibility. These techniques are designed to help advisors achieve their goals. Just knowing that the counterpart wants something is not enough. Whether it is time off, new weapons, or money, advisors have a limited supply of stuff they can promise; they can only bend so much on rules, regulations, and laws. It is at this point that the advisor enters the realm of negotiations.

Determine the Goal

10-22. Advisors influence others to achieve some purpose. To be successful at exerting influence, advisors have an end state or goal in mind. Sometimes the goal will be very specific. Many goals are less distinct and measurable but are still valid and meaningful.

Determine Who to Influence

10-23. Advisors develop a mental list of those whom they might need to influence. Getting a FSF platoon to take action may be as simple as influencing the commander, but the advisor may have to influence the soldiers and commander of the platoon. When practical, advisors consider many variations of whom to influence. They look for circumstances in the immediate environment that significantly affect their counterparts. Individual or group characteristics to consider can include—

- Gender.
- Religion.
- Age.
- Socioeconomic status.
- Ethnicity.
- Political affiliation.

- Level of education (important in determining how to access and persuade).
- Occupation.

10-24. The challenge is to determine which characteristics impact emotional or behavioral responses and under what circumstances. Ethnicity may significantly affect behavior and emotions regarding voting and politics but may have almost no effect on the decision to enlist in the military. Individual and group characteristics may provide additional insight on motivations.

Determine and Prioritze Motives

10-25. Motives are anything that is required or desired (needs and wants). Motives are the reasons behind the decisions people make. Motives come from an inner desire to meet a need or want. Advisors use the knowledge of what motivates others to influence them. Knowing one's counterpart and others who may be influenced gives advisors insight into influencing their behavior. The desires to fulfill, alleviate, or eliminate a need or want provides the motivation to change behavior.

Basic Motives

10-26. Basic motives involve physiological needs—food, water, and air—and safety needs—security and shelter. When people feel hunger, they are motivated to eat; when people feel pain, they are motivated to reduce the source of the pain. Such basic motives are extremely powerful in driving behavior and overwhelm psychological needs and wants.

Social Motives

10-27. Social motives associate access to other tangibles (money, goods, education, infrastructure, and healthcare) or more complex psychological motives with basic motives. For example, people learn to want money because it can be exchanged for food and other desired goods. Power, achievement, reassurance, escape, justice, and acceptance illustrate types of learned social motives. Examples include:

- Wanting better educational opportunities for their children.
- Wanting better paying jobs.
- Wanting their interests represented by the government.
- Wanting revenge for perceived wrongs.
- Wanting self rule.

Prioritize Motives

10-28. Advisors prioritize motives by immediacy of the need or want. Critical motives are immediate needs. Short-term motives are currently being satisfied or active efforts are being made to satisfy them in the near future. Long-term motives are desired but are not immediately important; satisfaction may be delayed until some point in the future.

Determine Beliefs, Values, and Attitudes

10-29. Although often difficult to derive, beliefs, values, and attitudes can prove quite effective in persuasion. Beliefs are what is thought to be true. People do not believe in something if they think it is not true. People vary from being gullible to skeptical. Values prioritize what one believes. Values can be ethical, doctrinal, social, aesthetic, or economic in nature. Attitudes reflect whether one likes or dislikes something. Likes and dislikes can often be traced back to a person's beliefs and values. This includes loves, hates, frustrations, and fears. Frustrations occur when a want is not met. By examining causes of frustration, the advisor may find another motive to address.

Compare Beliefs, Values, and Attitudes

10-30. The predominant cultural beliefs, values, and attitudes; traditions; and norms may differ greatly from the group the advisor is trying to influence. Many Americans act differently in different cultures. This change in behavior relates directly to their predominant culture.

Determine Susceptibility

10-31. Susceptibility is the likelihood that the individual or group will be open to persuasion. Determining susceptibility aids in prioritizing efforts. If susceptibility is moderate to high, then advisors may not need to focus as much effort as they would if the susceptibility is low. Advisors determine susceptibility by assessing the strength of the underlying source of the motivation and the perception of the desired corresponding behavior. The strength of the motivation depends on the perceived risk and the perceived reward, while the perception of the behavior directly links directly to how consistent the behavior is with the individual or groups' values and beliefs.

Perceived Risks and Rewards

10-32. Risks may include the threat of physical harm, financial loss, or disapproval of peers. Rewards may include financial gain, political power, approval of peers, or increased self-esteem. Risks and rewards can vary from immediate- to long-term. Immediate risks and rewards are stronger than long-term risks and rewards. For example, few people will walk a tightrope (immediate risk), but many people smoke (long-term risk). The higher the perceived rewards and the lower the perceived risks, the more likely the behavior will occur.

Consistency with Values and Beliefs

10-33. Advisors determine if the desired behavior is consistent with values and beliefs. Generally, behavior is consistent with the relative moral or ethical, religious, political, and cultural beliefs, values, and norms. Behavior that is inconsistent or incongruent with values and beliefs often decreases over time. Alternately, if behavior continues to be incongruent with values and beliefs, then individuals are likely to justify this behavior by modifying their values and beliefs. The more consistent a desired behavior change is with the individual's values and beliefs, the more likely the behavior will occur.

10-34. The lower the susceptibility, the less likely influencing behavior will be successful, which does not mean that the person cannot still be influenced. The lower the susceptibility, the more conditions will need to be modified for behavior change to occur.

APPLICATION

10-35. Influencing is a dynamic process. Advisors make a plan before attempting to influence. If one technique fails, they have to be able to shift to other techniques. Advisors do not need to sit down and write a plan before attempting to influence people. However, even with apparently positive results, advisors review their tactics and techniques.

CHECKS

10-36. Advisors check measures of effectiveness, unintended consequences, and second- and third-order effects. Advisors assess if their influencing is achieving their immediate measures of effectiveness. However, with every action some unintended consequences may occur or the activity may not prepare counterparts for more complicated tasks later on.

CROSS-CULTURAL NEGOTIATING

10-37. The principal form of negotiation is likely to be cross-cultural negotiation. Cross-cultural negotiation brings with it a series of unique challenges and techniques. Culture fundamentally affects language and behavior. It also significantly impacts the way people handle conflict.

10-38. Negotiations among members of the same culture can be stressful; negotiating with members of other cultures can be exceedingly difficult. Logically, working with other cultures is a basic skill for the Soldier and an absolute requirement when acting in an advisory role. Understanding the components, perspectives, roles, and outcomes will help minimize the difficulty of cross-cultural negotiation. Chapter 4 of TC 31-73 discusses cross-cultural negotiations.

Influences on Negotiations

10-39. Cultural differences may lead to a conflict between what the parties expect and what their families and communities expect. This has an obvious impact on negotiating behavior. Negotiations are influenced by perceptions, individualism, punctuality and pace, relationship building, and language.

Perceptions

10-40. Most Americans have certain preconceived ideas about people from other cultures. These perceptions may not be factually based, but they exist and influence negotiations. Similarly, foreign negotiators have certain perceptions about Americans. Again, these perceptions may not be based on fact, but they exist. Advisors need to understand how other cultures perceive Americans to appropriately adjust their negotiating style. Successful advisors find ways to capitalize on the positive preconceptions of their counterparts and find ways to neutralize the negative preconceptions.

Individualism

10-41. The United States encourages a largely individualistic culture. As such, the typical American negotiator prefers to enter into negotiations alone or, if required, as part of a very small negotiating team (two or three people, at most). This behavior is largely a by-product of U.S. culture, which focuses on the individual, performance, initiative, and accomplishments. Most other cultures of the world place less emphasis on the individual and more on the group.

Punctuality and Pace

10-42. Americans place a great emphasis on punctuality, viewing it as an indication of an individual's basic ability and commitment. However, other cultures emphasize punctuality differently. Some emphasize punctuality more than Americans. In other cultures, time is considered relative or may not be important at all.

Relationship Building

10-43. Americans are very competitive during negotiations and are inclined to stress short-term results. Interpersonal relationships are of minimal importance to American negotiators. Building long-term relationships only occurs after the successful completion of negotiations.

Language

10-44. U.S. negotiators must be considerate of others' with respect to language. Even if the counterpart is fluent in English, they may expect negotiators to speak their language or use an interpreter. Members of some cultures do not like to speak English as they are very proud of their language. When U.S. negotiator counterparts do not speak English, they will need an interpreter. Counterparts may also expect American negotiators to present detailed written proposals in the local language. Many cultures appreciate foreigners who take the trouble to learn a few simple courtesy words. However, some prefer that foreigners who are not fluent refrain from using more than a few, basic courtesy words.

Elements for Conducting Cross-Cultural Negotiations

10-45. The basic elements common to all negotiations must be applied in ways that allow for the differing cultures of the participants. Elements that must be considered include opening strategies, directness, strengthening behaviors, movement, power, face-saving, formal agreements, and mediation.

Components of Negotiating

10-46. Two opposite types of negotiating exist: distributive and integrative. They reflect a philosophy towards negotiating in general. Distributive negotiating is win-lose while integrative is win-win. Distributive negotiating involves dividing a fixed amount of a thing to the parties involved. This negotiation normally involves people who have never had a relationship before. Integrative negotiating involves joining two parties to achieve something together. Successful negotiations can involve a mix of both strategies.

10-47. When using distributive negotiating, advisors know the point at which counterparts will walk away. Advisors push them almost to that point by manipulation and withholding information. The distributive negotiator believes—

- The disputants are adversaries and the goal is victory.
- Fixed resources are divided so one side receives more than the other after making concessions.
- One person's interests or position opposes the other's interests or position.
- The dominant concern is usually maximizing one's own interests.

10-48. Integrative strategies used by advisors include cooperation, sharing information, and mutual problem solving. The integrative negotiator believes—

- The disputants are joint problem solvers seeking a wise decision.
- An understanding of needs will allow resources to be divided and both sides can win.
- The relationship between the parties is what is critical.
- The dominant concern is to maximize joint outcomes.

PERSPECTIVES AFFECTING NEGOTIATIONS

10-49. Four perspectives impact negotiations: time, environment, self, and nonverbal communication. Advisors consider how each party views these conditions when negotiating.

Time

10-50. Some cultures and individuals see time as a micro view. It is all about here and now. Other cultures and individuals see time in a macro view. Now is a point that bridges the past with the future. For example, imagine a person with the first scheduled event has a deadline to make an agreement now. The next scheduled event is 1500 so that person has two hours to come to an agreement. The second person might agree that they can come to an agreement now; but to the second person, now could be within the next month or two.

Environment

10-51. Some cultures believe they can impact, if not control, their environment. If something should go wrong, somebody is to blame. Every accident is preventable. Other cultures believe the environment impacts the person. Accidents just happen and no one could have done anything about it. One side may be quite ready to commit to something in the future *knowing* that nothing stands in the way, while the other side may be reluctant to talk in certainties of the future *knowing* that factors beyond their control may prevent them from fulfilling their commitments.

Self

10-52. In some cultures, when a person makes a mistake, blame can be isolated to that person. In other cultures, the individual is seen as an element of the group. So a mistake is made not by the individual but rather the whole group. In negotiations, one might accept a personal slight to get the goal achieved, whereas the counterpart represents the group and cannot allow shame to fall on them even though one might accept it personally.

Nonverbal Communication

10-53. Since advisors will often have to work through interpreters, nonverbal communication becomes more important. Some cultures respect people who mean just what they say and maintain their composure. Other cultures respect the eloquence in which something is said or the emotion used to convey it. Where words may be translated improperly, the advisor or counterpart can hear tone and see emotion.

Roles of Negotiation

10-54. When entering a negotiation, advisors must know what their role. The three major roles associated with negotiation are negotiator, mediator, and arbitrator. Each role has its own tactics or techniques and separate functions.

Negotiator

10-55. Advisors most often find themselves in the role of a negotiator. A negotiation is the process by which parties exchange commitments or promises to resolve disagreements and reach a settlement. Advisors prepare for each negotiation so they will not be at a disadvantage during the negotiation. To prepare for negotiations advisors examine their options, discover what motivates the parties, list all positive scenarios, and dispute resolutions.

10-56. Advisors examine their options. They question their goals, the importance of the goals, and the amount of leverage they have. They determine who is in power and what to do if no agreement is reached. Advisors uncover what motivates their counterparts. They delve into their counterparts' underlying needs.

10-57. By listing all the scenarios that satisfy their needs, advisors can maximize the gain for all parties. Advisors can add features to the scenarios that increase the opportunities for all parties to be satisfied. When advisors dispute resolutions, they often find some form of compromise. One of the best outcomes of compromise is when all parties leave the bargaining table satisfied but no single party feels like they have won or lost the argument. A skilled negotiator should strive to achieve a win-win result.

10-58. The negotiator represents one side and has a vested interest in the outcome of the negotiation. It is best to avoid situations where there is a conflict of interests. By blindly pushing the coalition agenda, the advisor could lose the trust of the counterpart. Considerations for a negotiator should include:

- If the foreign security forces ask the advisor to negotiate for them, they expect the advisor, as their negotiator, to have their best interests in mind. The coalition may try to use the advisor to push their agenda and influence rather than negotiate.
- If the foreign counterparts take illegal action, advisors cannot use commitment techniques. They need to use some form of compliant technique.
- If the foreign counterpart is not violating any laws, the advisor should focus on commitment techniques. Success will not be achieved all at once.

Mediator

10-59. Advisors may find themselves in the role of a mediator. The mediator has three general roles to perform: facilitator, formulator, and manipulator. Table 10-3 identifies the roles and tactics advisors use.

Table 10-3. Mediator roles and tactics

Roles	Tactics
Facilitator	• Make contact. • Arrange for interaction. • Clarify the situation. • Supply missing information. • Transmit messages between parties. • Ensure all interests are discussed.
Formulator	• Control the pace, formality, and physical environment of the meetings. • Highlight common interests. • Help parties save face. • Keep process focused on the issues. • Make substitutive proposals. • Suggest concessions parties could make.
Manipulator	• Keep parties at the table. • Change parties expectations. • Highlight costs of nonagreement. • Reward concessions. • Promise resources. • Threaten the withdrawal of resources. • Offer to verify compliance. • Threaten to withdraw mediation.

Arbitrator

10-60. The foreign security force may try to put the advisor in the position of an arbitrator. An arbitration is the process by which the parties to a dispute submit their differences to the judgment of an impartial person or group selected by mutual consent or statutory provision. They want the advisor to decide between two positions. This is a dangerous position for the advisor. When a win-win solution can be found, the advisor can improve rapport with both parties, but when it is a win-lose situation, the advisor could alienate one or both groups.

Expected Outcomes of Negotiation

10-61. The outcome of negotiations sometimes can determine the outcome. Advisors consider the desired outcome to determine which confrontation preference to use. Negotiation can focuses on physical outcomes of relationships. How advisors combine their preferences to their focus affects their perceptions of the outcome.

Confrontation Preferences

10-62. Negotiators will have one of three preferences for confrontation: avoidance, collaboration, or competition. Those that prefer avoidance will consider accommodation as a viable means to resolve an issue. Negotiators that prefer collaboration consider compromise as the preferred outcome. Those negotiators who use competition want to win. Intentionally losing in a negotiation to avoid confrontation seems counterintuitive; however, it tends to protect the current relationship. Collaboration requires a mutual respect of the existing relationship, while competition is generally indifferent. It would be fairly easy to take advantage of a negotiator who prefers to avoid confrontation. At the same time, expecting to be able to compromise with a highly competitive negotiator may become a point of frustration. For the competitor, winning may be more important than the deal itself.

Negotiation Focus

10-63. Defining whether the negotiation is a win-win situation is not as simple as which side gained the most from the negotiation. Advisors often focus on physical outcomes whereas diplomats focus on relationships. Every negotiation varies on which factor is more important. Just as car dealers may accept less profit on individual services, they gain overall if the customer returns.

Perception of the Outcome

10-64. Comparing confrontation preferences leads to nine potential combinations. If advisors can determine a participant's preference—avoidance, collaboration, and competition—and focus, they can determine how the participant might feel based on the outcome of the negotiation. If one participant is a collaborator and one is a competitor, then a win-win situation is possible, but neither side will be completely satisfied with the outcome. The competitors will not have gotten a clear win, and the collaborators will not have gotten what they considered fair. Subsequently, they will both have some minor negative feelings toward the other.

Appendix A

Army Special Operations Forces Imperatives

This appendix outlines the imperatives for Army special operations forces conducting security force assistance and foreign internal defense operations. Table A-1 lists these imperatives.

Table A-1. Army special operations forces imperatives

Imperatives	*General Descriptors*	*Specific Actions*
Understand the Operational Environment	• Knowing the U.S. command relationships (military and interagency) • Knowing the population (who, what, when, where, why, and how) • Knowing the enemy and other antigovernment forces (crime, militias, gangs) • Knowing the higher commander's intent • Knowing that the press is a part of the operational environment (news reports of events reach the U.S. population quickly)	• Define relationship as partnership or augmentation • Conduct analysis using ASCOPE (areas, structures, capabilities, organizations, people, events) • Analyze the enemy order of battle • Protect the population
Recognize Political Implications	• Identifying the political (and legal) restrictions and implications of actions by U.S. and foreign security forces • Identifying political implications of establishing and enforcing rules of engagement policy • Identifying short and long term political implications of security force assistance • Understanding the importance of establishing human rights (critical for tactical and strategic success) • Acknowledging the effects of public opinion on the mission locally, in the United States, and internationally • Understanding that insurgents conduct operations to have political effects among the local populace, external actors, and the U.S. and multinational civilian public	• Include rules of engagement in the planning and briefs of operations • Include local government leadership in operations • Reinforce human rights briefings and training to all foreign security forces • Prepare all U.S. and foreign security forces to talk with the press • Prepare information engagements for all operations • Anticipate and neutralize the methods enemy forces pass propaganda

Imperatives	General Descriptors	Specific Actions
Facilitate Interagency Activities	• Conducting security force assistance as both a military and an interagency effort; taking advantage of the greater impact of collective efforts • Coordinating the scope and limitations of each agency's influence and programs • (as opposed to a purely military program) • Obtaining the active cooperation of the foreign security forces • Including military and civilian agencies working together for unity of effort	• Include interagency members in planning and training as appropriate • Provide assistance to support limited agency resources, allowing agencies to prioritize their primary roles • Conduct regular coordination meetings with civil agencies to synchronize security efforts • Include host-nation agencies whenever including U.S. agencies
Engage the Threat Discriminately	• Avoiding the use of overwhelming firepower—to prevent alienating the population and turning them away from the host-nation government • Giving training and advice to foreign security forces based on the mission and threat • Targeting insurgents, criminal activities (lawlessness), or subversion—to avoid alienating the populace • Minimizing the use of force in populace and resources control measures—weighed for potential gain versus potential cost • Ensuring that rules of engagement and escalation of force are balanced between force protection and safeguarding civilians	• Take civilians into account when considering weapons systems and techniques for operations • Include civilians in risk assessment during mission planning • Train and rehearse rules of engagement and escalation of force to prepare Soldiers to react in accordance to policy • Publicize reasons for use of populace and resource control (for example, curfews protect civilians); avoid enemy propaganda or rumors dictating the psychological response of the populace • Update rules of engagement and escalation of force periodically • Rules of engagement or escalation of force reflects engaging targets that are visually verified and not simply suppressive fire in populated areas
Consider Long-Term Effects	• Building up the foreign security forces and alleviating the root causes of the current situation over time—not just treating the symptoms • Solving problems in a broader political, military, social and economic context • Making policies, plans, and operations consistent with national and theater priorities and the objectives they support	• Understand local grievances and underlying issues • Interact with all aspects of the population a government to identify root causes • Ensure the commander's intent remains focused on higher goals and policy related to addressing root causes, and protecting the local population • Monitor and update rules of engagement as local government and police gain control of rule of law. Conduct operation that adhere to local laws as possible • Understand legal and political constraints (rules of engagement) to avoid strategic failure while achieving tactical success

Imperatives	*General Descriptors*	*Specific Actions*
Ensure Legitimacy of Special Operations	• Reinforcing and enhancing the legitimacy and credibility of the foreign security forces and government • Gaining trust and credibility to ensure advisors and U.S. forces obtain legitimacy with the local population • Obtaining legitimacy in theater, in the United States, and internationally to ensure ongoing support	• Conduct joint and multinational patrols and operations. • Ensure information engagement is common in all operations and contributes to maintaining credibility • Engage the media with the host nation to inform as well as to justify actions • Focus on what the population considers the root causes, not the U.S. perspective • Develop long-term solutions to restore or improve basic services with host-nation leadership
Anticipate and Control Psychological Effects	• Considering the psychological effects of all operations (such as populace and resources control measures, combat operations, and civic action) • Avoiding negative psychological effects that can negate tactical victories • Appreciating the power of people's perceptions versus reality	• All operations are developed, analyzed and executed with information engagement • Continue to gauge the psychological effects with the local population, local government, and security forces as well as the enemy (propaganda and interrogation)
Apply Capabilities Indirectly	• Teaching host-nation forces to plan instead of giving them tasks • Supporting U.S. efforts enhancing the legitimacy and credibility of the host-nation government and foreign security forces • Balancing control and pressure on foreign security forces to ensure they conduct successful tactical operations and accomplish missions • Fulfilling primary role to advise, train, and help foreign security forces take the lead and stabilize the host nation • Influencing a foreign security force without commanding it; not accepting a lack of effort or a high level of corruption	• Conduct training of foreign security forces only to teach basics • Transition to mentor or advisor relationship during planning and operations to encourage the host-nation to take the lead • Use discretion when advising counterparts as not to embarrass foreign security forces leadership • Use host-nation methods whenever possible

Imperatives	General Descriptors	Specific Actions
Develop Multiple Options	• Considering the complex and irregular environment • Planning to use the full range of capabilities • Anticipating that an operational environment may dictate a change of rules of engagement or type of operations • Ensuring foreign security forces are capable of shifting from one option to another before and during mission execution	• Continue to refine and train on battle drills and contingencies • Develop an understanding of skills and abilities of the members of the unit and community • Elicit the input of foreign security forces on how to address situations to widen the options beyond conventional techniques • Include government agencies, special operations forces, nongovernmental organizations, local police, and the host-nation government to develop additional methods or options for solving problems
Ensure Long-Term Sustainment	• Avoiding advising and training foreign security forces in techniques and procedures beyond their capabilities to sustain • Modifying tactics, techniques, and procedures; training; operations; and sustainment to fit the culture, education, and technological capability of the foreign security forces • Developing host-nation sustainment before handing over economic, social, political, military, and security initiatives and projects • Avoiding host-nation dependency on U.S.-funded programs	• Understand and utilize the equipment and manpower present in the foreign security forces unit's inventory • Assess the capabilities of foreign security forces for leadership, training, sustainment (logistics, administration, and equipment), and ongoing development of a professional security force • Recognize the programs that are durable, consistent, and sustainable by the host nation • Do not begin programs beyond the economic or technological capacity of the host nation
Provide Sufficient Intelligence	• Ensuring detailed, near real-time, all-source tactical intelligence products support all activities and programs • Ensuring detailed and comprehensive intelligence on all aspects of the operational environment and its internal dynamics support operations • Ensuring human intelligence, urban recon, and internal security elements assess internal threats, warn the government, and take action to penetrate and neutralize instability • Considering aspects of the society not directly related to the tactical combat situation when assessing the enemy threat • Using information from intelligence assets to advise, train, and help foreign security forces counterparts and ease interagency efforts	• Ensure host nation forces are conducting operations using accurate, real-time intelligence • Establish priority of effort when identifying intelligence requirements • Assist foreign security forces to prioritize and focus counterintelligence on the commander's critical information requirement to receive timely and specific intelligence • Assess other organizations that impede the rule of law and government functions (such as gangs, criminals, smugglers, militias, mercenaries, drug lords, and narcoterrorists)

Imperatives	*General Descriptors*	*Specific Actions*
Balance Security and Synchronization	• Maintaining unity of effort and avoiding compartmentalizing key security forces from planning and intelligence contingencies even when security concerns exist • Avoiding either excessive or insufficient security • Anticipating reasons foreign security forces may not provide intelligence, such as lack of internal security or lack of U.S. intelligence support	• Encourage combined planning, preparation, and operations with foreign security forces' military, police, and governmental units • Resolve conflicting demands on mission planning and execution • Employ measures to mitigate risk of operations security instead of dismissing the host-nation government, police, or military needed for the overall effort • Strive to gain the trust of the local population to gain the most human intelligence and support from the foreign security forces

Appendix B
Legal Considerations

GENERAL LEGAL BASIS

B-1. Law and policy govern the actions of U.S. forces in all military operations, including security force assistance. For U.S. forces to conduct operations, a legal basis must exist. This legal basis profoundly influences many aspects of an operation. It affects the rules of engagement, how U.S. forces organize and train foreign forces, the authority to spend funds to benefit the host nation, and the authority of U.S. forces to detain and interrogate. Under the Constitution of the United States, the President is the Commander in Chief of U.S. forces. Therefore, orders issued by the President or the Secretary of Defense to a combatant commander provide the starting point in determining the legal basis. Laws are legislation passed by Congress and signed into law by the President and treaties to which the United States is party. Policies are executive orders, departmental directives and regulations, and other authoritative statements issued by government officials. This appendix summarizes some of the laws and policies that bear upon U.S. military operations in support of security force assistance. This summary does not replace a consultation with the unit's supporting staff judge advocate.

LEGAL AUTHORITY FOR SECURITY FORCE ASSISTANCE

B-2. If the Secretary of State requests and the Secretary of Defense approves, U.S. forces can participate in security force assistance. The request and approval can go through standing statutory authorities in Title 22, U.S. Code. Title 22 contains the Foreign Assistance Act of 1961, the Arms Export Control Act of 1976, and other laws. It authorizes security assistance, developmental assistance, security force assistance and other forms of bilateral aid.

B-3. The request and approval for security force assistance might also occur under various provisions in Title 10, U.S. Code. Title 10 authorizes certain types of military-to-military contacts, exchanges, exercises, and limited forms of humanitarian and civic assistance in coordination with the U.S. ambassador to the host nation. In such situations, U.S. forces may be granted status as administrative and technical personnel based on a status-of-forces agreement or an exchange of letters with the host nation. This cooperation and assistance is limited to liaison, contacts, training, equipping, and providing defense articles and services. It does not include direct involvement in operations. Assistance to police by U.S. forces is permitted but, generally, Department of Defense (DOD) does not serve as the lead government department. Without receiving a deployment or execution order from the President or Secretary of Defense, U.S. forces may be authorized to make only limited contributions during operations that involve security assistance.

EXISTING UNITED STATES LAW

B-4. The Constitution divides the power to wage war between the executive and legislative branches of government. Under Article I, Congress holds the power to declare war; to raise and support armies; to provide and maintain a navy; and to make all laws necessary and proper for carrying out those responsibilities. Balancing that legislative empowerment, Article II vests the executive power in the President and makes the President the Commander in Chief of the Armed Forces. This bifurcation of the war powers created an area in which the coordinate political branches of government exercise concurrent authority over decisions relating to the use of Armed Forces overseas as an instrument of U.S. foreign policy.

B-5. The Supremacy Clause of the Constitution (Article VI) states, in part, that all treaties made by the United States are the supreme law of the land. Therefore, ratified treaties, such as the United Nations (U.N.) Charter (see paragraphs B-7 through B-8) and the Geneva Conventions, create legal obligations on U.S. forces regarding their ability to perform various types of missions and functions.

B-6. The War Powers Resolution of 1973 ensures the collective judgment of both the executive and legislative branches to deploy U.S. forces by requiring consultation of and reports to Congress, in any of the following circumstances:

- Introduction of troops into actual hostilities.
- Introduction of troops, equipped for combat, into a foreign country.
- Greatly enlarging the number of troops, equipped for combat, in a foreign country.

B-7. The President is required to make such reports within 48 hours of the triggering event, detailing—

- The circumstances necessitating introduction or enlargement of troops.
- The constitutional or legislative authority upon which he bases his action.
- The estimated scope and duration of the deployment or combat action.

B-8. Since the War Powers Resolution was passed over the veto of President Nixon, no President has either conceded its constitutionality or complied fully with its mandates.

International Law and Treaties

B-9. The U.N. Charter became effective on 24 October 1945 after being ratified by the United States and a majority of other signatories. The U.N. Charter mandates that all member states resolve their international disputes peacefully. It requires that member states refrain in their international relations from the threat or use of force. The U.N. Charter also provides that all nations have the right to use self-defense to combat acts of aggression against them until such time as the Security Council takes action.

B-10. The U.N. Charter provides the essential framework of authority for the use of force, effectively defining the foundations for a modern *jus ad bellum*. Inherent in its principles are the requirements for necessity, proportionality, and timeliness. Necessity involves considering—

- The exhaustion or ineffectiveness of peaceful means of resolution.
- The nature of coercion applied by the aggressor State.
- The objectives of each party.
- The likelihood of effective community intervention.

B-11. Proportionality refers to limiting force in magnitude, scope, and duration to that which is reasonably necessary to counter a threat or attack. Timeliness refers to situations in which the delay of a response to an attack or the threat of attack attenuates the immediacy of the threat and the necessity to use force in self-defense.

B-12. U.S. forces obey the law of war. The law of war is a body of international treaties and customs, recognized by the United States as binding. It regulates the conduct of hostilities and protects noncombatants. The main law-of-war protections come from The Hague and Geneva Conventions. They apply at the tactical and operational levels and are summarized in ten rules shown in table B-1.

Table B-1. Summary of the law-of-war rules

• Fight only enemy combatants.
• Do not harm enemies who surrender; disarm them and turn them over to the chain of command.
• Do not kill or torture detainees.
• Collect and care for the wounded, whether friend or foe.
• Do not attack medical personnel, facilities, or equipment.
• Destroy no more than the mission requires.
• Treat all civilians humanely.
• Do not steal; respect private property and possessions.
• Do one's best to prevent violations of the law of war.
• Report all violations of the law of war to superiors.

B-13. During security force assistance operations, commanders must be aware of Common Article 3 of the Geneva Conventions and the status of insurgents under the laws of the host nation. One article contained in all four of the Geneva Conventions—Common Article 3—is specifically intended to apply to internal armed conflicts:

> *In the case of armed conflict not of an international character occurring in the territory of one of the High Contracting Parties, each Party to the conflict shall be bound to apply, as a minimum, the following provisions:*
>
> *(1) Persons taking no active part in the hostilities, including members of armed forces who have laid down their arms and those placed "hors de combat" by sickness, wounds, detention, or any other cause, shall in all circumstances be treated humanely, without any adverse distinction founded on race, colour, religion or faith, sex, birth or wealth, or any other similar criteria. To this end, the following acts are and shall remain prohibited at any time and in any place whatsoever with respect to the above-mentioned persons: (a) Violence to life and person, in particular murder of all kinds, mutilation, cruel treatment and torture; (b) Taking of hostages; (c) Outrages upon personal dignity, in particular humiliating and degrading treatment; (d) The passing of sentences and the carrying out of executions without previous judgment pronounced by a regularly constituted court, affording all the judicial guarantees which are recognized as indispensable by civilized peoples.*
>
> *(2) The wounded and sick shall be collected and cared for. An impartial humanitarian body, such as the International Committee of the Red Cross, may offer its services to the Parties to the conflict. The Parties to the conflict should further endeavor to bring into force, by means of special agreements, all or part of the other provisions of the present Convention. The application of the preceding provisions shall not affect the legal status of the Parties to the Conflict.*

Host-Nation Law and Status-of-Forces Agreements

B-14. After considering the type of baseline protections represented by fundamental human rights law, the military leader must be advised in regard to the other bodies of law that leader should integrate into planning and execution. This includes consideration of host-nation law. Because of the nature of most noninternational armed conflict missions, commanders and staffs must understand the technical and pragmatic significance of host-nation law within the area of operations. Although in theory, understanding the application of host-nation law during military operations is perhaps the simplest component, in practice it is perhaps the most difficult.

B-15. Judge advocates must recognize the difference between understanding the technical applicability of host-nation law and that law's application to control the conduct of U.S. forces during operations. In short, the significance of host-nation law declines in proportion to the movement of the operation toward violent conflict. Judge advocates should understand that U.S. forces enter other nations with a legal status that exists anywhere along a notional legal spectrum.

B-16. Judge advocates and planners must be familiar with any status-of-forces agreements or status-of-mission agreements that may be applicable. In any given country, there may be agreements short of status-of-forces agreements, such as diplomatic notes, on point. Relevant international documents affecting military operations may be difficult to locate. Several sources are available for locating applicable international agreements governing the status of U.S. forces or affecting military operations. Department of State (DOS) publications, such as *Treaties in Force,* contain unclassified international agreements. Both the relevant geographic combatant command's legal office and the defense attaché or military assistance group at the embassy normally have access to host-nation or international agreements impacting a military operation.

B-17. Status-of-forces and other forms of agreements frequently exist. They are essentially contractual agreements or treaties between two or more nations that establish the legal status of military personnel in foreign countries. Topics that are usually covered in a status-of-forces agreement include criminal and civil jurisdiction, taxation, and claims for damages and injuries. In the absence of an agreement or some other arrangement with the host nation, DOD personnel in foreign countries may be subject to its laws. Commanders ensure that all personnel understand status of U.S. forces in the area of operations and are trained accordingly.

LEGAL CONSTRAINTS ON THE SECURITY FORCE ASSISTANCE MISSIONS

B-18. U.S. law and regulation play a key role in establishing the parameters by which military forces may conduct security force assistance missions. These parameters tend to constitute constraints on the activities of military units. They range from the rules of engagement in combat situations to the ability to spend government funds in furtherance of a training or support mission.

GENERAL PROHIBITION ON ASSISTANCE TO POLICE

B-19. Historically, DOD is not the lead government department for assisting foreign governments. DOS is the lead when U.S. forces provide security assistance—military training, equipment, and defense articles and services—to host-nation governments. The Foreign Assistance Act of 1961 specifically prohibits assistance to foreign police forces except within specific exceptions and under a Presidential directive. When providing assistance to training, DOS provides the lead role in police assistance through its Bureau of International Narcotics and Law Enforcement Affairs. The President, however, may delegate this role to other agencies. This was done in 2004, when the President signed a decision directive granting the commander, United States Central Command, authority to train and equip Iraqi police.

TRAINING AND EQUIPPING FOREIGN SECURITY FORCES

B-20. All training and equipping of foreign security forces are specifically authorized. U.S. laws require Congress to authorize expenditures for training and equipping foreign forces. The laws of the United States also require DOS to verify that the host nation receiving the assistance is not in violation of human rights. The Secretary of Defense may authorize deployed U.S. forces to train or advise host-nation security forces as part of the mission. In this case, DOD personnel, operations, and maintenance appropriations provide an incidental benefit to those security forces. Numerous other programs to assist foreign forces are paid for with funds appropriated by Congress for that purpose. Consultation with a staff judge advocate or legal advisor early in the planning ensures the availability of funds for missions to train and equip foreign forces.

RULES OF ENGAGEMENT

B-21. *Rules of engagement* are directives issued by competent military authority that delineate the circumstances and limitations under which United States forces will initiate and/or continue combat engagement with other forces encountered (JP 1-02). Often these directives are specific to the operation. If there are no operation-specific rules of engagement, U.S. forces apply standing rules of engagement. When operating with a multinational force, commanders must coordinate the rules of engagement thoroughly.

FISCAL LAW CONSIDERATIONS

B-22. In security force assistance missions, like all operations, commanders require specific authority to expend funds normally found in the annual DOD appropriations act. As a general rule, operation and maintenance funds may not be used for security force assistance missions. In recent years, Congress appropriated additional funds to commanders for conducting more complex stability operations that were not typically covered by operation and maintenance funds. Examples include the Commander's Emergency Response Program (CERP), the Iraq Relief and Reconstruction Fund, Iraq Freedom Fund, and Commander's Humanitarian Relief and Reconstruction.

B-23. DOS has the primary responsibility, authority, and funding to conduct foreign assistance on behalf of the U.S. government. Foreign assistance encompasses any and all assistance to a foreign nation, including security assistance (assistance to the internal police forces and military forces of the foreign nation), development assistance (assistance to the foreign government in projects that will assist the development of the foreign economy or their political institutions), and humanitarian assistance (direct assistance to the population of a foreign nation). The legal authority for DOS to conduct foreign assistance is found in the Foreign Assistance Act of 1961.

B-24. There are two exceptions to the general rule requiring the use of Title 22 funds for foreign assistance. The first exception is for interoperability, safety, and familiarization training. DOD may fund the training (as opposed to goods and services) of foreign militaries with operation and maintenance dollars only when the purpose of the training is to enhance interoperability, familiarization, and safety training. Operation and maintenance funds may not be used for security force assistance training. This exception applies only to interoperability training. The second exception is congressional appropriation or authorization to conduct security foreign assistance. This exception allows DOD to fund foreign assistance operations if Congress has provided a specific appropriation or authorization to execute the mission.

CONSTRAINTS ON FUNDING

B-25. The Foreign Operations Appropriation Act (an annual DOS appropriations act) was enacted in 1997. Known as the Leahy Amendment, it prohibits the U.S. government from providing funds to a unit of the security forces of a foreign country if DOS has credible evidence that the unit has committed gross violations of human rights. The Leahy Amendment contains additional constraints on government funding of security force assistance missions. The law restricts funding until the Secretary of State determines and reports that the government of such country is taking effective measures to bring the responsible members of the security forces unit to justice.

B-26. Congress specifically appropriates funds for foreign assistance. The United States Agency for International Development (USAID) expends such funds under the legal authorities in Title 22. Provisions of Title 10 may also authorize amounts of money for these purposes. Standing funding authorities are narrowly defined and generally require advance coordination within DOD and DOS.

B-27. Effective foreign forces need training and equipment. U.S. laws require Congress to authorize such expenditures. U.S. laws also require DOS to verify that the host nation receiving the assistance is not in violation of human rights.

FUNDING AUTHORIZATION FOR SECURITY FORCE ASSISTANCE

B-28. All training and equipping of foreign security forces must be specifically authorized. Military and civilian personnel, operations, and maintenance appropriations should typically provide only an incidental benefit to those security forces. All other weapons, training, equipment, logistic support, supplies, and services provided to foreign forces must be paid for with funds appropriated by Congress for that purpose. Examples include the Iraq Security Forces Fund and the Afghan Security Forces Fund. Moreover, the President must give specific authority to DOD for its role in training and equipping foreign security forces. In May of 2004, the President signed National Security Presidential Directive (NSPD) 36 that made the commander, U.S. Central Command, under policy guidance from the chief of mission, responsible for coordinating all U.S. government efforts to organize, train, and equip Iraqi Security Forces, including police. Absent such a directive, DOD lacks authority to take the lead in assisting a host nation to train and equip its security forces.

KEY SECURITY ASSISTANCE AND FOREIGN ASSISTANCE FUNDING PROGRAMS

B-29. Congress has appropriated funds to allow DOS to conduct its foreign assistance mission, including the following activities:

- Foreign Military Financing Program.
- International Military Education and Training Program.
- Economic Support Fund.
- Peacekeeping operations.
- Antiterrorism assistance.
- Global humanitarian demining.

- Refugee assistance.
- Personnel details.

B-30. DOS administers the following programs but no funds are appropriated to sustain them:

- Foreign Military Sales Program.
- Foreign Military Lease Program.
- Economy Act Security Assistance.
- U.S. Government Commodities and Services Program.
- Direct Commercial Sales Program.

B-31. Additional special support created by law for the foreign assistance mission includes—

- Excess defense articles.
- Presidential drawdowns.

B-32. DOS directly, or indirectly through USAID, finances numerous development assistance activities to address the following needs:

- Rule of law.
- Governance.
- Agriculture and nutrition.
- Population control.
- Health.
- Education.
- Energy.
- Environment improvement.

B-33. CERP (see paragraph B-19) is an example of a targeted humanitarian assistance fund program. CERP's primary purpose is to enable military commanders to respond to urgent humanitarian relief and reconstruction requirements within their fields of interest by carrying out programs that will immediately assist the host-nation populace. CERP was originally funded with seized Iraqi assets, but Congress later appropriated U.S. funds for the purpose. CERP is not applicable to missions outside Iraq and Afghanistan. Future missions, though, may have similar funding sources established to facilitate a humanitarian assistance mission. Commanders consult with the servicing judge advocate to determine the availability of funds.

Appendix C
Information Engagement

INFORMATION SUPERIORITY

C-1. Information shapes a commander's operational environment at every level. Twenty-first century conflicts occur in operational environments with instant communication. Because information systems are everywhere and exposure to news and opinion media is pervasive, an individual's words and actions can have immediate operational and strategic implications. Gaining and maintaining the trust of key publics has never been more important. Army forces conduct operations in the public eye and must maintain public confidence by being consistent in word and deed.

C-2. *Information superiority* is the operational advantage derived from the ability to collect, process, and disseminate an uninterrupted flow of information while exploiting or denying an adversary's ability to do the same (JP 3-13). Advisors build partner capacity in this mission area. Advisors ensure that host-nation counterparts integrate information superiority into operations as carefully as fires, maneuver, protection, and sustainment. Army information tasks contribute to information superiority. Information tasks are used to shape the operational environment (see FM 3-0). The information tasks are—

- Information engagement.
- Command and control warfare.
- Information protection.
- Operations security.
- Military deception.

C-3. Advisors prioritize the five information tasks based on the host nation's capabilities and strengths. This appendix discusses some information engagement capabilities that advisors can expect to employ while conducting security force assistance, as well as considerations when working with the media. For a further discussion of the other information tasks, see FM 3-0.

INFORMATION ENGAGEMENT CAPABILITIES

C-4. *Information engagement* is the integrated employment of public affairs to inform U.S. and friendly audiences; psychological operations, combat camera, and U.S. government strategic communications and defense support to public diplomacy, and other means necessary to influence foreign audiences; and leader and Soldier engagements to support both efforts (FM 3-0).

C-5. In general, advisors focus on information engagement because land operations occur among civilian populations. Advisors can have a significant impact by focusing on this task. Advisors work with host-nation counterparts to communicate information to the populace, build their trust and confidence in friendly forces, promote popular support for friendly operations, and positively influence public perceptions and behavior. Advisors integrate information engagement throughout the operations process.

C-6. Advisors ensure host-nation counterparts focus their information engagement activities on achieving desired effects locally. However, land operations always occur in a regional and even global context. Therefore, local information engagement activities must support and complement higher headquarters information engagement activities, host-nation and U.S. government strategic communications guidance (when available), and broader host-nation and U.S. government policy where applicable.

C-7. Three information engagement capabilities deserve particular attention as the advisor can directly influence them. These are public affairs, combat camera, and leader and Soldier engagements.

Public Affairs

C-8. First, public affairs can have a profound and lasting effect on public perceptions of operational success or failure. This effect may have strategic consequences, even when related to tactical operations. Therefore, advisors ensure that public affairs efforts are integrated at tactical, operational, and strategic levels. (See paragraphs C-10 through C-20 for more information to help advisors working directly with the media.)

Combat Camera

C-9. Second, combat camera is used to document a wide range of tactical and operational successes and to counter enemy propaganda. If there is no dedicated host-nation combat-camera capability, the advisor recommends that a host-nation soldier be designated, equipped, and trained for this function. If that is not possible, the advisor should obtain a digital camera and, if necessary, photograph relevant host-nation activities personally. Positive photographs can have a tremendous benefit, and digital photos are easily taken and transmitted.

Leader and Soldier Engagements

C-10. Third, leader and Soldier engagements are a fundamental and highly effective method to influence people. They involve face-to-face interaction by leaders and Soldiers to influence the perceptions of the local populace. Military leaders engage key communicators and civilian leaders (often referred to as key leaders) through carefully planned meetings. Examples of key leaders include local tribal elders, elected or appointed officials, military and police commanders, and nongovernmental organization leaders.

WORKING WITH THE MEDIA

C-11. The appropriate public affairs officer normally manages media contacts. However, this is not always possible. A deployed advisor team seldom has a public affairs officer representative.

C-12. Occasions may arise when the host-nation press wants to question advisors. Even if an advisor team lacks a public affairs officer representative, advisors must avoid refusing to speak to the media. Apparent refusal to speak with accredited members of the media, for whatever reason, may inadvertently create strong negative impressions with strategic implications. Therefore, advisor teams designate a media spokesperson and keep this individual well informed. This individual coordinates media contacts for the advisor team. This individual ensures that official, published public affairs officer guidance and command policies regarding media relations, including anonymity and operations security, are observed. The following paragraphs provide general media relations guidance for advisor teams.

Enhancing Relationships with the Media

C-13. Advisors strive to convey a positive impression and enhance relationships between the media and the military. If deemed acceptable within the public affairs guidance and command policies, the advisor team's designated media spokesperson—

- Maintains a list of reliable reporters and editors that cover operations in the team's operational area and keeps them informed of significant activities.
- Answers media inquiries promptly, accurately, and courteously.
- Finds out reporters' deadlines and uses them to the friendly force's advantage.
- Stresses the human aspects of a story that readers, viewers, or listeners can identify with, including the negative impact of opposing operations on the people.
- Points out how advisors and host-nation counterparts are working to address the needs of the unfortunate.
- Encourages the media to observe constructive work the host nation and advisors are doing and to interview them about their efforts.
- Identifies a location in advance for the press to take photos and videos.

Correcting Public Misconceptions

C-14. If information is withheld, journalists often fall back on speculation. Such speculation is usually inaccurate. However, it is often near enough to the truth that it is easily accepted by large sections of the public and possibly by established governments. In addition, partisan sources may find it advantageous to leak part of a story to the press to build up public support for their position. On occasion, such activities can grow into fully orchestrated press campaigns.

C-15. With modern satellites and communications technology, media are able to distribute reports and photographs extremely quickly. Incidents, sometimes fabricated or slanted toward a partisan viewpoint, may be aired in living rooms across the globe the same day.

C-16. However, advisors and host-nation counterparts frequently can correct public misconceptions and provide accuracy, adequate context, and proper characterization promptly. They can turn a potential liability into a major advantage by remaining prepared to coordinate and communicate with the media. They can take advantage of the media's powerful and far-reaching ability to communicate with critical audiences.

C-17. If the media report inaccurate or damaging information, advisors use caution in deciding on a course of action. Advisors (usually through the team's designated media spokesperson) inform their chain of command. Advisors consider the following questions:

- Is it important enough to correct or is it a detail that—in the long term—is not really important?
- How damaging is the charge, criticism, or error? Will a correction simply give greater visibility to an unfavorable point of view?
- Is a correction worth a restatement of the entire problem to new audiences who did not read or see it the first time?
- Is it possible to target only the audience originally exposed to the story?
- Can significant gain result from pointing out an error?

C-18. When incorrect or misleading information is released to the public through host-nation government or media sources, the advisor team determines a course of action appropriate to the situation. It must include a plan to ensure that relevant host-nation government officials and media recognize the inaccuracy of the information. This may prevent future incidents. Operational and combat experience since 2001 shows that in the struggle for information superiority, friendly forces must be candid and honest. Additionally, they must not allow lies disseminated by enemies to stand unchallenged, particularly in areas of great international interest.

Interviews With Advisor Team Members

C-19. Advisor teams sometimes arrange for media interviews with selected advisor team members. Properly conducted interviews can support the mission and even improve morale. When preparing for an interview, the team's designated spokesperson—

- Provides all team members with a single and simple theme to convey to the press, sometimes in the form of a mission narrative (a succinct account of the mission as a whole, explained consistently with the host nation's cultural view).
- Conducts rehearsals in front of video cameras, using mock interviews to practice communication of the theme.
- Selects interviewees based on their understanding of the theme covered, not their comfort level in front of a reporter. (Shyness is a normal trait that may add a human touch to a sensitive situation.)

C-20. If an advisor team member does not know the answer to a reporter's question, the individual should attempt to find out and get back to the reporter or refer the reporter to an appropriate source. This may establish the team as a helpful information source and develop a relationship for future balanced coverage.

C-21. Communicating with the media is the most difficult aspect of information engagement. Anything advisors say or do publicly can impact the mission. Therefore, advisors ensure consistency with national policy, professionalism, and clear communication. Advisors must know and follow the policies of the geographic combatant commander, the American Embassy, and the appropriate public affairs officer regarding media interviews. Additionally, advisors follow the specific guidelines shown in table C-1.

Table C-1. Specific guidelines for media interviews

- Maintain a friendly, yet professional attitude.
- Remain positive, greet interviewers, and welcome questions. (This is difficult under stressful or tragic conditions; however, a calm, mature appearance earns respect from audiences.)
- Use language that is clear and easy to understand.
- Avoid military jargon or terminology that others can misinterpret.
- Never enter an interview unprepared.
- Prepare notes and study them carefully; use facts to back up statements.
- Ensure the facts supporting their position are up to date and come to mind easily.
- Anticipate possible questions and think about various responses.
- Rehearse before interviews on location if possible and, if time permits, role-play tough questions.
- Learn and use reporters' names.
- Research interviewers' organizations, including—
    - The organizations' views or views they defend.
    - Their underlying motivations.
    - How members of the organizations previously conducted interviews.
- Keep the interests of the local nationals or other beneficiaries of the mission at the forefront.
- Avoid talking from the point of view of U.S. interests.
- Finish answering one question before answering another if two or more questions are asked at the same time.
- Avoid speculation and give only factual information that can be verified.
- Use caution in quoting statistics as such data may be easily disputed or reinterpreted.
- Avoid repeating a question that contains incorrect information or inflammatory language (this may result in a misquotation).
- Strive to tell the truth.
- Avoid saying *no comment*. (It gives the appearance the advisor has something to hide.)
- If unsure of something, admit it rather than mistakenly lie.
- If questioned about another agency's activities, refer to that agency for comment.
- Do not speak for other organizations.
- Avoid exaggeration or claims that cannot be backed up.
- Refuse all off-the-record discussions.
- Expect any statement they make to be quoted publicly.
- Avoid criticizing other groups or organizations, and avoid committing information fratricide.
- Remind reporters that photographing recognizable dead Soldiers, charts, maps, supply depots, or electronic warfare assets is off limits.
- Remind reporters that their personal security is not a primary military concern.
- Remain objective and calm.
- Demonstrate conduct befitting a representative the U.S. government
- Never give personal opinions and beliefs.
- Avoid reacting emotionally when interviewers appear skeptical or hostile.
- Maintain focus on the mission.

Glossary

SECTION I – ACRONYMS AND ABBREVIATIONS

ABCA	American, British, Canadian, Australian, and New Zealand Armies Program
AR	Army regulation
BCT	brigade combat team
CCIR	commander's critical information requirement
CERP	Commander's Emergency Response Program
COA	course of action
COIST	company intelligence support team
DA	Department of the Army
DOD	Department of Defense
DODD	Department of Defense directive
DOS	Department of State
DOTMLPF	doctrine, organization, training, materiel, leadership and education, personnel, facilities
FM	field manual
FSF	foreign security forces
IDAD	internal defense and development
ISR	intelligence, surveillance, reconnaissance
JP	joint publication
METT-TC	mission, enemy, terrain and weather, troops and support available, time available, civil considerations (mission variables)
MTT	military transition team
NSPD	national security Presidential directive
PIR	priority intelligence requirement
S-2	intelligence staff officer
S-3	operations staff officer
S-4	logistics staff officer
SFA	security force assistance
TC	training circular
UN	United Nations
US	United States
USAID	United States Agency for International Development

SECTION II – TERMS

foreign internal defense

(joint) Participation by civilian and military agencies of a government in any of the action programs taken by another government to free and protect its society from subversion, lawlessness, and insurgency. (JP 3-05)

foreign military sales

(joint) That portion of United States security assistance authorized by the Foreign Assistance Act of 1961, as amended, and the Arms Export Control Act of 1976, as amended. This assistance differs from the Military Assistance Program and the International Military Education and Training Program in that the recipient provides reimbursement for defense articles and services transferred. (JP 1-02)

***foreign security forces**

Forces, including but not limited to military, paramilitary, police, and intelligence forces; border police, coast guard, and customs officials; and prison guards and correctional personnel, that provide security for a host nation and its relevant population or support a regional security organization's mission.

host nation

(joint) A nation that receives the forces and/or supplies of allied nations, coalition partners, and/or NATO organizations to be located on, to operate in, or to transit through its territory. (JP 3-57)

humanitarian and civic assistance

(joint) Assistance to the local populace provided by predominantly US forces in conjunction with military operations and exercises. This assistance is specifically authorized by Title 10, United States Code, section 401, and funded under separate authorities. Assistance provided under these provisions is limited to (1) medical, dental, and veterinary care provided in rural areas of a country;
(2) construction of rudimentary surface transportation systems; (3) well drilling and construction of basic sanitation facilities; and (4) rudimentary construction and repair of public facilities. Assistance must fulfill unit training requirements that incidentally create humanitarian benefit to the local populace. (JP 3-57)

internal defense and development

(joint) The full range of measures taken by a nation to promote its growth and to protect itself from subversion, lawlessness, and insurgency. It focuses on building viable institutions (political, economic, social, and military) that respond to the needs of society. (JP 3-07.1)

international military education and training

(joint) Formal or informal instruction provided to foreign military students, units, and forces on a nonreimbursable (grant) basis by offices or employees of the United States, contract technicians, and contractors. Instruction may include correspondence courses; technical, educational, or informational publications; and media of all kinds. (JP 1-02)

security assistance

(joint) Group of programs authorized by the Foreign Assistance Act of 1961, as amended, and the Arms Export Control Act of 1976, as amended, or other related statutes by which the United States provides defense articles, military training, and other defense-related services, by grant, loan, credit, or cash sales in furtherance of national policies and objectives. (JP 3-57)

security force assistance

The unified action to generate, employ, and sustain local, host-nation, or regional security forces in support of legitimate authority. (FM 3-07)

security sector reform

The set of policies, plans, programs, and activities that a government undertakes to improve the way it provides safety, security, and justice. (FM 3-07)

References

Field manuals and selected joint publications are listed by new number followed by old number.

REQUIRED PUBLICATIONS

These documents must be available to intended users of this publication.

FM 1-02 (101-5-1). *Operational Terms and Graphics*. 21 September 2004.

JP 1-02. *Department of Defense Dictionary of Military and Associated Terms*. 12 April 2001.

RELATED PUBLICATIONS

These documents contain relevant supplemental information.

JOINT AND DEPARTMENT OF DEFENSE PUBLICATIONS

Most joint publications are available online: <http://www.dtic.mil/doctrine/jpcapstonepubs.htm.>

DOD 5105.38-M. *Security Assistance Management Manual*. 3 October 2003. <http://www.dsca.osd.mil/samm/>

DODD 3000.05. *Military Support for Stability, Security, Transition, and Reconstruction (SSTR) Operations*. 28 November 2005. <http://www.dtic.mil/whs/directives/corres/html/300005.htm>

JP 3-0. *Joint Operations*. 17 September 2006.

JP 3-05. *Doctrine for Joint Special Operations*. 17 December 2003.

JP 3-07.1. *Joint Tactics, Techniques, and Procedures for Foreign Internal Defense (FID)*. 30 April 2004.

JP 3-07.4. *Joint Counterdrug Operations*. 13 June 2007.

JP 3-57. *Civil-Military Operations*. 8 July 2008.

ARMY PUBLICATIONS

Most Army doctrinal publications are available online: <http://www.army.mil/usapa/doctrine/Active_FM.html>. Army regulations are produced only in electronic media. Most are available online: < http://www.army.mil/usapa/epubs/index.html>.

AR 11-31. *Army International Security Cooperation Policy*. 24 October 2007.

AR 12-1. *Security Assistance, International Logistics, Training, and Technical Assistance Support Policy and Responsibilities*. 24 January 2000.

FM 3-0. *Operations*. 27 February 2008.

FM 3-05.137. *Army Special Operations Forces Foreign Internal Defense*. 30 June 2008.

FM 3-05.202 (FM 31-20-3). *Special Forces Foreign Internal Defense Operations*. 2 February 2007.

FM 3-07. *Stability Operations*. 6 October 2008.

FM 3-24. *Counterinsurgency*. 15 December 2006.

FM 3-34.400 (FM 5-104). *General Engineering*. 9 December 2008.

FM 4-0 (100-10). *Combat Service Support*. 29 August 2003.

FM 4-02 (FM 8-10). *Force Health Protection in a Global Environment*. 13 February 2003.

FM 6-0. *Mission Command: Command and Control of Army Forces*. 11 August 2003.

FM 6-22 (22-100). *Army Leadership*. 12 October 2006.

FM 7-0. *Training for Full Spectrum Operations*. 12 December 2008.

FM 7-15. *The Army Universal Task List*. 27 February 2009.

TC 31-73. *Special Forces Advisor Guide*. 2 July 2008.

OTHER PUBLICATIONS

National Defense Strategy of 2008. Washington, DC: U.S. Government Printing Office, 2008. <www.defenselink.mil/news/2008 National Defense Strategy.pdf>

National Military Strategy of the United States of America. Washington, DC: U.S. Government Printing Office, 2004. <www.defenselink.mil/news/Mar2005/d20050318nms.pdf>

National Security Strategy of 2006. Washington, DC: U.S. Government Printing Office, 2006. <www.marforres.usmc.mil/docs/nss2006.pdf>

Quadrennial Defense Review of 2006. Washington, DC: U.S. Government Printing Office, 2006. <www.defenselink.mil/qdr/report/Report20060203.pdf>

Armed Forces. Title 10, U.S. Code. <http://www.loc.gov/law/help/guide/federal/uscode.php>

Arms Export Control Act of 1976. <http://www.loc.gov/law/help/guide.php>

The United States Constitution. <http://www.house.gov/house/Constitution/Constitution.html>

Foreign Relations and Intercourse. Title 22, U.S. Code. <http://www.loc.gov/law/help/guide/federal/uscode.php>

Foreign Operations Appropriation Act of 1997. <http://www.loc.gov/law/help/guide.php>

Foreign Assistance Act of 1961. <http://www.loc.gov/law/help/guide.php>

The Geneva Conventions of 1949. <http://www.loc.gov/rr/frd/Military_Law/MLR_site-map.html>

National Security Presidential Directive 36. "United States Government Operations in Iraq." 11 May 2004.

Treaties in Force 2007. <http://www.state.gov/s/l/treaty/treaties/2007>

Charter of the United Nations. <http://www.un.org/aboutun/charter/>

War Powers Act of 1973. <http://www.thecre.com/fedlaw/legal22/warpow.htm.>

OUTSIDE RECOMMENDED READING

von Steuben, Friedrick Wilhelm. *Regulations for the Order and Discipline of the Troops of the United States*. Philadelphia: Styner and Cist, 1779.

PRESCRIBED FORMS

None

REFERENCED FORMS

DA Form 2028. Recommended Changes to Publications and Blank Forms

Index

Entries are by paragraph number.

Entries are by paragraph number.

Entries are by paragraph number.

Entries are by paragraph number.

LAW ENFORCEMENT INVESTIGATIONS

Headquarters, Department of the Army

DISTRIBUTION RESTRICTION: Approved for public release; distribution is unlimited.

Field Manual
No. 3-19.13

Headquarters,
Department of the Army
Washington, DC, 10 January 2005

Law Enforcement Investigations

Contents

---

*This publication supersedes FM 19-20, 25 November 1985.

Page

Page

Preface

This field manual (FM) is a guide for military police, military police investigators (MPIs), and United States (US) Army Criminal Investigations Command (USACIDC) special agents operating in all levels of tactical and garrison environments. This manual makes no distinction between the various levels of investigation, whether it is done by a uniformed military policeman, an MPI, or a USACIDC special agent. Where appropriate, this manual describes nationally recognized methods of investigation and evidence examination adopted from the Department of Justice (DOJ); Federal Bureau of Investigation (FBI); Bureau of Alcohol, Tobacco, and Firearms (BATF); National Association of Fire Investigators (NAFI); National Institute of Justice; and US Army Criminal Investigation Laboratory (USACIL). In addition to the techniques and procedures described in this manual, Army law enforcement personnel are encouraged to seek guidance on police and investigative matters from other approved official law enforcement sources. Special terms used are explained in the glossary.

The proponent for this publication is HQ, TRADOC. Send comments and recommended changes on *Department of the Army (DA) Form 2028 (Recommended Changes to Publications and Blank Forms)* directly to Commandant, US Army Military Police School (USAMPS), ATTN: ATSJ-DD, 401 MANSCEN Loop, Fort Leonard Wood, Missouri 65473-8926.

Appendix A complies with current Army directives, which state that the metric system will be incorporated into all new publications.

Unless this publication states otherwise, masculine nouns or pronouns do not refer exclusively to men.

PART ONE

Criminal Investigations and Testimonial Evidence

Part One provides information that is critical in understanding the role and responsibilities of the Army law enforcement investigator. *Chapter 1* establishes the foundation for understanding the objectives of criminal investigations and describes the investigator's responsibility to victims and witnesses. *Chapter 2* provides an understanding of trial preparation and describes proper courtroom testimony. *Chapters 3 and 4* describe the many challenges associated with obtaining information from victims and witnesses and interrogating suspects. These chapters offer the reader the most current techniques and procedures for conducting interviews and interrogations (I&I).

Chapter 1

Criminal Investigations

Military criminal investigations are official inquiries into crimes involving the military community. A criminal investigation is the process of searching, collecting, preparing, identifying, and presenting evidence to prove the truth or falsity of an issue of law.

OVERVIEW

1-1. For criminal investigations to be successful, the investigator must understand the general rules of evidence; provisions and restrictions of the *Manual for Courts-Martial, United States, 2000 (MCM 2000);* and the *Uniform Code of Military Justice (UCMJ).* Army law enforcement investigators must also be familiar with the capabilities and limitations of the USACIL. As investigators adopt a more scientific approach to criminal investigations and rely more on tangible evidence than on the confessions of suspects or eyewitness accounts, the relationship between the investigator and evidence examiners becomes crucial to the success of the investigation.

1-2. Most criminal investigations begin at the scene of a crime (the actual site or location in which the incident took place). It is important that the first officer on the scene properly protects the evidence. The entire investigation hinges on the initial law enforcement responder being able to properly identify, isolate, and secure the scene. Crime scenes should be secured by establishing a restricted perimeter. The purpose of securing the scene is to restrict access and prevent evidence destruction. There are many factors that dictate how a crime scene should be protected. However, nationally recognized standards for crime scene protection suggest the following three-layer or tier perimeter:

- An outer perimeter (established as a border larger than the actual scene to keep onlookers and nonessential personnel safe and away from the scene).
- An inner perimeter (allows for a command post and comfort area just outside of the scene).
- The core (actual scene).

OBJECTIVES OF CRIMINAL INVESTIGATIONS

1-3. In reality, the objectives of a criminal investigation are not as simple as just solving cases. Occasionally, a case is unsolvable, yet every lead must be exhausted. Criminal investigation is both an art and a science. In science, the absolute truth is often achieved. Experience has shown that in criminal investigations a less decisive hypothesis may sometimes be all that is possible to achieve.

1-4. Objectives of criminal investigations are as follows:

- Determine if a crime was committed.
- Collect information and evidence legally to identify who was responsible.
- Apprehend the person responsible or report him to the appropriate civilian police agency.
- Recover stolen property.
- Present the best possible case to the prosecutor.
- Provide clear, concise testimony.

ARMY LAW ENFORCEMENT INVESTIGATORS

1-5. Investigators conduct systematic and impartial investigations to uncover the truth. They seek to determine if a crime was committed and to discover evidence of who committed it. Investigators' efforts are focused on finding, protecting, collecting, and preserving evidence discovered at the crime scene or elsewhere. Their professional knowledge and skills include crime scene photography, development of latent fingerprints, and recording crime scene impressions. They are skilled in the techniques and methods used to interview witnesses and interrogate suspects.

1-6. Army law enforcement investigators document their actions and relevant details of an investigation in an investigator's notebook and use various methods of crime scene photography and sketches to capture the facts of a case. They ensure that evidence is accounted for by maintaining a complete chain of custody to allow it to be admissible in court. They must be skilled in providing professional testimony. An investigator's charter is to impartially find, examine, and make available evidence that will clear the innocent and allow prosecution of the guilty. As professional fact finders, investigators maintain unquestionable integrity during a criminal investigation.

1-7. Army law enforcement investigators frequently perform drug suppression, surveillance, and undercover operations. These operations are designed to gather critical information and police intelligence and stop illegal drug and contraband traffic on military installations. These efforts greatly increase the security of military communities by identifying criminals and their associates. Investigators network with other police and intelligence agencies to report and share information.

1-8. Less experienced investigators can gain valuable experience by reviewing cases, consulting with peers, and working with experienced investigators. They also expand their knowledge through formal military and civilian law enforcement courses designed to teach nationally recognized investigative techniques.

NEW INVESTIGATIVE CONSIDERATIONS

1-9. As criminal behavior and investigative techniques evolve, even experienced investigators must learn new investigative methods. Protecting and retrieving electronic evidence continues to be a challenge for the law enforcement community (refer to *Chapter 11*). Advances in technology have helped serology and blood pattern analysis to become an established part of the criminal justice procedure. Traditional blood and saliva testing have been rendered almost obsolete. The development of forensic deoxyribonucleic acid (DNA) testing has expanded the types of useful biological evidence. DNA is found in all body tissues and fluids. Because DNA is more sensitive than traditional serologic methods and is able to withstand far harsher environmental conditions, DNA testing is often successful when traditional testing is not (refer to *Chapter 24*).

SAFETY CONSIDERATIONS

1-10. Personal safety is a priority for investigators. Responding to a crime scene often places the investigator in danger of exposure to hazards or unsafe conditions. Personal protection against biohazards, chemical hazards, and physical hazards often requires special training and equipment. Law enforcement managers should seek every opportunity to ensure that all police personnel receive hazardous material (HAZMAT) and critical incident management training. Numerous federal, state, and local agencies provide such training. Emergency management agencies often provide HAZMAT and critical-incident training at no cost to police, fire, and emergency medical personnel.

UNITED STATES ARMY CRIMINAL INVESTIGATION LABORATORY

1-11. The successful investigation and prosecution of crimes requires (in most cases) the collection, preservation, and forensic analysis of evidence, which can be crucial to demonstrations of guilt or innocence. The USACIL provides forensic and technical services to military criminal investigators. Analyses of physical evidence ranging from blood and other biological materials to explosives, drugs, questioned documents, fingerprints, electronic evidence, firearms, and imaging are conducted by the USACIL. Nationally accredited by the American Society of Crime Laboratory Directors (ASCLD) and the laboratory accreditation board (LAB), the USACIL also serves as a continual source of new scientific techniques. Laboratory examiners provide expert witness testimony in court cases regarding the results of forensic examinations.

1-12. In addition to many innovative forensic techniques, the USACIL contributes to the Combined DNA Index System (CODIS). CODIS is an FBI program to database DNA profiles. It allows federal, state, and local crime labs to compare DNA profiles and to generate investigative leads. All military criminal investigators should become thoroughly familiar with CODIS (refer to *Chapter 24* for a complete discussion on CODIS).

1-13. Other investigative databases accessed by USACIL that may be of use to the investigator include the Automated Fingerprint Identification System (AFIS) and the National Integrated Ballistics Information Network (NIBIN). These important databases should be considered any time fingerprint and firearms evidence, such as cartridge casings, are recovered in an investigation.

VICTIM AND WITNESS ASSISTANCE

1-14. Army law enforcement investigators are required to inform victims and witnesses of the services available to them. Particular attention should be paid to victims of serious and violent crimes, including child abuse, domestic violence, and sexual misconduct. This information can be viewed on-line at <www.DOD.mil/vwac/> or a copy of the policy can be obtained by writing to Department of the Army: Criminal Law, Office of the Judge Advocate General (OTJAG), 2200 Army Pentagon Washington, DC 20310-2200.

1-15. *Department of Defense Directive (DODD) 1030.1* and *Department of Defense Instruction (DODI) 1030.2* implement statutory requirements for victim and witness assistance and provide guidance for assisting victims and witnesses of crime from initial contact with offenders through investigation, prosecution, and confinement. Together, the directive and instruction provide policy guidance and specific procedures for all sectors of the military to follow for victim and witness assistance.

1-16. The directive includes a bill of rights that closely resembles the federal crime victims' bill of rights. Department of Defense (DOD) officials are responsible for ensuring that victims of military crimes enjoy these rights. Victim's rights include the following:

- The right to be treated with fairness and respect.
- The right to be reasonably protected from the offender.
- The right to be notified of court-martial proceedings.
- The right to be present at court-martial proceedings.
- The right to confer with the government attorney.
- The right to available restitution.
- The right to know the outcome of an offender's trial and release from confinement.

1-17. DOD victim and witness assistance programs cover the entire military justice process from investigation through prosecution and confinement. In providing services and assistance to victims. DOD programs emphasize an interdisciplinary approach involving the following:

- Law enforcement personnel.
- Criminal investigators.
- Chaplains.
- Family advocacy personnel.
- Family service center personnel.
- Emergency room personnel.
- Equal opportunity personnel.
- Judge advocates (JAs).
- Unit commanding officers.
- Corrections personnel.

DOD victim and witness assistance programs use the following forms to advise victims and witnesses of their rights during all stages of a case:

- *Department of Defense (DD) Form 2701 (Initial Information for Victims and Witnesses of Crime).* This form provides notice to victims and witnesses about their rights and information on the military justice system.
- *DD Form 2702 (Court-Martial Information for Victims and Witnesses of Crime)* and *DD Form 2703 (Post-Trial Information For Victims and Witnesses of Crime).* This form provides notice to victims about their rights during court martial proceedings and information about the court-martial process.
- *DD Form 2704 (Victim/Witness Certification and Election Concerning Inmate Status)* and *DD Form 2705 (Victim/Witness Notification of Inmate Status).* These forms provide information to victims about the offender's sentence, confinement status, clemency and parole hearings, and release from confinement.
- *DD Form 2706 (Annual Report on Victim and Witness Assistance).* This form provides statistical information about assistance rendered to victims and witnesses.

1-18. An Interdisciplinary DOD Victim and Witness Assistance Council provides a forum for the exchange of information and the coordination of policy recommendations. The council helps to foster the implementation of consistent and comprehensive policies and procedures to respond to crime victims and witnesses in all of the military services. A senior program specialist with the US DOJ, Office for Victims of Crime, serves as a liaison member.

LEGAL CONSIDERATIONS

1-19. All military personnel are subject to the provisions of the *UCMJ*. The *UCMJ*, as established by Congress, provides one basic code of military justice and law for all services. The code authorizes the President of the United States to set rules of evidence; pretrial, trial, and posttrial procedures; and maximum punishments for violations of the *UCMJ*. Under this authority, the President has issued the *MCM 2000*.

THE MANUAL FOR COURTS-MARTIAL

1-20. The *MCM 2000* is a primary source document for matters relating to military justice. This executive order (EO) implements the provisions of the *UCMJ*. It establishes the military law of evidence. The *MCM 2000* is divided into four parts: a table of contents; the body, which is divided into parts; the appendixes; and the index.

The Body

1-21. The body of the manual is divided into the following five parts:

- **Part I, Preamble.** Part I is a brief discussion of military jurisdiction and the nature of military law.
- **Part II, Rules for Courts-Martial.** Part II outlines the steps that must be taken to hold a proper military court-martial. It covers matters ranging from military jurisdiction to posttrial appeals. It discusses both the rights of the accused and the obligations of the government.
- **Part III, Military Rules of Evidence.** Part III is mainly for the use of trial lawyers in the courtroom. But certain provisions of the rules impact heavily on the investigator's everyday activities. In particular, Section III of Part III discusses the rules and related matters concerning self-incrimination, search and seizure, and eyewitness identification.
- **Part IV, Punitive Articles**. Part IV contains a thorough discussion of crimes punishable by the military. Each punitive article is discussed and illustrated separately. This part is particularly important to investigators, because it offers a guide for the investigation, showing what facts are important and need to be determined. Part IV contains the text of the article, lists the elements of the offense, gives an explanation of the offense and examples, lists any lesser-included offenses, lists the maximum punishment, and gives a sample specification.
- **Part V, Nonjudicial Punishment**. Part V deals with the rules of nonjudicial punishment under the *UCMJ, Article 15*.

The Appendixes

1-22. The two most important appendixes in the *MCM 2000* are *Appendixes 1 and 2.*

- ***Appendix 1***. This appendix contains *The Constitution of the United States,* which sets the bounds within which the federal government must operate.
- ***Appendix 2.*** This appendix contains the *UCMJ*. The *UCMJ* creates the law while the *MCM 2000* implements and defines the law. If the *MCM 2000* conflicts with the *UCMJ* as interpreted by the Court of Military Appeals, the code must be followed. Federal rules of evidence, as developed in federal courts, are used to assist in the interpretation of the manual when military law is silent on the question.

PARTICIPANTS IN A CRIME

1-23. The primary participants in a crime are the "principals." The *UCMJ* discusses principals in detail. The person who directly commits an offense is a principal and so is an aider and abettor of a crime.

1-24. The aider and abettor share the criminal intent of the perpetrator. Being present at the scene of a crime or failing to prevent a crime does not make someone an aider and abettor. But someone who counsels, commands, induces, or procures another to commit a crime is an aider and abettor. An aider and abettor is a principal even if he is not present at the scene of the crime and even if the person he solicits to commit the crime does so by a means other than what was planned.

1-25. An aider and abettor, if his intent or state of mind is more culpable than that of the perpetrator, may be guilty of an offense of greater seriousness than the perpetrator. The reverse is also true. If, when a homicide is committed, the actual perpetrator acts in the heat of sudden passion caused by adequate provocation, he may be guilty of manslaughter. But the aider and abettor who hands a weapon to the perpetrator during this encounter with shouts of encouragement for him to kill the victim may be guilty of murder. On the other hand, two persons may agree to commit robbery by snatching purses in a particular place. If one acts as a lookout and the other, without the lookout's knowledge, seizes a victim and rapes her after taking her purse, the perpetrator will be guilty of rape and robbery, but the aider and abettor will only be guilty of the robbery.

1-26. An investigation of any given crime may also reveal the criminal liability of an accessory after the fact. A person who is an accessory after the fact is someone who, knowing that an offense under the *UCMJ* has been committed, receives, comforts, or assists the offender to prevent his apprehension, trial, or punishment. An accessory after the fact is also someone who, knowing that a crime has been committed, helps conceal the crime. Mere knowledge that a crime has been committed does not make someone an accessory after the fact. The person must have had a legal duty to report it or must have committed some overt act designed to prevent the punishment of the criminal. Conviction of the perpetrator of the offense to which the accused is allegedly an accessory after the fact is not a prerequisite to the trial of the accessory.

LEGAL PROTECTION OF JUVENILES

1-27. Most job contact with juveniles occurs when investigating minor offenses, such as disturbing the peace. Sometimes contact is made when juveniles are seen committing acts that could be harmful to people or property. Usually, police personnel stop the misconduct and, when needed, refer the incident to the parents. The investigation into the causes of the misconduct and the collection of background data is limited to essential information. Investigators may extend the investigation to include the conduct of the child's military sponsors if that conduct is dangerous or harmful to the child.

1-28. Investigative steps for the gathering of evidence in juvenile offenses are the same as those used in cases involving adult suspects. Ensure that the juvenile is processed according to *Chapters 401 and 403, Part IV, Title 18, United States Code (18 USC)*. Children are protected from unwarranted treatment.

1-29. If juveniles are to be detained, remember that detaining juvenile suspects in confinement facilities, detention cells, or hospital prisoner wards is strictly forbidden. Juveniles may be temporarily detained in the offices of the post commander or provost marshal (PM), but check with the office of the Staff Judge Advocate (SJA) to ensure that proper conditions exist. Unless a juvenile is taken into custody for serious offenses, do not take any fingerprints or photographs of him without written parental or judicial consent. Contact the office of the SJA to ensure proper judicial authority. Do not release any names or pictures of juvenile offenders to the public.

1-30. All records of juvenile offenders must be secured and released only on a need-to-know basis. During juvenile proceedings, data on the juvenile and the offense may only be given to the court, the juvenile's counsel, and others having a need to know. Others may include courts or agencies preparing presentence reports for other courts, or they may be police agencies requesting the information for the investigation of a crime. Records may also be released to a treatment facility that a juvenile has been committed to by the courts. The director of that facility must submit an inquiry in writing. Records may be released to an agency considering the subject for a position involving law enforcement or national security.

1-31. Records should give detailed listings of regulations that the juvenile has violated. They should include the disposition made by civilian authorities. Permanent records are not made for nonessential minor incidents or situations resolved in conference with parents of the juvenile. If a juvenile is found innocent, all records of the offense (including fingerprints) must be destroyed, sealed by the court, or disposed of according to local directives.

GENERAL RULES OF EVIDENCE

1-32. Military law enforcement investigators must develop skills and techniques to recognize, collect, evaluate, process, and preserve evidence. Evidence is the source from which a court-martial or jury must form its conclusions as to the guilt or innocence of an accused. Evidence is the means by which any alleged matter of fact is proven or disproved. Evidence includes all matters, except comment or argument, legally submitted to a court.

1-33. Military investigators conduct inquiries to find evidence and make it available for presentation in court. Something more than a mere collection of evidence is required of a successful investigation; the evidence obtained must be admissible in court.

1-34. A basic knowledge of the rules governing admission and rejection of evidence is fundamental to an investigation. This knowledge is needed to conduct inquiries and to prepare cases that will present to the court enough admissible and reliable information upon which to base a proper decision. Only evidence that satisfies the rules of admissibility is admitted.

1-35. Evidence from a search or a seizure is not admissible in a court-martial or in a federal judicial proceeding if it was obtained as a result of an unlawful search of the accuser's property. This is called the "exclusionary rule." Evidence that is derived from an exploitation of an illegal act may also be inadmissible under what is known as the "fruit of the poisoned tree" doctrine.

1-36. To be admissible, evidence must be relevant. Relevancy requires that the particular item of evidence have some tendency to prove or disprove a fact to be decided at the trial. All relevant evidence is admissible at trial unless some rule of law forbids its consideration by the court.

1-37. When submitting evidence, the main concern is the admission of facts and pertinent materials and not their "weight." The weight accorded a particular item of evidence is a question for the court-martial or jury to determine. The weight of evidence is its relative importance among differing items of evidence in a case. For example, an alibi being established by a defendant accused of murder would have more weight as evidence if a physician testifies he was attending the defendant in his home at the time of the alleged offense than if the defendant's mother testifies he was at home in bed at that time. The testimony of both is admissible, but it is apparent that greater weight would be given to that of an impartial witness than to testimony from a mother favoring her son.

CRIMINAL INTELLIGENCE

1-38. Army law enforcement agencies are the primary liaison representatives of the Army to federal, state, local, and host nation (HN) agencies for exchanging police intelligence. They collect criminal intelligence (CRIMINT) and human intelligence (HUMINT) within the provisions of applicable statutes and regulations. Army law enforcement agencies have the capability to analyze and disseminate collected time-sensitive information concerning a criminal threat against Army interests. The value of such intelligence cannot be overemphasized; it not only serves the purpose of police investigations, it also contributes to the development of countermeasures to safeguard Army personnel, materials, information, and other resources.

DEFINITIONS

1-39. The definitions for police intelligence operations (PIO) and CRIMINT, as defined by the Army law enforcement PIO critical task selection board, and other intelligence sources are provided in the following paragraphs:

- **Police intelligence operations.** PIOs are those operations conducted by law enforcement, security, and intelligence organizations. PIOs are designed to collect, analyze, fuse, and report intelligence regarding threat and/or criminal groups for evaluation, assessment, targeting, and interdiction.
- **Criminal intelligence.** CRIMINT is the result from the collection, analysis, and interpretation of all available information concerning known and potential criminal threats and vulnerabilities of supported organizations. Refer to *Army Regulation (AR) 525-13*.
- **Commander's critical information requirement (CCIR).** CCIR is a comprehensive list of information requirements identified by the commander as being critical in facilitating timely information

management and the decision-making process. The CCIR affects successful mission accomplishment. The two key subcomponents are critical friendly force information requirements (FFIR) and priority intelligence requirements (PIR). Refer to *Joint Publication (JP) 1-02*.

- **Friendly forces information requirement.** The FFIR is information that a commander and his staff need about which friendly forces are available for operations.
- **Priority intelligence requirement.** The PIR is based on those intelligence requirements for which a commander has an anticipated and stated priority in the task of planning and decision making to gain current battle information.
- **Essential elements of friendly information (EEFI).** EEFI are the critical aspects of a friendly operation that, if known by the enemy, would subsequently compromise, lead to failure, or limit success of the operation and, therefore, must be protected from enemy detection. According to *FM 3-0*, although the EEFI are not part of the CCIR, they become a commander's priorities when he states them. The EEFI help commanders understand what enemy commanders want to know about friendly forces and why.
- **Measurement and signature intelligence (MASINT).** MASINT is scientific and technical intelligence information obtained by the quantitative and qualitative analysis of data (metric, angle, spatial, wavelength, time dependence, modulation, plasma, and hydromagnetic) derived from specific technical sensors for the purpose of identifying any distinctive features associated with the source, emitter, or sender and to facilitate subsequent identification and/or measurement of the same.
- **Human intelligence.** HUMINT is a category of intelligence derived from information collected and provided by human sources.
- **Criminal intelligence cycle.** The CRIMINT cycle is effective intelligence results from a series of interrelated activities. The intelligence cycle is a continuous process of collecting and converting data into intelligence products to be integrated into operations. The cycle consists of five phases with continuous evaluation and feedback at each stage and at the end of each cycle. With the minor modification of adding "reporting" to the second phase, the following text outlines the intelligence cycle as discussed in *AR 525-13*:
  - CRIMINT planning and directing.
  - CRIMINT collection and reporting.
  - CRIMINT processing.
  - Analysis and production.
  - Dissemination and integration.

PLANNING AND DIRECTING

1-40. Planning and directing identifies CCIR, sets priorities for collection or interdiction, and provides guidance for the management of collection or interdiction assets. In this phase, planners should consider the following:

- **What.** The activities or indicators that will confirm the threat event.
- **Where.** The probable locations of the threat event including the points of vulnerability or target value, such as mission-essential vulnerable areas (MEVA).
- **When.** The time that the threat event may occur. Indicators and warnings may predict this.
- **Why.** The justification of the intelligence requirement. This is important for prioritizing collection and interdiction efforts.
- **Who.** The persons or agencies who need the results.

1-41. Once priorities have been set, a collection plan is prepared based on the intelligence requirements, available collection assets, and other factors including time and self-protection. The Army law enforcement version of the reconnaissance and surveillance (R&S) plan as discussed in *FM 101-5* is typically used in the patrol distribution plan. The patrol distribution plan will do the following:

- Synchronize intelligence requirements with collection by prioritizing collection tasks.
- Assign law enforcement assets for collection and interdiction coverage, such as patrols, checkpoints, and access control points, with special emphasis on vulnerable areas or other highlighted areas.
- Manage collection assets (such as patrols, contraband detectors, and access control teams), special services (such as investigations, K-9, physical security, and school resource officers), or external support (such as operationally controlled units, aviation, and explosive ordnance disposal [EOD]).
- Provide indicators and special instructions.

COLLECTION AND REPORTING

1-42. Collection involves the gathering of relevant data and raw intelligence products needed to produce actionable intelligence. To be effective, collection should be planned, focused, and directed based on the CCIR. Analysts must assist commanders and staffs to identify those gaps in information requirements that will help complete the threat picture. Commanders and staffs must also prepare collection assets based on their capabilities and limitations. The important addition of "reporting" to this already existing phase captures the critical activity performed by Army law enforcement personnel. This addition stresses the importance of reporting during all operations and is not PIO specific. Reporting integrates intelligence with other essential police activities, such as interdiction, legal adjudication, or numerous other activities (such as investigations, physical security, and vulnerability assessments).

Collection

1-43. Army law enforcement officers and agents are natural collectors; they are highly adaptable to any collection plan. They can provide effective collection as an independent mission while conducting any of the other four core military police functions or while providing protection program services. Missions that can result in collection opportunities include (but are not limited to) tactical or nontactical patrols, criminal investigations, route reconnaissance, checkpoints, dislocated civilian population operations, physical security inspections, or first-responder services.

1-44. Army law enforcement can choose from an array of existing and emerging technologies to collect intelligence. In addition to the collection of CRIMINT and HUMINT, intelligence managers should leverage available image intelligence (IMINT), signal intelligence (SIGINT), and MASINT assets to improve collection results. Today, Army law enforcement must be prepared to request and use satellite imagery, aviation assets for air scan or search-and-rescue support, cell phone blocking and intercepting technology, EOD services for postblast analysis, or any other available assets. Intelligence managers can also use police intelligence networks and databases as an important source for collecting raw data and completed intelligence products.

1-45. Collection can occur at any level. While military police patrols in both tactical and nontactical environments collect more raw information for input into data streams than any other Army law enforcement element, ascending levels of collectors in both military police and Criminal Investigation Division (CID) headquarters (HQ) gather both critical information and existing intelligence products. For example, when conducting an auto theft investigation, a special agent in charge (SAC) may query and receive from local civilian law enforcement completed auto theft cases with a similar modus operandi (MO). In this case, the SAC (as a collector) has not collected mere data but rather a developed intelligence product for entry into the product stream for further analysis and fusion. The results of collection are submitted through reports.

Reporting

1-46. Reports can be transmitted verbally or in writing, as hard copy reports, or by using Internet services. There are numerous Army law enforcement reports used for a variety of activities such as military police reports (MPRs), reports of investigation, situation reports (SITREPs), debriefs, and spot reports (SPOTREPs). While the means for transmitting reports can vary, it is important that reporting is standardized and that reporting through data entry points is fully understood by all collectors to ensure report timeliness. Data entry points may include:

- Military police patrols.
- Military police stations or substations.
- Installation or camp access/entry control points (gates).
- CID resident agency.
- Joint law enforcement operation centers.

- Tactical operations center.
- Special or proponent staff members.
- Network forums.

PROCESSING

1-47. Through this phase, intelligence is processed into data and product streams from subordinate units and external agencies at each level of the organization. Data and product streams are the natural flow of information or intelligence through established systems or protocols such as communication links, reports, and database sequencing.

1-48. Intelligence processing occurs at set points to reduce raw data into manageable portions for analysis and production. Processing is similar to what was traditionally considered collation and involves an initial evaluation to ensure that the intelligence is valid, reliable, and sufficient. During processing, intelligence is prioritized according to current collection and production requirements. During processing, intelligence managers should do the following:

- Prioritize incoming data according to collection and production requirements.
- Organize intelligence by category, such as crime or threat, and by the particular product or end user.
- Enroll information into databases.
- Collate actionable intelligence into interim products; for example, the battalion Intelligence Officer [US Army] (S2) consolidates patrol-debriefing reports into a single intelligence summary (INTSUM).

ANALYSIS AND PRODUCTION

1-49. In this phase, criminal intelligence analysts accomplish formal analysis and production using the standard established techniques outlined below. It is important to remember that informal analysis and other forms of production occur throughout all phases of the intelligence cycle. As discussed under collection, examples of informal analysis result from intuition, experience, and training.

Analysis

1-50. In analysis, intelligence from multiple sources is synthesized and interpreted to produce a more complete picture than may be evident from just a summary of intelligence data. Similar to collection, analysis is prioritized to address CCIR through production requirements.

Production

1-51. Intelligence production includes products developed to provide situational awareness or convey meaningful information regarding the threat picture. There is a multitude of intelligence products provided by law enforcement, force protection and antiterrorism managers, and intelligence agencies. Army law enforcement, for example, produces monthly crime surveys, physical security vulnerability assessments, crime analysis surveys, criminal intelligence reports, and computer crime vulnerability assessments. While some products are developed only once, others are produced on a recurring basis. Regardless of the product, an effective intelligence production should provide actionable intelligence for planning and directing intelligence collection, threat countermeasures, or interdiction operations. More specifically, intelligence products should provide the following:

- Identity of the person, group, organization, and nation.
- Timelines indicating either an immediate or future threat.
- Location of potential attack points, safe houses, and so forth.
- Motivation and goals.
- Priority targets for further collection or interdiction.
- Triggers to improve operations, tactics, techniques, and procedures (TTP), and standing operating procedures (SOPs).

1-52. Computer automation and software have dramatically improved production of intelligence products. Ideally, products will use standardized formats and software. Members of the PIO network should understand the intent and message of shared intelligence production. Final production should be closely coordinated with the office of the SJA, security managers, and other relevant representatives to promote efficient and effective dissemination and integration. Include the lowest appropriate classification, appropriate routing systems, and other restrictions that may affect secondary dissemination or integration.

DISSEMINATION AND INTEGRATION

1-53. Dissemination and integration takes analysis and production "off the shelf," and places them in the hands of the operators. If activities in this phase are not properly conducted, the result is an ineffective link between planning, collection, and production requirements and the subsequent execution of collection or interdiction. Accuracy and timely dissemination are essential to achieving effective integration of intelligence products into operations.

Dissemination

1-54. The objective of dissemination is to provide intelligence to meet user requirements. Systems used for dissemination must ensure effective intelligence by matching intelligence products with the appropriate end user at the correct place and by using the correct media at the correct time. No matter what the quality of the product, it serves little purpose if it is not disseminated to the commander in time to use in the decision-making process. This can only be achieved by using standard distribution systems that are readily accessible to all appropriate intelligence users. A security level appropriate for the content must be provided, yet caution must be taken to not overclassify the information.

1-55. Distribution systems include both push and pull mechanisms. Dissemination has traditionally been accomplished between producers and customers by using push mechanisms such as those distributed using Secret Internet Protocol Router Network (SIPRNET) e-mail, hard-copy distribution, or a secure fax. This can be extremely efficient for recurring products when supported by effective interagency liaison or networks. The push system requires a scheduled exchange where both the producer and user are available and, therefore, may not be as effective for unscheduled production. Push systems are also less efficient than pull systems because they distribute intelligence through a planned distribution network of users regardless of their needs.

1-56. Intelligence dissemination using pull mechanisms has become increasingly more popular as advances in technology have allowed production to be positioned on standard Internet sites that are accessible to intelligence requestors using standard but secure retrieval protocols. The pull system is equally effective for both recurring and situational or event-specific products. As products are developed, they can be positioned on the site for ready access. Such sites also provide a platform for organizing products by threat category, level, or location that allows users to efficiently browse and select intelligence relevant to their needs. It is important when using this system that producers update the site as appropriate to ensure that posted intelligence products are current. It may help to establish a special marking system to delineate between current and dated intelligence that still may have value.

Integration

1-57. Integration is the ultimate recognition that intelligence supports operations. Integration enables intelligence managers to support both mission planning and execution, ensuring that intelligence is effectively linked to operations. Effective integration can occur either as part of a deliberate planning process or as part of self-generating processes or production cycles. The subsequent discussion of this phase clearly identifies the formal processes that Army law enforcement routinely conducts everyday.

INVESTIGATIONS

1-58. Conducting a successful investigation is often the result of having a wide range of knowledge and using common sense in its application. There are certain actions that apply to all investigations. Investigators follow these intelligent and logical steps to ensure that an investigation is conducted systematically and impartially. There are certain actions that, over time, have proven useful for specific investigations. It is a wise investigator who understands and applies the knowledge, skills, and techniques learned for a particular investigation and uses them wherever they are most useful in any investigation. This means that, in order to conduct a successful larceny investigation, the investigator must do more than just follow the investigative process and the guidelines for investigating larcenies. Knowing and using a technique usually used for investigating a robbery may be just what is needed to help solve a larceny case.

HYPOTHESIS DEVELOPMENT

1-59. Success on any case is always a function of intellect and experience. Develop a hypothesis that serves as the framework for the case. The hypothesis is based on a survey of the crime scene. It is simply a set of reasoned assumptions of how the crime was committed and the general sequence of acts that were involved.

HYPOTHESIS MODIFICATION

1-60. Reassess the hypothesis as new facts and leads are uncovered. Investigators must overcome a natural tendency to make contradicting information fit a set of existing assumptions. For example, if there is substantial evidence that a murder was committed at the place where the body was found, it is tempting to ignore a fact or a lead that does not fit that assumption. Often, the lack of some item or event is just as important as its presence. As an investigator obtains new information, he must be willing to modify or change his initial ideas about how a crime was committed. Only through constant reassessment can the full value of his experience be realized.

EVIDENCE GATHERING

1-61. Generally, the art of an investigation lies in gathering and evaluating information and evidence, both testimonial and physical. Testimonial evidence, like sworn statements of eyewitness accounts and admissions of guilt, is obtained through communication with people. Physical evidence, like identified weapons and fingerprints, is obtained by searching crime scenes, tracing leads, and developing technical data.

1-62. Investigators must always be evidence conscious. The scene of any crime is evidence in and of itself, as is the testimony of trained investigators regarding observations and findings. Both physical and testimonial evidence are vital to the successful prosecution of an investigation.

TESTIMONIAL EVIDENCE

1-63. Obtaining testimonial evidence requires skillful interpersonal communication (IPC) with human sources of information, particularly with the persons directly involved in a case. Questioning victims, witnesses, complainants, suspects, and sources is the investigative method most often used to obtain testimonial evidence. It is also the method used to obtain background information that gives meaning to the physical evidence collected. The solution to many crimes is the direct result of leads and testimonial evidence developed through interviews and interrogations.

1-64. All law enforcement personnel must be skilled in IPC to elicit useful information. They must know how and when to ask the "right" questions. An investigator's attitude and method of questioning, as much as the questions asked, can elicit the leading information and testimonial evidence needed to bring an investigation to a successful conclusion.

1-65. Victims and witnesses are questioned to gain information that will help show the facts of the crime. Investigators should do the following:

- Question victims and witnesses to gather information on what they saw, know, or did in regard to an offense.
- Check information received from one person against information received from another.

- Question sources for information pertaining to the case under investigation.
- Ask questions to obtain observations and develop descriptions that will identify suspects.
- Question suspects to remove suspicion from the innocent and give the guilty an opportunity to confess.
- Record information obtained from interviews and interrogations. From this information, develop statements that may become documents admissible in court as evidence when sworn to under oath and signed by the swearer (see *Chapter 4*).

EVIDENCE EVALUATION

1-66. Frequently, the successful outcome of a case depends on an accurate evaluation of the evidence. Evaluation of evidence begins with the first information received about the occurrence of a crime. Evaluate evidence in light of the circumstances and conditions found at the crime scene and the information obtained by questioning persons connected with the event. Evaluate each piece of evidence individually and collectively in relation to all other evidence. If doubt exists about the evidentiary value of an item, secure and process it as evidence. Later evaluation can determine the worth of the item as evidence to the investigation.

1-67. After evaluating evidence and statements of expected testimony gathered during the preliminary investigation, decide what facts are still required to establish the elements of proof of the offense being investigated. Coordinate with other agencies and commands to gain the information or documents needed to support the investigation. Make sure administrative action is started early to secure help from and refer undeveloped leads to others agencies. Take early action to give other agencies time to comply with requests. Exploit every available local source of information while awaiting replies or action.

1-68. Carefully use selected sources and seek out reliable persons who possess information that is material to the case. Check with the criminal information office or the joint police information team. The information needed can often be obtained from a central location. Also, contact the US Army Crime Records Center (USACRC) to see if suspects have a past record or if victims have ever been victims of another crime. If new information is found, ensure that it is widely disseminated.

1-69. Evaluate evidence again in light of all new information. Support the evaluation with common sense and sound judgment enhanced by your past experience. Discuss the evaluation of the evidence with supervisors, other investigators, technicians, other experts in a given field, or the office of the SJA.

1-70. Continue this evaluation process until the investigation has been concluded. Prepare a final report to document the findings when an investigation has been completed. The report must reflect the who, what, where, when, why, and how of the offense. A final report must be a thorough, timely, and objective evaluation of the findings.

Chapter 2

Testimony

The final and most severe test of an investigator's efficiency is often as a witness before a court of law. The effectiveness of the evidence can be directly affected by the impression the investigator makes as a witness. Testimony is only effective when it is credible. Credibility is established when the investigator articulates his testimony with sincerity, knowledge of the facts, and impartiality. Although the substance of his testimony is of great importance, equal significance is attached to his conduct on the stand and to the manner in which he presents the facts discovered during the course of the investigation.

RULES OF EVIDENCE

2-1. For an investigator to be better equipped to provide testimony, he should—

- Guide the investigation so that evidence relevant to the issues receives primary consideration.
- Acquire a general knowledge of what is occurring in the courtroom.
- Understand what the defense and prosecuting attorneys are trying to achieve through a particular strategy.
- Recognize relevant and material evidence with an understanding of the rules as they apply to hearsay, confessions, polygraph results, and documents.
- Testify only to the facts that he acquired firsthand, through his own senses. This means that he cannot be prompted to provide an opinion unless he is recognized as an expert witness. It also means that it is permissible for him to pause before providing a seemingly inadmissible answer, which allows the opposing counsel an opportunity to object.

TRIAL PREPARATION

2-2. The final result in bringing a successful investigation to a close is often the investigator's testimony in the courtroom. The investigator must prepare carefully and ensure that all known case facts are in order. Whenever the investigator is preparing for trial, he should coordinate with the counsel, who is calling him as a witness, in order to prevent misunderstandings and surprises. The investigator should develop a close working relationship with the trial counsel (TC) so both parties will clearly understand the questions that will be asked and the answers that will be provided. The investigator must be professional in every way and never give the impression that he is attempting to conceal information from the court. The accused has the right to a fair trial regardless of the investigator's opinion.

COURTROOM TESTIMONY

2-3. Military police training seldom devotes sufficient time to trial preparation and testimony. A good police witness must be positive and firm in answering all questions. Military police and investigators must be taught to defend against the tactics used by lawyers in the courtroom. They must use intelligible and understandable language when talking to the panel or members of the court. Military police have been trained to recognize that providing good first impressions and giving solid concluding statements in testimony give them an advantage in the courtroom.

FIRST IMPRESSIONS IN THE COURTROOM

2-4. Impressions that the investigator first gives in a courtroom are critical to his testimony. The following guidelines will assist the investigator in setting a good first impression with the members of the court:

- Know what to do when entering the courtroom. This may require the investigator reporting ahead of time to get familiar with the layout of the room.
- Walk confidently. The investigator should appear secure without looking cocky or arrogant.
- Carry any notes or reports in a clean manila folder in your left hand if they are taken into the courtroom. This leaves the investigator's right hand free for taking the oath; however, any notes taken into the courtroom can, and likely will, become a record of trial.
- Stand straight and face the administrator of the oath with the palm of your right hand facing that person.
- Answer the question, "Do you swear to tell the truth, the whole truth, and nothing but the truth, so help you God?" in a clear, firm voice. The investigator should not look at anyone other than the administrator of the oath.
- Sit in the chair with your back straight when instructed to take the stand, but be comfortable. The investigator should keep his hands folded in his lap or rested on the arms of the chair. He should avoid becoming rigid, fidgeting in the chair, or swiveling the chair around. This is distracting to the testimony and may be perceived as indications of deceit or anxiety.
- Hold any notes and/or reports in your lap while on the stand. If the investigator must refer to them, he should do so before responding to the question and should not wave them around when referring to them during testimony.

INVESTIGATOR'S TESTIMONY

2-5. The prosecutor will ask the investigator for his full name, social security number (SSN), unit, and duty position. Before appearing in court, the TC and the investigator should discuss whether the investigator will be requested to disclose his military rank and, if so, what the appropriate response should be. Additionally, depending upon the nature of the offense, the investigator may be requested to provide a summary of his investigative training relating to the investigation that he intends to testify about. The investigator must be cognizant of inadmissible statements, which include information related to—

- Privileged communication.
- Hearsay.
- Polygraph results or declinations.
- Opinions (unless the investigator is a qualified and recognized expert in the related field of questioning).
- Statements about the character and reputation of the defendant, to include a past criminal record (unless the statements establish a pattern of conduct).

2-6. The investigator testifies only to what he saw, heard, or did. He does not testify to what he thinks or believes, heard about, or was told about unless the information falls within one of the exceptions to the hearsay rules of evidence as described in the *MCM*.

2-7. The verbal and nonverbal factors of the investigator are equally important to the success of an effective testimony.

Verbal Factors

2-8. The following are verbal factors that are important to the success of an effective testimony:

- Speak slowly and deliberately, with expression, and loud enough to be heard. Do not use profanity or vulgarity unless asked to provide the exact words of the suspect, victim, or witness. If profanity or vulgarity is present in those words, forewarn the court.
- Pause briefly to form answers before answering each question. If an opposing counsel makes an objection, stop speaking until the court rules on the objection. Never blurt out answers to a question objected to by the counsel.
- Only refer to notes for clarification of exact details; do not rely on them. Note dependence weakens the testimony and gives the appearance of not being prepared. The investigator's notes may be introduced into the record.
- Answer questions from either counsel in a polite, courteous manner. When a question is not understood, ask that it be repeated or clarified. If the investigator does not know the answer to a question, he should respond with "I do not know." He should not speculate or guess.

Nonverbal Factors

2-9. The following are nonverbal factors that are important to the success of an effective testimony:

- **Appearance.** Present a clean and well-kept appearance, dress in appropriate business attire, and ensure that personal grooming standards are according to *AR 670-1*, unless an exception for ongoing covert activities is authorized. In such instances, ensure that the TC is aware of this authorization before any anticipated testimony. If an exception to policy is authorized, the investigator's appearance should still be neat and professional.
- **Posture.** Present a comfortably straight posture when standing and sitting. Refrain from slouching, as this may suggest contempt or a lack of confidence.
- **Gestures.** Use meaningful gestures to emphasize what is being stated, but avoid waiving and using excessive hand movements.
- **Eye contact.** Make eye contact with the person for whom the information is intended. The investigator must be aware that the primary audience of his testimony is the panel. It is appropriate for the investigator to look at the counsel who is asking the question while it is being asked, but he may look at both the counsel and the panel when he provides his response. Common sense must be used when answering a series of short questions.
- **Rates and tone of speech.** The rate at which the investigator speaks may have an effect on the panel's perception of his credibility. If he gives his testimony too quickly, the panel may not hear him.
- **Movement.** Avoid movements that are annoying and distracting.

AGGRESSIVE DEFENSE COUNSEL

2-10. During examination by the defense counsel (DC), it is not uncommon for him to become overbearing or aggressive. The DC may demand "Yes" or "No" answers to complex questions or to questions that present parts that could be answered "Yes" while other parts could be answered "No." The best approach for investigators dealing with a DC who employs these tactics is to turn to the judge and advise him that he cannot accurately respond to the question with a "Yes" or "No" answer, and seek his guidance and intervention. In most cases, the judge will ask the DC if he wants the question answered and then issue instructions to him on how to resolve the conflict. It is paramount that investigators never take an adversarial tone with the DC. If the DC is hostile while the investigator is calm and composed, the loss of credibility and respect will impact the DC, not the investigator. Conversely, if the investigator loses his temper or becomes hostile, his professionalism and credibility are diminished in the eyes of the panel.

CROSS-EXAMINATION

2-11. The most difficult part of testifying is usually the cross-examination. The investigator must remain calm during cross-examination and avoid arguing. The DC will use a variety of questioning techniques to establish possible inconsistencies or prejudice. He may attempt to cast doubt on the investigator's testimony in an attempt to get an acquittal. It is important for the investigator to remember that the DC is merely doing his job the best way he can and that such attacks are not personal, they are merely a tactic used to create doubt. The investigator must be familiar with the methods of attack for cross-examination to avoid falling prey to such tactics. These tactics include the following:

- Acting overly friendly or brutal.
- Attacking the investigator's skills or knowledge of investigative procedure.
- Using rapid-fire questioning.
- Employing the silent treatment after the investigator's response.
- Demanding a simple answer to a complex question.
- Asking leading questions.
- Misquoting the investigator and declaring him incompetent or inconsistent. The investigator should not become defensive, but may merely restate his previous testimony in correcting the perception of inconsistency. Investigators should not argue with the DC.
- Attempting to force contradictions. The DC repeats the question using slightly different verbiage in an attempt to create disparity in the investigator's answers. The investigator should merely say, "As I previously stated," and repeat his earlier response.

TESTIMONY CONCLUSION

2-12. The investigator has completed his testimony. This does not mean he should let his guard down. He must maintain the same presence that he had when he entered the courtroom. He should leave the stand when directed to do so. While leaving the stand, he should not direct his attention to the DC, defendant, prosecutor, judge, or panel members. The investigator should do the following:

- Determine if he has been temporarily or permanently excused. If he has been permanently excused, he may observe the remainder of the proceedings or leave the court building and return to work. If he has been temporarily excused, he should return to the waiting area until he is permanently excused.
- Show neither approval nor disapproval of a verdict rendered to a defendant.
- Do not discuss testimony with anyone other than the TC until the case has been adjudicated.

Chapter 3

Observations, Descriptions, and Identifications

Investigators rely on their five senses when performing the supporting investigative skills of observing, describing, and identifying persons, places, events, and objects. While sight and hearing are most often depended upon, other senses, such as smell and touch, may occasionally be used advantageously to further the success of an investigation. When honed, the skills of observing, describing, and identifying become invaluable investigative tools.

OVERVIEW

3-1. Observations help investigators build descriptions of persons, places, events, and objects so that who or what was seen may later be identified. Descriptions help an individual relate what he saw to others. They may be either written or oral. They include signs, gestures, sketches, and other means to convey information about what was seen by an observer. Accurate observations and descriptions add to credible identifications of persons, places, and objects in an investigation.

FACTORS INFLUENCING OBSERVATION

3-2. The ability to observe accurately is developed through practice and experience. Most people are not trained in remembering and evaluating what they see, so the observations and descriptions they provide may not be as detailed or as objective as those of a trained observer. Trained observers, such as investigators, know that their observations can be affected by lack of sleep, illness, perceptions, or other outside influences. Influences include the following:

- Environmental factors, such as weather and light.
- The presence of unrelated, distracting events. These events cause an individual to focus his attention in a particular direction; for example, a spectator watching an exciting play on a football field may fail to note the actions of a person sitting next to him.
- The passage of time between when an event is seen and when it is recalled. This time delay can cause the observer to forget or confuse the details of the event, thus influencing his description of what he saw.
- The location of an observer at the time he sees an event influences what he sees. It is unlikely that more than one or two people will view an event from exactly the same place, therefore, a difference in location may account for a difference in their observations. Someone observing an event from a great distance may be able to give a good

overall description of what took place, but he might not be able to see and give the details that someone at close range could give. Likewise, the closer person may be unsure of the overall picture.

- Psychological, physiological, and experiential factors. These factors influence what people see and how they retain the information. People tend to evaluate and interpret what they observe by their past experiences with like incidents. They are inclined, for instance, to compare the size of an object with the size of another object with which they are familiar. A very short or very tall person may fail to judge another's height correctly. Someone 6 feet tall may seem "very tall" to an observer only 4 feet 10 inches tall. The same 6-foot person would appear to be "normal height" to a person 5 feet 10 inches tall.
- Unfamiliar sounds, odors, tastes, and other perceptions. A bartender can relate to and accurately describe these factors and his relationship to a brawl that occurred in his place of business. However, a young 21-year-old man who made his first visit to the bar on the night of the brawl probably could not recall much about any of these factors because they were new and unfamiliar to him. Stimuli, which cannot be easily compared to a past experience, is often mistakenly interpreted in terms of familiar things, and a wrong interpretation of a past experience may influence the perception of a present experience.
- Personal interests. Special interest training may increase one's power of observation, but it may also limit the focus of attention, causing the loss of other details. Specialists often have acute perception within their own field, but fail to be observant in other fields. An artist may take special note of color, form, and proportion, but fail to discern or properly interpret sounds or odors. Conversely, a mechanic may quickly note the sound of a motor or an indication of the state of repair of a car, but fail to clearly discern the appearance and actions of the driver.
- Pain, hunger, fatigue, or an unnatural position of the body. Discomfort may cause an observer to fail to correctly interpret things he would normally comprehend. The senses of taste and smell are often distorted by physical illness and external stimuli. These senses are generally the least reliable basis for interpretation. The presence of a strong taste or odor may completely hide the presence of other tastes or odors.
- Emotions (fear, anger, or worry) and mind sets (prejudice or irrational thinking patterns, such as a victim of a robbery may have been in great fear of the weapon used by the criminal). He may only be able to recall the size of the bore of the weapon and not be able to describe the offender. Such a person might be expected to exaggerate the size of the bore. Sometimes an observer may have great prejudice against a class or race of people. For example, a person who dislikes police may unwittingly permit this prejudice to affect his view of the actions of a night watchman or a security guard. How he interprets what he sees may be wrong, even if his senses recorded a true report of what occurred.

OBSERVATIONS AND DESCRIPTIONS BY INVESTIGATORS

3-3. Investigators must use a systematic approach to observations and descriptions in conjunction with the use of notes, photographs, videos, and sketches. These help the investigator to remember what he observed and also help to improve the accuracy of descriptions. Generally, accuracy is most assured if a set pattern is followed. The pattern used most often for observations starts with general features and moves to specific features. For example, when observing a person to develop a description, look first at general features like sex, height, and race. Then check features like color of hair and eyes, unusual scars, and specific body weight. Next, note changeable characteristics like clothing, eyeglasses, hats, and hairstyles. Lastly, note the mannerisms and behavior of the person.

3-4. When an investigator observes a person to try to match him to a description, he may change or reverse his pattern of observation. This is most likely if the person he is looking for has some very noticeable feature. For example, if a man with a limp were being looked for, the first feature one would look for would be the limp, followed by his general features, and then his specific features. Even when reversed, a pattern using a systematic approach is being followed.

PERSONS

3-5. When observing and describing a person, start with the person's general features, beginning at the head and progressing downward to the feet. This process should continue until all characteristics have been noted through the changeable characteristics.

General Features

3-6. When describing the general features of a person, include the following:

- Gender.
- Ethnic background or race.
- Skin color.
- Height.
- Build and posture, such as stout, slim, or stooped.
- Weight.
- Age.
- Complexion, such as flushed, shallow, or fair.

Specific Features

3-7. Every individual has features or a combination of features that are unique or distinguishing to only him. Because these features set him apart from others, they are the most important part of the description of a person. To describe the specific features of a person, begin by describing the size and shape of the head and then move to the profile, mentally divide it into three parts. Describe each third in separate detail and in relation to the whole. (The profile, unless it has a peculiarity, is not as useful as the shape of the face for identifying people.) See *Table 3-1* for descriptive features. This table is not intended to be all inclusive.

Table 3-1. Descriptive Terms for People

Term	Description
Face	Long, round, square, fat, or thin. Had scars or acne.
Hair	Color (to include fad colors), thick, thin, long, short, straight, wavy, curly, groomed, unruly, style, texture, partially bald, or completely bald. Wore a wig or hairpiece.
Forehead	High, low, or wide.
Eyebrows	Thin, bushy, or average thickness.
Eyes	Color, small, large, wide-set, close together, piercing, oval shape, or round shape. Wore glasses.
Nose	Large, small, long, short, pug-nosed, broad, narrow, straight, crooked, large nostrils, deep-pored, or hairy.
Ears	Large, small, close to the head, protruded away from the head, and number of piercings, if pierced. Wore hearing aids.
Mustache	Color, short, long, thick, thin, handlebars, or pointed ends.
Mouth	Large, small, straight, upturned, or drooping.
Lips	Thick or thin.
Teeth	Large, small, protruding, bucked, spaced, close-set, gold, broken, or missing. Wore a retainer or other observable dental work.
Chin	Broad, narrow, short, long, square, pointed, round, or double-chinned.
Beard	Color, long, short, bushy, thin, groomed, or free-flowing. Wore a goatee
Neck	Thin, thick, long, or short.
Shoulders	Broad, narrow, square, drooped, or muscular.
Arms	Length (compared to the rest of the body).
Hands	Slim, thick, well-groomed, dirty, missing, or mechanic's hands. Had crooked digits.
Distinguishing marks	Tattoos, scars, cuts, moles, birthmarks, or amputations.
Personal mannerisms	Calm, nervous, male with feminine traits, or female with masculine traits. Scratched his nose, ran his hand(s) through his hair, jingled his keys, or flipped coins.
Individualities	Limp, unusual gait, muscle twitch (of the eye, mouth, or the like), or smile (such as one that would not be forgotten). Spoke with a stutter.
Voice and speech	Tone (low, medium, or loud), soft, gruff, nasal twang, pronounced drawl, foreign accent, or mute condition. Spoke in cultured, vulgar, clipped, fluent, or broken English.
Clothing	General description (include hat, scarf, and gloves), military or civilian uniform, well-groomed, or bloodstained. Wore a certain type of footgear.

Table 3-1. Descriptive Terms for People (Continued)

Term	Description
Jewelry	Any that was observed, such as earrings, watches, toe rings, necklaces, or bracelets.
Makeup	Type of makeup worn and whether it appeared to disguise the natural face, such as, wore lipstick that made the lips look larger.
Vehicle	Color, type, make, model, year, and damages (location and extent). The location where the vehicle was observed.
Weapon	Location and type. How the weapon was carried or who left the scene with it.

Changeable Features

3-8. These features are those that the person who is being observed can readily change. It may be his clothing, hairstyle, use of cosmetics, or any other item that he uses to portray his identity. Investigators must look for deceptive ploys while observing changeable characteristics, such as the person having a false limp or wearing a hairpiece or glasses.

DECEASED PERSONS

3-9. When an investigator is observing and describing a dead person, possibly the victim of a crime, he is describing traits that are permanent, that a person does to "create" himself, and that an offender creates on the deceased.

Permanent Traits

3-10. These traits include a description of those characteristics that the person was born and died with and are not readily changeable. For example, an Asian female dies at age 35 and is 4 foot 9 inches tall and weighs 100 pounds. Her gender, race, skin color, eye shape, height, and number of teeth cannot be changed.

Personal Appearances

3-11. When describing what a person does to himself, an investigator is concerned with the clothing, footgear, jewelry, makeup, hair (such as, the color and style and whether it is real or artificial), and any other item that a person can change to make himself appear different than his natural self. This includes cosmetic changes, such as implants, hairpieces, body piercing, and tattoos.

Traits Created by an Offender

3-12. When the body of a deceased is being observed for the traits that an offender created or changed, the investigator is looking for those things specifically altered from their normal state, such as any teeth knocked out, cuts to the neck and abdomen, trauma to the skin, stab or close-contact wounds, or any hair or limbs chopped off.

PLACES

3-13. To show the exact scene of an incident or crime, an investigator will have to make detailed observations of places and locales. The purpose may be to connect the place to an incident or to connect the place to information given by a witness.

3-14. Descriptions should cite the elements the investigator observes. The goal is to give a concise and easily understood word picture of the scene. Sketches and photographs will support the word description. The pattern of observation will depend on whether the investigator is looking at an outdoor or indoor scene.

Outdoor Scenes

3-15. While observing and describing outdoor scenes from their general to specific characteristics, look for natural or man-made landmarks and do the following:

- Note the general scene and its relation to roadways, railways, and/or shore lines. Use these features to pinpoint the general site.
- Pinpoint exact site locations in relation to fixed or semifixed features, such as buildings, bridges, or power line poles.
- View outstanding objects or features within the scene.
- Check the details of the scene and any items of high interest. Some outdoor sites may not have such landmarks, so they will need to be marked for reference.
- Mentally assign boundaries to the area. Use boundaries that are neither too far apart nor too close together.

Indoor Scenes

3-16. Indoor scenes have obvious and definite boundaries like walls, hallways, and basements and are easier to observe and define than outdoor scenes. Because an indoor area often contains many objects, it is very important to use a methodical pattern of observation. Investigators must do the following in the right order:

- Note the location of the place to be observed, and state if it is at the front or rear of the building and at what floor level.
- Check the distances to stairways, entrances and exits, and elevators.
- Get the room number or other designation. Observe details near the entries to the area that is the specific point of concern, and note the objects located within the area.
- Get the exact location as it relates to other objects of interest or concern.

EVENTS

3-17. If an investigator is present when an unlawful event occurs, he must observe and remember it systematically and quickly. Take in the important factors of time, place, persons, objects, and actions involved and the immediate results of the event. These factors are involved in the essential questions of who, what, when, where, why, and how. In most cases, the investigator arrives at the scene of an incident after the crime has occurred. He seldom sees an event as it takes place. His observation of connected actions after an event may give him major clues in whole or part to what did take place.

3-18. Small, but important actions or events often provide an important lead for an investigation. Verbal remarks, emotional states of excitement, gestures, looks of concern, and unlikely claims of lack of knowledge can all be clues. An investigator may get leads from such things as the way a fire burned, the presence of certain fumes or odors, the sound of a voice, or the warmth of a body. These deductions may aid in reconstructing the cause, start, or progress of an incident or offense. Investigators must recognize related acts or conditions and understand them correctly.

3-19. The description of an event must be as complete as the circumstances allow. It should contain the following facts: time, place, order of action, objects and persons involved, and what happened because of these factors. To get a thorough and logical description of an event, an investigator must think about it in terms of his observations and from the point of view of possible witnesses or victims. He must consider statements made by witnesses, victims, and suspects and evaluate the physical evidence from the crime scene. A description of an event must be supported with sketches, photographs, and collected evidence.

OBJECTS

3-20. The pattern of observation an investigator uses to describe objects is like the pattern used to describe people; go from the general features to the specific features. This same pattern is used when trying to find objects to match an already built description.

3-21. Start with the general features that clearly define the broad category of the object. This prevents it from being confused with objects of other classes. In identifying objects—

- Use a noun to describe it, such as a car, gun, or club.
- Note its type, size, and color.
- Look for other general features that are easy to discern and that may help give quick, sure recognition.
- Describe the specific features of the object that set it off from all other like items, such as whether or not the car has a sunroof or whether or not the radio or computer is portable.
- Search for any damage or alterations that make the item unique.
- Look for serial numbers or other identifying marks and labels.

3-22. Distinguishing marks, scratches, alterations, damaged parts, worn areas, signs of repair, faded paint, serial numbers, identifying markings, and missing parts must be noted in detail. For example, when observing and describing a computer, begin with the brand name. Then list it as "nonportable, large technological wall model, light gray in color, serial number NM97JT02." Include remarks like "No previous damage or markings" or identify the damage that was caused. Follow the same procedure for all other objects.

OBSERVATIONS AND DESCRIPTIONS BY WITNESSES

3-23. An investigator must ask a witness specific questions about the general, specific, and changeable features of a person as with the investigator's systematic approach to observations and descriptions. A witness is not trained to commit details to memory about things he has observed and may not be able to without specific questions, such as, "What color eyes did he have?" Asking specific questions from the items listed in *Table 3-1*, without suggesting clues or hints, can trigger the recall of a witness. Normally, a witness will not have to provide a description associated with a deceased person. The witness may be required to identify that person based on previous observations.

3-24. A witness must be spoken to as soon as possible after he has made his observations and before he has time to talk to others or to change his observations, consciously or unconsciously, to fit a pattern of other things he may have seen or heard. Imaginative persons often use conjecture to fill in the gaps in their knowledge of an incident. This is particularly true if they later learn that the incident is important in an investigation. Investigators must evaluate the information of a witness and compare it with all related data before using it to investigate further. Investigators must also be aware of, and make allowances for the many factors that may influence a person's understanding and retention of the details he relates.

3-25. When obtaining a description from a witness, the investigator must learn of any influences affecting the understanding of what the witness observed. The investigator must determine if there are influences that might cause a witness to give false answers or to purposely withhold information so that he does not become involved. To help put the witness at ease, the investigator can talk to him briefly before questioning him. The investigator should ask the witness to repeat his descriptions to reveal discrepancies made on purpose or by incomplete observation. These flaws should be clarified in an attempt to get a better description. The investigator's questions may lead a witness to admit that he distorted the truth. Usually, a witness who lies or hides information often makes unconscious slips that can be detected by the investigator.

IDENTIFICATION

3-26. When a witness or victim identifies a person, place, object, or event, it helps the investigator to relate that identity to an incident. Before a witness makes an identification, ensure that he has made as complete a description as he can. This will help avoid false identifications and reduce the chance for error. Let the witness identify a person or an object from among a group of like persons or objects. Showing a witness one weapon or one person for identification may confuse him. The witness may give a false identification because the investigator presented a particular weapon or person to him. A witness can identify a person through means of field identification, mug shot identification, and photograph or lineup identification. When a witness identifies a place, have him describe its general location in relation to known landmarks and then describe it in detail. The witness should then be asked to take the investigator to the scene.

3-27. Field identification should be used when the person (suspect) is apprehended while committing a crime or is apprehended in the general vicinity of the crime scene. Witnesses and victims should be asked to make on-the-scene identification of the suspect as soon as possible after the occurrence of a crime.

3-28. Mug shots should be used if the witness saw the suspect and it is believed that the suspect has a previous police record with a photograph. If the witness or victim can provide substantial information concerning a suspect's appearance, a composite sketch should be drawn.

3-29. When a field lineup cannot be conducted or a suspect is not in custody, photographic identification should be used. A suspect does not have the right to a lawyer, does not need to be present, and does not need to be informed of when his picture is used in photographic identification. When a witness or victim is viewing the lineup, stress the point that he need not identify anyone from the photographs. There should be at least six photographs in the lineup with comparable similarities. At a minimum, height, age, race, weight, and general appearance must be considered. The investigator documents the photographs of the suspects with an individual number (1 through 6) and their position. The lineup should be completed a minimum of three times with each witness or victim, with the position of the photographs varied each time. The investigator documents the location and identity of the individual used in the photographic lineup within his notes and report, in addition to taking a photograph or copying the image of the lineup. When having an object identified, the lineup should be conducted in a like manner. The items should be similar, such as, if a particular name-brand revolver is in question, then all pictures should be of a revolver of that name brand.

3-30. When a suspect is in custody and there were witnesses to the crime, lineup identification may be used. The requirements are basically the same as those of photographic identification. At least six people of the same height, age, race, weight, and general appearance must be used. If one person is asked to perform a movement, then all who participate in the lineup must be required to perform the movement. The same is true with verbal statements. If one person is directed to make a certain statement, then all persons must make the same statement. Witnesses and victims must be told that they do not have to identify anyone from the lineup.

3-31. Each person in the lineup is given a numbered card. The name and number of each person's position in the lineup is recorded. The lineup is photographed to verify the location of each person in it. Military suspects are required to participate in lineup identifications. They are entitled to counsel but do not have to be advised of this unless they are in pretrial confinement. At this time, the attorney assigned to represent the military suspect must be afforded the opportunity to observe the lineup. Attorneys who are assigned to observe physical lineups may not issue guidance, participate, and/or interfere in anyway. They are restricted to an observer role.

COMPOSITE SKETCHES

3-32. Composite photographs or sketches are often used to help identify persons. Composites are developed from separate photographs, templates, or sketches of foreheads, eyes, noses, mouths, chins, or other facial features. The witness selects the example that most nearly looks like the particular facial feature of the person to be identified. Do not show a witness a photo lineup before having him help develop a composite. It may influence his memory of the subject.

3-33. Commercially manufactured kits can be used to make composite drawings or photos from verbal descriptions. The drawing from such a kit can resemble a person so closely it removes others from suspicion. Kit models that use true photos of facial features, hairstyles, eyeglasses, and hats produce realistic photo-like composites.

3-34. If photographs or sketches of separate features are not available, many photographs of different persons or objects may be used. Have the witness pick out the features that most closely look like the person or object to be drawn, or have an artist sketch a likeness of persons or objects from descriptions given by one or more witnesses. Even this kind of drawing or portrait may be useful to an investigation.

3-35. Another consideration for making composites is the use of computer programs. These composite-generating programs allow the user to make an initial sketch based on information provided by the witness. The sketch is carefully created based on multiple-choice and free-form questions. The witness views the sketch and recommends changes that can be quickly made.

Chapter 4

Interviews and Interrogations

I&I is an investigator's means of obtaining testimonial evidence from or about persons connected with an incident. Interviewing skills are among the most basic and essential tools at an investigator's disposal. Although extensive investigations may reveal extremely incriminating physical evidence that will certainly lead to a conviction, there is only one way to ever truly know the full extent of the criminal acts committed by an individual. In almost all cases, the only place from which all of the answers can be obtained is from the person who committed the offenses.

OVERVIEW

4-1. The goal of any investigator should be to determine what happened, which can only be fully revealed by discovering all of the facts pertaining to an event. One of the most difficult types of evidence to obtain can be testimonial evidence, especially from the offending party in the form of a confession.

4-2. The key to obtaining information from a suspect is the style of questioning and the approaches used during questioning to get to the truth. Years of research and experience have clearly established that criminals are more likely to confess to someone they like, trust, and respect than to someone who does not allow them to retain their dignity. Additionally, rapport-based interview techniques provide investigators with greater protection against obtaining false confessions than more confrontational interrogation styles.

4-3. Rapport-based interrogation techniques are substantially and consistently more effective in gaining cooperation and truth than other more confrontational styles. Additionally, rapport-based interview techniques provide investigators with greater protection against obtaining false confessions when safeguards that will ensure that false confessions can clearly be distinguished from valid ones are implemented.

4-4. Evidence and facts developed from a professionally processed crime scene can often provide sufficient information to identify, prosecute, and convict the offender. However, it is the testimonial evidence from victims, witnesses, and the offender that can fully document and explain the events being scrutinized. Using the proper skills to interview all persons present at the scene, those having knowledge of the events surrounding the crime, and those with knowledge of the participants will reveal exactly what took place.

TESTIMONIAL EVIDENCE

4-5. Testimonial evidence is the result of interviews conducted during an investigation. The testimony can be written or verbal and may stem from any of the various types of interviews conducted. Interviews include those resulting from an area canvass or taken from a complainant, victim, witness, suspect, or subject. Investigators must continuously strive to be objective while conducting interviews and try to put all of the collected information together like pieces of a puzzle. This will help investigators determine if something is out of place or missing. Testimonial evidence can never stand alone. Therefore, it should always be compared to all other aspects of the investigation, including physical or forensic evidence, circumstantial evidence, alibis, and motives.

4-6. Although testimonial evidence can be the most beneficial evidence in many investigations, it is also the least reliable form of evidence. It does require investigators to maintain a greater level of objectivity and skill than many other forms of evidence identification, collection, and preservation. There are several factors that contribute to the lesser degree of reliability in testimonial evidence. Some of these factors are the fragility of human memory and the fact that people have the ability and, on occasion, the inclination to lie. The observations or perceptions of others may conflict due to the fact that people observe a single event from various vantage points. Although peoples' observations are factually accurate, they may be skewed by perception. The guidance provided in this chapter is designed to limit the effects of these factors and aid investigators in discovering the truth.

4-7. Understanding human nature can be extremely helpful in thwarting the collection of contaminated information while conducting an investigation. With the passage of time, memory fades and details will be lost, which can result in the loss of valuable information. Similarly, by nature, people generally want to be as helpful as possible. They do not want to be considered less than observant, which can result in the contamination of evidence. If witnesses are allowed to discuss their observations with one another before being interviewed, they tend to incorporate statements and observations (provided by other witnesses) into their account of the incident. This is generally not intended (in any way) to derail the investigative process. However, several studies have proven that erroneous information inserted into a scenario is frequently incorporated in future witness accounts by the individuals who were provided such information. Being cognizant of this and recognizing the fragility of human recollection will aid the investigator in avoiding the collection of contaminated information.

INTERVIEW TYPES

4-8. There are various types of interviews that an investigator may have to conduct. The type used depends on the situation and the person to be interviewed. Interview types consist of canvass, victim, witness, and suspect.

CANVASS INTERVIEW

4-9. Canvass interviews are conducted in areas surrounding the locations where criminal acts are committed. During a canvass interview, nobody (in particular) is a target of the interview. Canvass interviews are designed in a web-like fashion to capture information or identify witnesses from people who may not come forward on their own or who may be otherwise linked to the offense under investigation. Frequently, individuals who possess material facts that can aid in the resolution of a criminal investigation are not aware that a crime was committed. They would not think to report something they observed just because it may have been suspicious. For example, if someone observed a stranger in his neighborhood, he may mentally take note of their observations but not report it to the authorities unless informed of a crime and asked if he heard or saw anything suspicious during the time frame in question.

4-10. Canvass interviews are most effective if conducted within 24 hours of the incident being investigated. Human memory fades with time, and people only tend to recall information for long periods of time if it is significant to them. Merely seeing a person or vehicle in a neighborhood or hearing a loud bang that awakened them from sleep would probably not be a significant event. As time passes, it becomes increasingly difficult to obtain details, such as what time an event occurred; the color, make, and model of a vehicle; and the physical descriptions of any suspicious people. Although canvass interviews are most effective in the early stages of an investigation, they can also yield leads and provide evidence much later in the investigative process. Consequently, they should never be overlooked as an investigative tool in any appropriate situation.

4-11. Canvass interviews are conducted when investigators physically walk through the immediate area where a crime was committed, such as a residence, business district, or public gathering place. The investigators should talk to every individual they can locate, such as parents, children, passersby, or other persons likely to have been in a position to see or hear something that may have evidentiary value. Each person talked to should be fully identified and recorded in the investigative case file by name, address, and phone number. A brief description as to what he observed or did not observe should also be recorded in the file. Frequently, a more detailed interview of persons identified during a canvass interview will be warranted, and it is important to know what he previously reported. In some cases, it may be beneficial to conduct the canvass interviews at the same time of day or night that the crime occurred. Additionally, some instances may require conducting canvass interviews on the same day of the week that the crime occurred. Some witnesses may only be in a particular area on a certain day or at a specific time as part of their routine, such as truck drivers and Army National Guard (ARNG) or US Army Reserve (USAR) personnel. This is especially helpful in locating potential witnesses who work shifts or are only in the area during specific times of the day, such as postal workers or people who jog or walk pets past the area.

VICTIM INTERVIEW

4-12. The victim of a crime (the person who suffered an injury or loss) should be interviewed as soon as possible. The victim will describe items used in the commission of the offense that may not otherwise be identified during a crime scene examination. Although victim interviews and other investigative actions often occur simultaneously, it is important that the scene not be released until after the victim is thoroughly interviewed.

4-13. When interviewing a victim, the investigator must always be cognizant that the victim may be traumatized and emotionally affected. The investigator must be aware of this fact and approach the victim with compassion and concern for his well-being. Although conducting interviews around the emotional state of a victim can be time-consuming, the victim will be much more cooperative for longer periods of time if he is treated with dignity and respect. When a victim is severely traumatized, his emotions can cause him to provide partial or fragmented information or to focus only on the most traumatic aspects of the incident. In these cases, it is generally a good idea to obtain as much information as possible in a verbal format. Conduct a follow-up interview at a later date when the victim can provide a detailed, logically written statement.

4-14. Investigators should be familiar with the provisions of *DODD 1030.1* and *DODI 1030.2*. These directives implement statutory requirements for victim and witness assistance and provide guidance for assisting victims and witnesses of a crime. Together, the directive and instruction provide policy guidance and specific procedures to be followed for victim and witness assistance in all sectors of the military. These procedures are described in *Chapter 1*.

4-15. It is important to note that a large percentage of people lie or omit information when interviewed by law enforcement officials. This phenomenon is not restricted to suspects or individuals making false complaints. Genuine victims frequently omit information that they feel is embarrassing or that causes other people to question their decency. This may go as far as a direct lie to prevent embarrassment or shame from certain actions that they think may be scrutinized. Identifying the fact that a purported victim omitted information or lied does not necessarily indicate that he is not a bona fide victim—this is normal human behavior. It can be helpful to explain to the victim that everyone tends to cast himself in the best light possible when providing an account of an adversarial situation. Addressing this fact up front and letting the victim know that he will not be judged will frequently lesson this factor. It is also important to keep in mind that a victim or the parent of a child victim may exaggerate aspects of the reported crime. This may be done in order to overcome perceived law enforcement indifference to the crime or in an effort to ensure rapid responses. It is the investigator's responsibility to obtain accurate information and ensure that he does not commit victims to false or misleading statements. This is particularly important when interviewing the victim of a sexual assault. Unless the investigator establishes good rapport and obtains the victim's complete trust, it is very likely that the victim will omit information, which he believe shows complicity in the incident.

False Report

4-16. On occasion, individuals who are not victims of a crime will make a false report. Generally, they do this in order to conceal other illicit behaviors they are involved in. Even when an investigator is suspicious as to the veracity of a victim's complaint, it is generally a good idea to accept the report at face value and thoroughly investigate the allegation, while being objective and open-minded. If it is determined at the conclusion of the investigation that the complaint was false, the purported victim's status will change to that of a suspect. This is the time the individual should be interrogated, not when he initially makes the suspicious complaint. When investigators allow their objectivity to become jaded by a previous experience or by the lack of "logic" in the reported crime, they develop tunnel vision. This causes them not to see an investigation for what it is. Investigators should remember that if someone makes a false report, there is plenty of time for the investigation to expose this fact. However, if he is the victim of a legitimate crime, time is a critical component of solving the investigation and a lack of objectivity can result in catastrophic failures in protecting the public.

Self-incrimination

4-17. If information is obtained during an interview with a victim indicating that he committed a criminal offense (such as drinking under age or adultery) in the course of events related to the offense, then he must be advised of his legal rights before being asked any questions about the suspected misconduct. It is generally advisable to complete the victim's interview without addressing the offenses for which he is suspected before executing the rights warning and asking about his criminal conduct. However, if the interview cannot be completed without addressing these issues, appropriate rights warnings must be rendered. If the victim is in the military, he must be advised of his legal rights. If the victim is a civilian, then a noncustodial warning would be appropriate, and he should be advised that his suspected actions constitute a crime. If the offense is of a nature that is not under the investigative purview of the organization conducting the investigation, investigators may refer that aspect to the appropriate authority, to include the unit commander. However, the conduct cannot simply be ignored.

WITNESS INTERVIEW

4-18. There are several categories of witnesses, including an eyewitness, significant party, and expert witness. For the purpose of this chapter, a "witness" is defined as anyone having direct knowledge of criminal activities. An "eyewitness" is defined as anyone who observes criminal activities.

4-19. Significant parties are defined as anyone having other than direct knowledge or observations, but bearing information that would tend to prove or disprove aspects of the crime being investigated. These people may be able to verify or refute an alibi or may have been present when an individual who was involved in the crime made statements about what they did, but was not present when the crime was committed.

4-20. An expert witness is a person who can provide insight as to the significance of the evidence collected and what that evidence indicates with regard to guilt or innocence. An expert witness can interpret crime scene evidence, such as how the crime was committed.

4-21. Witnesses should always be interviewed, one at a time, to obtain facts about the incident under investigation. Witnesses should be separated before the interview and not allowed to discuss the case before being interviewed. If witnesses are traumatized as a result of what they observed, they will often want to be with people that they feel comfortable with, such as family or friends who also may have witnessed the incident. A case-by-case determination as to how to handle each situation should be made. In some instances, it may be appropriate to accommodate such requests. However, the investigator should instruct the witnesses not to discuss the incident and explain why discussions could contaminate what each individual actually observed or heard. The investigator should have the witnesses wait in the presence of another investigator to ensure that these discussions do not occur.

4-22. When conducting witness interviews, it is generally acceptable to take notes from the onset of the interview. When witnesses appear to be traumatized from the incident, investigators should take a more sympathetic approach, allowing them to tell their story for the first time without interruption. While listening to the entire account, investigators should be supportive and understanding of the witnesses' emotional states. Once the story is fully disclosed, it becomes easier for the witness to talk about it. This will provide the investigator the opportunity to seek out details and take adequate notes.

4-23. A written sworn statement should be obtained from all witnesses. Caution should be taken in providing witnesses with copies of their statements. The release of these documents can create difficulty later in the investigation, especially when a purported victim or witness becomes a suspect at a later point in the investigative process. Additionally, witnesses who are provided with a copy of their statement tend to read them again before subsequent interviews. This makes it difficult for investigators to identify inconsistencies in stories over a period of time. It is normal for a person's memory to fade over a period of time. However, people who read their previous statement before subsequent interviews will tend to provide the same level of details months after the incident. This raises suspicion on the part of investigators who may be unaware that the individual may have a copy of their previous statement.

4-24. Not releasing copies of statements is an accepted practice. It is done to protect the integrity of the investigation. If the person who rendered a statement requests a copy of it, consideration must be given as to whether or not releasing the statement will hamper future investigative efforts. In many of these situations, it may be better to provide copies of these statements only after the completion of the investigation. If a formal Freedom of Information Act (FOIA) request is received that requests the release of investigative products, it should be coordinated with a USACIDC-designated representative or an installation release authority.

4-25. Occasionally, an investigator may encounter a victim or witness who is reluctant or refuses to cooperate with the investigation and provide a statement. The victim or witness may be afraid of retribution or may be a friend of the suspect and does not want to see the suspect prosecuted. The investigator should make an attempt to understand why the victim or witness does not wish to cooperate. With that understanding, the investigator can attempt to alleviate any fears and explain the importance of his cooperation. Some victims or witnesses are under the mistaken impression that the "right to remain silent" applies to them. If necessary, military members can be ordered to cooperate with investigators. This should be used as a last resort. In the case of civilians, the prosecutor can issue a subpoena and compel them to testify. In either case, the investigator should consult with the appropriate prosecutor to determine the best course of action when dealing with uncooperative victims and witnesses.

SUSPECT INTERVIEW

4-26. The first thing that must be addressed in determining whether to interview or interrogate a suspect is to recognize the difference between an interview and an interrogation. An interview is generally unstructured and takes place in a variety of locations, such as a residence, workplace, or police station. It is conducted in a dialogue format where investigators are seeking answers to typically open-ended questions, and the guilt or innocence of the person being interviewed is generally unknown. An interrogation is planned and structured. It is generally conducted in a controlled environment free from interruption or distraction and is monologue-based. The investigator must be reasonably certain of the suspect's guilt before initiating an interrogation.

4-27. When questioning a suspect for the first time, if the investigator is not reasonably certain of guilt, an interview is the preferred investigative tool. During this process, a suspect will be asked questions about his actions during the time of the incident. Investigators should obtain detailed descriptions of the individual's activities for the periods of time before, during, and after the crime occurred. By obtaining this information, investigators will be able to determine the accuracy of any alibis and whether or not there is evidence that is in conflict with the purported actions of the suspect. Through observation of the suspect's verbal and nonverbal responses, the investigator can assess if any indications of deception are present, which may cause the investigator to transition to an interrogational setting.

4-28. When conducting an interview, the suspect is asked questions and is encouraged to speak freely. The investigator should take notes throughout this process, recording the information provided and behavioral responses to both nonthreatening (not case-related) and relevant (case-related) questions. Interviews are not confrontational in nature, which means investigators do not directly accuse the suspect of committing the crime under investigation; however, questions pertaining to the suspect's possible involvement, motivation, or other issues that could connect the individual to the crime may be explored.

4-29. If the investigator is reasonably certain that the suspect is responsible for the crime under investigation, it is generally appropriate to initiate an interrogation without conducting an interview first. However, due to the need to develop rapport and themes, interrogations should still begin with an open discussion. Interrogation techniques are discussed later in this chapter. This discussion may be related to nonthreatening issues or may include minor aspects of the offense under investigation. If the investigator is reasonably certain as to the guilt of the suspect, he should try to avoid asking the suspect questions that will likely result in denial. This will make getting to the truth more difficult during the interrogation. An interrogation is confrontational in nature, which means the suspect will be directly confronted with his involvement in the offense. The confrontation should not be adversarial. It should provide the suspect with dignity and respect. An interrogation is not an open two-way communication. If the suspect is allowed to interrupt and provide false denials, he will be entrenched into his lie, making it progressively more difficult to obtain the truth during the interrogation.

INTERVIEW OR INTERROGATION SETTING

4-30. It is important to determine what physical setting or environment will be most conducive to gaining the trust and confidence of an interviewee and will produce the most truthful and meaningful information. Interviews or fact-finding explorations can be and frequently are conducted in a myriad of settings, locations, and environments. Generally not planned or structured, they consist of free-flowing information to and from the investigator. An interview can be effectively conducted in both suspect-supportive and -nonsupportive environments. It is completely acceptable to conduct an interview at a suspect's work, home, or other location where he may feel more comfortable. Comfort sometimes allows a suspect to talk more openly and freely, which can greatly benefit the investigative process.

4-31. An interrogation needs to be strictly planned and controlled. An interrogation should rarely, if ever, be conducted in a suspect-supportive environment. The location selected for an interrogation should be supportive to the interrogator and provide absolute privacy (free from distraction or disruption). Interrogation rooms should not be equipped with phones, outside windows, wall ornamentation, and so forth. In addition to these requirements, the room should be strategically arranged to ensure the most practical and conducive environment. If the room is equipped with a two-way mirror, the suspect should not face directly toward it. This serves as a constant reminder that someone may be monitoring the interview from another room and allows the interviewee to become distracted by his own reflection.

4-32. It is generally a good idea to have a small table in the room to complete paperwork, such as interview worksheets and rights warning certificates. It is paramount to position the table and chairs in a manner that does not allow the table to become a physical barrier once the interrogation stage begins. Ideally, there should be two different types of chairs in the room. One should be a standard four-legged chair that is comfortable but not mobile. Prevented from rolling around the room, the subject and interrogator will not become distracted by movement. The second chair should be equipped with rollers, which allows the interrogator to move back and forth from the table to the suspect as paperwork is being completed. See *Figure 4-1* for a depiction of a favorable interrogation setting and *Figure 4-2* for a depiction of a poor interrogation setting. Additionally, a mobile chair will aid the interrogator in applying proximal (physical proximity) and haptic (physical contact) techniques. See *paragraph 4-87*, page 4-31.

GENERAL RAPPORT BUILDING

4-33. Establishing a good rapport with those involved in an investigation will aid the investigator immensely in conducting a thorough and timely investigation. Rapport building begins immediately upon first contact with a victim, witness, suspect, commanders, or other law enforcement agencies and continues throughout the course of the investigation. Investigators must

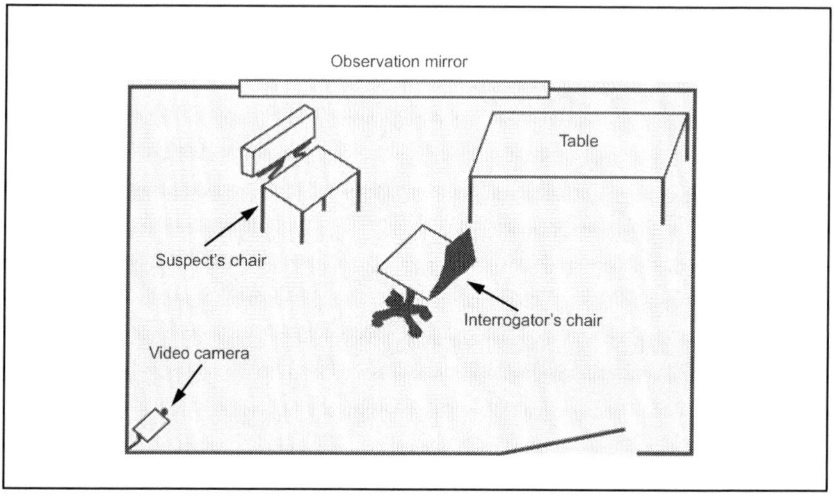

Figure 4-1. Correct Arrangement of Interrogation Room

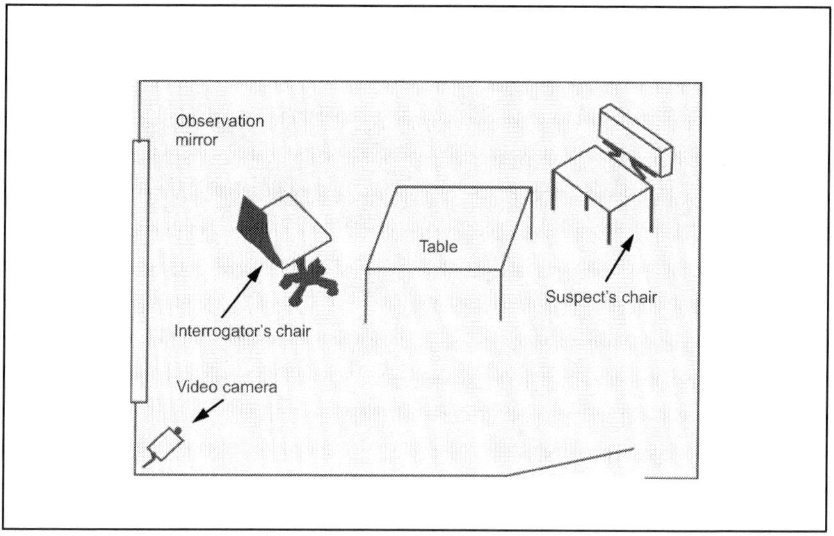

Figure 4-2. Incorrect Arrangement of Interrogation Room

always be cognizant of their behavior and conduct when in the presence of these persons. Victims, witnesses, and suspects will immediately begin sizing up the investigator. An investigator who is perceived as unprofessional, rude, lazy, unprepared, or uncaring will not gain the trust and confidence of the victim or witness and is more likely to be challenged by a suspect. An investigator who presents a professional appearance and attitude and conveys a genuine concern for the interviewee will obtain a much higher level of cooperation. He will have a higher resolution rate during assigned investigations. Each person must be treated with dignity and respect regardless of the investigator's personal views or opinions. Investigators displaying these positive attributes will see a greater level of cooperation from commanders and other law enforcement agencies. Rapport is the cornerstone of testimonial evidence and must be maintained throughout the investigative process.

CUSTODIAL VERSUS NONCUSTODIAL SETTINGS

4-34. According to federal law and *CID Regulation 195-1*, investigators may conduct noncustodial interviews of suspects who are not subject to the *UCMJ*. According to the decisions and case law stemming from Arizona versus Miranda, law enforcement personnel who question suspects who are in custody must issue a "Miranda" warning. Accordingly, suspects not in a custodial status may be interviewed without issuing a "Miranda" warning; however, the standard for whether a person is in custody is measured by the suspect's perception regarding his ability to leave or terminate the interview (not from the police officer's perspective). The most effective means for investigators to ensure that the suspect knows his custodial status is to advise him by saying, "You are not under arrest, you are free to leave at any time, and you do not have to answer any questions." Once the suspect acknowledges this warning and the interview is conducted, the investigator should document this warning on the first line of the suspect's statement, which serves as a written record of the warning. If a suspect's ability to leave is in any way impeded, he is reasonably in custody, and must be issued the "Miranda" warning. If a civilian suspect is interviewed in a custodial setting, he must be advised of his "Miranda" rights. These rights are on the reverse side of *DA Form 3881 (Rights Warning Procedure/Waiver Certificate)*. Properly executing *DA Form 3881* in such instances is the appropriate mechanism for administering this warning.

4-35. According to *UCMJ, Article 31(b)*, anyone subject to the *UCMJ* may not elicit self-incriminating statements from anyone else subject to the manual until that person has been advised of his rights indicated therein. *UCMJ, Article 31(b)*, requires the suspect to be informed as to the nature of the accusation, that he does not have to make any statements regarding the offense, and that any statements made by him may be used as evidence against him. *DA Form 3881*, which incorporates the warning requirements from both the *UCMJ* and "Miranda," should be used whenever feasible in conducting suspect interviews of military personnel or civilian personnel in a custodial setting.

4-36. There is no relief from the requirements of *UCMJ, Article 31(b)*, and failure to comply will result in the suppression of all statements. The *UCMJ* reads, "No statements obtained from any person in violation of this article, or through the use of coercion, unlawful influence, or unlawful inducement may be received in evidence against him [or her] in trial by court-martial."

DEPARTMENT OF THE ARMY FORM 3881

4-37. This section describes the procedures for preparing a *DA Form 3881*. See *Figures 4-3, 4-4,* and *4-5,* pages 4-12 through 4-14. This form is filled out each time an accused or suspected person is questioned. It is best to fill out the administrative data on the form first.

Block 1. Enter the geographic location; such as the city or installation and the state in which the form will be executed. Do not enter specifics such as a building number or military police station.

Block 2. Enter the date (day, month, and year). Have the suspect initial above the date.

Block 3. Have the suspect write in the time and initial above it after he has been administered the rights warning.

Block 4. Enter the CID or military police case file sequence number.

Block 5. Enter the suspect's last name, first name, and middle initial.

Block 6. Enter the suspect's SSN.

Block 7. Enter the military or civilian pay grade of the person being advised; for example, E3, O3, GS09, or enter "Civ" if there is no military affiliation. If the person is in the military, indicate his status as either active duty (AD), USAR, or US National Guard (USNG).

Block 8. Enter the suspect's complete military or governmental organization including unit, installation, state, and zip code or Army Post Office (APO) or Fleet Post Office (FPO). If interviewing a civilian who does not have any military affiliation, enter the his current home address including the city, state, and zip code or APO or FPO.

Part I, Section A. Enter the applicable organization and status of the person issuing the rights warning, such as "Criminal Investigation Command as a Special Agent" or "Provost Marshal's Office as a Military Police Investigator." A formal accusation can only be accomplished through the referral of charges. Investigative personnel will always line through the word "accused" to ensure that the suspect understands that he is merely a suspect. However, if interviewing someone subsequent to the referral of charges, the word "suspect" should be stricken. The investigator must then orient the suspect toward the nature of the offense for which he is suspected. Plain language should be used (not a statute or other citation). For example, if suspected of murder, the word "murder" would be entered in this section, not *UCMJ, Article 118.* After all suspected offenses have been indicated, three slashes (///) should be placed behind the last offense cited. Having the suspect draw in the slashes and initial both where the word "accused" was stricken and again following the slashes will help to show the suspect's understanding of his status in the interview and specifically what offenses he is being interviewed for.

Blocks 1 through 4. Administer the rights warning advisement by reading the information, starting with number 2, that is located in the

RIGHTS WARNING PROCEDURE/WAIVER CERTIFICATE

For use of this form, see AR 190-30; the proponent agency is ODCSOPS

DATA REQUIRED BY THE PRIVACY ACT

AUTHORITY:	Title 10, United States Code, Section 3012(g)
PRINCIPAL PURPOSE:	To provide commanders and law enforcement officials with means by which information may be accurately identified.
ROUTINE USES:	Your Social Security Number is used as an additional/alternate means of identification to facilitate filing and retrieval.
DISCLOSURE:	Disclosure of your Social Security Number is voluntary.

1. LOCATION Fort Leonard Wood, MO	2. DATE JDW 1 Jan 02	3. TIME JDW 1930	4. FILE NO. 0012-02-CID901
5. NAME *(Last, First, MI)* WRIGHT, John D.	8. ORGANIZATION OR ADDRESS HHC, 26th Infantry Battalion Fort Leonard Wood, MO 65473		
6. SSN 124-76-9834	7. GRADE/STATUS E5/RA		

PART I - RIGHTS WAIVER/NON-WAIVER CERTIFICATE

Section A. Rights

The investigator whose name appears below told me that he/she is with the United States Army Criminal Investigation Command As A Special Agent and wanted to question me about the following offense(s) of which I am suspected/accused: JDW Robbery, Kidnapping/// JDW

Before he/she asked me any questions about the offense(s), however, he/she made it clear to me that I have the following rights:

1. I do not have to answer any question or say anything.
2. Anything I say or do can be used as evidence against me in a criminal trial.
3. *(For personnel subject to the UCMJ)* I have the right to talk privately to a lawyer before, during, and after questioning and to have a lawyer present with me during questioning. This lawyer can be a civilian lawyer I arrange for at no expense to the Government or a military lawyer detailed for me at no expense to me, or both.

   \- or -

   (For civilians not subject to the UCMJ) I have the right to talk privately to a lawyer before, during, and after questioning and to have a lawyer present with me during questioning. I understand that this lawyer can be one that I arrange for at my own expense, or if I cannot afford a lawyer and want one, a lawyer will be appointed for me before any questioning begins.
4. If I am now willing to discuss the offense(s) under investigation, with or without a lawyer present, I have a right to stop answering questions at any time, or speak privately with a lawyer before answering further, even if I sign the waiver below.

5. COMMENTS *(Continue on reverse side)*

Sample

Section B. Waiver

I understand my rights as stated above. I am now willing to discuss the offense(s) under investigation and make a statement without talking to a lawyer first and without having a lawyer present with me.

WITNESSES *(If available)*	
1a. NAME *(Type or Print)* SA Josephine P. Garrett	3. SIGNATURE OF INTERVIEWEE John D. Wright
b. ORGANIZATION OR ADDRESS AND PHONE Fort Leonard Wood Resident Agency, CID Fort Leonard Wood, MO 65473	4. SIGNATURE OF INVESTIGATOR Regina R. Brown
2a. NAME *(Type or Print)*	5. TYPED NAME OF INVESTIGATOR SA Regina R. Brown
b. ORGANIZATION OR ADDRESS AND PHONE	6. ORGANIZATION OF INVESTIGATOR Fort Leonard Wood RA, CID Fort Leonard Wood, MO 65473

Section C. Non-waiver

1. I do not want to give up my rights

   ☐ I want a lawyer ☐ I do not want to be questioned or say anything

2. SIGNATURE OF INTERVIEWEE

ATTACH THIS WAIVER CERTIFICATE TO ANY SWORN STATEMENT *(DA FORM 2823)* SUBSEQUENTLY EXECUTED BY THE SUSPECT/ACCUSED

DA FORM 3881, NOV 89 EDITION OF NOV 84 IS OBSOLETE USAPA 2.01

Figure 4-3. Sample 1 of DA Form 3881

RIGHTS WARNING PROCEDURE/WAIVER CERTIFICATE

For use of this form, see AR 190-30; the proponent agency is ODCSOPS

DATA REQUIRED BY THE PRIVACY ACT

AUTHORITY:	Title 10, United States Code, Section 3012(g)
PRINCIPAL PURPOSE:	To provide commanders and law enforcement officials with means by which information may be accurately identified.
ROUTINE USES:	Your Social Security Number is used as an additional/alternate means of identification to facilitate filing and retrieval.
DISCLOSURE:	Disclosure of your Social Security Number is voluntary.

1. LOCATION	2. DATE	3. TIME	4. FILE NO.
Fort Leonard Wood, MO	1 Feb 02	10:36	0911-02-CID901

5. NAME *(Last, First, MI)*		8. ORGANIZATION OR ADDRESS
SMITH, Everett R.		A CO, 1/55th Engineer Battalion Fort Leonard Wood, MO 65473
6. SSN 229-87-0653	7. GRADE/STATUS E4/RA	

PART I - RIGHTS WAIVER/NON-WAIVER CERTIFICATE

Section A. Rights

The investigator whose name appears below told me that he/she is with the United States Army Office of the Provost Marshal as a Military Police Investigator and wanted to question me about the following offense(s) of which I am suspected/accused: Assault///

Before he/she asked me any questions about the offense(s), however, he/she made it clear to me that I have the following rights:

1. I do not have to answer any question or say anything.
2. Anything I say or do can be used as evidence against me in a criminal trial.
3. *(For personnel subject to the UCMJ)* I have the right to talk privately to a lawyer before, during, and after questioning and to have a lawyer present with me during questioning. This lawyer can be a civilian lawyer I arrange for at no expense to the Government or a military lawyer detailed for me at no expense to me, or both.

   - or -

   (For civilians not subject to the UCMJ) I have the right to talk privately to a lawyer before, during, and after questioning and to have a lawyer present with me during questioning. I understand that this lawyer can be one that I arrange for at my own expense, or if I cannot afford a lawyer and want one, a lawyer will be appointed for me before any questioning begins.
4. If I am now willing to discuss the offense(s) under investigation, with or without a lawyer present, I have a right to stop answering questions at any time, or speak privately with a lawyer before answering further, even if I sign the waiver below.

5. COMMENTS *(Continue on reverse side)*
Waived legal rights, but declined to initial or sign DA Form 3881. INV Metelski brought in to witness waiver.

Section B. Waiver

I understand my rights as stated above. I am now willing to discuss the offense(s) under investigation and make a statement without talking to a lawyer first and without having a lawyer present with me.

WITNESSES *(If available)*	3. SIGNATURE OF INTERVIEWEE
1a. NAME *(Type or Print)* DAM INV Donald A. Metelski Donald A. Metelski	Declined to Sign
b. ORGANIZATION OR ADDRESS AND PHONE Law Enforcement Command Fort Leonard Wood, MO 65473	4. SIGNATURE OF INVESTIGATOR
2a. NAME *(Type or Print)*	5. TYPED NAME OF INVESTIGATOR INV Chance C. Davis
b. ORGANIZATION OR ADDRESS AND PHONE	6. ORGANIZATION OF INVESTIGATOR Law Enforcement Command Fort Leonard Wood, MO 65473

Section C. Non-waiver

1. I do not want to give up my rights
   - ☐ I want a lawyer
   - ☐ I do not want to be questioned or say anything
2. SIGNATURE OF INTERVIEWEE

ATTACH THIS WAIVER CERTIFICATE TO ANY SWORN STATEMENT *(DA FORM 2823)* SUBSEQUENTLY EXECUTED BY THE SUSPECT/ACCUSED

DA FORM 3881, NOV 89 EDITION OF NOV 84 IS OBSOLETE USAPA 2.01

Figure 4-4. Sample 2 of DA Form 3881

PART II - RIGHTS WARNING PROCEDURE

THE WARNING

1. WARNING - Inform the suspect/accused of:
   a. Your official position.
   b. Nature of offense(s).
   c. The fact that he/she is a suspect/accused.
2. RIGHTS - Advise the suspect/accused of his/her rights as follows:
   "Before I ask you any questions, you must understand your rights."
   a. "You do not have to answer my questions or say anything."
   b. "Anything you say or do can be used as evidence against you in a criminal trial."
   c. *(For personnel subject to the UCMJ)* "You have the right to talk privately to a lawyer before, during, and after questioning and to have a lawyer present with you during questioning. This lawyer can be a civilian you arrange for at no expense to the Government or a military lawyer detailed for you at no expense to you, or both."

   \- or -

   (For civilians not subject to the UCMJ) You have the right to talk privately to a lawyer before, during, and after questioning and to have a lawyer present with you during questioning. This lawyer can be one you arrange for at your own expense, or if you cannot afford a lawyer and want one, a lawyer will be appointed for you before any questioning begins."
   d. "If you are now willing to discuss the offense(s) under investigation, with or without a lawyer present, you have a right to stop answering questions at any time, or speak privately with a lawyer before answering further, even if you sign a waiver certificate."

Make certain the suspect/accused fully understands his/her rights.

THE WAIVER

"Do you understand your rights?"
(If the suspect/accused says "no," determine what is not understood, and if necessary repeat the appropriate rights advisement. If the suspect/accused says "yes," ask the following question.)

"Have you ever requested a lawyer after being read your rights?"
(If the suspect/accused says "yes," find out when and where. If the request was recent *(i.e., fewer than 30 days ago)*, obtain legal advice whether to continue the interrogation. If the suspect/accused says "no," or if the prior request was not recent, ask him/her the following question.)

"Do you want a lawyer at this time?"
(If the suspect/accused says "yes," stop the questioning until he/she has a lawyer. If the suspect/accused says "no," ask him/her the following question.)

"At this time, are you willing to discuss the offense(s) under investigation and make a statement without talking to a lawyer and without having a lawyer present with you?" *(If the suspect/accused says "no," stop the interview and have him/her read and sign the non-waiver section of the waiver certificate on the other side of this form. If the suspect/accused says "yes," have him/her read and sign the waiver section of the waiver certificate on the other side of this form.)*

SPECIAL INSTRUCTIONS

WHEN SUSPECT/ACCUSED REFUSES TO SIGN WAIVER CERTIFICATE: If the suspect/accused orally waives his/her rights but refuses to sign the waiver certificate, you may proceed with the questioning. Make notations on the waiver certificate to the effect that he/she has stated that he/she understands his/her rights, does not want a lawyer, wants to discuss the offense(s) under investigation, and refuses to sign the waiver certificate.

IF WAIVER CERTIFICATE CANNOT BE COMPLETED IMMEDIATELY: In all cases the waiver certificate must be completed as soon as possible. Every effort should be made to complete the waiver certificate before any questioning begins. If the waiver certificate cannot be completed at once, as in the case of street interrogation, completion may be temporarily postponed. Notes should be kept on the circumstances.

PRIOR INCRIMINATING STATEMENTS:

1. If the suspect/accused has made spontaneous incriminating statements before being properly advised of his/her rights he/she should be told that such statements do not obligate him/her to answer further questions.
2. If the suspect/accused was questioned as such either without being advised of his/her rights or some question exists as to the propriety of the first statement, the accused must be so advised. The office of the serving Staff Judge Advocate should be contacted for assistance in drafting the proper rights advisal.

NOTE: If 1 or 2 applies, the fact that the suspect/accused was advised accordingly should be noted in the comment section on the waiver certificate and initialed by the suspect/accused.

WHEN SUSPECT/ACCUSED DISPLAYS INDECISION ON EXERCISING HIS OR HER RIGHTS DURING THE INTERROGATION PROCESS: If during the interrogation, the suspect displays indecision about requesting counsel (for example, "Maybe I should get a lawyer."), further questioning must cease immediately. At that point, you may question the suspect/accused only concerning whether he or she desires to waive counsel. The questioning may not be utilized to discourage a suspect/accused from exercising his/her rights. (For example, do not make such comments as "If you didn't do anything wrong, you shouldn't need an attorney.")

COMMENTS *(Continued)*

REVERSE OF DA FORM 3881 USAPA V2.01

Sample

Figure 4-5. Reverse of DA Form 3881

warning section on the reverse side of *DA Form 3881* (see *Figure 4-5*). The investigator should provide the rights waiver certificate that is located on the front of *DA Form 3881* to the suspect for him to follow. Upon completion of the rights warnings advisement, the investigator administers the rights waiver advisement by reading verbatim the questions in quotations that are located in the waiver section on the reverse side of another *DA Form 3881* (see *Figure 4-5*). Proper use of *DA Form 3881* will help investigators ensure that only knowing and voluntary waivers are obtained.

Block 5. Use the comments section to record clarification questions posed to the suspect and his responses. If additional space is required, use the comments section located on the reverse side of the form. Because of the "McOmber Rule" being abolished, there is no longer a requirement to ask if the suspect has requested an attorney after being read his rights within the past 30 days. However, if the suspect has an attorney, the attorney must be present for any interviews.

Part I, Section B, Blocks 1 through 2. All witnesses to a rights warning, which may include military police and/or investigators, parents or guardians of juveniles, and military sponsors, should print their names in the blocks dedicated to witnesses (blocks 1a or 2a). Additionally, the witness should record his unit or organization, address, and telephone number in blocks 1b or 2b. It is also a good idea to have the witness initial adjacent to his name (see *Figure 4-4,* page 4-13).

Block 3. If the suspect states the following: (1) I understand my rights, (2) I have not requested an attorney after being read my rights, (3) I do not want an attorney, and (4) I am willing to make a statement, he has waived his rights and should sign his name in this block. If the suspect has questions or does not answer the questions to the satisfaction of the investigator to show that he has knowingly and voluntarily waived his rights, the investigator must provide clarification or ask additional questions. He must do this before having the suspect sign the waiver certificate (see *Figure 4-3,* page 4-12).

Blocks 4 through 6. The special agent or military policeman administering the rights advisement signs in block 4 and prints or types his name in block 5. The investigator's unit of assignment should be entered in block 6.

Part I, Section C, Blocks 1 and 2. If the suspect invokes his rights by requesting counsel or declines to discuss the offenses at that time, he should then check the appropriate boxes in block 1 and sign in block 2. If the suspect invokes his rights and declines to sign the document, the investigator may write, "declined to sign" in block 3 (see *Figure 4-4,* page 4-13). Once the suspect invokes his legal rights, the investigator must stop all discussions that are designed or are likely to cause the suspect to be incriminated. If the suspect did not request an attorney, but related he did not desire to be interviewed at that time, he may be reapproached by investigators after a reasonable "rest bit," which has normally been held to as little as two hours. However, if the suspect requested counsel, the investigator may not reapproach him until he is afforded a "reasonable opportunity" to seek out and confer with counsel, which is normally several working days.

4-38. In the instance that a suspect waives his rights, but declines to sign the *DA Form 3881,* this fact may be recorded both in Part I, Section A, Block 5 (comments) and in Part I, Section B, Block 3 (signature of interviewee). Whenever possible, such waivers should be witnessed by at least one other party. This can be accomplished by simply bringing a third party into the interview room and asking the suspect if he was advised of his rights, understood his rights, and is willing to discuss the offenses under investigation but did not want to sign the form. If the suspect affirms this to be true and accurate, the witness should sign his name in Part I, Section B, Block 1a to confirm his presence during this portion of the interview (see *Figure 4-4,* page 4-13).

TRICKERY AND DECEIT

4-39. The use of trickery, deceit, ploys, and lying is legally permissible during the course of an interrogation under the following conditions. However, no form of trickery or deceit may be used to gain a knowing and voluntary waiver of legal rights, and no such lies, ploys, trickery, or deceit can be of a nature to shock the conscience of society or to cause an otherwise innocent person to confess. For instance, threatening a mother with the loss of her children may likely cause her to choose her children over the truth and consequently confess even if she is not guilty. Conversely, it would be permissible to tell a suspect that his fingerprints were found on evidence or that a witness observed him at the scene because an innocent person would know that he was not there and refute the statement, whereas a guilty person may feel trapped by the overwhelming evidence and confess.

4-40. Although it is legally permissible to employ deceitful tactics when conducting interrogations, it is not advised or recommended for several reasons. First, the suspect knows more about the crime than the investigator so it is easy for him to catch the investigator in a lie. Once caught in a lie by the suspect, the investigator loses all credibility. Secondly, juries and judges do not like to place their trust and confidence in police officers who appear to manipulate case facts and/or lie as part of their duties. Although lying rarely results in a confession being thrown out, it is frequently a factor used in a deliberation for panel members and judges who are not certain they can completely trust the officer who they know to be a convincing liar. This can result in reduced sentences and occasional acquittals. Defense attorneys have become very adept at bringing out lies told during interrogations in courtroom settings and at turning these lies into credibility issues for the panel.

4-41. Historically, the reason police officers have felt the need to lie during the conduct of interrogations was to apply pressure to the suspect. Convincing a suspect that the evidence clearly links him to the crime may pressure him into telling the truth. Although this pressure is essential to the interrogation process, investigators do not need to resort to lying to pressure the suspect into confessing or telling the truth. The presentation of "potential evidence" is far superior to deceit in not only creating evidentiary pressure for a suspect, but it allows the investigator to do so with absolute integrity and credibility.

4-42. Rather than telling the suspect that his fingerprints were found on an item of evidence, the investigator should consider telling the suspect, "We have recovered several latent fingerprint impressions from the crime scene, and I am going to send them to the lab with your record fingerprints." Remind the suspect that he may find it very difficult to explain how his fingerprints were discovered at the scene if he continues to insist that he was never there.

4-43. Presenting potential evidence to cause a positive reaction from the suspect works well for many different situations. For example, reminding a suspect that a parking lot or store is equipped with surveillance cameras may cause him to believe he was recorded on tape. If the investigator does not know if the location was equipped with cameras, he can send another investigator to check. He can then tell the suspect that he just sent someone over to obtain any videotapes that may exist. Although the investigator does not know whether or not this evidence exists, he presents the potential for the recovery of evidence that will link the suspect to the crime.

4-44. In just about any situation where an investigator can develop a lie to apply evidentiary pressure to a suspect, he can just as easily identify potential evidence to apply the same pressure without lying. If challenged in court, the investigator can explain potential evidence to a panel without losing credibility, but it is difficult to explain deceit and lying as a "legitimate" interrogation technique without losing credibility.

SELECTION OF AN INTERVIEW OR INTERROGATION STYLE

4-45. This section describes the two basic styles (indirect and direct) used to conduct I&I. Choosing the correct style can be a challenge even for the most experienced investigator. Human quirks of behavior or personality can affect the success of getting a person to honestly and effectively communicate with an investigator. An investigator's failure to establish rapport and gain the suspect's trust, confidence, and respect may greatly bear on his ability to glean a truthful statement from the suspect. When selecting an interview style, consideration should be given to whether or not the investigator is reasonably certain of the suspect's guilt. When the suspect's guilt is certain, a direct style may be the proper choice. If he was previously interviewed, find out what he was told by previous investigators and what he said during those interviews. In all cases where the investigator is not reasonably certain that the suspect committed the offense, he should be interviewed using an indirect style.

INDIRECT STYLE

4-46. If a suspect's guilt is uncertain or doubtful, indirect questioning is used to gain a detailed account of the suspect's activities before, during, and after the time the offense took place. The indirect style of questioning is exploratory in nature and is generally initiated by merely asking the suspect to tell the investigator his story or to explain his actions on the date in question. The investigator listens and takes detailed notes making observations of both verbal and nonverbal mannerisms that may provide insight as to the likelihood of deception. Questions are generally asked for clarification or to explore apparent inconsistencies. Questions posed in this style are generally open-ended and are not inclined to lead the suspect in any direction with his answer. Facts that are definitely known or strongly suggest the suspect's guilt are used to formulate questions designed to evaluate his verbal response and physical reaction, which will in turn provide insight into the inclination of a suspect to lie. The indirect style is not designed to cause perpetrators to confess. It is designed to arm the investigator with an extremely detailed account of the suspect's activities, which can be verified or refuted by facts developed through the investigative process. Once sufficient evidence is established to warrant an interrogation, the investigator will transition to the direct style.

DIRECT STYLE

4-47. The direct style of interviewing is designed to prevent the suspect from getting entrenched in a lie, which will make getting to the truth more difficult. In this style, the investigator tries to maintain an atmosphere of monologue. The investigator lays out case facts. He works on the emotions, logic, and reasoning of the suspect in an attempt to get him to see his situation in a more realistic light, without enabling him to build a wall of lies. This style of interviewing is very structured and should be well-planned before initiating the interrogation. In addition to providing case facts to the suspect to make him feel he was identified as the person responsible for the crime, the investigator must decide what evidence will be withheld from the suspect to identify or corroborate admissions or confessions. In order for this style to work effectively, the investigator must present no uncertainty as to the suspect's guilt, but he should seem relatively uncertain as to motivation. If the investigator seems uncertain, the suspect will be significantly more likely to hold out hope that his involvement in the crime will go undetected. Caution should be used when employing this style with weak-minded individuals or people with a less than average intelligence quotient (IQ) because they are more susceptible to rendering false confessions. Safeguards must be used to ensure that confessions are the result of guilt, not mental fatigue. All investigators conducting interrogations should ensure that a suspect who confesses is able to provide information he would not otherwise know unless he was involved in the commission of the crime. Investigators should also ensure that investigative personnel did not provide the information in the suspect's confession. Additionally, when an investigator suspects that an interviewee is making admissions merely to stop the interview process, he should consider providing information that is not true to see if the suspect incorporates this information into his statement. If a suspect attempts to incorporate this information into his statement, the investigator should recognize the propensity of the individual to accept responsibility for the things he did not do and should implement strong safeguards to prevent false confessions.

OBSERVATION OF BEHAVIOR (VERBAL AND NONVERBAL)

4-48. Interviewers observe verbal and nonverbal responses to recognize whether an individual is being truthful or deceptive. The nonverbal responses will either compliment or contradict the verbal response. Although an individual may provide verbal responses to direct questions, as much as 70 percent of all communication takes place at the nonverbal level. Through a comparison of the two, investigators are much more successful in determining if an individual is attempting to deceive them. Because nonverbal responses are typically generated at a subconscious level, they are more reliable than the spoken word. Observation and active listening will assist in determining if the interviewee is being truthful or deceptive. Very few people are so adept at lying that they can control both their verbal and nonverbal responses.

VERBAL INDICATORS OF DECEPTION

4-49. Signs of verbal deception include the following:

- **The chosen word.** Normally, an innocent person will use realistic terms such as rape, robbery, and murder. A guilty person will generally avoid harsh terms and use more socially acceptable terms, such as had sex, took, or died when discussing alleged offenses.
- **Vocal characteristics.** A nondeceptive person will normally have a clear, steady voice and will remain this way regardless of the stress applied during the interview process. His vocal characteristics will be consistent with and support what he is saying is saying verbally. A guilty person tends to lose this quality and clarity when responding to questions that cause him to lie. His voice may change pitch or waver, and the speed of his response is usually drawn out as he takes additional time to formulate his response. Anticipate follow-up questions and look for consistency in their story. Once he has developed his response, he may speed up his responses, displaying confidence in his answer.

4-50. It is extremely important to note that assessing verbal and nonverbal responses for indicators of deception is accomplished by evaluating clusters of behavior. One or two observed behaviors do not necessarily mean that the person is practicing deception; however, a constant comparison of responses and reactions to nonthreatening and threatening questions is essential to this process.

EVALUATION OF VERBAL RESPONSES

4-51. When analyzing verbal responses, there are specific reactions that will assist investigators in recognizing truthful and deceptive responses. It is important to remember that these reactions are just general guidelines, but they are effective and accurate most of the time. The accuracy of these reactions depends on the mental and physical state of the suspect. Specific reactions are as follows:

- **Truthful response.** Suspects generally answer questions with direct and spontaneous answers. Speech clarity, rate, and pitch are more understandable and smoother than that of a guilty suspect. Although truthful suspects will sometimes use clipped word endings to emphasize their innocence and your understanding, such as, "I...did not...do it," they will much more frequently use contractions, which are more free flowing, such as "I didn't do it."
- **Deceptive response.** Suspects normally delay in answering questions by using vocalized pauses, asking for questions to be repeated, answering with vague responses, or repeating questions before providing answers (some people do this out of habit). Evaluation for these behaviors should begin during the opening stages of the interview and continue during nonthreatening questions. This will help to delineate between someone employing deceptive practices and someone reacting out of habit.

4-52. It is important to remember that often times, what the suspect "does not" say is as important as what he "does" say. Be cognizant of gaps in timelines and specific questions that the suspect provides only evasive answers to or just does not answer. A suspect may be very committed to stating that he did not break into a building but may entirely avoid answering whether or not he stole contents from the building. This may indicate that the building was unlocked or that another suspect broke into the building. This way he uses a specific denial to adamantly assert his innocence without technically lying.

NONVERBAL INDICATORS OF DECEPTION

4-53. Observe the position (posture) of the entire body. Try to determine if the person has an open or closed posture. An innocent person will generally sit upright, appearing more relaxed and casual. In most cases, he will go as far as to lean toward the interviewer inviting the questions and demonstrating an eagerness to resolve the issue (see *Figure 4-6*). A guilty person will appear to be stiffer and more rigid. He may lean back in his chair or slouch. He will generally avoid a frontal alignment with the interviewer, which presents a closed posture. The crossing of arms or legs, holding elbows close to the body, or tucking legs under the chair are good indicators of a deceptive person (see *Figure 4-7*). Normally, these body positions will change frequently and be very erratic.

Figure 4-6. Portrayal of an Innocent Person

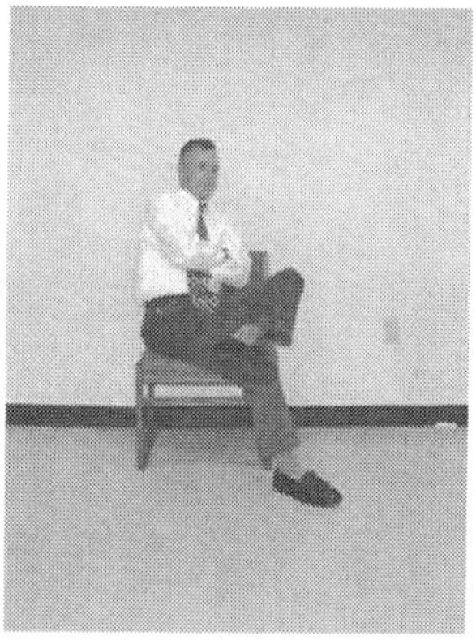

Figure 4-7. Portrayal of a Guilty Person

4-54. Signs of nonverbal indicators of deception include the following:

- **Physical movements.** Initial contact and the introduction of nonthreatening questions will allow the interviewer to observe the natural movement of the interviewee's hands. As the interview progresses, the interviewer can observe if there is any change in the smooth, fluid motion of the interviewee's hands when asking threatening questions. A sign of deception could be if the interviewee engages in grooming gestures or closes his previously open posture or if his bodily reactions become erratic.
- **Gestures.** Certain gestures are good indicators of deception. Investigators should look for the following two basic types of gestures:
  - Grooming gestures. Examples of grooming gestures are when a person removes lint from his clothing, rubs or wrings his hands, taps his fingers on a tabletop, scrapes at the surface of the tabletop with his fingernails, or rearranges his clothes or jewelry.
  - Supportive gestures. Supportive gestures are when a person places his hands over his mouth or eyes when speaking, places his hands or fists under his chin, holds his forehead with his hands, or places his hands under or between his legs.

All of these bodily reactions and postures occur as a subconscious protection mechanism and are generally more consistent with deception than truthfulness. They are the result of a guilty person subconsciously attempting to release the stress and anxiety of lying. In these cases where verbal and nonverbal responses do not complement one another, the physical responses are typically more reliable than the verbal responses.

- **Facial expression.** This is another subtle indicator of deception. A guilty person may use a variety of facial expressions resulting from fear of detection. The variation of facial expressions may be suggestive of untruthfulness or the lack of such variation may be suggestive of truthfulness. Truthful individuals should exhibit expressions that coincide with the subject of the conversation and should not be obviously exaggerated. Deceptive individuals tend to feign shock, surprise, or overly exaggerate emotional responses.

4-55. Nonverbal responses will either support or contradict the spoken word. If the person answers a question while leaning forward in an upright posture without accompanying movements and using direct eye contact, normally, this is considered to be the actions of a truthful person. However, if the person gave the same reply and also crossed his legs, folded his arms, looked at the floor, or shook his head in agreement while answering in the negative, this would tend to indicate deception.

INTERROGATION PROCESS

4-56. Establishing and maintaining effective rapport with suspects during an interrogation is instrumental in obtaining their cooperation and getting them to expose facts they would not otherwise feel comfortable disclosing. The key to establishing rapport lies almost entirely within the mind of the interviewer. If the investigator cannot relate to the suspect as an innocent person, he will not be able to effectively develop rapport or display sincerity. The investigator must look for characteristics or commonalities between himself and the suspect in order for the investigator to see the suspect as an individual. It is important for the investigator to remember that no one is inherently evil and that good people sometimes do bad things. If an investigator is able to separate the crime from the person being interrogated, he is more likely to get the suspect to like, trust and respect him. Guilty persons are substantially more likely to confess to an investigator they like, trust, and respect; whereas, innocent suspects are provided the greatest safeguards from false confessions through rapport-based interrogation techniques.

4-57. Sincerity is commonly the most difficult component of rapport building for most interrogators. Investigators often believe they can "pretend" to care about the suspect's situation in order to gain his trust. However, sincerity cannot be faked; it is either sincere or fake. Sincerity is the best tool to use to emotionally move the suspect. Separating the crime from the suspect, and finding something within the suspect that the investigator can relate to is the most effective means by which investigators can develop genuine concern for the suspect. This is the key to developing and maintaining rapport with the suspect and gaining his confidence.

INTERROGATION PHASES

4-58. The interrogation consists of four phases. Each phase serves a distinct purpose and allows the investigator to move in a structured and meaningful manner through the interrogation process. The four phases include—

- Preparation.
- Opening.
- Body.
- Closing.

PREPARATION

4-59. Probably the most neglected phase of the interrogation process is the preparation phase. By nature, interrogations are structured; however, it is difficult to develop this structure without adequate planning. There are several key components of preparation that provide the greatest advantage to the investigator. There are several components of preparation. The following key components provide the greatest advantage to the investigator:

- Researching the suspect's background.
- Researching case data.
- Planning the interrogation.
- Mentally preparing for the interrogation.
- Establishing the most appropriate interrogation environment.

4-60. Researching the suspect's background involves collecting as much background information on the subject, as possible. This will aid the investigator in analyzing the suspect's training, education, financial situation, upbringing, and other factors that may be useful during the interrogation. The more background information an investigator has at his disposal, the better. This information will have a dramatic psychological impact during an interrogation. Presentation of this background information during the interview causes the suspect to realize how thorough the investigation is and will continue to be. This is extremely beneficial in increasing anxiety at key points of the interrogation process. Investigators should consider numerous sources when seeking background information on a suspect. Sources include the following:

- Employment or military records.
- Evaluation reports.
- Service record briefs.
- Medical records.
- National Crime Information Center (NCIC).
- Financial Crimes Enforcement Network (FinCEN).
- Education records.
- Internet sources.

4-61. Another key component of the preparation phase is to research case data. A complete understanding of the incident under investigation is paramount to a successful interrogation. This should include physically visiting the crime scene, viewing all crime scene photos and sketches, reading all statements collected during the investigation to date, and discussing the investigation in detail with the case agent and other investigative personnel involved in investigating the offense. While conducting this review, the investigator should identify all actual and potential evidence that could link the suspect to the scene, victim, or crime. Evidence is the pressure plate used in the interrogation that forces a suspect to face the fact that at the end of the investigation there will be no doubt as to who committed the offense. Actual evidence is any evidence that is known to exist, such as physical evidence collected; surveillance tapes seized; testimonial evidence from victims, witnesses, or co-conspirators; or other items that the investigative organization already knows to exist. Potential evidence is evidence that could be identified during the investigation and link the perpetrator to the crime.

4-62. Planning the interrogation and mentally preparing for the interrogation involves a thorough analysis of all known case facts, the suspect's personal background, and what trends indicate why others have committed similar crimes. The investigator will be able to speculate what the suspect's motive was for committing the crime with a relatively high degree of accuracy. Once the investigator understands the likely motive for the crime, he should determine the most effective means of approaching the suspect. If the investigator asks himself, "If I did this, how would I have to be approached, and what would I have to be confronted with in order for me to tell the truth about this?" By placing himself in the mental state of the suspect and answering these questions, an investigator can then develop the approach he should use in getting the suspect to see his situation in the most realistic light, while not making him defensive. Causing a suspect to take a defensive posture places barriers between him and the interrogator. This hampers the investigator's ability to maintain rapport. Mapping out the interrogation process in this manner provides the structure required for an interrogation. It will enable the investigator to more effectively control the process, ultimately resulting in full disclosure of the truth by the suspect being interrogated.

4-63. Establishing the most effective interrogation environment is a very important component of the interrogation process. Generally, the most effective interrogation environment is one controlled by the investigator, such as a police station. If possible, a room specifically designed for interviews should be used; one that is free of exterior windows and wall ornamentation so as not to distract the suspect. There should be one small table situated in a such a manner so that the suspect cannot use it as a barrier to hide behind. The chair intended for the suspect should be a four-legged stationary chair, which does not lend itself to swiveling or other movement. This should keep the suspect from moving around in the interview room and causing a distraction during the interrogation process. The interrogator's chair should be equipped with wheels and a swivel mechanism, allowing the investigator to move around subtly during the interrogation process.

4-64. Where feasible, there should be a two-way mirror installed in the interview room that allows other investigative personnel to observe the interrogation and take copious notes instead of being physically located in the interview room serving as a distraction. In complex interrogations, it may be advisable to establish a means of communication between personnel observing the interrogation and the interrogator. This allows the observers to point out issues that create anxiety in the suspect, point out conflicts in the suspect's stories, or provide the interrogator with questions that should be asked. One of the best mechanisms for this is to place a computer monitor in the interview room, which is split from a computer system in the observation room, and place it only where the interrogator can observe the information displayed on it. Using this setup, the interrogator can read comments placed on the monitor at his convenience. This would not be as much of a distracter as using an earpiece to relay verbal commands.

OPENING

4-65. The opening portion of the interview is essential in establishing the tone of the interrogation. Before the interrogation, the investigator should attempt to assess where the power base and anxiety level of the suspect will be, based on his status in the community. Power base refers to how much control and power the suspect believes he has as compared to the interrogator. For example, if a senior military officer were being interviewed, he would likely feel that he has greater power and control over the interview process than someone of lesser rank. Likewise, the anxiety level of a senior officer would likely be lower. The investigator's introduction is designed to adjust the power base and anxiety level appropriately. When the power base is anticipated to be high and the anxiety low, the investigator should seek to reduce the power base, while increasing the anxiety level. Whereas, when the power base is low and the anxiety is high, the investigator should seek to increase the power base, while reducing the anxiety. In an optimal situation, a suspect's power base, as he may see it, should be about equal to that of the investigator. However, the investigator should, in actuality, control the structure and conduct of the interrogation. If a suspect's anxiety level is too low, there is a lack of concern on the part of the suspect, which reduces the possibility of emotionally affecting him. However, if the anxiety level is too high, the investigator will not be able to accurately assess the suspect's nonverbal communication or get him to focus on key issues.

4-66. When an investigator introduces himself to a suspect with a high power base, he should not subordinate himself to the suspect. For example, if the suspect is a senior officer who outranks the investigator, the investigator should use the suspect's rank and last name when he initially introduces himself. He should then refer to himself by using his official title and last name (such as, "Good morning Lieutenant Colonel Johnson. My name is Special Agent Smith. I am glad you could come in today. Could you please follow me?"). It should be noted that generally, the investigator always walks behind the suspect for safety reasons; however, when interviewing senior personnel, walking in front of them subtly diminishes the perception of subordination. The investigator may then ask the suspect if he would like to take off his coat. If he does, show him where a hanger is and instruct him to hang it up. The investigator may then ask the suspect if he would like some coffee. If he accepts, the investigator should then point out where the coffee pot is and tell him to help himself. During this process, the investigator should be polite, courteous, and professional; however, he should not refer to the suspect as "sir" or "ma'am." He should just use the rank and/or the rank and last name. This approach has proven itself in modifying the power base and anxiety level to the desired levels.

4-67. Suspects who have a lower power base are often uncomfortable and very apprehensive. Consequently, when the investigator introduces himself to the suspect, he should call him "sir" or "ma'am," even if the person is a janitor. He should use the person's first name if he is a lower enlisted soldier. The investigator should then introduce himself using only his first and last name in a very friendly and congenial manner. The investigator should then thank the person for coming in and tell him that he understands that he is very busy, but it is good to see that he wants to take care of this issue. The investigator should cater to the suspect by offering to hang his coat up, pouring him a cup of coffee, and putting cream and sugar in his coffee for him. This makes the suspect feel that he is important in the process and that the investigator sees him as an individual, not by the status that others may see him as having in the community.

4-68. The opening portion of the interrogation is where investigators begin the rapport-building process. In building rapport, the investigator should explore the suspect's interests, looking for common ground that he can relate to on a personal level. When those interests are identified, the investigator should spend a reasonable period of time discussing these topics from a purely personal vantage point. This not only helps the suspect to see the investigator as a person with whom he can relate, but also aids the investigator in a similar fashion and is paramount to successful interrogations.

4-69. After establishing the initial tone of the interview and rapport, the investigator should complete a biographical data sheet. While completing the form, the investigator should maintain the previously established environmental tone, continuing to build rapport. Being friendly and polite will reduce anxiety; whereas, being more rigid and professional will increase anxiety.

BODY

4-70. The body of the interview is the heart of the interrogation process. This is where all actual discussion pertaining to the offense itself transpires. It is during this phase of the interrogation that the investigator strives for the suspect to tell what happened in a detailed statement, which includes the who, what, where, when, why, and how of the crime. The presentation of themes and alternative questions, along with the observation of the suspect for deceptive behavior, all occur in this phase of the interrogation process. The end result should be a complete description of events, people, places, things, and thoughts (planning and intent), and all elements of proof for each offense should be clearly established.

CLOSING

4-71. It is important to close the interview in a manner that leaves the suspect feeling good about himself and the investigator. After a suspect confesses, many investigators succumb to the desire to tell the suspect that he is going to jail or what he did was disgusting, and a week later found themselves in a situation wherein the same suspect had to be reinterviewed. Unfortunately, since the suspect felt taken advantage of, or ridiculed at the conclusion of the previous interview, he was no longer willing to cooperate. Letting the suspect leave the interview with a sense of dignity and the perception that the investigator cared about him as a person, leaves the opportunity for positive future contact with the suspect. Honesty should always be rewarded with gratitude and respect, which serves to encourage further honesty. After a suspect renders his statement, it is advisable for the investigator to escort him out of the interview room and explain the fingerprinting process. He should explain why it is being done, what happens to the fingerprints, and why they will be released to the military police desk sergeant. The impact of the investigator explaining these things before they happen continues to make the suspect feel that he matters as a person and that he is important to the investigator. This opens the door for future contact.

INTERROGATION THEMES

4-72. Themes are essentially the vehicle used by interrogators to deliver the interrogation approaches. Themes are used to help the suspect see and appreciate his situation more clearly without making him defensive in the process. Themes can be as simple as the development of a suspicion of the suspect's motive (following the determined approaches) or as complex as a story that allows the suspect to draw his own conclusion about the right thing to do. Story type themes are very effective in getting suspects to participate in the interview process in a philosophical manner without actually discussing the crime in question, which keeps the suspect from becoming defensive and putting up barriers.

4-73. Other story themes can be used to prevent the initial lie from being told and to humanize the investigator. If an investigator is reasonably certain of a suspect's guilt, it is significantly easier for investigators to stop the lie before it is told. Once a suspect lies and the more he is allowed to lie, the more difficult it is to get beyond the lie to the truth. If a suspect knows that he is going to be interviewed, he goes over in his mind what he will be confronted with and what evidence may exist that can link him to the crime. Then he explores his options. One aspect of this exploration is, almost always, what can he say to explain the evidence away in an attempt to get away with the crime. Themes can frequently help get beyond the preparation that the suspect has done by the investigator distracting him from his prepared story and attempting to get him to modify his view of the situation to a more realistic and practical one.

4-74. A theme may be designed to pry at those things most important to the suspect, which is why it is vital during the rapport-building stage for investigators to seek out the things that will help a suspect better recognize the situation for what it is and to influence the suspect to be honest. For instance, if a suspect has a strong relationship with his mother, investigators may want to have him reflect on how his mother would feel about the situation. This could also be effective when used with how he handles himself subsequent to the incident. Other themes may be designed to change the suspect's perspective of the investigator. Frequently, in interrogation settings, the suspect has a predisposed depiction of the interrogator as the enemy or someone who does not care about him but just wants to solve the investigation and put someone in jail. Themes of this nature are designed to show the suspect that the investigator is a human being who cares about doing the right thing for the right reason. These types of themes attempt to show the suspect that the investigator is impartial, understanding, and trying to ensure that the most thorough efforts are undertaken to protect the interests of all involved in an ethical manner. Allowing the suspect to recognize these facts helps him to relate to the investigator as an individual, cultivate trust, and build respect between himself and investigator.

APPROACHES

4-75. Various approach methods are used in interrogation. The following paragraphs discuss the different types of approach.

FLATTERY

4-76. One approach is to use flattery. This is used to build self-esteem within the suspect or feed his ego. It is accomplished by making favorable observations, telling the suspect positive things that his command, friends, or loved ones said about him. By complimenting his appearance, prior positive conduct, character, patriotism, or other attributes, the investigator makes the suspect feel better about himself. As long as this approach is delivered with sincerity, it will help the investigator build and maintain rapport and a sense of mutual trust and respect. After a suspect makes an admission, he may feel depressed or uncertain that this was a good course of action. The use of flattery at this point helps reinforce his decision to be honest and enables him to continue telling the truth. The subject knows he did something wrong and he does not need condemnation for his actions. He just needs understanding and respect. Flattery can help him maintain his self-respect, ease inner turmoil, and maintain a positive interview environment.

SUSPECT VERSUS SUSPECT

4-77. When investigating a crime involving more than one suspect, it is often effective to pit the suspects against each other. Seat the suspects where they can see each other through the door to the interview room, but where they cannot communicate. When a suspect realizes that other possible suspects to the crime have been identified and are also being interviewed, he will likely realize that the evidence against him is mounting and that another suspect may attempt to place himself in the best light by attributing the greater responsibility to one of the other suspects. This approach can have a dramatic psychological impact on a suspect and compel him to relate his story first in order to cast himself in the most credible and understandable situation possible.

SYMPATHY

4-78. This technique is designed to let the suspect know that the investigator understands what he did and why he did it. However, the investigator should be careful not to condone or condemn the offense, merely to understand the suspect's emotional state. To some extent, people want pity and understanding. The investigator should try to truly understand the suspect's motivation and emotional state, which will allow him to show sympathy and sincerity when talking to the suspect. The suspect will quickly identify attempts at sympathy that are void of sincerity as hollow words, resulting in more of an obstacle than a tool.

LOGIC AND REASONING

4-79. This approach is generally the favored mechanism in cases with a great deal of evidence or when interviewing someone who is street-smart (someone who has committed numerous crimes and dealt with police on numerous occasions). The logic and reasoning approach is more effective with regard to hardened or career criminals. In delivering this approach, the investigator should be rational, tell the suspect that the facts prove the case, and lay out those facts. At the end of the interview, the only thing left to understand is the why of the offense. The facts already show who, what, when, where, and how.

COLD SHOULDER

4-80. This approach is used best on an egotistical individual after a lot of flattery. After building up the suspect's self-esteem, the investigator begins the interview process. However, when the suspect offers a denial that is clearly untrue, the investigator does not verbally respond to the statement. He merely realigns himself in his chair, leans back, and looks at the suspect in blank silence. Frequently, the suspect will realize how implausible his statement was and attempt to clarify it or, in some cases, retract it. The investigator may follow up with a plea to be treated with the same respect that he has provided the suspect and request that his intelligence not be insulted. This can, in many cases, result in an agreement to provide mutual respect and honesty.

TRANSFERENCE OF BLAME

4-81. In some cases, especially theft, a suspect will feel that the victim's actions justified his crimes. For instance, a suspect may feel that the government owed him compensation for the many long hours he worked and attempt to gain restitution by taking items of value from the Army. Sometimes, the suspect feels that if a victim really wanted the items, he would have secured them better. Occasionally, it is appropriate for investigators to allow and encourage this justification. However a suspect might use this as a crutch to legally justify his actions, such as self-defense in an assault situation. The investigator must evaluate all case facts to determine if the claims being asserted by the suspect are reasonable and indisputable.

HYPOTHESIS

4-82. This approach is used after a thorough evaluation leads to the how and why a crime was committed. The investigator may be able to speculate as to the suspect's motivation and present a hypothesis. This should only be done if the investigator reasonably believes the hypothesis to be accurate based on case facts.

ALTERNATIVE QUESTIONS

4-83. Alternative questions are asked of suspects who assume a defeated posture and are ready to tell the truth but have not yet made a disclosure. The alternative question is designed to help the suspect feel that the investigator understands and does not judge him. It does not provide an excuse for the suspect that justifies his actions. Many disciplines within the interview and interrogation arena have been criticized by experts in the field of "the false-confession syndrome" because of the way they present alternative questions. These professionals often testify against police officers for several reasons. The primary reason is, in their opinion, that police officers "provide" suspects with their story without regard for the truth to obtain an admission. An investigator's purpose and function is to discover the truth; these tactics should not be employed if the investigator does not believe what he is suggesting to the suspect to be an accurate hypothesis regarding his motivation. When suggesting the acceptable motivation, the investigator must have evaluated all evidence and case facts, and the speculation should be as accurate as possible. Then the investigator should contrast the acceptable motivation against a less acceptable reason for committing the same offense.

4-84. Speculation as to the suspect's true motivation should be as close as humanly possible to the suspect's actual motivation, which helps the suspect feel understanding from the interrogator as opposed to judgment. For example, when interviewing a suspected child molester, do not approach the suspect's motivation as a one-time incident. Historical evidence strongly supports that these types of crimes are of a multiple nature and, if the suspect admits to assaulting one child based on this approach, it would be difficult to get him to talk about other abuses. The interrogator may want to address the motivation issue using supported case facts, such as, I know you truly care about the child or children, you tried to be a good role model, and you were taught right from wrong, but your feelings for the child manifested themselves in the form of sex instead of nurturing. This approach could be contrasted with someone who waits in parks and abducts children with the intent of hurting them. The interrogator could ask the suspect, "You're clearly not this guy are you? You do care about these kids, and you never intended to hurt them, did you"?

4-85. Based on what is known about the majority of pedophiles, this speculation is accurate. However, when the suspect is approached in this manner, he does not feel that he is being judged or condemned by the investigator, who earnestly wants to understand "why" this happened.

4-86. The presentation of alternative questions in this fashion will lead to more complete disclosures by the suspect. It will also help interrogators retain more credibility during courtroom presentation.

PROXIMAL AND HAPTIC TECHNIQUES

4-87. It is important to realize that within the structure of rapport-based interrogations, the object is to help the suspect become comfortable with the interrogator. The proper use of proximal and haptic techniques reduces psychological barriers and improves rapport building. With the intent to build rapport, rather than to increase anxiety or to intimidate, distance between the interrogator and suspect can be adjusted as the interrogation progresses. Similarly, touching can be extremely effective; however, caution should be used when employing this technique. Frequently, investigators misunderstand what a touch accomplishes during the interrogation process. A touch should generally be executed at a key point in the interrogation—when the suspect becomes emotional, when he is about to make an admission, or immediately after he has made an admission. The message a touch sends is that the investigator understands and can relate to the suspect and does not judge him. The delivery of a touch is the most critical aspect of this technique. The investigator should not reach out and grab or pat the suspect, as it may be interpreted as condescending and insincere. A touch should be delivered using a couple of fingers on the outer portion of the suspect's leg or arm. The more unintentional a touch appears, the more effective it is in relaying the desired message.

FALSE CONFESSIONS

4-88. The vast majority of false confessions result from some form of coercion, such as being threatened with the death penalty or the loss of their children or being forced into a situation where they felt the need to protect someone close to them. Interrogators must be cognizant that false confessions can be produced and know how to prevent and/or delineate them from viable confessions. People who have low IQs (generally considered to be under 80) and youths are typically more susceptible to providing a false confession induced by extreme psychological pressures.

4-89. Although false confessions are rare, there have been several instances where people who confessed to a crime and were subsequently convicted were later proven to be innocent through forensic evidence. Because juries tend to place a great deal of weight in confessions when deliberating a case, it is paramount that investigators and interrogators implement safeguards to prevent false confessions. A confession should never stand alone on its own merit; it must be collaborated. Therefore, it is important that interrogators not provide the suspect with details from the investigation that only the perpetrator would know. These details should also be shielded from the media. A suspect who confesses should be able to provide information and details of the crime not known or only known by authorized police personnel. If interrogators show the crime scene photos to the suspect or take him to the crime scene before getting this information, this critical safeguard can be negated. Additionally, videotaping the entire interrogation from start to finish can be an indispensable tool in preserving the integrity of the process. This can be extremely useful in refuting claims of police misconduct in interrogations. Videotaped interrogations can show a lack of coercion and can clearly depict whether the suspect provided details without having been provided by the interrogator. Therefore, videotaping all interrogations should be strongly considered as an investigative tool when practical.

DOCUMENTATION OF STATEMENTS

4-90. Statements from suspects are important documents. They must be recorded and made a permanent part of the investigation.

4-91. There are forums that can be appropriate mechanisms in recording information obtained through the interview process. However, they all have strengths and weaknesses that make some more appropriate than others based on the situation.

AUDIOTAPING AND/OR VIDEOTAPING

4-92. Investigators should consider whether recording the interrogation is feasible and practical. If a determination is made to record interviews, the suspect's permission must be obtained unless there is Army general counsel approval for recording and/or intercepting. Permission can be obtained verbally or in writing. It is best to include the verbal permission as part of the recording. The rights advisement of suspects and subjects should also be included in the recording if the decision is made to videotape and/or audiotape the interview. Recording an interrogation should start with an introduction of the parties being recorded, the time and date, and the agreement of each party to the recording. The interrogation should not be conducted without recording all of it. Turning on a video camera to record just the confession negates nearly all of the safeguards sought from the recording process.

4-93. Breaks may be taken during the recording. However, when a break is taken, the investigator conducting the interview should state that a break is being taken and the time and date of the break. The recording equipment should be turned off during the break. Once the interview resumes, the investigator should annotate on the tape the parties that are present along with the time and date and a brief description stating what occurred during the break, such as ate lunch or took a rest room break.

4-94. When interviewing children, the investigator should use open-ended questions instead of closed-ended or leading questions. Using leading questions, especially with child victims, can cause severe contamination of the child's statement. The tape generated from the interview should be maintained with the case file and the transcripts.

WRITTEN STATEMENTS

4-95. Written statements are permanent records of the pretrial testimony of accused persons, suspects, victims, complainants, and witnesses. They may be used in court as evidence attesting to what was told to investigators. They also are used to refresh the memory of the persons making the statements.

Typewritten Sworn Statements

4-96. Typewritten sworn statements are the preferred method for recording and preserving statements from victims and witnesses. However, when taking confessions from a suspect, defense attorneys have grown very adept at attacking the credibility of these statements. They suggest that the words in the statement are the investigator's and not the suspect's. Defense attorneys will try to establish that the investigator manipulated what the suspect said. This defense tactic is easily defended because the suspect is being asked clarification questions throughout the writing of the statement and is able to read the statement as it is being typed. The suspect is allowed and encouraged to change any aspect of the statement that did not accurately capture what he was trying to say. After the statement is completed, the suspect reads it again, then signs, and swears it to be true and accurate. The last question on the statement is, "Is there anything you would like to add to or delete from this statement at this time?" The suspect's response is recorded and appropriate action to delete and/or add information is taken at that time. The benefit of this type of statement is that it is generally brief and concise, well written, and address the elements of proof in a very clear manner. These statements can be taken in conjunction with handwritten statements, interview sketches, letters of apology, and/or audio recording or video recording.

4-97. A second option for obtaining typewritten statements from a suspect who has admitted to criminal culpability is to ask the suspect if he can type. If he states that he can, the investigator may ask him to type his own statement. This approach bears the same protections against defense assertions than if the investigator edited the statement. If this technique is employed, the first question asked in the question and answer portion of the statement should be, "Did you type the statement above?" Additional questions should be used to clarify or explore aspects of the suspect's typed narrative.

Handwritten Sworn Statements

4-98. Handwritten sworn statements are tools that can be used by investigators to record statements. This tool is most appropriate for suspect interviews. These types of statements are very effective in recording confessions. They are written by hand and in the words of the suspect, making it very difficult for defense attorneys to suggest manipulation of the facts by investigators. This has become a defense attorney's routine tactic when typewritten sworn statements are presented in court. As with any tool, there are some advantages and disadvantages to using the handwritten statement. One disadvantage is that the person providing the statement may not be able to write the statement in a clear and concise manner based on his educational level. His handwriting may be so bad that it is difficult to determine exactly what was written. However, after the suspect has written his statement, it is the responsibility of the investigator to carefully read over the statement and ensure that all elements of proof are adequately addressed, and all reasonable questions are answered. The question and answer section is no different than in the typewritten sworn statement. The investigator will ensure that the question and answer section clears up any doubt or confusion in the suspect's statement. He ensures that all the elements of proof have been addressed appropriately. This type of statement will frequently require many more clarification questions than would be required of typewritten statements.

4-99. Handwritten statements lend themselves to a technique known as statement analysis. Statement analysis should only be conducted by trained personnel and should only be considered when the circumstances are conducive to such analysis. Once a suspect is interviewed, the content of any statements taken will likely have been influenced and will not be conducive to true statement analysis. Consequently, the need for and value of statement analysis must be addressed before any interviews of the suspect. Generally, the suspect whose statement is to be analyzed is suspected of a criminal offense. Therefore, he should generally be advised of his legal rights before being asked to write such a statement.

4-100. In some cases, there may be a need to conduct a statement analysis of a victim's statement. Not because the victim is suspected of lying or criminal complicity, but to identify aspects of his story that may be embellished or omitted out of embarrassment or other factors. This is especially useful in sexual-assault complaints. Even bona fide victims frequently omit information they feel could cause them to be judged or otherwise scrutinized. Statement analysis on a victim's statement should be considered on a case-by-case basis and used only when there is an articulable benefit. However, if the investigator suspects that the victim is making a false report, he must advise the victim of his legal rights according to his military and custodial status before being requested to make such a statement. It should be noted that an accurate statement analysis cannot be conducted on a statement typed by a third party or written by a suspect after he has been questioned about the matter under investigation. These activities contaminate the interviewee's articulation of the incident.

Department of the Army Form 2823

4-101. Sworn statements are recorded on *DA Form 2823 (Sworn Statement),* and rights warnings are recorded on *DA Form 3881.* To permit written statements to be admissible in court, they must be carefully and completely prepared. Follow the instructions listed below when preparing a *DA Form 2823.* See *Figures 4-8, 4-9,* and *4-10,* pages 4-35 through 4-37.

Block 1. Enter the geographic location; such as the city or installation in which the statement is rendered.

Block 2. Enter the date of the interview. Have the interviewee initial above the date after he signs the sworn affidavit located on the last page of the statement.

SWORN STATEMENT

For use of this form, see AR 190-45; the proponent agency is ODCSOPS

PRIVACY ACT STATEMENT

AUTHORITY: Title 10 USC Section 301; Title 5 USC Section 2951; E.O. 9397 dated November 22, 1943 (SSN).

PRINCIPAL PURPOSE: To provide commanders and law enforcement officials with means by which information may be accurately identified.

ROUTINE USES: Your social security number is used as an additional/alternate means of identification to facilitate filing and retrieval.

DISCLOSURE: Disclosure of your social security number is voluntary.

1. LOCATION	2. DATE (YYYYMMDD)	3. TIME	4. FILE NUMBER
Fort Lewis, WA	2002/09/11	1943	0033-02-CID018-32656

5. LAST NAME, FIRST NAME, MIDDLE NAME	6. SSN	7. GRADE/STATUS
FISHAL, Art Edward	232-33-4544	E7/AD

8. ORGANIZATION OR ADDRESS
B Co, 3/47 Infantry Battalion, Fort Lewis, WA 98433

9.
I, Art E. Fishal, WANT TO MAKE THE FOLLOWING STATEMENT UNDER OATH:

Around 1800, 10 Sep 02, I received a telephone call from my friend, Bobby J. STELLS, (NFI), who asked me to meet him at Young's Bar and Grill, Tacoma, WA 98435. I was told there was a new disco band playing at the club and some of the band members were asking if anyone in the area had some marihuana for sale. I told STELLS that I still had plenty of marihuana to sell if the band members were willing to pay a fair price. STELLS told me the band members said that money was no object and they were willing to pay whatever the local price was. Around 1900, 10 Sep 02, I opened my safe and removed approximately 80 grams of marihuana and placed it inside a small trash bag. I then placed the trash bag containing the marihuana under the passenger seat of my pickup truck in order to transport it to Young's Bar and Grill. About 1945, 10 Sep 02, I departed my safe and drove directly to Young's Bar and Grill in order to meet the band members and sell them the marihuana. I was met by STELLS at the front door of the bar, prior to meeting any of the band members. STELLS asked me how much marihuana I brought and how much it would cost. I told STELLS that I brought enough marihuana to last the band throughout the entire week they would be in Washington, and that it would cost $500.00, which was not negotiable. STELLS told me the price was fair and the band should have no problem paying that amount. Around 2300, 10 Sep 02, I met with one of the band members, who introduced himself by the name "Big Joe" (NFI). Big Joe told me that he and his other band members needed marihuana in order to continue playing, as they were starting to lose their sanity. I handed Big Joe the garbage bag containing the marihuana and he handed me a piece of paper, which indicated he would make restitution after his band completed their performance later that night. Initially, I told Big Joe that would not be acceptable; however, I felt intimidated by him and realized I should do what he said. Big Joe asked me if I wanted to go back to the dressing room and smoke some of the marihuana with him and his partner prior to the concert. We waited a few minutes, but the other band members never showed up. So STELLS, Big Joe, and myself went back to the dressing room and smoked three joints, which he told me to roll. Around 0010, 11 Sep 02, Big Joe departed our company stating he needed to go do what it was he came to do, which I assumed he meant to perform with the band. A few minutes later, Big Joe came back to the room and took the bag of marihuana. I stayed in the dressing room enjoying my buzz and I must have passed out. When I woke up, the police were asking me about the joint remnants in the ashtray next to me. I told them I didn't know what they were talking about. They told me they would check it for fingerprints, at which time I knew I was already in trouble and told them that it was mine. While being transported to the Police Station, I asked the officers why they were at the club, and they informed me that an anonymous male caller had notified them that I was in the dressing room and that I was in possession of marihuana. I asked about the band, and they told me that the bar was closing when they arrived and there was no band there. It was at that time, that I realized that Big Joe had made the phone call and skipped out to avoid paying me for the drugs.

Q: What was the name of the Band "Big Joe" was playing in?
A: I'm not sure, but the sign on the door said "The Boogie Brothers" were playing live that night, so I think that was the name of Joe's band.
Q: Where is the band from?
A: Again I don't know, but I remember Big Joe saying something about New Orleans.
Q: How long were they in Washington State?
A: I don't know, STELLS told me they were going to be here for a week, but I think that either he lied to me, or they lied to him.
Q: Do you know where they were going next?
A: I don't have any idea, but I know they're a traveling band.
Q: Describe the man you call "Big Joe".
A: He was a black male, approximately 73 inches tall, 300 lbs, short black and grey hair, and he was around 49 years old.
Q: Did you observe any identifying marks, scars or tattoos on Big Joe?
A: Yes, he had a pierced left ear, from which he wore a trumpet earring; he had a tattoo of a really cool cross on his right hand.
Q: Do you know what Big Joe did in the band?

10. EXHIBIT	11. INITIALS OF PERSON MAKING STATEMENT	PAGE 1 OF 3 PAGES
	AEF	

ADDITIONAL PAGES MUST CONTAIN THE HEADING "STATEMENT OF ______ TAKEN AT ______ DATED ______

THE BOTTOM OF EACH ADDITIONAL PAGE MUST BEAR THE INITIALS OF THE PERSON MAKING THE STATEMENT, AND PAGE NUMBER MUST BE BE INDICATED.

DA FORM 2823, DEC 1998 — DA FORM 2823, JUL 72, IS OBSOLETE — USAPA V1.00

Figure 4-8. Sample DA Form 2823, Page 1

USE THIS PAGE IF NEEDED. IF THIS PAGE IS NOT NEEDED, PLEASE PROCEED TO FINAL PAGE OF THIS FORM.

STATEMENT OF Art E. Fishal TAKEN AT Fort Lewis RA DATED 2002/09/11

9. STATEMENT *(Continued)*

A: Based on the trumpet earring and a comment he made about blowing notes, I assume he played the trumpet.
Q: Describe STELLS?
A: He is a white male, approximately 71 inches tall, 180 pounds, about 22 years old. He has severe acne, a nose ring, short black hair, kind of tan, and he has a tattoo of a rat on his right shoulder. He also walks with a slight limp, I think he might have hurt himself recently.
Q: How do you know STELLS?
A: We inprocessed together at the Fort Lewis Replacement Center. We kind of became friends, and I see him downtown about once or twice a week.
Q: Where can we find STELLS?
A: He was in the Army, I think he was a medic or something. I don't know exactly when he came in, but he was a private when we inprocessed together in April of 2000. I think he got chaptered out of the Army for using drugs around July of 2002. I don't know exactly where he lives, I think it is on Freak Street, but I've never been to his house.
Q: Did you ever meet any other members of the band?
A: No.
Q: It is my understanding that you took 80 grams of marihuana to Young's Bar and Grill for the purpose of selling it to a party you did not know for the amount of $500.00; that you provided the marihuana to the party in question (Big Joe), but you did not receive the money for this transaction because you passed out, he skipped out, and you were arrested. Is this accurate?
A: Yes - he got my dope, but I didn't get paid.
Q: Since you entered the Army, how many times have you consumed controlled substances?
A: I've been smoking marihuana off and on since I was about 16 years old. I would say I have smoked marihuana about twice a week for the past 10 years.
Q: Have you consumed any other types of drugs while serving in the Army?
A: Yes, I have used cocaine periodically.
Q: What do you mean by periodically.
A: I would say only at parties when other people have it. Maybe once a month or so.
Q: Who do you get your drugs from?
A: I buy all of my marihuana in bulk, as I sell most of it, and just use what I want. I get it from a guy by the name of Stan DELEVERS, who is a SGT in the 1/48th FA, on Fort Lewis. He has a cousin who has a plantation somewhere in Oregon. I don't have a steady source for cocaine, but I only do it with people I trust. The only ones I remember doing it with are MAJ Lieks PAYNE, and SSG Major CHAPTER, both assigned to the Ranger Bat. They always provide it to me.
Q: What do you mean when you say "plantation"?
A: I mean an area where he cultivates marihuana plants, harvests them, then sells the marihuana.
Q: Do you know this cousin's name?
A: No, SGT DELEVERS just calls him Johnny.
Q: Do you know where this plantation is located?
A: No, but I think it is located in the southern part of Oregon. Somewhere near Mount Ashland.
Q: Where did you get the marihuana you provided to Big Joe?
A: I got that from SGT DELEVERS about three days ago.
Q: How much marihuana did you obtain from SGT DELEVERS on this occasion?
A: He provided me with three pounds. I sell it, then give him the money before he gives me my next shipment.
Q: How often do you get shipments from him?
A: Generally about once every two weeks.
Q: How much do you pay him for these shipments.
A: Unless I specify otherwise, he always gives me about three pounds. For this I pay him $4,000.00, which is the price I get from his due to our friendship.
Q: How much do you sell the marihuana for?
A: I generally sell it for $2,000.00 a pound; or about $150.00 to $300.00 per ounce, depending on if I know the person I'm selling it to.
Q: Who do you sell marihuana to?
A: I don't know many of their names off the top of my head, but I keep a ledger in my safe which contains all of the names, dates and amounts of these transactions.
Q: Do you have any other controlled substances in your safe?
A: Yes, I have about 2 pounds of marihuana in the safe right now.
Q: Where do you keep this safe?
A: I didn't think it was safe at my house if I got raided, so being smart, I keep my safe at the storage unit I rent at World Wide Storage, located on Loser Ave, in Tacoma, WA. My storage unit is number C-76.
Q: How did you know you were providing 80 grams of marihuana to Big Joe?
A: The only thing I use my storage unit for is my drug trade. I have scales, bags, coffee and other items in there that I use to evade detection by the police. I used the scales to measure out how much I was giving to them.
Q: In your statement you said you were intimidated by Big Joe, why?

INITIALS OF PERSON MAKING STATEMENT AEF PAGE 2 OF 3 PAGES

Figure 4-9. Sample DA Form 2823, Page 2, Continuation Page

STATEMENT OF Art E. Fishal TAKEN AT Fort Lewis, WA DATED 2002/09/11

9. STATEMENT *(Continued)*

A: He was a big man, and when I told him that later was not acceptable, he just looked at me like he was going to snap my neck.
Q: Did he say anything to you that was threatening in nature?
A: No, I changed my mind before he showed me what that look was all about.
Q: Who else knew you were making this transaction?
A: Nobody, I'm pretty smart and I don't talk about my business.
Q: You used the word "joint" in your statement. What did you mean by this?
A: I meant a marihuana cigarette.
Q: Do you make transactions at Young's Bar and Grill frequently?
A: Yes, it's a pretty shady place. The owner is a lady by the name of Diane YOUNG, I think. I'm not sure about her first name, but I'm pretty sure it starts with a "D". She has a live and let live attitude and doesn't mind if we do business in there as long as she's kept happy.
Q: What do you mean when you say "she's kept happy"?
A: I pay her a little bit of money for the use of her business and she watches my back.
Q: How much do you pay her?
A: Generally about $100.00 per week, but there's no set fee - it's on the honor system to pay her commensurate to what I make on a deal.
Q: Does she know you're selling drugs in her business?
A: We've never specifically discussed what the money is for, but she ain't stupid, she knows what's going on.
Q: Describe Ms. YOUNG?
A: She is a white female, about 56 inches tall, about 130 pounds, she has blonde hair, and she is about 35 years old.
Q: Do you know where Ms. YOUNG lives?
A: I think she lives in the upstairs portion of the bar, but I'm not sure. I just see her take guys up there all of the time, and assume she has an apartment up there.
Q: How long have you been selling drugs out of Ms. YOUNG's bar?
A: For about four years. I feel pretty safe there, plus I'm hoping to hook up with her one of these days.
Q: When you say hook up with her, do you mean to go into business with her?
A: No, I mean hook up in the romantic sense.
Q: How were you treated by CID today?
A: You were persistant, but courteous and respectful.
Q: Why did you decide to tell us the truth about your drug related activities?
A: Because I knew I was busted and I figured it was better just to lay it all out.
Q: Is there anything you would like to add or delete from this statement at this time?
A: Yes, I would like to apologize to my commander and first sergeant, I know I've let them down, but I hope they will still support me.
Q: Is there anything else you would like to add to this statement?
A: NO///END OF STATEMENT/// AEF

AFFIDAVIT

I, Art E. Fishal, HAVE READ OR HAVE HAD READ TO ME THIS STATEMENT WHICH BEGINS ON PAGE 1, AND ENDS ON PAGE 3. I FULLY UNDERSTAND THE CONTENTS OF THE ENTIRE STATEMENT MADE BY ME. THE STATEMENT IS TRUE. I HAVE INITIALED ALL CORRECTIONS AND HAVE INITIALED THE BOTTOM OF EACH PAGE CONTAINING THE STATEMENT. I HAVE MADE THIS STATEMENT FREELY WITHOUT HOPE OF BENEFIT OR REWARD, WITHOUT THREAT OF PUNISHMENT, AND WITHOUT COERCION, UNLAWFUL INFLUENCE, OR UNLAWFUL INDUCEMENT.

Art E. Fishal
(Signature of Person Making Statement)

WITNESSES: Shaun M. Cummins
SA SHAUN M. CUMMINS
Fort Lewis RA, CID
FORT LEWIS, WA 98643
ORGANIZATION OR ADDRESS

ORGANIZATION OR ADDRESS

Subscribed and sworn to before me, a person authorized by law to administer oaths, this 11th day of September, 2002 at Fort Lewis, WA 98643

Noel A. Thompson
(Signature of Person Administering Oath)

SA Noel A. Thompson
(Typed Name of Person Administering Oath)

Article 136 UCMJ
(Authority To Administer Oaths)

INITIALS OF PERSON MAKING STATEMENT AEF | PAGE 3 OF 3 PAGES

PAGE 3, DA FORM 2823, DEC 1998 USAPA V1.00

Figure 4-10. Sample DA Form 2823, Page 3, Affidavit or Final Page

Block 3. Have the interviewee write in the time and initial above it after he signs the sworn affidavit located on the last page of the statement.

Block 4. Enter the CID or military police case file sequence number.

Block 5. Enter the interviewee's last name, first name, and middle initial.

Block 6. Enter the interviewee's SSN.

Block 7. Enter the military or civilian pay grade of the person being advised; for example, E3, O3, GS09, or "Civ" if there is no military affiliation. If the person is in the military, indicate his status as either AD, USAR, or USNG.

Block 8. Enter the interviewee's complete military or governmental organization including unit, installation, state, and zip code or APO or FPO. If interviewing a civilian who does not have any military affiliation, enter the his current home address including the city, state, and zip code or APO or FPO.

Block 9. In most cases, the statement will consist of a narrative section followed by a question and answer portion. The narrative format is where the interviewee provides his version of events in a logical story-based format in his own words (but not necessarily verbatim). After the entire story is laid out in the narrative format, questions and answers will be used to draw out inconsistencies, gaps, and other issues that are not clear. If the elements of proof were not adequately addressed in the narrative, they must be addressed by specific questions that will draw out these details. Seasoned investigators may use several questions to resolve one element of an offense, as opposed to formulating one question directly from the *UCMJ*. For example, do not ask the interviewee "When you struck the man in the head with the brick, did you intend to cause grievous bodily harm or death?" because this almost compels him to lie. Instead the investigator should ask the following:

- "Why did you hit the man with the brick?"
- "What did you think would happen when you hit him in the head with the brick?"
- "Did you think that hitting him in the head could seriously hurt him?"
- "Why didn't you just use your fist?"
- "Did you want to hurt this man?"
- "Do you think striking someone with a brick could kill them?"

The responses to these broken down parts of the question are much more likely to result in honest answers, because it would sound implausible to say, "No, I just wanted to get his attention." The entirety of the statement must answer the who, what, where, when, why, and how of the crime. All elements of proof for the crime must be addressed. Additionally, the statement must specify the times and dates of specific acts and the methods used to complete the crime. It must identify suspects, accomplices, witnesses, and persons who knew of the crime. It must account for stolen property and instruments used in the crime. It must tie the evidence to the victim and/or to the interviewee. The interviewee must be given the chance to edit the statement when it is completed. The first step in this editing process is to ask, "Is there anything you would like to add or delete from this statement at this time?" This question is repeated until an answer of "No" is received. After the last word in the body of the statement, write "End of Statement" to close it out. If the statement will not fit on the front, back, and top of the final page of *DA Form 2823,* use continuation pages. Begin the statement on the front of *DA Form 2823.* Line out the reverse side with one diagonal line drawn from corner to corner if the page is not needed. Do not include the lined-out side of the form in the page count; it does not need to be initialed. Use white bond paper for your continuation pages. Each page must have a heading giving the same information as the heading of *DA Form 2823* and bearing the word "continued." The bottom of each continuation page must show the initials of the person making the statement and cite the page number in relation to the total pages of the statement. Instructions for using continuation pages can be found at the bottom of page 1, *DA Form 2823* (see *Figure 4-8,* page 4-35). Conclude the statement on page 3 (final page) where the affidavit is printed. Before administering the oath of affirmation, the interviewee must read the entire statement, or have it read to him. He must line through and correct all misstatements, errors, and corrections as he edits the statement. It is generally an accepted practice to program several easy to identify typographical errors into the statement, which the interviewee should identify, correct, and initial during his review. This helps to establish that the interviewee read the statement before signing it and that he felt comfortable changing errors or misstatements.

4-102. If preparing a statement on a typewriter and a mistake is made and noticed while the statement is being typed, make slash marks over each letter and leave a space for the interviewee's initials. This way the wrong word can still be seen, and it will not cause any doubt when and if the statement is introduced into court. If a mistake is found after the statement is completed, line the word out, write the correction above the mistake, and have the interviewee initial it. Do not use correction tape or white correction fluid to correct errors.

4-103. The affidavit is the last section of *DA Form 2823.* It states that the information was given voluntarily and that all mistakes and corrections on the statement were corrected and initialed by the interviewee. It shows that the number of pages in the statement was verified. Before administering the oath, the investigator must have the interviewee read the affidavit out loud, which will allow the investigator to refute later claims that the interviewee could not read and was too embarrassed to tell the investigator. The interviewee could also claim that, even though he could not read, he signed the statement to avoid ridicule. The investigator should then ask the interviewee what the oath means to him. Having the interviewee state in his own words what the oath means will allow the investigator to testify to the interviewee's level of reading comprehension. Administering the oath is accomplished when the investigator raises his right hand and has the interviewee raise his right hand and asks, "Do you swear or affirm that this statement is true and correct to the best of your knowledge, so help you God." If the interviewee objects to the use of the word God, the investigator deletes it from the oath and then reads the oath again. After the interviewee answers the oath in the affirmative, the investigator has him place his initials in

Block 2 and write in the current time in Block 3 along with his initials. The investigator should also have the interviewee initial beside the first and last word on the page and on the bottom of each page of the statement. The interviewee signs the affidavit on the line above "signature of person making statement" (page 3). The investigator signs the affidavit on the line above "signature of person administering oath" (page 3). Additionally, the investigator must type or print his name on the line above "typed name of person administering oath" and the authority on the line above "authority to administer the oath." If the interviewee is subject to the *UCMJ*, the authority is *UCMJ, Article 136 (b)4*. If the interviewee is a civilian, not subject to the *UCMJ*, the authority is *Section 303, Title 5, USC (5 USC 303)*. The affidavit page is included in the page count of the statement even if it contains no text from the statement. It is an integral part of the statement. See *Figure 4-8*, page 4-35.

INTERVIEW SKETCHES

4-104. After completing the body of the interview or interrogation of the suspect and before the closure, an interview sketch should be obtained. Interview sketches are very effective in capturing routes that the suspect took in committing the crime and fleeing from it. They also capture actions taken in the commission of the crime, which include locations along the route where each action occurred. The sketch should be hand drawn by the suspect and should outline the area of the crime scene. The investigator can ask the suspect questions pertaining to the "who, what, where, when, and how" of the crime and then ask him to annotate his response on the sketch. The interview sketch is a free-hand drawing of a scene or object relevant to the crime and describes the actions taken during the crime. The sketch is only obtained after full disclosure is reached. The suspect must draw the sketch as he remembers it. Do not provide information that is pertinent to the crime; however, the investigator can guide the suspect in the completion of the sketch. Relevant facts, such as buildings, structures, and other pertinent features should be included in the sketch, if possible, along with the entrance and exit points, avenues of approach, and evidentiary items. The suspect should use the word "I" when annotating his location in relation to the sketch and when writing directions or routes he used during the commission of the crime. Once the sketch is completed, the investigator should have the suspect sign and date it. The investigator's printed name and signature should also be on the sketch as a witness. See *Figure 4-11*.

LETTERS OF APOLOGY

4-105. Letters of apology are extremely effective in interrogations involving crimes against persons, but can be applied to other offenses as well. This is especially effective when interviewing someone suspected of sexually abusing multiple children but has confessed to only one incident with one child. The investigator has the choice of taking a sworn statement from the suspect to record his first confession or to go on with the interrogation in the hope that the suspect will continue to cooperate.

4-106. At the point when a suspect has confessed to crimes against a particular person, but the investigator needs to continue the interrogation for

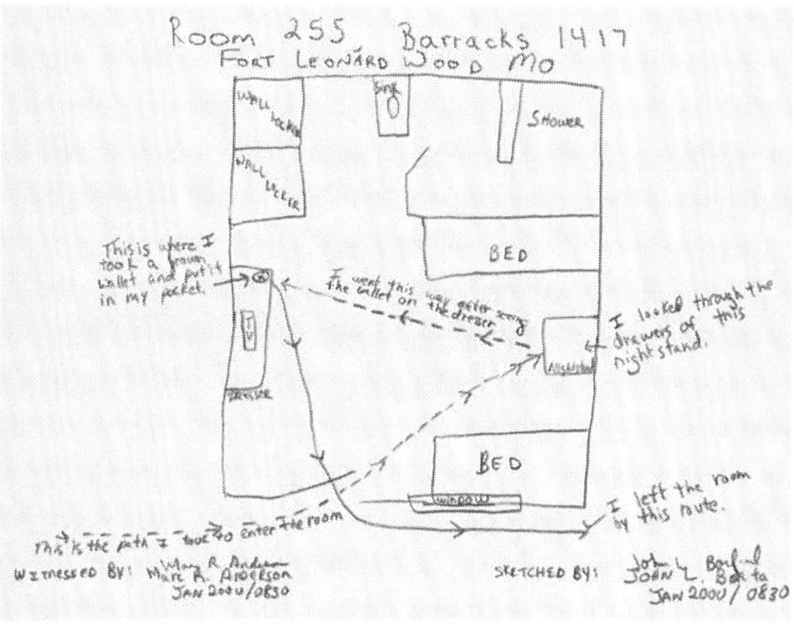

Figure 4-11. Completed Sketch

other issues or to explore other potential victims, the investigator can explain that it took a lot of courage to take responsibility for his actions. The investigator can tell the suspect that it would be very beneficial if he would tell the victim how he feels about what he did. The investigator should then suggest that the suspect write the victim a letter taking responsibility for what he did and telling the victim how sorry he is for hurting him.

4-107. Once the suspect has written the letter, if he has not specifically addressed what he did to the victim, the investigator can urge him to come straight out and tell the victim what he did, that it was wrong, and how he feels about having done it. Writing this letter serves two very significant purposes. First, it serves as a confession handwritten by the suspect. Second, it places the suspect in the victim's state of mind and helps him to view his actions from the victim's prospective.

4-108. Once the suspect is in this state of mind, he is significantly more emotionally vulnerable, creating the perfect opportunity to transition into the exploration of additional victims and/or crimes. Once the letter is obtained, it is evidence in the investigation just as any other statement; however, a copy of the letter should be provided to the victim's advocate for consideration as to whether or not the letter should be provided to the victim. The advocate may determine that the letter will only serve to further traumatize the victim or decide that it will help the victim in the healing process. The decision whether or not to provide the letter to the victim should be made by trained clinical personnel.

SPECIAL CONSIDERATIONS

4-109. When conducting interviews, an investigator may encounter individuals or situations that require special consideration. When this occurs, the investigator should maintain his focus of discovering the truth. Distinct regulations or guidelines may govern the conduct of certain interviews, but they should not take away from the basic guidelines on conducting interviews.

SENIOR PERSONS

4-110. Whenever an allegation of criminality or impropriety against a senior person is received, the local PM and USACIDC offices will be notified immediately of the allegation. A determination will be made whether the investigative responsibility is kept locally or sent to the Department of the Army Inspector General (DAIG) to investigate the allegation. Under no circumstances will a senior person be interviewed as a suspect or subject without specific prior authorization from the designated investigative command.

4-111. When a senior person reports to either the military police or USACIDC field element that he is a victim of (or witness to) a crime, he is interviewed to begin the investigative process. Whenever a senior person needs to be interviewed as a victim or witness to an investigation, the field element concerned will conduct the interview according to normal investigative procedures. Senior persons are identified as the following:

- A general or flag grade officer (includes active and reserve component).
- A general officer or flag grade officer designee.
- A DA civilian in the grade of GS16 or higher.
- Any other person of equivalent rank who occupies a key position as designated by the appropriate commander.

JUVENILES

4-112. A juvenile is identified as a person who is less than 18 years of age at the time of the incident being investigated and who is not AD military. Certain precautions must be observed when questioning juveniles regardless of whether they are the victim, witness, suspect, or subject.

4-113. To get a complete and accurate story, investigators must avoid alienating the juvenile. The use of abusive language, threats, or otherwise abrasive behavior toward juveniles will in almost all instances only make it more difficult to relate to them in general. Do not use condemning terms like juvenile delinquent, thief, or liar. An investigator must be in control of his emotions and temper at all times. In some situations, feigning anger momentarily may be appropriate in adjusting an juvenile's power base and anxiety. In most cases, these behaviors will only create distance between the juvenile and the investigator.

4-114. Avoid coercive practices or any acts that might push an innocent, but frightened or emotionally troubled, juvenile to falsely confess. What most juveniles, especially troubled juveniles, want more than anything is to be respected, listened to, and understood. One of the greatest frustrations in young people is that nobody listens to them, and most authoritative figures talk down to them. Strong-arm tactics will reinforce this frustration and cause the juvenile to retreat into his feelings of being misunderstood and oppressed.

4-115. Investigators can gain a juvenile's confidence by treating him with respect and dignity and encouraging him to take care of his immediate needs and to reciprocate the respect being offered to him. A juvenile will normally be interviewed in a noncustodial setting. He should not be interviewed in front of his peers because he will feel the need to maintain a hard-line stance with authority figures. Consider questioning a juvenile in his home if the parents are cooperative and willing to work with investigators. However, if the parents are making excuses for the juvenile without him offering any denials, investigators will have to gain the support of the parents before conducting an effective interview. To get the interview started, establish a bond of mutual interest, respect, and experience. Treat a juvenile with consideration and be friendly. When getting personal information, tell him why the information is needed. While obtaining information from a juvenile, the investigator may offer information about himself that will help the juvenile relate to him and keep the interview going.

4-116. An interview or interrogation of a juvenile by an investigator of the opposite sex should be conducted in the presence of a witness. When possible, the witness should be of the same sex as the juvenile being interviewed. If the juvenile to be interviewed is considered a suspect, the parent or guardian of the juvenile should be present during the rights advisement process. The juvenile should be advised of his rights against self-incrimination with the consent of the parent or guardian. Further, the juvenile should be advised that the interview could be terminated at anytime. In some cases, the presence of the parents may be beneficial to the interview process. However, in many instances, juveniles are reluctant to discuss misconduct in the presence of their parents, and consideration should be given to asking the parents to observe the interview in another room.

4-117. Avoid contacting juveniles at school, if possible. If circumstances require that interviews be conducted at a school, the investigator must observe certain precautions. Contact the school principal first and explain the circumstances related to the interview and why it must be conducted at school. Do not contact the student's teacher or the student first. Ask if there is a room in which the student can be interviewed. Never enter the classroom to question or apprehend a student. Do not contact students at school during hours when a number of other students may see investigators with the juvenile. Remove a juvenile from school only as a last resort. If a student is to be apprehended, investigators must explain this to the school officials and get permission to take the juvenile from the school, unless specific court instructions mandate otherwise.

4-118. Ensure that the juvenile's parents have been notified if the student must be removed from school. When an investigator detains or takes a juvenile into custody for an offense, the following actions should be taken:

- Immediately advise the juvenile of his rights in language that he understands.
- Promptly notify his parent or guardian and the local office of the SJA.

4-119. When juvenile subjects are in custody, they will not be fingerprinted or photographed without the written permission of a federal judge responsible for juvenile cases. This permission is required, regardless of consent by the juvenile, parent, or guardian.

INTERPRETERS

4-120. Investigators may have to question suspects who do not speak English. Although in an ideal scenario, the interrogator should be able to conduct the interview in the suspect's native language. This is not always possible or practical. An investigator should not attempt to conduct an interrogation by himself in a foreign language, unless he is fluent in the required language.

4-121. When using interpreters, there are several considerations that should be planned before the interview. It is best if the interpreter is a member of the US armed forces or is a US citizen. If this is not possible, a qualified local inhabitant can be used; however, it is essential that the interpreter being used is completely trustworthy and understands that he should not edit what the interrogator or the suspect says. Interpreters should be intelligent and articulate. They should be able to quickly learn the habits, methods, and ways of different investigators. They should be educated in the use of their own language and English. Social and educational levels are often discernible by speech habits or peculiarities. Interpreters must be able to express themselves clearly and intelligibly to all persons whom they are likely to question. They must be honest and free from criminal tendencies or affiliation. If they are native to the area, they should be free from unfavorable notoriety among the local populace. A good reputation and standing in the community prevents an interpreter from being intimidated by persons of higher rank or standing. They must be willing to accept a subordinate or supportive role in the actual questioning of a suspect, which means they must be willing to permit the investigator to ask the questions and to receive and evaluate the answers. Additionally, they should understand that they are not to ask questions on their own accord unless told to do so by the investigator. If the investigator has an understanding of the language being used in the interrogation, he will be better equipped to check the accuracy, loyalty, and obedience of the interpreter. If the investigator does not have any familiarity with the language used, it is good practice to have another linguist observe the interview and report their observations to the investigator.

INTERPRETERS TRANSLATING QUESTIONS AND ANSWERS

4-122. The interviewer asks the suspect the questions and the interpreter translates the questions as close to verbatim as possible. The interpreter then translates the answer back to the interviewer who evaluates it and formulates the next question. Investigators should prepare the questions they intend to ask before initiating the interview. Questions should be clear and concise and designed to invoke brief, factual answers. Direct the interpreter to make his translation of long statements at regular and convenient pauses during the suspect's utterances. The interruptions must come at the end of complete thoughts, but not after extended statements, as the interpreter will naturally abbreviate the response to only include those portions he believes are most relevant. This may be challenging if the subject is allowed to give extensive narrative versions of his account. Therefore, investigators should try to avoid questions that take long explanations and invite digressions.

4-123. Whenever possible, the investigator should give the questions to the interpreter in writing before the interview, so he can adequately prepare and clarify questioning approaches before the interview. Additionally, the interpreter may need to research specialized vocabulary in relation to the specific offense. The investigator should be frontally aligned with the suspect and look at him while issuing questions and listening to the responses. The interpreter should generally observe the suspect, but try to limit eye contact, which will encourage him to look at the investigator during the interview process.

4-124. The interpreter should stand or sit to the side and slightly forward of the investigator. This will allow the investigator to converse with the interpreter, but still portray to the suspect that the investigator is conducting the interview. The interpreter should not move around or do anything that will distract the suspect's attention from the investigator. The investigator should address the suspect directly, looking him in the eye to hold his attention. Ask questions slowly and clearly in concise and unambiguous English. Do not use slang words or expressions unique to a region, which could confuse the interpreter. Do not tell the interpreter to ask the suspect a question. For example, do not say to the interpreter, "Ask him if he knows John Doe." Instead, say the question directly to the suspect in English, "Do you know John Doe?" The interpreter translates the question into the language of the suspect. The interpreter should do this promptly in a clear and well-modulated voice. He should try to reproduce the tone and emphasis used by the investigator. When the suspect answers the question in his native language, the interpreter should repeat the answer in English, translating it as literally as possible. Insist that the interpreter translate the suspect's answer directly. Do not permit the interpreter to reply, "He says he does." Investigators should get the answer through the interpreter as though the suspect answered it in English, such as, "Yes, I know him," using the first person. If the investigator feels an explanation is warranted as to the word selection of the suspect, he should ask the interpreter at a later time. If fact clarification is required, additional questions should be posed directly to the suspect by the investigator.

4-125. Hold digressions to an absolute minimum. Do not allow the suspect and interpreter to begin an extensive conversation or argument. If the investigator wants to take detailed notes, he needs to instruct the interpreter to speak slowly and distinctly and to pause while notes are being recorded. Use of a stenographer or a recording device may be warranted and used within appropriate guidelines. If the stenographer speaks both languages, he should record all statements made in both languages. This gives a means of cross-checking the accuracy of the translation. The Status of Forces Agreement (SOFA) is different in every country and effects how local nationals are interviewed. Appropriate coordination must be made before conducting these interviews.

NATIONAL GUARD MEMBERS

4-126. Personnel who commit criminal acts while on duty as a member of the USNG may or may not be subject to the *UCMJ*. Therefore, before interviewing a soldier in the USNG, determine if he was activated under *Title 10, USC (10 USC)* or *Title 32, USC (32 USC)*. Whenever USNG personnel are deployed overseas, they must be activated under *10 USC*, which means they are subject to the *UCMJ*. Personnel serving in the continental United States (CONUS) are likely activated under *32 USC*, meaning they are not subject to the *UCMJ*. Before interviewing a USNG soldier, verify his status. The interviewer should be familiar with the appropriate elements of proof under the statute that the soldier can be charged with to determine if he may be interviewed in a noncustodial setting. Because USNG soldiers activated under *32 USC* are not subject to the *UCMJ*, they may be interviewed without giving the *UCMJ, Article 31(b)*, warnings required of investigators who are subject to the manual.

JOINT AND COLLATERAL CRIMINAL INTERVIEWS

4-127. The *Memorandum of Understanding (MOU) (January 1985)* between the DOD and the DOJ, was implemented by *DODD 5525.7* (which can be found in *AR 27-10*). The MOU defines the respective investigative responsibilities of the FBI and military criminal investigative organizations. It establishes three categories of coordination with the FBI to determine primary investigative responsibility in situations where dual responsibility exists. Local military police and USACIDC units will conduct investigations covered by those categories unless a US attorney directs otherwise or the FBI makes a positive assertion of its investigative authority. When the FBI undertakes an investigation in which there is an Army interest, it should be undertaken as a joint investigation with military police and USACIDC participation.

4-128. Questions concerning the types of coordination to be affected in any particular case should be referred to the major subordinate command office of the SJA. When investigative purview is in question, coordination of investigations with the FBI must be conducted before any interview. When conducting interviews or interrogations in conjunction with other investigative organizations of interest, conduct them according to the guidelines of the lead investigative organization. However, if such interviews (especially with foreign organizations) deviate from military guidelines, regulations, or laws, military investigators may not participate and may choose to completely disassociate from these activities even if they are legally permitted to do so for the other organization.

CONFESSION SUPPRESSION COUNTERMEASURE

4-129. In recent years, criminal defense attorneys and the military DC have adopted numerous standardized tactics designed to suppress confessions. An analysis of this trend indicates that confessions are very difficult obstacles for the DC to overcome with panel members. Especially when evidence and investigator testimony implicates the accused, along with a confession. Consequently, panel members tend to be more ready to convict suspects who admit to criminal misconduct.

4-130. In an attempt to gain acquittals, the DC will frequently attack the veracity of confessions so he may more effectively argue away other evidence presented during the trial process. The following are some examples of tactics used by the DC to impugn confessions: (1) coercive tactics were employed to gain a confession, (2) the suspect was denied his basic human rights, (3) the suspect was not allowed to eat, sleep, or take breaks, (4) the suspect did not actually read the statement, (5) the investigator put words into the suspect's statement that were not derived from the suspect, and (6) the investigators were overbearing and/or abusive.

4-131. Confession suppression countermeasures are designed to thwart future claims made by the DC. Many of these countermeasures have been discussed throughout this chapter.

4-132. One of the most basic countermeasures is to program easy to identify typographical errors within the body of the statement. When the suspect corrects the errors, it demonstrates that he read the statement and felt comfortable making changes to it before swearing that it was true and correct.

4-133. If the rapport-based interview style is used appropriately during interrogations, suspects will like, respect, and trust the investigator at the time of the interview. However, after speaking to counsel, the suspect will often claim to have been maltreated. Therefore, asking questions within the body of the statement designed to explore the suspect's perception of his treatment at the conclusion of the interview are very effective in refuting or preventing future claims of abuse or coercion. Examples are as follows:

- "How were you treated by CID and/or MPI today?"
- "Why did you decide to tell me the truth today?"
- "You were questioned for 10 hours today, were you allowed bathroom, smoking, or meal breaks?"

4-134. Suspects who are treated with dignity and respect will respond in a similar fashion. They will frequently answer with the following:

- "I was treated with respect."
- "You were very professional and nice."
- "You were courteous but persistent."

4-135. Although DC may try to attack sworn statements typed by the investigator, interview sketches, letters of apology, and handwritten statements will diminish future claims that investigators put words in the suspect's mouth or that the suspect did not read his statement.

4-136. Videotaping I&I can also be instrumental in disproving claims of coercion, overbearing investigators, and other alleged acts of misconduct on the part of investigators. In the rapport-based interview process, investigators are kind, caring, and compassionate toward suspects, which is in direct contradiction to claims of abuse. However, the DC will use television interview and interrogation tactics to reinforce these claims in the mind of the panel. Recording the interview can serve to impugn or prevent these claims from ever being made.

PART TWO

Crime Scene Processing and Documentation

Part Two describes numerous legal and scientific requirements concerning the collection, preservation, and documentation of evidence and crime scene processing. Guidance on many of the technical aspects of crime scene photography is in Part Two. The chapters of Part Two provide investigators and first responders with crime scene control measures and safety considerations.

Chapter 5

Crime Scene Investigations

There is no steadfast rule that can be applied to defining the dimensions of a crime scene. Areas that are well removed from the actual crime scene frequently yield important evidence. The scene of a crime is, in itself, evidence. Valuable physical evidence is normally found at or near the site where the most critical action was taken by the criminal against the victim or property. Just as it is likely to discover and develop critical evidence in the immediate area surrounding the body in a homicide, the site of forcible entry into a building often has the greatest potential for yielding evidence. Technological advances in forensic science, coupled with the ability of crime laboratories to successfully analyze evidence, have greatly increased the investigator's responsibility. Items not previously thought to be of evidentiary value may hold the key to the successful identification or elimination of a suspect or lead to the identification and conviction of the offenders. The investigator is responsible for conducting a thorough investigation of the crime scene and ensuring the proper collection, preservation, documentation, and transportation of evidence.

LEGAL AND SCIENTIFIC REQUIREMENTS

5-1. The legal requirements concerning evidence are met when the investigator can identify each piece of evidence, describe the exact location of the item, indicate when the item was collected, and maintain and show a proper chain of custody. The investigator or a trained crime laboratory technician must also describe changes that may have occurred in the evidence between the time it was collected and its introduction as evidence before the court.

5-2. An investigator's notes, photographs, and sketches become the legal record of his actions and aid immeasurably with introducing physical evidence in court. The written and photographic records of the scene also put the presented evidence into perspective and help laboratory experts determine what tests are needed. Finally, the investigator's notes, photographs, and sketches provide him with a valuable reference of his actions during the processing of the crime scene. Investigators must be thoroughly familiar with the techniques and procedures described in *Chapter 6*.

5-3. Crime scene processing in a deployed area can prove to be more challenging, simply because of the lack of support and a possible hostile environment. *Appendix B* contains a predeployment list which will assist in preparing agents for conducting criminal investigations in a forward-deployed area.

5-4. Scientific requirements in handling and processing evidence are met when the investigator takes every precaution to minimize any change or modification to the evidence. Biological materials always undergo some change. Weather and other unavoidable circumstances may induce change in some types of materials. The use of clean, spill-proof containers and proper packing materials reduce spillage, evaporation, or seepage of a sample or alteration of an item by accidentally scratching or bending it.

FIRST MILITARY POLICE RESPONDER

5-5. It is important for the first responder at the scene to be observant when approaching, entering, and exiting a crime scene. He must engage every sense to detect and determine the presence of potential threats, such as people, chemicals, or explosive devices.

INITIAL RESPONSE

5-6. One of the most important aspects of securing the crime scene is to preserve the scene with minimal contamination and disturbance of physical evidence. The initial response to an incident should be expeditious and methodical. Upon arrival, military police should assess the scene and treat the incident as a crime scene.

5-7. Military police shall promptly, yet cautiously, approach and enter a crime scene, remain observant of any people, vehicles, events, potential evidence, and environmental conditions. Military police will—

- Note or log dispatch information, such as the address or location, time, date, type of call, and parties involved.
- Be aware of any individuals or vehicles leaving the crime scene.
- Approach the scene cautiously, scan the entire area to thoroughly assess the scene, and note any possible secondary crime scenes. Look for any individuals and vehicles in the vicinity that may be related to the crime.
- Make initial observations (look, listen, and smell) to assess the scene and ensure officer safety before proceeding.
- Remain alert and attentive. Assume that the crime is ongoing until it is determined otherwise.
- Treat the location as a crime scene until it has been assessed and determined to be otherwise.

SAFETY PROCEDURES

5-8. The safety and physical well-being of military police and other individuals, in and around the crime scene, is the first priority of the first responding military policeman. The control of physical threats will ensure the safety of military police and others that are present. Military police arriving at the scene should identify and control any dangerous situations or individuals. They should do the following:

- Ensure that there is no immediate threat to other responders by scanning the area for sights, sounds, and smells that may be dangerous to personnel, such as HAZMAT, gasoline, and natural gas. If the situation involves a clandestine drug laboratory, biological weapons, or weapons of mass destruction (WMD), contact the appropriate personnel or agency before entering the scene.
- Approach the scene in a manner designed to reduce the risk of harm to other military police while maximizing the safety of victims, witnesses, and others individuals in the area.
- Survey the scene for dangerous individuals and control the situation.
- Notify supervisory personnel and call for backup and other appropriate assistance.

EMERGENCY CARE

5-9. After controlling any dangerous situations or individuals, the next responsibility of the first responding military policeman is to ensure that medical attention is provided to injured individuals while minimizing contamination of the scene. Assisting, guiding, and instructing medical personnel during the care and removal of injured individuals will diminish the risk of contamination and loss of evidence. The first responding military policeman should ensure that medical attention is provided with minimal contamination of the scene. The first responding military policeman should do the following:

- Assess the victim(s) for signs of life and medical needs and provide immediate medical attention.
- Call for medical personnel.
- Guide medical personnel to the victim to minimize contamination or alteration of the crime scene.
- Point out potential physical evidence to medical personnel, and instruct them to minimize contact with the evidence. For example, ensure that medical personnel preserve all clothing and personal effects without cutting through bullet holes or knife tears, and document any movement of individuals or items by medical personnel.
- Instruct medical personnel not to "cleanup" the scene and not to remove or alter items originating from the scene.
- Obtain the name, unit, and telephone number of attending personnel and the name and location of the medical facility where the victim was taken if medical personnel arrived first.
- Attempt to obtain a "dying declaration" if there is a chance that the victim may die.

- Document any statements or comments made by the victims, suspects, or witnesses at the scene.
- Send a military policeman with the victim or suspect to document any comments made, and preserve evidence if the victim or suspect is transported to a medical facility. If military police are not available to accompany the victim or suspect, stay at the scene and ask medical personnel to preserve evidence and document any comments made by the victim or suspect.

CRIME SCENE SECURITY

5-10. Controlling, identifying, and removing individuals at the crime scene are important functions of the first responding military policeman in protecting the crime scene. Controlling the movement and limiting the number of persons who enter the crime scene are essential in maintaining scene integrity, safeguarding evidence, and minimizing contamination. Military police should—

- Restrict individuals' movement and activity to maintain safety at the scene and prevent them from altering or destroying physical evidence.
- Identify all individuals at the scene.
  - Suspects. They must secure and separate suspects from one another.
  - Witnesses. They must secure and separate witnesses from one another to prevent collaboration of their stories.
  - Bystanders. They must determine if bystanders are witnesses. If they are witnesses, treat them as stated above; if not, remove them from the scene.
  - Victims, family, and friends. They must control these individuals while showing compassion.
  - Medical and other assisting personnel.

5-11. Military police must exclude unauthorized and nonessential personnel from the scene. This includes military police not working the case, politicians, media, and so forth.

BOUNDARY ESTABLISHMENT

5-12. Identifying perimeters assists in the protection of the scene. Establishing the boundaries is a critical aspect in controlling the integrity of evidentiary material. The perimeters of a crime scene may be easily defined when an offense is committed within a building. When the perpetrator takes the fruits of his crime away from the crime scene to another area, the perimeter becomes more difficult to establish. If an offense is committed in the open, such as in the desert, the perimeter of the offense becomes even more vague. Regardless of the ease in which perimeters may be established, this is a measure that must be determined at the beginning of the investigation. The location and types of crime determine the number of crime scenes and the boundaries. Boundaries are established beyond the initial scope of the crime scene with the understanding that the boundaries can be reduced in size when deemed necessary, but cannot be easily expanded.

5-13. The first responding military policeman at the scene should conduct an initial assessment to establish and control the crime scene and its boundaries. Military police should—

- Establish the boundaries of the scene. Start at the focal point and extend outward to include the following:
  - Location where the crime occurred.
  - Potential points and paths of the exit and entry of suspects and witnesses.
  - Places where the victim or evidence may have been moved. Military police must be aware of and protect trace and impression evidence while assessing the scene.
- Set up physical barriers to include ropes, cones, crime scene barrier tape, available vehicles, personnel, or other equipment. Existing boundaries, such as doors, walls, and gates can be used.
- Document the entry and exit of all individuals entering and leaving the scene once the boundaries have been established. Maintain a log of each individual's name, rank, SSN, unit, and telephone number.
- Control the flow of individuals and animals entering and leaving the scene to maintain the integrity of the scene.
- Implement measures to preserve and protect evidence that may be lost or compromised due to the elements (rain, snow, or wind) or footsteps, tire tracks, sprinklers, and so forth.
- Document the original location of the victim or objects that are being moved. If possible, insist that the victim and objects remain in place until the arrival of the investigation team.
- Consider search and seizure issues to determine the necessity of obtaining consent to search and/or obtain a search warrant. The investigation team will generally perform this measure.
- Ensure that no one—
  - Smokes.
  - Chews tobacco.
  - Uses the telephone or bathroom.
  - Eats or drinks.
  - Moves any items (including weapons) unless it is necessary for the safety and well-being of individuals at the scene.
  - Adjusts the thermostat or opens windows or doors (maintain the scene as it was found).
  - Touches anything unnecessarily (note and document any items moved).
  - Repositions moved items, litters, or spits within the established boundaries of the scene.

DOCUMENTATION OF ACTIONS AND OBSERVATIONS

5-14. The first responding military policeman must produce clear, concise, documented notes attesting to his actions and observations at the crime scene. However, all military police must record their activities at a crime scene and note any observations they make as soon as possible after the event occurs to preserve information. These notes, especially those of the first responding military policeman, are vital in providing information to substantiate investigative considerations. All notes must be maintained as part of the permanent case. The first responding military policeman should record the following:

- Observations of the crime scene, to include the location of persons and items within the crime scene.
- The appearance and conditions of the crime scene upon arrival, such as smells, ice, liquids, movable furniture, weather, temperature, or personal items and whether the lights are on or off; the shades are up, down, open, or closed; or the doors and windows are open or closed.
- Personal information, statements, or comments from witnesses, victims, and suspects.
- The actions of others and any of his own actions.

5-15. Once the investigation team arrives to take control of the scene, the first responding military policeman briefs the investigator in charge and relinquishes responsibility of the scene. The military policeman then assists the investigation team as necessary.

INITIAL NOTIFICATION TO THE INVESTIGATOR

5-16. Upon initial notification of an investigation, the investigator begins making crime scene notes. The investigator annotates the date and time of notification and how the complaint was received. He also obtains full identification of the reporting party and the details pertaining to the initial discovery. He must ask the following six "W" and "H" questions:

- "Who is involved (subject, victims, and witnesses)?"
- "What occurred?"
- "When did the incident occur?"
- "Where did the incident occur?"
- "Why this type of crime, victim, and location?"
- "How was the crime committed?"

5-17. The investigator determines the route of response and, upon arrival, parks the vehicle in a safe place. The first investigator on the scene approaches with caution to consider his safety.

INVESTIGATION TEAM ARRIVAL

5-18. This section covers the use of the investigation team concept. When an investigator is the only one to respond to a scene, he must make adjustments to these procedures to ensure that no steps are missed.

CONDUCT A SCENE ASSESSMENT

5-19. A scene assessment allows the investigator in charge to determine the type of incident to be investigated and the level of investigation to be conducted. The investigator can develop a plan for the coordinated identification, collection, and preservation of physical evidence and for the identification of witnesses. During the scene assessment, an exchange of information between law enforcement personnel can occur. The investigator in charge identifies specific responsibilities, shares preliminary information, and develops investigative plans according to CID policy; federal, state, and local laws; and the laws of the HN when applicable. The investigator in charge should—

- Talk with the first responding military policeman regarding his activities and observations.
- Evaluate safety issues, such as blood-born pathogens and chemical hazards that may affect personnel entering the scene.
- Evaluate search and seizure issues to determine the necessity of obtaining consent to search and the requirement of obtaining a search warrant.
- Evaluate and establish a path of entry and exit to the scene to be used by authorized personnel, such as medical, HAZMAT, and other law enforcement agencies, if not already accomplished by the first responders.
- Evaluate the initial scene boundaries.
- Determine and prioritize the number and size of the scenes.
- Establish a secure area within close proximity to the scenes for the purpose of consultation and equipment staging.
- Establish and maintain communication with personnel when multiple scenes exist.
- Establish a secure area for temporary evidence storage according to the rules of evidence and chain of custody.
- Determine and request additional investigative resources as required, such as personnel, specialized units, legal consultation, prosecutors, and equipment.
- Ensure continued scene integrity by documenting the entry and exit of authorized personnel and other related actions and by preventing unauthorized access to the scene.
- Ensure that witnesses to the incident are identified and separated. Obtain a valid identification from each.
- Ensure that the surrounding area is canvassed and the results documented.
- Ensure that the preliminary documentation, such as notes, photographs, and sketches of the scene, is complete. The initial photographs should include any injured individuals and vehicles involved in the incident.

CONDUCT THE PRELIMINARY SURVEY OF THE CRIME SCENE (SCAN THE SCENE)

5-20. The investigator should consider the crime scene unstable and fragile in the sense that the evidentiary value of items can be easily degraded or lost. Normally, there is only one opportunity to search the scene properly;

therefore, it is important that it is done right. Conduct the preliminary survey by—

- Performing a cautious "walk-through" of the scene.
- Developing a general theory of the crime.
- Recording the general condition of the scene.
- Taking preliminary photographs and making videos.
- Determining the extent of the search area.
- Identifying and protecting fragile physical evidence.
- Determining personnel and equipment needs.

5-21. If it is determined that the search could be lengthy, set aside an area that is close (but outside the critical area) to use as a collection point for trash generated during the search. Equipment not in immediate use should be placed in this area. This area is also the designated break area for police personnel. Using this type of area reduces the chance of contaminating the scene.

5-22. By the end of the preliminary survey of the scene, the investigator should have noted what obvious items of evidence should be collected. He must decide the processing order and collect them. The order of collection is predicated on how best to move through the crime scene without destroying the value of the evidence. Some forms of evidence are easily collected, such as a beer can or broken padlock. Others, such as a handprint on a sheetrock wall, may require assistance from an outside agency. If this is the case, protect the item of evidence (for example, with a plastic dome covering), check the item for trace evidence; secure the covering over the surface containing the evidence with evidence tape or paper-packaging tape; and note the time, date, and initials of the investigator. Note the time, method of protection, and the results of the check for trace evidence in the crime scene notes. The time and method of protection must also be reflected in the evidence control document when collecting the item.

5-23. If the scene is very large or more than one individual is going to search, decide what should be searched and how the tasks and the area is going to be divided. If the search must extend beyond the immediate crime scene, use military police personnel for additional assistance.

CONDUCT A SCENE "WALK-THROUGH"

5-24. The scene "walk-through" provides the investigator in charge with an overview of the entire scene. The walk-through provides the first opportunity to identify valuable and fragile evidence and determine initial investigative procedures. It provides a systematic examination and documentation of the scene. The walk-through also identifies any threats to scene integrity. The written and photographic documentation of the investigation team, along with the notes of the first responding military policeman, provide a permanent record of the crime scene. The investigator in charge conducts a walk-through of the scene with the individuals who are responsible for processing the scene. During the scene walk-through, the investigator in charge—

- Uses the established path of entry to avoid contaminating the scene.
- Prepares the preliminary documentation (notes, photographs, and sketches) of the scene as it is observed.
- Identifies and protects fragile and perishable evidence, considering elements such as wind, rain, and snow. He also considers crowds, hostile environments, and so forth.

EVALUATE PHYSICAL-EVIDENCE POSSIBILITIES

5-25. Evidence evaluation begins upon arrival at the scene and becomes detailed based on observations from the preliminary survey. The investigator determines what evidence is likely to be present. He concentrates on the most transient evidence and works to the least volatile material.

5-26. The investigator focuses his efforts on easily accessible areas in open view and progresses eventually to possible out of view locations. He considers whether the evidence appears to have been moved inadvertently and evaluates whether the scene and evidence appears to be intentionally contrived.

INITIATE THE NARRATIVE

5-27. The investigator starts a narrative to record his findings. A narrative is a running description of the crime scene and should include the following:

- The case identifier number.
- The date, time, and location.
- The weather and lighting conditions.
- The condition and position of the evidence.
- The assignments of support personnel.

5-28. Nothing is insignificant to record if it catches the investigator's attention. Under most circumstances, the investigator does not collect evidence during the recording of the narrative, but rather, he makes photographs and sketches to supplement, not substitute for, the narrative.

CRIME SCENE PROCESSING

5-29. When processing the crime scene, the investigator in charge has multiple responsibilities. He must determine team composition, establish the requirements for the contamination control of the evidence, verify the requirements for scene documentation, conduct the search, and prioritize the evidence collection. The investigator enforces the collection, preservation, inventorying, packaging, transporting, and submission of evidence. As a result of the scene assessment that the investigator in charge conducted earlier, he can determine the number of personnel required and how responsibilities will be assigned.

DETERMINE THE TEAM COMPOSITION

5-30. Based on the type of incident and the complexity of the scene, the investigator in charge determines the team composition and decides what specialized resources are required. Only trained personnel will perform scene processing. Following the walk-through, the investigator in charge should do the following:

- Assess the need for additional personnel. Additional personnel may be needed in cases with multiple scenes, victims, witnesses, or other circumstances.
- Assess forensic needs and request the assistance of forensic specialists at the scene for their expertise and/or equipment.
- Ensure that scene security and entry and exit documentation continue.
- Select qualified personnel to perform specialized tasks, such as taking photographs, making sketches, lifting latent prints, and collecting evidence.
- Document team composition by name and assignment.

MINIMIZE CONTAMINATION

5-31. The investigator in charge minimizes the contamination of evidence by implementing safe, clean, and careful procedures. These procedures are implemented to ensure the welfare of all individuals at the scene and maintain the integrity of the evidence. Other team members to include investigators and other responders should do the following:

- Limit access to the scene to those individuals directly involved in scene processing.
- Follow established entry and exit control routes at the scene.
- Identify the first responders and collect elimination samples from them.
- Designate a secure area for trash and equipment.
- Use personal protective equipment (PPE) to prevent contamination of personnel and minimize scene contamination.
- Clean, sanitize, or dispose of tools, equipment, and PPE between evidence collections and scenes.
- Use single-use equipment when directly collecting biological samples.

DOCUMENT THE SCENE

5-32. A well-documented scene ensures the integrity of the investigation and provides a permanent record for later evaluation. It also assists in the final written report and court testimony. A team member should do the following:

- Review the assessment of the scene to determine what type of documentation is needed.
- Coordinate the notes, photographs, sketches, videos, and measurements.

- Take photographs of the following:
  - The scene using overall, medium, and close-up coverage.
  - The evidence collected with and without a measurement scale and/or other evidence identifiers.
  - The victims, suspects, witnesses, crowds, and vehicles.
  - Any additional perspectives, such as aerial shots and the area under the body once it is removed.

5-33. A team member may also videotape the crime scene as an optional supplement to the photographs. Additionally, a team member prepares the preliminary sketches and measures the following:

- The immediate area of the scene, noting case identifiers (numbers) and indicating which way is north on the sketch.
- The relative location of items of evidence and the correlation of the evidence items with the evidence records, such as the log and *DA Form 4002 (Evidence/Property Tag).*
- The evidence before it is removed.
- The rooms, furniture, and other objects.
- The distance to adjacent buildings or other landmarks.

5-34. The same team member, or another, generates the investigator's notes at the scene being sure to document circumstances that deviate from usual procedures. (See *Chapter 6* for investigator's notes.)

CONDUCT THE SEARCH

5-35. A crime scene search is a planned, coordinated, and legal search conducted by law enforcement officials to locate physical evidence. The search is conducted in an efficient, systematic manner and covers the entire crime scene area. The method used depends on the nature of the scene and the characteristics of the crime. When an investigator decides his approach to the crime scene search, he—

- Establishes the perimeters of the crime scene search area.
- Determines the search patterns.
- Briefs the personnel.
- Coordinates information.
- Terminates the search.

Establish the Perimeters of the Crime Scene Search Area

5-36. The perimeters of a crime scene may be easily defined when an offense is committed within a building. When the perpetrator takes the fruits of his crime away from the crime scene to another area, the perimeter becomes more difficult to establish. If an offense is committed in the open, such as in the desert, the perimeter of the offense becomes even more vague. Regardless of the ease with which perimeters may be established, this is a measure that an investigator must determine at the beginning of the investigation. Identifying the perimeters assists in the protection of the scene.

Determine the Search Patterns

5-37. A competent search of a crime scene demands close attention to detail. Items and materials that may seem unimportant at first may later prove to be critical to the case. For this reason, begin the search of a crime scene with determination and alertness.

5-38. A successful crime scene search produces a comprehensive and nondestructive accumulation of all available physical evidence within a reasonable period of time. The search should minimize the movement and avoid the unneeded disturbance of evidence. Conduct a crime scene by using one or more of the following methods:

- Circle search.
- Strip search.
- Grid search.
- Zone or sector search.

5-39. The search method is determined by the intent of the search and the area to be covered. In rooms, buildings, and small outdoor areas, a systematic circle search is often used. In large outdoor areas a grid search is most useful. After mentally dividing the area into strips about 4 feet wide, the searcher begins at one corner of the main area and moves back and forth from one side to the other, each trip being made within one strip. Then the searcher moves from end to end. Both indoor and outdoor areas may be searched using the zone or sector method. Refer to *Figures 5-1, 5-2, 5-3.*

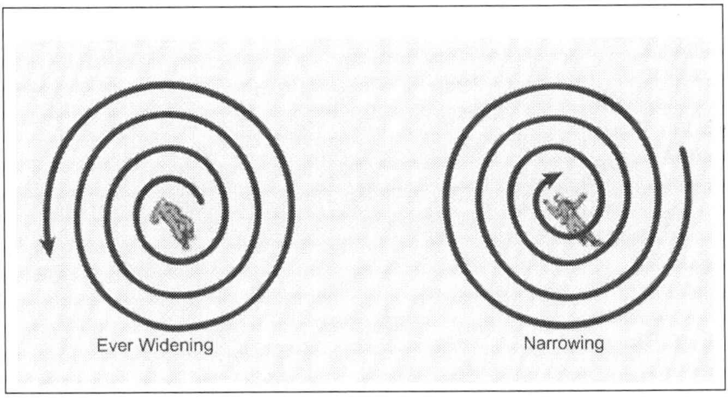

Figure 5-1. The Circle Search Method

Brief the Searchers

5-40. The searchers must be thoroughly briefed. They must be provided with a full description of the evidence being looked for. Tell the searchers that the evidence may have been hidden or discarded and what to do when they find a

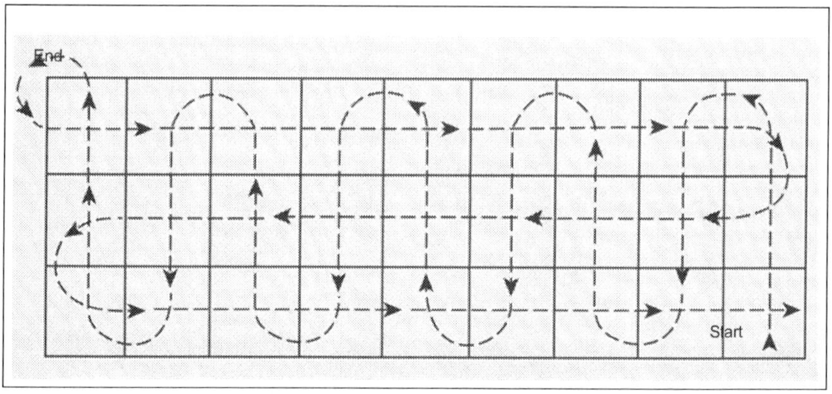

Figure 5-2. The Grid Search Method

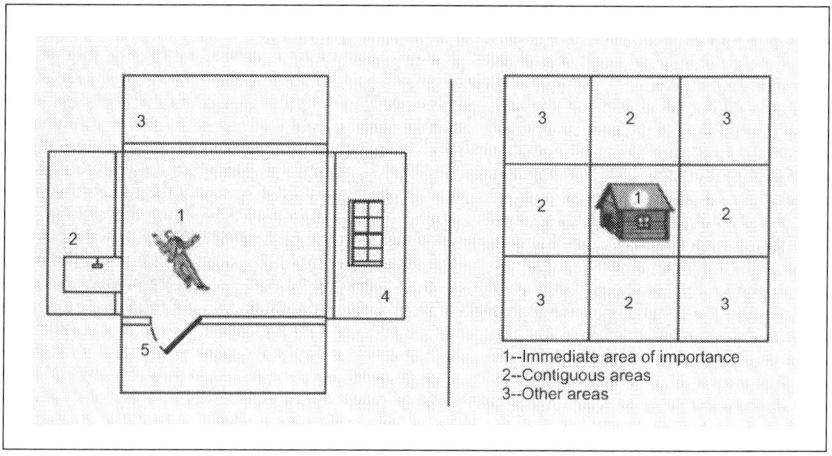

Figure 5-3. The Zone or Sector Search Method

piece of evidence or an item believed to be the one being looked for. They must—

- Not touch or move the item.
- Immediately report it to the person in charge of the search.
- Protect the area until an investigator arrives.

Coordinate Information

5-41. The investigator in charge of a crime scene must integrate the investigative efforts and findings with the requirements and findings of lab examiners, morticians, and other technicians. The coordination involves the lateral sharing of information discovered as it pertains to a crime. This is particularly true of technicians on the scene. A delay in the sharing of information may affect the successful recovery of physical evidence. Likewise, a delay in providing information to investigators may delay the apprehension of a suspect.

Terminate the Search

5-42. An investigator must determine when to terminate the search of a scene. The extent and nature of the scene impacts the amount of time required on that scene. Once an investigator believes that a scene has been fully examined, he can terminate the search. If the investigator must temporarily stop the search, perhaps to protect fragile evidence, he must secure the scene.

PRIORITIZE THE COLLECTION OF EVIDENCE

5-43. Prioritize the collection of evidence. The timely and methodical preservation and collection of evidence prevents its loss, destruction, or contamination. The investigator in charge and team members determine the order in which to collect evidence. Team members should—

- Conduct a careful and methodical evaluation considering all physical evidence possibilities, such as biological fluids, latent prints, and trace evidence.
- Focus first on the easily accessible areas that are in open view and proceed to the out-of-view areas.
- Select a systematic search pattern to collect evidence based on the size and location of the scenes.
- Select a progression of processing and collecting methods so that initial techniques do not compromise subsequent methods. The team members must concentrate on the most transient evidence and work toward the least transient. They must move from the least intrusive to the most intrusive processing and collection methods.
- Assess environmental and other factors that may affect the evidence.
- Be aware of multiple scene victims, suspects, witnesses, vehicles, and locations.
- Recognize other methods that are available to locate, technically document, and collect evidence, such as an alternate light source, blood pattern documentation, and projectile trajectory analysis.

PROCESS CRIME SCENE EVIDENCE

5-44. The integrity of a crime scene should always be maintainted.Evidence at crime scenes should be protected from contamination or other harmful change. During the processing of the scene, evidence should be appropriately packaged, labeled, and maintained in a secure, temporary manner until final packaging and submission to a secured evidence storage facility or the

USACIL. See *Chapter 19* for details on secure storage of evidence. The handling of physical evidence is one of the most important factors of the investigation; therefore, the team members must—

- Maintain scene security throughout the processing and until the release of the scene.
- Document the collection of evidence by recording and photographing its location at the scene, and by recording the name of the person who collected it and the date.
- Collect each item identified as evidence.
- Establish the chain of custody for each piece of evidence.
- Obtain standard or reference samples from the scene.
- Obtain control samples.
- Obtain elimination samples.
- Secure electronically recorded evidence immediately, such as answering machine tapes, surveillance camera videotapes, and computers from the vicinity.
- Identifying and securing evidence in containers at the crime scene. This requires a team member to label, date, and initial the container. The case number should be annotated, if known, when marking the container. Different types of evidence require different types of containers, such as porous, nonporous, and crushproof.
- Package items to avoid contamination and cross-contamination. (See *Chapter 19*).
- Document the condition of firearms and weapons before rendering them safe for transportation and submission.
- Avoid excessive handling of evidence after it is collected.
- Maintain evidence at the scene in a manner designed to diminish degradation or loss.
- Transport and submit evidence items for secure storage. (See *Chapter 19*).

CRIME SCENE INVESTIGATION DOCUMENTATION AND COMPLETION

5-45. The investigator in charge establishes a crime scene debriefing team, performs a final survey of the crime scene, and compiles the crime scene case file to complete the investigation. The investigator in charge oversees the debriefing team and the compilation of the case file while verifying accuracy and completion.

ESTABLISH A CRIME SCENE DEBRIEFING TEAM

5-46. The crime scene debriefing is the best opportunity for law enforcement personnel and other responders to ensure that the crime scene investigation is complete. It enables law enforcement personnel and other responders to share information regarding particular scene findings before releasing the scene. It provides an opportunity for input regarding a follow-up investigation, special requests for assistance, and the establishment of postscene responsibilities.

When participating in a crime scene debriefing, law enforcement personnel and other responders should—

- Ensure that the crime scene debriefing team includes the investigator in charge of the crime scene, other investigators and evidence collection personnel, such as photographers, evidence technicians, latent print personnel, specialized personnel, and the first responding military policeman (if still present).
- Discuss what evidence was collected.
- Discuss preliminary scene findings with the team members.
- Discuss potential, technical forensic testing and the sequence of tests to be performed.
- Initiate any actions identified during discussions that are required to complete the crime scene investigation.
- Brief the person or persons in charge upon the completion of the assigned crime scene tasks.

CONDUCT A FINAL SURVEY OF THE CRIME SCENE

5-47. Conducting a final survey of the crime scene ensures that all evidence has been collected and the scene has been completely processed before release. The investigator in charge directs a walk-through at the conclusion of the crime scene investigation and ensures that the scene investigation is complete. He ensures that—

- Each area identified as part of the crime scene is visually inspected.
- All evidence collected at the scene is accounted for.
- All equipment and materials generated by the investigation are removed.
- Any dangerous materials or conditions are reported and addressed.
- The crime scene is released according to jurisdictional requirements.

COMPILE A CRIME SCENE CASE FILE

5-48. Reports and other documentation pertaining to the investigation are compiled into a case file by the investigator in charge of the crime scene. This file is a record of the actions taken and of the evidence collected at the scene. This documentation allows for independent review of the work conducted. The investigator in charge obtains the following for the crime scene case file:

- First responding military policeman's documentation.
- Emergency medical personnel documents.
- Entry and exit control logs.
- Photographs and videos.
- Crime scene sketches and diagrams.
- Evidence documentation.
- Documentation of other responders.
- Record-of-consent form or search warrant.
- Reports, such as forensic or technical reports. They should be added to the file when they become available.

Chapter 6

Notes, Photographs, and Sketches

An investigator should know the requirements necessary to document a crime to include notes, photographs, and sketches. He must be skillful in developing these tools. He must consider himself the "artist" of the crime scene because all three of these tools are necessary to successfully reconstruct the scene. The better an investigator is in the development of these skills, the better the opportunity for solving a crime and convicting the perpetrator.

NOTES

6-1. Taking good notes may be one of the most important actions an investigator does while responding to and processing a crime scene. There are two main reasons why notes are important. They represent the basic source of information, or the raw material that will be used in the written report of the investigation, and they aid an investigator in the recall of events for testimony in court.

6-2. The type of notebook an investigator uses may seem to be immaterial, but it can be of significant importance. Unless a separate notebook is used for each case, a loose-leaf notebook is best for taking notes. A notebook may be examined in court. If notes from several cases are included in the same book, there is a chance of unauthorized disclosure of information on matters not being dealt with in the case being heard. If a loose-leaf notebook is used, the pages on other cases can be removed. Unauthorized disclosure of facts related to other cases should be avoided.

6-3. An investigator should place his name and pertinent business information on the inside cover of the notebook in case it is misplaced. At no time should an investigator annotate his home address or phone number in the notebook. The investigator's objective should be to make notes that will be meaningful months after the event. Notes that are unclear shortly after they are written may later be unintelligible. Investigators should never expect to rely on their memory of associated events to give single-word notes full meaning. This becomes especially important when the investigator has to write the report or testify in court. The formal written report may not require the level of detail that is required for a testimony. The details recorded in an investigator's notes should anticipate both the needs of the written report and the questions that an investigator may be called on to answer in court.

6-4. In cases involving a lot of physical material and a large crime scene, an investigator may want to use a portable audio and/or video device. By taping observations and findings, more details can be included in the notes. In all cases, the tapes should be transcribed into a written record that may be carried into court. An investigator should—

- Keep his notes in a safe place with the local office case file. Even after a criminal has been convicted and sent to prison, there is always a chance that an appeal or another civil action will require an investigator's appearance in court again.
- Print his notes if his handwriting is not easy to read. Use blue or black ink that will not smudge easily.
- Number each page of his notes and write his name, title or rank, the case number (when known), and the current date on each page.
- Record the time when an action is taken, information is received, and an event is observed.
- Not edit or erase notes. Line out an entry if a mistake is made, initial it, and then write the correct information.
- Include a detailed description of the scene and any item believed to be pertinent to the case. The description should be as complete as possible.
- Record the exact location (the measurements and triangulation of evidence) of where an item was found.
- Cite the relative distances separating various items.
- State the techniques used to collect evidence, and record identifying marks placed on the item or the package in which the evidence was placed.
- Tell what techniques were used to provide crime scene security and the methods used to search the scene.
- Include any actions taken that may have a bearing on the evidence obtained or that may significantly affect the investigation.

PHOTOGRAPHS

6-5. One of the most valuable aids to a criminal investigation is quality photographs. When properly taken, photographs and a photograph log supplement notes and sketches, clarify written reports, provide identification of personnel, and are a permanent record of fragile or perishable evidence. Photographs are of great value during every phase of the investigation and any subsequent judicial proceedings. They may also be of great value if a crime scene has to be reconstructed or questions arise concerning the sequence of events, such as who was present during the initial investigative phase or how did particular items of evidence appear at that time. Photographs may be a judge's or jurors' only view of a crime scene and prove most useful when they closely duplicate the exact scene as seen by the investigator.

6-6. Preserving and safeguarding images is important to the successful introduction of forensic images in a court of law. The original image should be stored and maintained in an unaltered state. This includes maintaining the original digital images. The duplicates or copies should be used for working images when applicable. Keep registered mail receipts and copies of work orders for film processing in the case file.

6-7. The following medias are recommended for the preservation of original images because of their quality, durability, permanence, reliability, and ease with which copies may be generated:

- Silver-based film negatives in 35-millimeter or larger format.
- Write-once compact disk recordable (CDR).
- Digital versatile disk recordable (DVD-R).

6-8. Time is an important consideration when photographing the crime scene. Fragile trace evidence (such as tire tracks in rain or fresh snow) that is subject to damage or destruction should be photographed before any other crime scene processing is conducted.

CAMERAS

6-9. All cameras, whether they are film-based or digital, share common characteristics. They all have a lens to form the image, some ability to focus the image, and a place in a light tight box to hold film. All cameras work on the same principle; light enters the lens and strikes either the film or the digital chip. In the case of film, the silver halide molecules in the film are "exposed." This makes some areas darker on the negative than other areas. The resulting negative can be printed. In the case of a digital camera, the light energy is converted to electrical energy and the information is converted to a usable computer-based file. The main components of the typical camera are the lens, camera body, exposure meter, and film.

LENS

6-10. The lens is usually made of glass and metal or plastic components. The lens is used to form an image on the film or chip. A 50-millimeter lens is considered a normal lens for 35-millimeter photography. This means that the 50-millimeter lens approximates what the eye sees when you look at a scene or object. There is less apparent distortion with the 50-millimeter lens than with a wide-angle (28-millimeter) or a telephoto (200-millimeter) lens. The lens usually has the following two aspects that require input from the photographer:

- **Focus.** Turning the front ring on the barrel of the lens changes the focus. Point-and-shoot cameras sometimes have a fixed focus, where anything from 8 feet to infinity is in the acceptable focus. Other point-and-shoot cameras may have a variable or automatic focus. Single-lens reflex (SLR) cameras have a variable or automatic focus. The ability to precisely control focus is one of the main benefits of the SLR.
- **Aperture or F-stop.** Most cameras have an aperture range from F-2.8 to F-22. The aperture should be viewed as a fraction of the length of the lens. A good rule of thumb is to make the aperture a fraction. Instead of F-16, view it as 1/16. Instead of F-2.8, view it as 1/2.8. That will assist in visualizing why F-16 is a smaller F-stop than F-2.8. The aperture controls the amount of light entering the lens/camera. A small aperture such as F-16 or F-22 allows less light to enter the camera over a given time than a large aperture such as F-2.8. The

aperture also controls the depth of field. The depth of field is the range at which objects being photographed appear sharp in the photograph. At F-16, a sharp focus may be obtained over a range from 3 to 100 feet. At F-2.8, sharpness may be obtained only from 3 to 5 feet. The larger the F-stop (F-2.8), the less depth of field. This becomes critical when doing close-up photography because the closer the object, the depth of field is less.

CAMERA BODY

6-11. The camera body is nothing more than a lighttight box that holds the lens on one end and the film on the other end. In the case of modern cameras, various mechanical and electrical devices and controls allow adjustments. Modern cameras also have shutters that allow the light to strike the film for a specific period of time. The main adjustment on the camera body that requires concern is the shutter speed. All modern cameras allow the user to adjust the speed or the duration or time the shutter stays open. Most cameras allow shutter speeds from 8 seconds to 1/4000 second. This is important because it is impossible to handhold a camera at a slow shutter speed and get a sharp print due to camera movement. The basic rule of thumb is to use the lens length to determine the lowest possible shutter speed that a camera can safely be handheld. With a 50-millimeter lens, a camera should not be handheld at a shutter speed that is less than 1/60 second. With a 200-millimeter lens, a camera should not be handheld at a shutter speed less than 1/250 second.

EXPOSURE METER

6-12. Exposure is determined by the amount of light falling on the object being photographed. A bright sunny day requires less exposure than a rainy, cloudy day. The meter in the camera automatically adjusts the camera exposure to provide acceptable pictures in most cases. The two adjustments that can be made are the aperture and shutter speed. They are interrelated and a change in one affects the other. If the exposure value is 10, the camera meter might give a proper exposure of F-8 at 1/60 second. If a greater depth of field were needed, the aperture would be changed, increasing the aperture from F-8 to F-16. The automatic exposure meter in the camera would automatically slow down the shutter speed to 1/15 second to maintain the correct exposure. This would cause a problem. The camera could no longer be handheld with a normal 50-millimeter lens and get sharp photographs. Therefore, a tripod must be used. The general rule is the slower the shutter speed, the smaller the aperture opening and the faster the shutter speed, the larger the aperture opening.

FILM

6-13. The selection of film is very important. Films come in various speeds. A film with a slow speed, such as 100 International Organization for Standardization (ISO), will have a finer-grain structure and better resolution than a faster film, such as 400 ISO. This allows photographs to be taken with greater resolution and detail. A faster film allows for a greater depth of field and works better in low-light conditions. However, the faster film has larger grains of silver embedded in the emulsion. While that makes the film more sensitive to light, it increases the grain structure and causes lower resolution. An investigator might not wish to use the faster film if the photographs are going to be enlarged to show fine details, such as those found in tire tracks, footwear, or fingerprints.

6-14. The following are three general types of film:

- **Color negative.** Used for most general and all violent crime scene photographs. Most color negative film has less contrast and resolving ability than black and white film. It generally comes in speeds of 100 ISO to 800 ISO. While the 800 ISO film may allow the camera to be handheld without a flash or tripod, the grain structure is large and results in low resolution (less detail).
- **Color reversal.** Also known as slide film and is used only when color slides for projection or presentation purposes are desired. Color reversal film is not to be used for crime scenes because color prints made from color "slides" do not have good resolution and generally have increased contrast levels.
- **Black and white film.** Used when high-resolution photographs are needed. Black and white film should be used when photographing items that may require examination at the laboratory. Tire tracks, footwear, and latent fingerprints should be photographed with black and white film. Black and white film is usually found in speeds of 100 ISO to 400 ISO.

CRIME SCENE PHOTOGRAPHY

6-15. The crime scene is photographed as soon as possible and before any evidence is disturbed. Photographs are not a substitute for crime scene sketches. All photographs are documented in a photograph log with camera positions indicated on a sketch. The crime scene photograph log records the photograph and describes the type of photograph. The photograph is precisely identified, and the identifying data is recorded as each shot is taken. See *Figure 6-1*, page 6-6.

6-16. The camera positions and distances are also recorded in the crime scene photograph log. This is achieved by measuring from a point on the ground directly below the camera lens to an object used as the focus point for the picture. In making overall crime scene photographs, it is best to keep the camera at about eye level. Take overlapping photographs of interior scenes intended to depict an area as a whole, moving in one direction around the room or area.

6-17. One of the most important elements in investigative photography is maintaining perspective. Photographs must reproduce with the same impression of relative position and size of visible objects (the scene as it would appear to someone standing in the photographer's shoes). Any significant distortion in perspective will reduce, or destroy altogether, the evidentiary

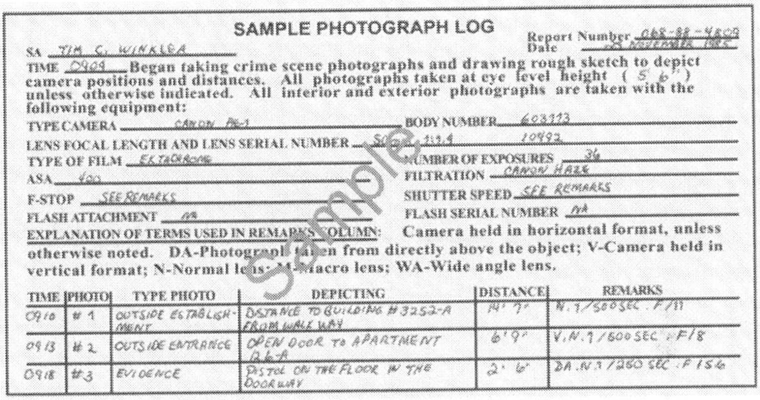

SAMPLE PHOTOGRAPH LOG

Report Number CAR-88-4808
Date 28 NOVEMBER 1985

SA TIM C. WINKLER

TIME 0909 Began taking crime scene photographs and drawing rough sketch to depict camera positions and distances. All photographs taken at eye level height (5' 6") unless otherwise indicated. All interior and exterior photographs are taken with the following equipment:

TYPE CAMERA CANON AE-1 BODY NUMBER 603373
LENS FOCAL LENGTH AND LENS SERIAL NUMBER 50mm 1:1.4 10992
TYPE OF FILM EKTACHROME NUMBER OF EXPOSURES 36
ASA 400 FILTRATION CANON HAZE
F-STOP SEE REMARKS SHUTTER SPEED SEE REMARKS
FLASH ATTACHMENT NA FLASH SERIAL NUMBER NA

EXPLANATION OF TERMS USED IN REMARKS COLUMN: Camera held in horizontal format, unless otherwise noted. DA-Photograph taken from directly above the object; V-Camera held in vertical format; N-Normal lens; M-Macro lens; WA-Wide angle lens.

TIME	PHOTO	TYPE PHOTO	DEPICTING	DISTANCE	REMARKS
0910	#1	OUTSIDE ESTABLISHMENT	DISTANCE TO BUILDING #3252-A FROM WALKWAY	14' 7"	N. 1/500 SEC. F/11
0913	#2	OUTSIDE ENTRANCE	OPEN DOOR TO APARTMENT 26-A	6' 9"	V. N. 1/500 SEC. F/8
0918	#3	EVIDENCE	PISTOL ON THE FLOOR IN THE DOORWAY	2' 6"	DA. N. 1/250 SEC. F/5.6

Figure 6-1. Sample Photograph Log

value of the photograph. An investigator taking basic crime scene photography should—

- Not assume that the confines of the crime scene have been properly identified or that they have been properly secured.
- Watch where he is walking, and consider that he might be walking on evidence while concentrating on obtaining photographs.
- Photograph evidence before recovery.
- Photograph a room or area from eye level to represent a normal view.
- Ensure that a progression of overall, medium, and close-up photographs of any objects of evidence are taken.
- Photograph the interior crime scene using a series of overlapping 360° photographs and a wide-angle lens.
- Photograph the exterior crime scene establishing the location of the scene by a series of overall photographs including landmarks. Photographs should have a 360° coverage.
- Consider aerial photography to capture large areas, roads, and other means of approach or departure. Coordinate with the laboratory on proper aerial-photography techniques.
- Use a ruler for size determination (scale). Always photograph the item without the ruler first.
- Photograph areas adjacent to the crime scene, such as points of entry, exits, windows, and attics.
- Photograph items and places to corroborate the statements of witnesses, victims, and suspects.

6-18. All evidence should be photographed three times. An evidence-establishing photograph should be taken to show the evidence and its position in relation to other evidence, and a close-up photograph that fills the frame should be taken with and without a ruler.

6-19. For a photograph to have the highest quality as evidence, it must depict the scene, person, or object precisely as it was found. Photography is an exclusive function of the crime scene and must not include people working or extraneous objects in the image.

6-20. Photographs are admissible in court if the investigator can testify that they accurately depict the area or item observed, and the personnel who captured the original image or who were present at the time the original image was captured can verify that the image is a true and accurate representation. The accuracy of a photograph relates to the degree that it represents the appearance of the subject matter as to form, tone, color (if applicable), and scale. A lens other than normal (50 millimeter) may not correctly portray distances between objects or show them in proper perspective. In these situations, crime scene sketches and notes play strong supporting roles.

6-21. Because scale, distances, and perspective are important in interpreting photographs taken at crime scenes, always include a ruler. Because some courts may not allow even this minor modification to the scene, an identical photograph without the ruler must first be taken.

FLASH PHOTOGRAPHY

6-22. A flash should be used when exposing photographs inside a building or whenever there is a general lack of light. Most electronic flashes are automatic and do not require adjustment from the operator. Remember the following rules when using an electronic flash for crime scene photography:

- Remove the flash from the camera and use a flash extension cord when taking close-up fingerprint and toolmark photographs. This prevents "hot spots" of light that can cause severe overexposure to the top part of the photograph. Do this by using the following procedures:
  - Place the camera on a tripod; do not attempt to handhold it.
  - Ensure that the camera back is parallel with the latent print. If the photograph is taken at an angle, the print will not be completely in focus.
  - Fill the viewfinder of the camera with the latent area. Get as close as possible. Focus carefully. There is an extremely shallow depth of field when focusing on a close object. Unless the camera back is kept parallel and a small F-stop (F-11 through F-22) is used, parts of the evidence will probably be out of focus in the photograph.
  - Hold the flash 1 to 2 feet away and at about a 45° angle from the evidence being photographed.
  - Expose several frames at different flashes to subject distances and angles.
- Remove the flash from the camera and use the flash extension cord to create shadows in the impression when taking footwear and tire track photographs. This will result in increased contrast or detail in the

negative, which the laboratory will enlarge for examination. Do this by using the following procedures:

- Place the camera on a tripod, and ensure that the legs of the tripod do not obscure the footwear or tire track impressions. Focus and fill the frame. Do not attempt to take usable examination type photographs from 6 to 10 feet away.
- Ensure that the back of the camera is parallel with the impression. If the impression is on sloping ground, ensure that the camera back slopes in the same direction and at approximately the same degree.
- Ensure that there is at least a 6-inch ruler visible in the photograph. The ruler should be at the same level as the bottom of the impression to allow enlargements to life-size photographic prints.
- Attach the flash to the off-camera flash synch cord. Take a series of photographs with the flash held about 3 feet away at a 45° angle to the impression. Expose numerous photographs varying the angle and distance of the flash from the impression. The minimum number of exposures should be two each from all four directions, front, back, and two sides of the impression. This will provide the laboratory with usable negatives for later examination and comparison.
- Shoot some photographs at the recommended meter setting, such as 1/250 at F-11 when photographing impressions in snow. Then shoot several photographs at a higher exposure, such as 1/125 at F-11 or 1/60. Bright objects can fool in-camera exposure meters.
- Photograph longer impressions, such as tire tread patterns, in segments of 1 to 2 feet in length. Ensure that the sections overlap. Place the camera on a tripod to assist in keeping the same camera to the subject distance. This is important when sizing and printing later in the darkroom.

SPECIALIZED PHOTOGRAPHY

6-23. Some crime scenes require specialized photography because of the type of film required, the type of shot required or the manner in which shots must be taken, or the uniqueness of the crime or scene. Specialized photography is required for arson scenes, death scenes, and autopsies.

Arson Scenes

6-24. All arson scenes should be shot with color film, especially if the fire is in progress. At arson scenes, an investigator will work with either an active or inactive fire scene. At an active fire scene, the investigator must use caution in photographing the fires due to the many dangers present. The investigator must always follow the orders of the fire chief. Do the following at an active fire scene:

- Photograph the fire in progress using a tripod from all angles or sides.
- Use color film. The color of the flame and smoke may provide information regarding the accelerant and the temperature of the fire.

- Photograph the entire fire at intervals, keeping note of times. This will show the speed of progression.
- Photograph the crowd in small segments so that faces are identifiable. Many times arson suspects will remain at the scene to watch the fire.

6-25. At an inactive fire scene, the investigator must use caution and coordinate with the fire chief before entering the scene. At an inactive fire scene, the investigator should do the following:

- Photograph the exterior from all sides and the interior using 360° overlapping photographs.
- Be aware that when photographing burned areas, 2 to 3 extra F-stops of exposure are needed. The investigator will have to override the camera's built in meter, take a reading with it on automatic, and switch it to manual and increase the exposure by 2 to 3 F-stops (decrease the shutter speed or increase the aperture setting). Because dark, charred areas do not reflect a great deal of light, extra exposure is required to show the charred patterns. Always use a flash and tripod to photograph all fire scenes.
- Have the fire chief point out the suspected point of origin. Photograph that area and its relationship to the rest of the area or building. Pay attention to an "alligator" pattern charring because it shows the intensity of the fire.
- Photograph anything of possible evidentiary value, such as cans, bottles, strings, wires, or matches.
- Photograph any suspected points of entry.
- Photograph areas that are adjacent to the fire scene to show the proximity of other structures, if present.
- Photograph any voided areas where items, such as, knick-knacks, televisions, or stereos could have been removed. This may indicate arson by the owner.
- Photograph electrical junction boxes and the main electrical box in the building.
- Watch for trail patterns where an accelerant could have been used.

Death Scenes

6-26. All death scenes should be photographed using color film. The investigator should do the following:

- Take an integrity photograph before entering the area where the deceased is located. This records the area before anyone enters and disturbs the scene. This action is secondary to rendering first aid to a living victim.
- Photograph the area of the deceased with 360° overlapping photographs.
- Photograph the deceased from all four sides.
- Take several identification photographs of the individual's face.
- Photograph all wounds, abrasions, or other marks on the deceased. Photograph with and without rulers.

- Photograph the overall scene and establish close-up photographs of all potential evidence, such as weapons or pill bottles.
- Photograph livor mortis patterns if present before moving the deceased.
- Photograph all bloodstains on the skin and clothing of the victim, with and without a ruler, before removal. Movement of the deceased can alter the bloodstain patterns.
- Photograph all signs of gunshot powder stippling or tattooing or the lack of it in all firearm-related deaths. Much of the powder pattern can inadvertently be washed off during the autopsy.

Autopsy

6-27. Always use color film to photograph autopsies. When using a scale in close-up photographs, use white, gray, or a combination of both to allow the person printing the photographs to balance the lighting. This also reduces the green color that is usually obtained when shooting film under fluorescent lights. The investigator should do the following:

- Take overall shots to record the entire body from all angles, to include the front and back. The use of a wide-angle lens may be required. Photographs must be taken both before and after the body is undressed and cleaned up.
- Photograph the face to include the front, profile, and any unique facial or dental features. Take several photographs of the face straight on using a normal focal length lens (50 millimeter) to avoid distortion. A ladder may be required to accomplish this.
- Photograph old scars (including those as a result of operations or previous injuries) and tattoos.
- Photograph wounds by taking orientation photographs and close-up or macrophotographs with and without a ruler. The ruler must be parallel with the camera back so that the photographs can be printed life-size, if necessary. This is especially important if impressions from belt buckles, threaded pipes, or bite marks are present.
- Photograph internal injuries, by taking orientation and close-up photographs.
- Photograph projectiles while still in the body and once they are removed.
- Photograph anything noted by the pathologist as being out of the ordinary or anything that proves or disproves the facts or witness testimony. Use the photographs to substantiate and document the findings, both pathological and investigative.

6-28. When photographing organs or objects removed from the body, ensure that the item is wiped off and there is a clean "field" behind it. A clean cloth works best. Bloody objects, such as a table, clothes, or other items in and around the area being photographed, are distracting and can confuse the viewer.

6-29. Investigators should expose all necessary identifying photographs of the genital area as soon as possible. If no defects exist in and around the genitalia, investigators should then cover that area with a small cloth.

SURVEILLANCE PHOTOGRAPHY

6-30. All surveillance photography should be taken using high-speed black and white film (400 ISO or higher). In surveillance photography, a longer length lens, such as 200 to 500 millimeter, is almost always required. The investigator should do the following:

- Select the lens remembering that the general rule to obtaining a recognizable image of a face is to use 2 millimeter of lens focal length for each foot of distance between the camera and the subject. A 200-millimeter lens will give a recognizable facial image at 100 feet.
- Use a tripod and a cable release. A long lens magnifies movement. If the camera or lens is not stable, the photographs will show movement and will result in blurry images. If no tripod is available, brace the lens against a wall or tabletop. In a vehicle, a side window can be rolled partway down with the lens resting on the top edge of the window. Ensure that the vehicle engine is turned off to avoid any vibration. Remember to always use a shutter speed that is faster than the length of the lens; for example, if using a 200-millimeter lens, use a 1/250 second shutter speed. Besides obtaining a facial image on the film large enough to identify, the most important consideration is to have a fast shutter speed, usually 1/250 second or faster to stop motion.
- Shoot several exposures of each subject and/or transaction.
- Recognize that the depth of field is less important when doing surveillance photography than the shutter speed. The lens can be opened to the widest aperture in order to gain a faster shutter speed. Focus carefully because the depth of field (range of acceptable focus) will be very shallow at wider apertures, such as F-2.8 or F-4.

PHOTOGRAPHY OF SCENES AND OBJECTS FOR EVIDENCE

6-31. Crime scene photographs of impressions or other evidence, which may require laboratory evaluation and analysis, must be recorded using 35-millimeter film. Digital images are currently not suitable for USACIL analysis of evidence. Use film to record latent fingerprints, blood patterns and stains, footwear impressions, toolmark impressions, and tire tracks. Digital images for this type of evidence may look satisfactory but the fine detail needed for laboratory analysis is not discernible.

6-32. Video imaging can be used in a supplementary capacity. It is suggested that the camera have the ability to disable on-camera audio to avoid unnecessary conversations during taping. Do not rely on video to record fine details.

CRIME SCENE VIDEO

6-33. In addition to photographing and sketching a crime scene, investigators should consider videotaping the scene. Videotaping is not a replacement for still photography of the crime scene but rather a supplement. Videotaping should be well-planned and follow the same basic principles as still photography (movement from the general to the specific). This will result in fewer mistakes and allow a smooth transition from one area of the crime scene to the next. A tripod or monopod will help steady the camera and provide a smooth video.

VIDEOTAPING (BEFORE)

6-34. Before taking a video camera to the crime scene, verify that there is extra videotape and batteries available. Ensure that the batteries are fully charged.

6-35. While at the scene, but before videotaping, play the tape for approximately 10 seconds before beginning the recording. This will prevent recording on the tape leader and possibly loosing the first few seconds of the recording. White balance the camera before shooting and anytime the lighting conditions change. Turn off or disable the audio record capabilities. This eliminates the possibility of making erroneous statements or picking up extraneous comments.

VIDEOTAPING

6-36. Identify the subject matter at the beginning of the tape, such as case number, investigator's name, or scene. Plan each shot in advance to provide sufficient information and prevent confusion. Do not record anything on the tape that is not of potential evidentiary value and do not delete anything from the tape. Use the video light if the light levels are low.

6-37. Use standard play (SP) to record. Image quality quickly degrades when using other than SP. Be careful not to over zoom because this can be distracting and it magnifies movement.

6-38. Try to leave the focus of the camera on normal to wide-angle and move in closer to the subject for close-ups. Be consistent and establish a routine. Always pan from the same direction (right to left or left to right).

VIDEOTAPING (AFTER)

6-39. Do not rewind the tape at the scene. Wait until leaving the scene when no additional recording is required. After videotaping the scene, always do the following:

- Make a master copy or working copy to use in making duplications and for viewing. Each time a videotape is played there is degradation.
- Place the date, time, and your initials on the tape, and preserve it the same as other physical evidence.

- Punch-out or break the antierase tabs.
- Store tapes in a place free of dust and equipment that produces magnetic fields. These items may reduce the tape quality or even destroy the tape.

SKETCHES

6-40. The two types of sketches that an investigator makes of a crime scene are the rough sketch and finished sketch. If an investigator wishes to provide an additional dimension to a sketch to demonstrate hidden walls or other areas, he may want to draw a cross-projection sketch. Sketches help to recreate the crime scene by depicting the location of evidence, the victim's body, points of entry and exit to a crime scene, and other details.

ROUGH SKETCHES

6-41. A rough sketch is the sketch that is drawn by an investigator while at the crime scene. Its purpose is to portray information accurately. A rough sketch is usually not drawn to scale, but it must show accurate distances, dimensions, and relative proportions. To eliminate excessive detail in a sketch, there may have to be more than one sketch drawn. For example, one sketch may be devoted to the position of the victim's body and one or two of the more critical evidence items. Other sketches might show the lay of evidence items with respect to the point of entry or to other critical points. See *Figure 6-2*, page 6-14, for an example of a rough sketch.

FINISHED SKETCHES

6-42. A finished sketch uses the information provided in the rough sketch, but it is more of a professional version. A finished sketch does not need to be drawn by the same person who drew the rough sketch, but whoever draws the rough sketch must verify the accuracy of the finished sketch. It is best if an experienced draftsman draws the finished sketch. The name of the person who drew the finished sketch is shown in the report and on the sketch. A copy of the smooth sketch is attached to each copy of the investigation report. Finished sketches are often drawn to scale from information in the rough sketch. By making a scaled drawing, the numbers showing distances can be left out. If the finished sketch is not drawn to scale, these distances must be shown. See *Figure 6-3,* page 6-15, for an example of a finished sketch drawn to scale.

CROSS-PROJECTION SKETCHES

6-43. Cross-projection is used to add another dimension to sketches. This dimension is useful when items or locations of interest are on or in wall surfaces in an enclosed space. The walls, windows, and doors in a cross-projection sketch are drawn as though the walls had been folded out flat on the floor with the surface projected up. The required measurements and triangulation of evidence are then entered on the sketch. Measurements in cross-projection have length, not height, and there are no tops or bases to measure. Measurements are made from corner to corner. Several items of information are considered essential in a cross-projection sketch. The sketch should include the following:

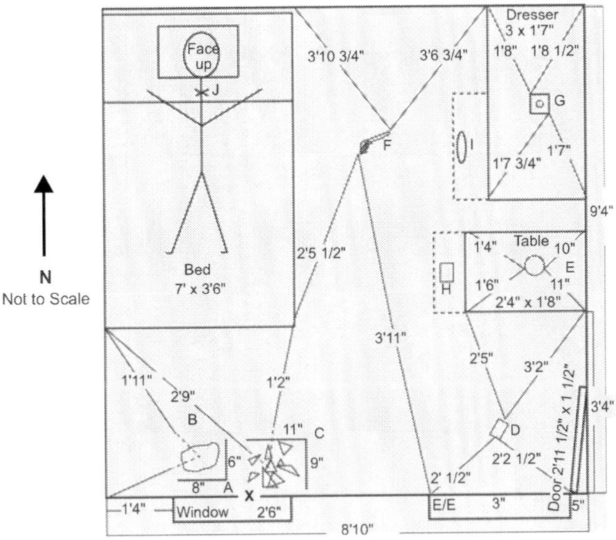

Legend

A. Hole
B. Red stain
C. Glass fragments
D. Shell casing
E. Glass
F. Pistol
G. Bottle
H. Canister
I. Cigarette
J. Neck injury

Title Block

Case number: 0123-02-CID037
Offense: Undetermined death
Scene portrayed: Room #C-33, Bldg #3203, troop barracks
Location: Fort Leonard Wood, MO 65473
Victim: SGT Janet Williams
Time & date began: 1115 2 Jan 02
Sketched by: SA William Mac
Verified by: SA John Friend

Figure 6-2. Example of a Rough Sketch Depicting Evidence, Measurements, and Triangulation (Inside Surface)

- Dimensions of the horizontal surface (length and width).
- Entry and exit way (length and width).
- Windows (length and width).
- Distances between objects, persons, bodies, entrances, and exits.
- Measurements showing the location of evidence.
- Compass orientation, scale, caption, and legend.

6-44. The information that is annotated for the legend and title block in a flat projection rough sketch must also be annotated for a cross-projection sketch. See *Figure 6-4*, page 6-16.

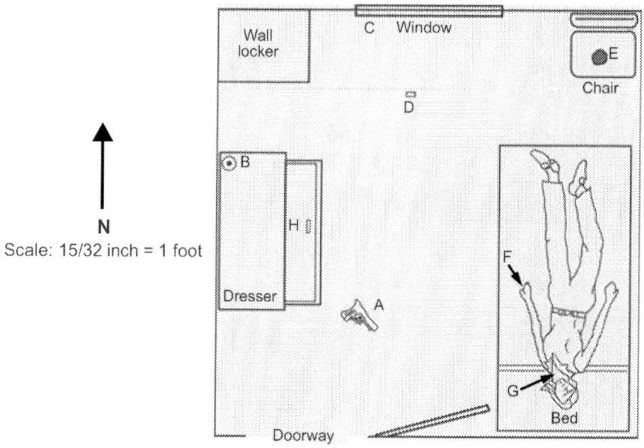

Legend

A. Pistol
B. Bottle
C. Letters "BR"
D. Shell casing
E. Red stain
F. Fibers
G. Gunshot wound
H. Cigarette inside opened drawer

Title Block

Case number: 063-02-4809
Offense: Death investigation
Scene portrayed: Room #7, Bldg 5-3251-B, troop billets
Location: Fort Leonard Wood, MO
Victim: SP4 Faye M. Miller
Time & date began: 1000 12 Jan 02
Sketched by: SA Tim C. Winkler
Verified by: SGT Betty C. Goodwin

Figure 6-3. Example of a Finished Sketch Drawn to Scale

PREPARATION OF A ROUGH SKETCH

6-45. Any kind of paper may be used for a rough sketch. However, bond or graph paper is best. It can be placed on a clipboard large enough to form a smooth area for drawing. At a minimum, the following items are required to prepare a rough sketch:

- A soft lead pencil.
- A 100-foot and a 25-foot steel tape.
- A straightedge ruler.
- Several thumbtacks to hold one end of the steel tape down when you are working alone.
- A magnetic compass.
- Crime scene templates.
- A clipboard.

6-46. Several items of information are considered essential in a crime scene (flat-projection) sketch. The sketch should include the following:

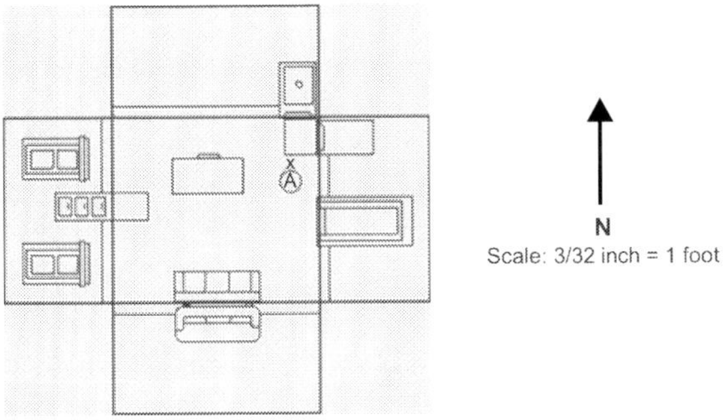

Legend
A. Crowbar

Title Block
Case number: 1426-02-CID063
Offense: Burglary investigation
Scene portrayed: Room #7, Bldg 5-3251-B, troop billets
Location: Fort Leonard Wood, MO
Victim: SP4 Betty M. Holmes
Time & date began: 1030 12 Jan 85
Sketched by: Mr. Frank L. Wright
Verified by: SA David L. Kelly

Figure 6-4. Example of a Cross-Projection Sketch

- Weather and lighting conditions.
- Dimensions of the rooms (length, width, and height); furniture (width, height, and depth); doors (width, height, and thickness); and windows (length and width).
- Distances between objects, persons, bodies, entrances, and exits.
- Measurements showing the location of evidence. Each object should be located by two measurements from nonmovable items, such as doors or walls.
- Compass orientation, scale, and legend.

6-47. The major constraint on detail in sketching is that the result must be completely understandable to a viewer without a detailed study. An investigator must not include too much detail in a sketch or the major advantage of a sketch over a photograph will be lost.

6-48. Each sketch should include the critical features of the crime scene and the major, discernible items of physical evidence. Evidence sketches must show accurate measurements of the crime scene. They must also show the location of evidence established by using the triangulation method. A photograph sketch must show camera positions and distances to focus points.

6-49. Each sketch should have a caption to identify the illustration. For example, a caption might read: "flat-projection rough sketch showing camera positions and distances." Each sketch must have a legend. The legend explains the symbols, numbers, and letters used to identify objects on the sketch. Numbers are used to explain items of interest and letters are used to explain items of evidence. Standard military symbols are used where practical.

6-50. The sketch must show the compass direction north. Scaled drawings must include a scale designation. If no scale is used, write: "not drawn to scale." Each sketch must have a sketch title block containing the following entries:

- Incident report number, such as an MPR, USACIDC sequence number, or report of investigation (ROI) number.
- Alleged offense.
- Name and rank or title of the victim.
- Scene portrayed, citing the room number, building number, and type of building (such as the post exchange [PX], commissary, house, or troop billets).
- Location, citing the complete name of the installation, city, state, and zip code.
- Time and date the sketch was started.
- Name and rank or title of the person who drew the sketch.
- Name and rank or title of the person who verified the sketch.

6-51. The measurements shown on the sketch must be as accurate as possible and should be made and recorded uniformly. Steel tapes are the best means of taking accurate measurements. It is important to verify the measurement with a tape measure or laser range finder. A measurement error on a sketch can bring the entire crime scene search into doubt. If one aspect of a sketch is inaccurate, such as the dimensions of a field in which a body was found and the position of an object within the field is only roughly estimated, the distortion introduced renders the sketch relatively useless. It is important that the coordinate distances of an item in the sketch be measured in the same manner. For example, one coordinate leg of the victim should not be paced and the other measured with a tape measure. It is also a mistake to pace off a distance and then show it on the sketch in terms of feet and inches. This implies a far greater degree of accuracy than the measurement technique could possibly produce. If the point arose in court, such imprecision could greatly detract from the value of the sketch.

EVIDENCE LOCATION SHOWN ON A SKETCH

6-52. Various sketch methods may be used to show evidence location and other important items at the scene. The simplest form of a sketch is a two-dimensional presentation of a scene as viewed directly from above. Evidence location is shown on this type of sketch by triangulation. Triangulation is used for indoor and outdoor sketches having fixed reference points. Items are located by creating a triangle of measurements from a single, specific, identifiable point on an item to two fixed points all on the same plane at the scene. If movable items are used as reference points, they must first be "fixed." Do not triangulate evidence to evidence. Do not triangulate under or through evidence. Do not take a line of measurement through space. Measure the line along a solid surface like a floor, wall, or tabletop. In the interest of clarity, keep the angle of triangulation measurements between 45° and 90° on the sketches. Ensure that the measurements in the notes and sketches do not conflict. Although height measurements are taken and annotated in the notes, they are not reflected on crime scene sketches.

6-53. Regular-shaped items are fixed by creating two separate triangles of measurements. Each originates at opposite points on the item and ends at two fixed points on the same plane at the scene. This is commonly known as the 2-V method of triangulation. See *Figure 6-5*.

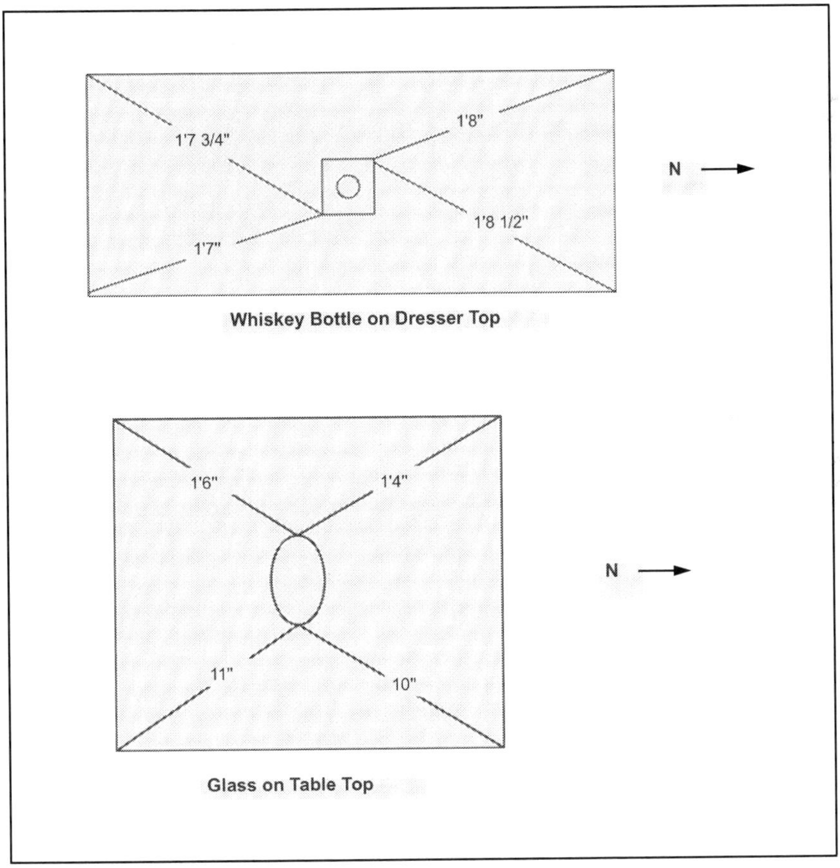

Figure 6-5. 2-V Method of Triangulation

6-54. Irregular-shaped items are fixed by creating a single triangle of measurements from the approximate center of mass of the item to two fixed points on the same plane at the scene. The longest and widest dimensions of the item are also measured. This is commonly known as the 1-V method of triangulation. See *Figure 6-6*.

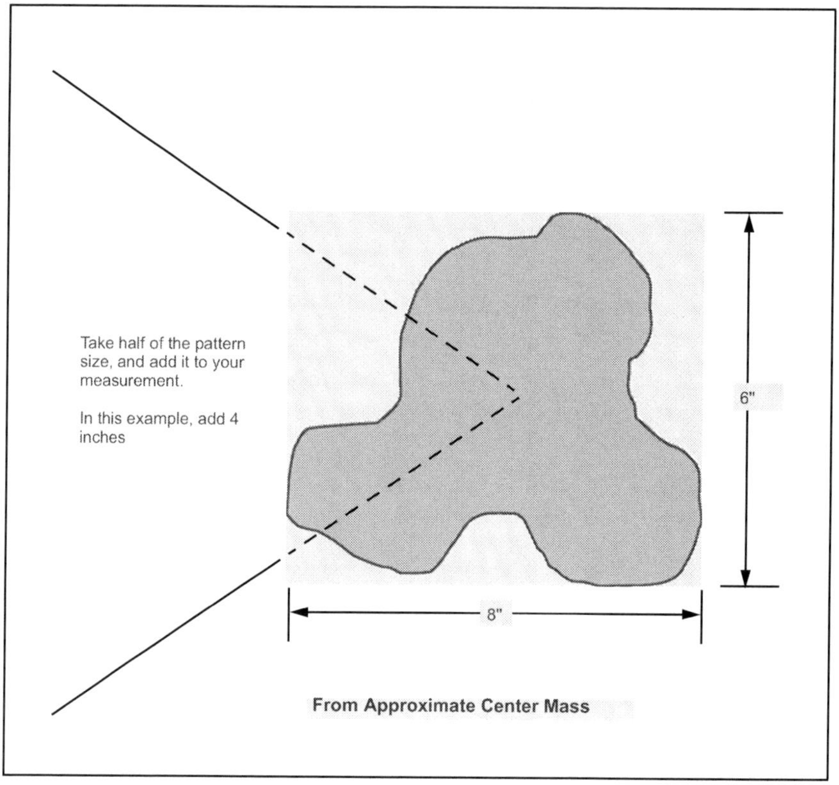

Figure 6-6. 1-V Method of Triangulation

6-55. Inhabited outdoor areas usually have easily defined, fixed reference points such as buildings, edges of roads, and sidewalks. When these are present, the triangulation method can be used to establish the location of objects. Uninhabited or remote areas may not have easily defined, fixed points within close range. Use a GPS or precision lightweight GPS receiver (PLGR) to fix the points. If not available, use engineers to stake a reference point. See *FM 3-25.26* for application of this method. Situations may occur where no logical point of reference exist. The investigator must then establish reference points by creating them. Care must be exercised in describing the location of these reference points so that accurate placement of the fixed item can be accomplished upon reconstruction of the crime scene.

6-56. Upon completing a sketch, the investigator must recheck the crime scene against his notes and sketch to verify that all items have been identified and recorded. If new evidence is discovered, it must be described, photographed, triangulated, and annotated on the appropriate sketch.

0PART THREE

Investigations of Specific Criminal Offenses

Part Three addresses the most common types of crimes investigated by Army law enforcement investigators. Each chapter describes the offense, provides legal considerations and, where appropriate, describes the nationally recognized method of investigation as established by the DOJ, FBI, BATF, and NAFI.

Chapter 7

Arson and Explosives Investigations

Similar investigative techniques are used to investigate arson and explosive incidents. The investigator must determine the nature and cause of the incident. Initially, he must distinguish between an accidental fire or explosion and one produced intentionally or by criminal negligence. Crimes involving fire and explosions present many difficulties, paramount among which is the partial or total destruction of evidence. Although the usual investigative steps are followed at the crime scene, great care must be taken in handling items of evidence and interpreting the pattern suggested by their location. Police intelligence gathered as a result of these investigations is reported and shared with other federal, state, local, and HN police agencies.

ARSON

7-1. Arson investigation is a difficult field for the investigator. Arson frequently involves the use of an accelerant (a material used to spread or increase the rate of burning). Typical accelerants include ignitable liquids, such as gasoline, mineral spirits, kerosene, and similar materials. The presence of such materials, if there is no legitimate reason for their presence, may indicate arson.

7-2. During a fire involving an accelerant, the accelerant will undergo change. The more volatile components will be lost to a much greater extent than the components of lower volatility and those that remain may be absorbed into the wood or carpeting of the structure. Laboratory analysis can determine if an ignitable liquid is present in the sample. The volatility range and general chemical composition can also be determined. There will be situations in which one sample may disclose the presence of an ignitable liquid and another will not, even though the samples were taken from areas quite close together.

FIRE CHEMISTRY AND BEHAVIOR

7-3. There are four types of fires: accidental, natural, incendiary, and undetermined. Most fires are accidental. To determine if a fire was accidental or incendiary, investigators must understand the basics of fire chemistry and behavior. To prove it was incendiary, they must first prove it was not accidental or natural.

Chemistry

7-4. Fire is a chemical reaction that takes place when fuel, heat, and oxygen combine in an uninhibited chain reaction. Fire can only exist when all four of these factors are present. Remove any one of the elements and the fire goes out because the continuing chemical reaction has been stopped.

7-5. Because only gases burn, solid and liquid fuels must be heated until they become vapor (gas) before they can burn. Heat chemically decomposes a fuel into its gaseous elements. This decomposition is known as pyrolysis. For example, when wood is heated, it pyrolyzes to form hydrogen, oxygen, ethane and methane gases, and methyl alcohol. It is these highly flammable vapors that burn. Fuel in vapor form in its normal state, like natural gas, does not need to be pyrolyzed.

7-6. Most fuels are compounds of carbon, hydrogen, and oxygen along with traces of mineral matter. When the fuels burn completely and freely in air, the carbon reacts with the oxygen to form carbon dioxide and the hydrogen combines with the oxygen to form water vapor. The mineral matter remains behind as ash. As the oxygen in the fuel is used up, oxygen is drawn from the air to continue the reaction. That is why drafts and air supplies directly affect the behavior of a fire. A fire started in a completely enclosed space soon dies. It uses up all of the available oxygen and generates noncombustible gases that smother it. On the other hand, the rate of burning is greatly increased if a chimney effect exists when the hot gases and flame contact combustible material. Disastrous fires result in large buildings where elevator shafts or stairways served as chimneys to direct the up-rushing flames and gases.

7-7. It takes heat to ignite the fuel and start the chemical reaction. Once ignition has taken place, the reaction (fire) produces its own heat and becomes self-generating as long as fuel and oxygen remain present.

7-8. It is understandable then, that fuels do not need to be touched by flame to begin burning. They simply need to be heated above their ignition points. It is for this reason that heat, not flame, is the greatest cause of fire spread.

Heat can be transferred from one place to another by convection, conduction, or radiation.

- **Convection.** Convection occurs when heat is transferred by a circulating medium like air or water. Heat convected by circulating air is the most common method of fire spread.
- **Conduction.** Conduction transfers heat by contact. Often, heat from a fire in one room is conducted throughout the structure by pipes or electrical conduit. If combustible material in another room is in contact with the pipe, it can become heated above its ignition point and start a second fire. Metallic objects are the most frequent conductors of heat. Sometimes, even brick walls can conduct enough heat to cause a second fire.
- **Radiation.** Radiation transfers heat in the form of energy waves through space. Heat radiates through any transparent medium like air, glass, or even water. No physical contact is needed. This is how the heat of the sun is transferred through the vacuum of space to the earth. Often, a fire in one building radiates enough heat to start a fire in another building, even though a curtain of water is sprayed between the buildings.

Behavior

7-9. Fires behave according to well-defined principles of burning. Fires produce heat, flame, smoke, and gases. These combustion by-products may or may not be readily seen. Flame includes both an open flame and a smoldering glow. Smoke is composed of very fine solid particles and condensed vapors. The composition of fire gases emitted by the burning materials depends on the chemical makeup of the burning material, the amount of oxygen available during burning, and the temperature of the fire. Most fire gases are highly toxic. They are the biggest cause of fire deaths. The biggest single killer is carbon monoxide, not because it is the most toxic, but because it is the most abundant. When large quantities are breathed in, carbon monoxide causes unconsciousness and eventually death. At less than lethal concentrations, it causes disorientation and confusion, subjecting victims to other hazards present in the fire. The second most dangerous gas produced by a fire is carbon dioxide. While not toxic in itself, a 2 percent increase in carbon dioxide in the air causes a 100 percent increase in a human's breathing rate. This doubles a victim's intake of other toxic gases.

7-10. Fire burns up and out. It leaves a V-shaped char pattern on walls and vertical structures. A fire that is hot and fast at the point of origin will leave a sharp V pattern. A slow fire will produce a shallow V pattern. If fire meets an obstruction, such as a ceiling, it will burn across the obstruction looking for a place to go up.

7-11. Fire travels with air currents. It never travels into the wind unless the entire fire load, the combustible material or fuel in an area, is on the windward side of the fire. If this happens, the fire slowly eats into the fire load as its tendency to follow the wind is overcome by its attraction for fuel.

7-12. Fire seeks oxygen. Because fire consumes great amounts of oxygen, it is always drawn toward new sources of oxygen when burning occurs indoors. It is not unusual to see a char pattern going up a wall to the ceiling and across the ceiling toward an open window. It is also common to find deeper charring and evidence of higher heat on window frames and doorways.

7-13. These principles of burning account for most of the fire and char patterns seen during and after a fire. But sometimes there are unusual patterns that are the result of a flashover or backdraft, natural conditions which only occur during fires when conditions are right.

7-14. A flashover occurs when flames instantaneously erupt over the entire surface of a room or confined area. Once a fire starts, it produces gases that rise and form a superheated gas layer at the ceiling. As the volume of this gas layer increases, it begins to move down to the floor, heating all objects in the room regardless of their proximity to the flaming objects. In a typical contained fire, the gas layer at the ceiling can rapidly reach temperatures in excess of its autoignition point. If there is enough existing oxygen, a flashover occurs and everything in the room becomes involved in an open flame all at once. This sudden eruption into flames generates a tremendous amount of heat, smoke, and pressure with enough force to push beyond the room of origin through doors and windows. This combustion process will accelerate more now because it has a greater amount of heat to move to objects.

7-15. Backdrafts occur when a structure burns with all doors and windows closed and the fire uses up all of the available oxygen. It then turns into a slow smoldering fire, generating huge amounts of superheated carbon monoxide gas. The hot gases rise to the top of the room and stay there. Because carbon monoxide is a flammable gas and is heated above its ignition point, it only needs more oxygen to burst into flames. Oxygen entering around cracks in doors and windows keeps the fire smoldering. This produces more and more superheated carbon monoxide. Then, when a door is opened or a window melts out, the in-rushing oxygen combines with the superheated carbon monoxide, causing an explosive fire.

7-16. Windows will blow out, and the explosion may be strong enough to damage the structure of the building. Damage caused by a backdraft may look similar to that caused by a low explosive. A backdraft produces an unusual char pattern. Most of the burn damage will be at the extreme top of the room. There will also be a rather sharp line of demarcation at the bottom of the char pattern on the wall. As with a flashover, accelerant residue may not be present.

7-17. When wood burns, it chars a pattern of cracks that looks like the scales on an alligator's back. The scales will be the smallest and the cracks the deepest where the fire has been burning the longest or the hottest. Most wood in structures chars at the rate of 1 inch in depth per 40 to 45 minutes of burning at 1400° to 1600° Fahrenheit, which is the temperature of most house fires. However, no specific time of burning can be determined based solely on the depth of charring since most fires vary with intensity and fuel load. A room fire chars only the upper one half to two thirds of the room. Ceiling damage in a normal structural fire is usually at least five times the floor damage. Sometimes a char pattern has a sharp line of demarcation on one side. This indicates that the fire quit spreading in that direction when a draft entered and blew it back.

7-18. When glass is exposed to fire, it begins to melt at about 1200° Fahrenheit. It becomes runny at about 1600° Fahrenheit. A lot can be learned about a fire from the glass at the scene. Remember, though, to examine all of the glass and not jump to conclusions from the appearance of just one piece. As a general rule, glass that contains many cracks indicates a rapid heat buildup. Glass that is heavily stained indicates a slow, smoky fire.

7-19. Bright metals, like the chromium on toasters, turn colors when heated. These colors may remain after the fire and indicate the temperature of the fire at that location.

FIRE INVESTIGATION

7-20. Investigators arriving at a fire scene should observe and mentally note the conditions and activities and, as soon as conditions permit, initiate permanent documentation of the information (such as notes, voice recordings, photographs, or videotapes). Investigators should document the following:

- Presence, location, and condition of victims and witnesses.
- Vehicles leaving the scene, bystanders, or unusual activities near the scene.
- Flame and smoke conditions (the volume of flames and smoke; the color, height, and location of the flames; and the direction in which the flames and smoke are moving).
- Type of occupancy and the use of the structure (such as a residential occupancy being used as a business).
- Condition of the structure (lights turned on; fire through the roof; walls standing; open, closed, or broken windows and doors).
- Conditions surrounding the scene (blocked driveways, debris, or damage to other structures).
- Weather conditions.
- Any unusual characteristics of the scene (the presence of containers, exterior burning or charring on the building, the absence of normal contents, unusual odors, or fire trailers).
- Fire suppression techniques used (including ventilation, forcible entry, and utility shutoff measures).
- Status of fire alarms, security alarms, and sprinklers.

7-21. Before entering the scene, the investigator should identify and contact the current incident commander (IC). He should request a briefing with the IC to determine who has jurisdiction and authorization (legal right of entry) and to identify others at the scene, such as law enforcement, firefighting, EOD, emergency medical services (EMS), HAZMAT, and utility services personnel.

7-22. Information obtained from the IC and the first responders will help the investigator determine the level of assistance required and whether additional investigative personnel are needed. Before entering the scene, investigators should determine initial scene safety by using observations and discussions with first responders. Consider environmental as well as personnel safety concerns, and assess changes in safety conditions resulting from suppression efforts.

Define the Extent of the Scene

7-23. To provide for the safety and security of personnel and to protect the evidence, the investigator should perform a preliminary scene assessment. He should determine the area in which the site examination will be conducted and establish or adjust the scene perimeter.

7-24. To determine the boundaries of the scene, the investigator should—

- Make a preliminary scene assessment to identify areas that warrant further examination, being careful not to disturb evidence. The preliminary scene assessment is an overall tour of the fire scene to determine the extent of the damage, proceeding from areas of least damage to areas of greater damage.
- Inspect and protect adjacent areas, even areas with little or no damage, that may include nonfire evidence, such as bodies, blood stains, latent prints, toolmarks, or additional fire-related evidence (such as unsuccessful ignition sources, fuel containers, or ignitable liquids).
- Mark or reevaluate the perimeter and establish or reassess the procedures for controlling access.

Identify and Interview Witnesses at the Scene

7-25. Persons with information about the scene, activities before the incident, the incident, and its suppression are valuable witnesses. Determine the identities and locations of witnesses and make arrangements to conduct interviews.

7-26. To develop a witness list, the investigator should—

- Contact the IC, identify first responders and first-in firefighters, and arrange to document their observations either in writing or through recorded interviews.
- Determine who reported the incident, and secure a tape or transcript of the report if available.
- Identify the owner of the building or scene, any occupants, and the person responsible for property management.
- Identify who was last to leave the building or scene and what occurred immediately before he left.
- Identify and interview other witnesses (such as neighbors, bystanders, people injured during the incident, or public agency personnel arriving later) and record their statements.

Determine Scene Security at the Time of the Incident

7-27. The investigator should determine whether the building and/or the vehicle was intact and/or secure. He should also determine if the intrusion alarms and/or fire detection and suppression systems were operational at the time of the fire. This information helps to establish factors, such as ventilation conditions, possible fire development timelines and scenarios, and whether vandalism of the property or systems occurred before the incident.

7-28. To determine the status of security at the time of the fire, the investigator should—

- Ask first responders where entry was made, what steps were taken to gain entry, and whether any systems had been activated when they arrived at the scene.
- Observe and document the condition of doors, windows, and other openings. Attempt to determine whether they were open, closed, or compromised at the time of the incident.
- Observe and document the position of timers, switches, valves, and control units for utilities, detection systems, and suppression systems, as well as any alterations to those positions by first responders.
- Determine if the security or suppression systems were available and contact the monitoring agencies to obtain information and available documentation about the design and functioning of the systems.

Identify Resources

7-29. Except in the most obvious cases, the determination of the origin and cause of a fire may be a complex and difficult undertaking that requires specialized training and experience as well as knowledge of generally accepted scientific methods of fire investigation. The investigator must either have the appropriate expertise or call on the assistance of someone with that knowledge. This is especially true in cases involving deaths, major injuries, or large property losses.

7-30. Based on the preliminary scene assessment and analysis of fire patterns and damage at the scene, the investigator should identify a distinct origin (the location where the fire started) and an obvious, fire cause (an ignition source, the first fuel ignited, and circumstances of the event that brought the two together). If neither the origin nor the cause is immediately obvious, or if there is clear evidence of an incendiary cause, the investigator should conduct a scene examination according to recognized national guidelines, such as the NAFI or National Fire Protection Association (NFPA). Investigative guidelines can be found on NAFI, NFPA, and other official government Web sites.

Obtain Equipment and Tools

7-31. Once arson is determined to be the cause of the incident, the investigator will need to obtain specialized equipment to assist with the investigation. These items should be kept on hand and available to the investigator at all times to prevent a delay in the investigation. Specialized equipment includes—

- Barrier tape.
- Clean, unused evidence containers (lined paint cans with friction lift lids, glass jars, or nylon or polyester bags specifically made for fire debris evidence).
- Compass.
- Decontamination equipment (buckets, pans, and detergent).
- Evidence tags, labels, and tape.

- Gloves (disposable gloves and work gloves).
- Hand tools (hammers, screwdrivers, knives, crowbars, and shovels).
- Lights (flashlights and spotlights).
- Marker cones or flags.
- PPE.
- Photographic equipment.
- Rakes, brooms, and spades.
- Tape measure (100 feet).

Document the Scene

7-32. Written descriptions of the scene, along with accurate sketches and measurements, are invaluable for focusing the investigation. Written scene documentation recreates the scene for investigative, scientific analysis, and judicial purposes and correlates with photographic evidence. Photographic documentation creates a permanent record of the scene and supplements the written incident reports, witness statements, or reports on the position of evidence. The investigator should create and preserve an accurate visual record of the scene and the evidence before disturbing the scene. Additional photography or videotaping should occur as the investigation progresses. *Chapter 6* describes the procedures for notes, photographs, and sketches.

7-33. The scene should be photographed before the disturbance or removal of any evidence and throughout the scene investigation. The investigator (or other individual responsible for evidence) should—

- Photograph and/or videotape the assembled crowd and the fire in progress.
- Remove all nonessential personnel from the background when photographing the scene and evidence.
- Photograph the exterior and interior of the fire scene (consider walls, doors, windows, ceilings, and floors) in a systematic and consistent manner. (Videotaping may serve as an additional record but not as a replacement for still photography.)
- Photograph any points or areas of origin, ignition sources, and first material ignited.
- Photograph any physical reconstruction of the scene.
- Maintain photograph and video logs.
- Determine whether additional photographic resources are necessary (aerial photography, infrared photography, stereo photography).

7-34. Written documentation of the scene provides a permanent record of the investigator's observations that may be used to refresh recollections, support his opinions and conclusions, and support photographic documentation. The investigator should—

- Prepare a narrative, written descriptions and observations (to include assessments of possible fire causes).
- Sketch an accurate representation of the scene and its dimensions, including significant features, such as the ceiling height, fuel packages (combustible contents of the room), doors, windows, and any areas of origin.
- Prepare a detailed diagram using the scene sketches, preexisting diagrams, drawings, floor plans, or architectural or engineering drawings of the scene. This may be done at a later date.

Process Fire Scene Evidence

7-35. Collecting evidence at a fire scene requires attention to documenting and maintaining the integrity of the evidence. The investigator should ensure that evidence collectors identify and properly document, collect, and preserve evidence for laboratory analyses, further investigations, and court proceedings.

7-36. To optimize the recovery and evaluation of physical evidence, evidence collectors should—

- Take precautions to prevent contamination and cross contamination.
- Document the location of evidence using written notes, sketches, photographs, photograph and video logs, an evidence recovery log, evidence tags, and container labels. When evidence is excavated, additional photographs may be of value.
- Take special care to collect evidence in any areas of origin (such as the first fuel ignited and ignition source) in cases where the fire is not accidental.

NOTE: In cases where the fire appears to be accidental, evidence should not be needlessly disturbed, but the property owner or insurer should be notified to avoid issues of spoliation.

- Place evidence in labeled containers for transportation and preservation. Evidence collected for laboratory identification of ignitable liquids must be immediately placed in clean, unused vapor-tight containers (clean, unused paint cans; glass jars; or laboratory-approved nylon or polyester bags) and then sealed.
- Label each container so that it is uniquely identified. Labeling may include the name of the investigator, date and time of collection, case number, sample number, description, and location of recovery.
- Collect and preserve comparison samples. Comparison samples are comprised of the same kind of materials or substances as the samples that are taken from possible points of origin for ignitable liquid analysis. These samples are not always available.

- Package evidence according to the policies and procedures of the laboratory.
- Recognize the presence of other physical evidence, such as bloodstains, shoe prints, latent prints, and trace evidence. Use proper preservation and collection methods or seek qualified assistance.

Prevent Contamination

7-37. Preventing contamination during evidence collection protects the integrity of the fire scene and evidence. The investigator should ensure that access to the fire scene is controlled after fire suppression and that evidence is collected, stored, and transported in a manner that will not contaminated it.

7-38. To prevent contamination, evidence collectors should—

- Establish and maintain strict control of access to the scene.
- Recognize that fuel-powered tools and equipment present potential contamination sources and should be avoided. When it is necessary to use these tools and equipment, the investigator should document their use.
- Wear clean, protective outer garments, including footwear.
- Use clean disposable gloves for collecting items of evidence. (To avoid cross-contamination, gloves should be changed between collection of unrelated items of evidence or when visibly soiled.)
- Use clean tools for collecting items of evidence from different locations within a scene. (Disposable tools also can be used.)
- Place evidence in clean, unused containers and seal immediately.
- Store and ship fire debris containers of evidence that are collected from different scenes in separate packages.
- Package liquid samples to prevent leakage, and ship them separately from other evidence.
- Store and ship fire debris evidence separately from other evidence.
- Follow any specific laboratory requests, such as submitting an unused sample container or absorbent medium for detection of any contaminants.

Package and Transport Evidence

7-39. Preventing changes in the condition of a sample after it has been collected ensures the integrity of the evidence and requires controlled packaging and transportation. The investigator should ensure that packaging, transportation, and storage procedures are followed to prevent any destructive changes in the condition of the samples.

7-40. To minimize changes in the condition of samples, the personnel responsible for packaging and transport should—

- Take precautions to prevent contamination.
- Package fragile items carefully.
- Freeze or immediately transport items containing soil to the laboratory.

- Transport all volatile samples to the laboratory in a timely manner.
- Comply with the shipping regulations

SCENE INVESTIGATION COMPLETION

7-41. The investigator should ensure that the scene is not released until reasonable efforts have been made to identify, collect, and remove all evidence from the scene for further examination and all physical characteristics of the scene have been documented. In addition, before releasing the scene to the receiving party, associated legal, health, and safety issues must be articulated and reported to public safety agencies, if necessary. Doing so minimizes the risk of any further incident or injury and the potential liability of the authority releasing the scene.

7-42. The investigator should complete the necessary tasks before releasing the scene. He should—

- Perform a final critical review.
  - Ensure that all evidence is inventoried and in custody.
  - Discuss preliminary scene findings with team members.
  - Discuss postscene issues, including forensic testing, insurance inquiries, interview results, and criminal histories.
  - Assign postscene responsibilities to law enforcement personnel and other investigators.
  - Address legal considerations.
- Verify that all scene documentation has been completed. This can be accomplished using an incident documentation checklist.
- Address structural, environmental, health, and safety issues.
- Remove all investigative equipment and material.
  - Recover and inventory equipment.
  - Decontaminate equipment and personnel.
- Document the following information:
  - The time and date of release.
  - The receiving party.
  - The authority releasing the scene.
  - The condition of the scene at the time of release (structural, environmental, health, and safety issues). The investigator may want to consider photographing and/or videotaping the final condition of the scene.
  - The cautions given to the receiving party on release (safety concerns, conditions, evidence, and legal issues).

7-43. Detailed fire information is collected, integrated, and disseminated through national and state databases. This data assists authorities in identifying fire trends and developing innovative procedures and equipment. The responsible agencies must file incident reports with the correct authorities for input into the appropriate databases.

7-44. The investigator should collect enough information to facilitate entering it into the following databases as appropriate:

- Arson and Explosives National Repository (BATF).
- Bomb Data Center (FBI).
- National Fire Incident Reporting System (NFIRS) (US Fire Administration [USFA]).
- National Incident-Based Reporting System (NIBRS) (FBI).
- State and local fire incident reporting systems.

EXPLOSIVES

7-45. Bomb threats (written or telephonic), bombings, and other explosive incidents on military installations are immediately reported to local police, fire, USACIDC, BATF, FBI, and other appropriate agencies. The examination of an explosion scene is generally conducted in the same manner as an arson investigation. In some cases, an explosion may lead to a subsequent fire. Like arson scenes, it is extremely important that evidence or debris is not moved before careful examination. The relationship between the various pieces may be critical to reconstructing the cause of the explosion.

CHARACTERISTICS OF EXPLOSIVES

7-46. An explosive is a substance that, through chemical reaction, violently changes into a gas, creating pressure and liberating heat. Explosives are divided into the following two classes:

- **Low-order explosives.** With low-order explosives, the rate of change to a gaseous state is relatively slow and must be in a compressed or enclosed state to explode. Low-order explosives tend to produce large chunks of debris. Examples of low-order explosives are black powder, smokeless powders, volatile flammable liquids, and flammable gases. These explosives can also be called deflagration agents. A dust or grain explosion can also be considered a low-order explosion.
- **High-order explosives.** With high-order explosives, the rate of change to the gaseous state is extremely rapid. They tend to pulverize everything nearby. Compressing or enclosing the explosive is not required. Such an explosion is said to detonate. High-order explosives include dynamite, military explosives, trinitrololuene (TNT), pentaerythrite tetranitrate (PETN), composition B (CB), composition 4 (C4), and mixtures of ammonium nitrate and fuel oil.

7-47. The effects of the two classes of explosives are different. Low-order explosions tend to "push" rather than shatter. Large chunks of debris can usually be found. Twisting and tearing of objects tend to occur. High-order explosives tend to shatter and fragment material near the center of detonation, and there is much evidence of impact by small, high-velocity missiles near the center of detonation. The resulting debris is in small fragments. Scientific and technical information on explosives can be found on official government Web sites.

EXPLOSIVES INVESTIGATIONS

7-48. Debris and soil close to the point of detonation are likely to bear residue from the explosive. Evidence of components used in the explosion may be found. Some examples are fragments of blasting caps, safety fuse fragments, wire, matches, match folders, fuse lighters, batteries or other sources of electric current, fragments of a timing device, delay mechanisms, or switches. Information of these accessories assists in the investigation. Containers or material foreign to the scene should be collected. The site should be checked for—

- Fingerprints.
- Footprints.
- Tire tracks.
- Toolmarks.
- Unusual odors.

7-49. The following procedures serve only as a guide. The ultimate determination of how to handle an explosive incident must be made by the investigating team responding to the scene. Availability of time, personnel, and other resources greatly influences what course of action to choose.

- The team leader—
  - Selects and assembles personnel and equipment and coordinates with other jurisdictions.
  - Conducts a scene overview.
  - Determines and establishes scene integrity, security, and safety.
  - Establishes command post and media control.
  - Conducts a scene walk-through with the explosives technician and forensic chemist.
  - Coordinates all personnel and search patterns.
  - Assigns immediate area search and investigative units.
  - Assigns the general area search and investigative units.
  - Manages, evaluates, and finalizes search and investigative actions.
  - Conducts final scene evaluation conferences.
- The photographer—
  - Selects and assembles equipment.
  - Photographs the immediate and general area including the victims, crowd, and vehicles.
  - Photographs team operations.
  - Photographs the blast seat and damage showing the measurements.
  - Photographs evidence as found
  - Photographs the immediate and general area from an aerial perspective.
  - Takes scene reconstruction photographic series.
  - Photographs blueprints, maps, and previous photographs of the scene, if necessary.

  - Photographs known or potential suspects.
  - Identifies additional photographic needs with all scene investigators.
- The evidence technician—
  - Selects, assembles, and distributes collection equipment to the search team members.
  - Prepares an evidence control log, and sets up an evidence collection point.
  - Coordinates and controls evidence collection techniques and procedures.
  - Records the receipt of all properly marked and packaged evidence from search teams on the evidence control log.
  - Categorizes collected evidence.
  - Maintains custody and control of collected evidence at the scene.
  - Coordinates with the team leader and other investigators.
  - Verifies collected evidence with the evidence control log before departing the scene.
  - Documents the chain of custody and provides temporary storage.
  - Prepares laboratory analysis requests, and transmits evidence to the laboratory.
- The schematic sketch artist—
  - Selects and assembles equipment.
  - Diagrams the immediate blast area.
  - Diagrams the general area.
  - Identifies the evidence found by indicating the assigned evidence numbers on the evidence control sketch showing the location where the evidence was found.
  - Shows the necessary measurements of heights, lengths, and widths.
  - Makes an artist's conception of the scene before the blast with the help of witnesses showing where furniture was arranged or how the structure was before the explosion.
  - Prepares a legend on the diagrams.
  - Inventories collected evidence with the evidence technician and ensures that all evidence is noted on the control sketch.
  - Marks and identifies the evidence control sketch and other diagrams for proper court presentation.
  - Coordinates with the team leader and other investigators.
- The immediate area investigative unit—
  - Selects and assembles investigative equipment.
  - Interviews local officers, firemen, and all possible witnesses at scene.
  - Determines the owner of the property, the victim of the explosion, and if any persons were injured in the blast.
  - Obtains names of any persons that are normally on the premises, such as employees, watchmen, or janitors.

  - Provides the general area investigative unit with the names and location of all persons or groups who should be interviewed. Ensures that the list includes the injured persons who were taken to a hospital or rescue workers who have departed from the scene.
  - Records the descriptions and times of the sounds, colors of the smoke, and any odors noticed by witnesses.
  - Identifies all persons at the explosion scene, and coordinates with the photographer to film the crowd and vehicles in the immediate and general area.
  - Questions the witnesses, and records the facts pertaining to the general activity at the scene before the explosion.
  - Questions the witnesses, and records the facts pertaining to anything unusual about the activity or any facts concerning unidentified packages, items, persons, or vehicles.
  - Reconstructs the immediate area activity, and coordinates with the team leader and other investigators.
- The immediate area search unit—
  - Selects and assembles equipment.
  - Stays alert for structural hazards, secondary devices, and entrapment devices before and after entering the blast area.
  - Locates the seat of the explosion or the point of origin of the fire.
  - Coordinates with the schematic artist and photographer before disturbing the crater or immediate blast area.
  - Measures and records the size, depth, and shape of the crater or damage.
  - Collects samples from the blast seat, and retains the necessary control samples.
  - Searches and sifts the seat of the explosion for device components.
  - Divides the immediate area into a search pattern, and makes a methodical search. Searches from the seat of the explosion to an expanded area that overlaps with the general area search unit.
  - Records and packages (individually) the evidence found, and follows routine procedures with the photographer, schematic artist, and the evidence technician.
  - Reconstructs the immediate area scene, and coordinates with the team leader and other investigators.
- The general area search unit—
  - Selects and assembles equipment, and coordinates the search pattern.
  - Stays alert for structural hazards, secondary devices, and entrapment devices.
  - Checks all surrounding buildings, vehicles, and objects for damage by missiles from the explosion, and marks them for the photographer and the schematic artist
  - Searches the area of ingress and egress for associative evidence, such as footprints, tire tracks, torn clothing, blood, hair, fingerprints, or other evidence that may relate to a suspect.

  - Searches the area for evidence from the explosion.
  - Searches rooftops and trees or other high places that may have caught debris from the explosion. Documents the blast effect and glass breakage in the surrounding area.
  - Determines the extent of the outer perimeter of thrown missiles and evidence. Indicates this finding to the schematic artist, photographer, and explosives technician
  - Adjusts the outer perimeter of the search pattern as necessary.
  - Records and packages (individually) the evidence found. Coordinates with the evidence technician, schematic artist, and photographer.
  - Reconstructs the general area scene, and coordinates with the team leader and other investigators.
- The general area investigative unit—
  - Selects and assembles investigative equipment.
  - Reviews the maps, evaluates the ingress and egress, and selects a methodical pattern for canvassing the area.
  - Determines the possibility of deliverymen being in the area, and makes a list of their names and addresses for follow-up interviews.
  - Canvasses the neighborhood for witnesses.
  - Canvasses business premises that may be related to ingress and egress (such as all-night service stations, cafes, taverns, and toll bridges).
  - Prepares a suspect list with necessary facts relating to the investigation.
  - Records descriptions of suspects, suspect vehicles, and suspect premises for future use.
  - Checks sources of device components and/or materials recovered at the crime scene.
  - Follows team leader assignments as developed from the scene, and makes an overall evaluation from the conferences.
  - Maintains communication with the team leader, and coordinates with other investigators.
- The forensic chemist—
  - Selects and assembles equipment.
  - Conducts a preliminary walk-through of the scene with the explosives technician and team leader.
  - Assists the team leader in evaluating the situation and discusses the method of approach.
  - Assists the general area search unit where appropriate.
  - Assists the immediate area search unit.
  - Coordinates with the state and/or local laboratory personnel, as appropriate.
  - Acts as a technical advisor for all laboratory-oriented questions arising at the scene.
  - Conducts field tests, where appropriate.

- Assists the evidence technician and team leader in the evaluation of collected evidence.
- Assists the evidence technician with proper packaging for submission to the laboratory.

- The explosives technician—
  - Selects and assembles equipment.
  - Accompanies the team leader on a walk-through in order to provide a technical evaluation and assessment of the fire or explosion scene.
  - Establishes scene parameters
  - Identifies the seat of the explosion or point of origin of the fire.
  - Evaluates investigative information and recovered materials.
  - Determines whether the incident was incendiary or accidental.
  - Reconstructs the sequence of occurrence and physical evidence.
  - Provides technical briefings.
  - Prepares a statement regarding the technical determination.

Chapter 8

Assault

Assault is among the most serious and feared criminal offenses because it involves threatened or actual violence to the victim. This offense occurs more frequently than either rape or homicide. Many assault victims suffer serious injuries ranging from broken bones to life-threatening gunshot or knife wounds.

DEFINITIONS AND LEGAL CONSIDERATIONS

8-1. The *UCMJ, Article 128,* identifies basic assaults. The *UCMJ, Article 134,* encompasses assaults involving intent to commit certain crimes of a civil nature.

ASSAULT

8-2. An assault is an attempt or offer with unlawful violence of force to do bodily harm to another individual whether or not the attempt or offer is accomplished *(UCMJ, Article 128).* An attempt type assault must contain an overt act as required under *UCMJ, Article 80.* An offer type assault must contain some overt act whereby a person is put in fear of bodily harm. To constitute an assault, the act of violence must be unlawful.

BATTERY

8-3. Battery is an assault in which the threat to do bodily harm is achieved. To constitute battery, it must be established that unlawful force or violence caused bodily harm to a certain person as alleged.

AGGRAVATED ASSAULT

8-4. There are two types of aggravated assault. The first is an assault with a dangerous weapon or another means of force likely to produce death or grievous bodily harm. It is not necessary that death or grievous bodily harm be actually inflicted. The second type of aggravated assault is that in which grievous bodily harm is intentionally inflicted with or without the use of a weapon. Grievous bodily harm does not include minor injuries, such as a black eye or a bloody nose, but does include fractured or dislocated bones, deep cuts, torn members of the body, serious damage to internal organs, and other serious bodily injuries. When it is difficult to determine the actual degree of assault (aggravated or simple), consult with the appropriate office of the SJA for advice.

ASSAULTS WITH THE INTENT TO COMMIT CERTAIN CRIMES

8-5. Assaults committed with the intent to commit certain other crimes are usually investigated during the investigative process of the intended crime. An assault with the intent to commit an offense is not necessarily the equivalent of an attempt to commit the intended offense. Assaults under this category are those perpetrated with the intent to commit murder, voluntary manslaughter, rape, robbery, arson, burglary, or housebreaking.

MAIMING

8-6. Maiming is committed when someone with the intent to cause injury to another inflicts injury that seriously disfigures, destroys, or disables any member or organ or seriously diminishes the physical vigor of another. According to the *UCMJ*, the following injuries are considered to be maiming:

- Putting out an eye.
- Cutting off a hand, foot, finger, or other appendage.
- Knocking out a tooth.
- Cutting off an ear.
- Scaring a face with acid.
- Causing other similar injuries that would destroy or disable any members or organs mentioned above or cause serious disfigurement.

8-7. A disfigurement does not necessarily mean that an entire member is mutilated; it may merely impair (perceptibly and materially) the victim's comeliness. Because maiming is a specific intent crime, it can only be associated with assault type investigations; however, it only requires intent to injure, not intent to maim. For example, if someone is seriously disfigured in a traffic accident, where there was no intent to cause injury to another, maiming cannot be associated with the incident. However, if two people engage in a fistfight and, during the course of the altercation, one of the parties bites off the ear or knocks out a tooth of the other party this would constitute maiming.

ASSAULT INVESTIGATIONS

8-8. Aggravated assault or assault in connection with other crimes may warrant the expertise of an investigator. When investigating an assault, the investigator must also be aware of techniques involved in investigating other crimes and offenses, such as robbery or murder. In cases of aggravated assault in which the victim has been badly injured, the investigator will normally follow many of the same techniques used to investigate a homicide. An assault investigation normally requires actions that include:

- Substantiating the allegation.
- Questioning the victim.
- Consulting with medical authorities to determine if the sustained injuries could have been inflicted as reported by the victim.
- Obtaining permission from the victim to photograph injuries.
- Providing the victim with necessary victim assistance information. (See *Chapter 1*.)

- Attempting to establish a motive for the assault.
- Conducting a careful, detailed search of the crime scene. (See *Chapter 5.*)
- Taking steps to trace weapons or other physical evidence, as appropriate.
- Seeking, locating, and questioning witnesses as soon as possible after the assault.
- Warning all suspects of their rights before questioning.

SUBSTANTIATING THE ALLEGATION

8-9. When an alleged assault has been reported, first establish that the offense did occur. Normally, this can be established by questioning the victim, an attending physician, and any available witnesses.

8-10. To investigate an assault—

- Interview the victim to substantiate the allegation and determine who the complainant was (if other than the victim).
- Search the scene.
- Interview witnesses, to include medical personnel, and suspects.

INTERVIEWING THE VICTIM

8-11. While interviewing the victim, make a determination as to whether or not the elements of proof for assault are included in the victim's account of the incident. For example, if the installation emergency room notifies the military police station that it is treating an individual who was struck by a baseball bat, this does not necessarily substantiate that an assault occurred. Upon interviewing the victim, it may be determined that the individual was accidentally struck with a bat during a friendly baseball game. Because there was no intent to do bodily harm, there is no assault.

8-12. Make attempts to record, as chronologically as possible, the events that took place. This allows the investigator to double check the victim's story. Many victims lie during initial interviews. This does not indicate that they are not victims; even individuals who are legitimate victims frequently lie about some aspect of the incident. Feelings of guilt, embarrassment, or the attempt to hide other facts from the investigator may cause a victim to lie. Investigators who learn to appreciate what can cause victims to omit information or lie to avoid embarrassment will likely gain greater cooperation from victims than those who make conclusions about conflicting information.

8-13. In unknown subject cases or complex investigations, it is generally advised that the victim be interviewed on at least two occasions. A victim may recall more information as time is spent in reflection of the incident. This is especially true when the victim has consumed excessive amounts of alcohol before the assault. Information received in the initial interview may be hazy and unclear. A second interview conducted a day or two later may produce additional details that the victim did not initially recall. Refer to *Chapter 4* for a complete discussion on I&I.

QUESTIONING AND PHOTOGRAPHING THE VICTIM

8-14. During initial questioning, obtain the victim's consent to take photographs of visible injuries. Photography techniques are described in *Chapter 6*. In general, use color film or color, digital photography during the preliminary investigation, and take another set of photographs about three days later to show the full extent of the injuries, because bruises may take up to 72 hours to completely form.

8-15. To avoid lengthy initial questioning of the victim, the investigator should prepare questions based on the information provided by military police. This action assists in conducting a timely crime scene search and may assist in the identification and capture of the assailant (if at large). If there is a possibility that the injuries are life-threatening, an effort must be made to obtain as much information and detail about the incident as possible. Interviews of a dying victim must be conducted with the cooperation of medical personnel. If the interview will likely result in a "dying declaration," it should be recorded via audio and video whenever possible. Such tapings require informed consent of the person being interviewed. A dying declaration consists of the utterances of a dying person and can be used in court as an exception to the hearsay rules of evidence. Investigators should ask a victim of an assault the following questions:

- What were you doing before the assault?
- What did you do after the assault?
- Did you have a confrontation or meeting with the suspect before the assault?
- Who witnessed the assault?

8-16. If the assailant used a weapon, ask the victim to describe it. During the initial interview, victims may be able to identify the assailant and provide insight as to the motivation of the assailant. In cases where the assailant is unknown to the victim, the investigator must obtain a composite sketch as soon as possible. It is extremely rare that a victim cannot provide some insight as to the motivation of the assailant, because most assaults result from an emotional outburst during which the assailant frequently makes statements to the victim as to why he is committing the assault.

8-17. Investigators must ask a victim other relevant questions to establish his actions leading up to the assault. In later interviews with the victim, information from previous interviews should be revisited.

8-18. If questioning a victim fails to identify the assailant, check the victim's background, associates, and activities. Check police and personnel records to see if the victim has been involved in previous incidents. Question relatives, members of his military unit, neighbors, and so forth. An investigator may find that the victim has a motive for withholding information, such as his involvement in drug or other illicit activities, which may have spawned the attack. However, telling the police about these activities would likely incriminate the victim in other offenses. Information gleaned through this process may aid the investigator in a reinterview of the victim to help reveal additional details of the reported offense.

DOCUMENTING THE VICTIM'S MEDICAL TREATMENT

8-19. When a victim receives medical treatment, the investigator obtains copies of his treatment record where the attending physician recorded the physical injuries and the victim's accounts of how the injuries were sustained. A written sworn statement from the attending physician can be of great assistance to the investigation, but may not always be practical due to the pace of the emergency room. An interview of the physician should be conducted to record his observations, the treatment provided, the prognosis, recommended follow-up treatment, and other information he may have obtained during the treatment process.

8-20. The investigator should ask the physician the estimated age of all injuries, which can be provided based on the coloration of bruising and the healing stage of the wounds. The treating physician should be able to determine if the injuries sustained were consistent with the accounts of the sequence of events provided and be able to articulate how the injuries logically occurred. This is especially important in suspected assaults on children where the parent or guardian, who may have caused the injuries, provides the only story as to how the injuries were sustained. The physician may also be able to help identify the type of weapon that could have caused the injuries.

8-21. In cases where a victim has been substantially traumatized or injured, use discretion. The investigator should assist the victim in seeking professional counseling from a member of the clergy or social work services.

EXAMINING PHYSICAL ASSAULT VICTIMS

8-22. When investigating a physical assault, investigators should follow the procedures described in *CID Pamphlet 195-5*. This pamphlet includes the following:

- Have the victim examined by a medical doctor.
- Seize any evidence recovered during the examination.
- Collect fingernail scrapings and the results of a head combing and plucking from the victim.
- Collect blood from the victim for comparison with blood at the scene or on clothes.
- Arrange for a change of clothing for the victim.
- Collect clothing worn by the victim at the time of the attack.
- Have the victim disrobe while standing on paper or a hospital sheet. Place each item of clothing into separate paper bags and properly seal. Be sure to collect the paper or sheet as evidence, taking care to fold it up with the relevant surface on the inside.
- Air-dry any bloodstained clothing at the office in an area separate from the suspect's clothing. Do not use a fan or similar method of speed-drying.
- Take color photographs (with and without a ruler) of the injuries. Use discretion. Have medical personnel or a medical photographer expose any photographs of the genitals and/or the anal or breast area if these areas are injured.

- Take photographs of individual injuries with the camera parallel to the body surface.
- Take all close-up photographs with and without a scale.
- Photograph injuries over a period of several days.
- Ask the doctor what type of weapon might have caused the injuries and determine if the injuries are consistent with the victim's account of the assault.
- Obtain a copy of all the associated medical reports.
- Consider attempting to develop latent fingerprint impressions on the skin, if appropriate.

CRIME SCENE SEARCH

8-23. After questioning the victim, search the crime scene according to *Chapter 5* of this manual. Ascertain if the scene has been disturbed and identify who has been in the scene before, during, and after the incident (to include medical and law enforcement personnel). Photograph the scene before the collection of evidence (see *Chapter 6*). Determine if the victim knows the type of weapon used or other specific items touched or disturbed during the commission of the crime. Attempt to locate these items. If the suspect took the weapon when he departed, get a complete description of the weapon for later searches. The investigator must place special emphasis on areas, such as a bed, closet, or latrine when the victim describes that location as the place in which the incident occurred.

8-24. The number of personnel in the crime scene must be kept to a minimum. Only those personnel working the crime scene and providing medical care should be allowed access. Witnesses and suspects should be removed from the crime scene and secured in another location for later questioning. The witnesses and suspects must be separated from one another so that they cannot confer with each other.

8-25. If the victim is not present during the search, the investigator may want to go to the scene with him after the initial search has been conducted. The investigator should ask the victim to walk him through the incident, describe the action, and show the location where the assault took place. This may give the investigator a better understanding of how the assault happened. It could also lead to additional evidence. It is good practice to return to the scene at the same time of day and on the same day of the week as when the incident occurred. This will allow investigators to make observations about lighting, traffic, deliveries, trash collection, and so forth.

IDENTIFICATION OF WITNESSES AND SUSPECTS

8-26. Investigators must identify all potential witnesses and suspects in an assault, either through the victim or other present witnesses. Additionally, evidence, motive, and opportunity may lead to possible suspects.

QUESTIONING WITNESSES

8-27. Investigators must locate (if necessary) and question the witnesses to an assault as soon as possible. Witness interviews must be conducted separately because each witness will see different aspects of the same event and will report them differently. Other witness accounts can skew the true observations of each individual. The investigator must—

- Conduct each witness interview independently.
- Allow each witness to describe what he saw. Ensure that the information he provides is the result of his personal observations and not what he was told by another witness.
- Ask the same questions pertaining to events leading up to, during, and after the assault to determine the credibility of the victim and/or witnesses.
- Determine what attracted the attention of the witnesses to the event, which will aid in developing a chronological account of the assault based on what people saw and when they began seeing it.
- Explore any gaps or differences in the description of events in an attempt to separate fact from speculation.
- Conduct canvass and witness interviews as soon as possible after the event, because human memory is volatile and will fade rapidly with the passage of time.
- Determine if any of the witnesses knew the parties involved and, if so, the nature of their relationship.
- Obtain full, physical descriptions of the persons the witnesses observed in the event (to include other potential witnesses). This includes the individuals' names (if known), what they were wearing, scars, and other individualities.
- Record all personal information of the witnesses (to include contact numbers) for follow-up interviews, as necessary.

UNDERSTANDING WITNESS ACCOUNTS

8-28. In almost all instances, witness accounts will differ to some degree. This does not necessarily indicate that any of the witnesses are lying, but that they have seen different aspects of the same event. In some cases, witnesses who have actually observed an attack will be reluctant to come forward out of fear; whereas, others who were not present may come forward in support of a friend or relative. Those who attempt to describe events differently than they actually occurred will stand out with time and can be refuted through supporting evidence. Those who did not actually observe the event, but attempt to assert information will not be able to provide detailed information about who was standing where, who said what to whom, where the other witnesses were, and so forth. The lack of detailed information should always alert investigators to use caution when trying to piece the incident together. These reasons support the importance of obtaining detailed, chronological statements from each witness.

QUESTIONING SUSPECTS

8-29. After a suspect has been identified, he must be advised of his rights. If the suspect has waived his rights, the investigator can begin interrogating him. The investigator should base his questions on information and evidence he developed throughout the investigation. Evidence considerations should be for both testimonial and potential physical evidence. The investigator should consider the following questions:

- Did the suspect have a motive?
- Did the suspect have the opportunity to commit the assault?
- Was the suspect in the area at the time of the assault?
- Did the suspect have access to the type of weapon used in the assault?
- Can any evidence found at the scene be tied to the suspect?
- Are the suspect's alibis supported?

PROCESSING SUSPECTS

8-30. Once an investigator has determined that a suspect is most likely the assailant, he must process the suspect. He must get fingerprints, photographs, and a written sworn statement when a suspect agrees to provide a statement. The investigator must obtain any possible evidence from the suspect, such as photographs of any cuts and abrasions, the weapon that was used to commit the assault, and bloody clothing. When there is any uncertainty to the involvement of a suspect in an assault, consideration should be given to photographic, physical, and voice lineups of the suspects with the witnesses and victims. To conduct these lineups (see *Chapter 3*), use the suspect and five other persons (not suspected of having involvement in the incident).

INSTRUCTING A VIEWER BEFORE THE LINEUP

8-31. Before the viewing process, the viewer should be instructed that the person who committed the crime may or may not be in the lineup and that there is no expectation for him to recognize anyone. He should view each individual or photograph for several seconds or for as long as he feels necessary to determine if he recognizes anyone. He should not make any statements until he has viewed all of the parties in the lineup. Once he has viewed all parties, he should tell the investigator if he recognizes any of the participants and why he believes he recognizes that person. Emphasis must again be made that the perpetrator may or may not be in the lineup, that the viewer should not feel any pressure to make an identification, and that no action will be taken solely on his ability or inability to identify an individual in the lineup. Pressure on a viewer to make identification can result in the identification of an individual who was not involved in the assault and must be avoided. Voice lineups are conducted in the same basic fashion with each individual reading words spoken during the crime.

PREPARING FOR A LINEUP

8-32. In photographic and physical lineups, all participants should meet the same basic physical characteristics as the suspect; this includes approximate height and weight, the wearing of glasses, facial hair, hairstyle, and so forth. In a physical lineup—

- The suspect should be allowed to select his number (1 through 6).
- Each participant will hold his number in front of his abdominal area during the lineup.
- All of the participants are directed to stand on a line in numerical order.
- A photograph of the lineup is taken with color film or digitally to capture the composition of the lineup.
- The victim views the lineup. This process is completed three times regardless of the viewer's (victim or witness) reaction, such as a positive identification of an individual or the inability to recognize any of the parties.

8-33. Photographic lineups are conducted in the same manner. However, the investigator should arrange the photographs under a template, which allows all of the photographs to be aligned the same way exposing the same portion of the lineup photographs. During a photographic lineup—

- The investigator photocopies the lineup composition between each viewing and has the viewer annotate his observations at the completion of each viewing.
- The viewer (victim or witness) is instructed to circle the photograph on the photocopy when he recognizes an individual in the lineup and is told to write in his own words how he believes this individual is familiar to him.
- The viewer annotates on the photocopy when he does not recognize anyone.

EXAMINING A PHYSICAL ASSAULT SUSPECT

8-34. The need to examine the suspect will depend on the possibility of finding incriminating evidence and the amount of needed corroborating evidence. When an examination of a physical assault suspect is necessary, the investigator should—

- Obtain legal authorization, as necessary.
- Look for any injuries on the suspect and any blood transferred from the victim.
- Recover the suspect's clothing, sealing each item into a separate paper bag, and collect known samples, if appropriate.
- Consider blood alcohol and drug tests.
- Search for weapons used in the crime.
- Search for other evidence that might link the suspect to the crime scene.
- Consider using an alternate light source.

Chapter 9

Black Marketing

Black marketing provides an opportunity for a seller to make money and a buyer to possess an item that is restricted from his possession. By definition, black marketing is the illegal business of buying or selling goods in violation of restrictions or price controls. Black marketing is punishable under the *UCMJ*, but because of its financial appeal, many soldiers, their family members, US government employees, and other personnel participate in its illegal activities. Those who are restricted from possessing black marketing items buy them because of the enjoyment, status, or sole desire of having the item. Both the seller and buyer are aware of the legal consequences associated with participating in black marketing activities.

DETECT BLACK MARKET ACTIVITY

9-1. Black marketing thrives when goods commonly available in developed nations are imported into developing nations on a restricted basis. The potential for black market activities exists whenever US forces are located in a HN. US forces introduce, through supply and PX channels and by HN allowances of tax-exempt mail and baggage, many items that are not available to the HN populace through commercial markets. Many of these items have a potential or an actual black market value.

9-2. The impact of black market activities on the economy of a HN can be devastating. The cost to the United States of replacing military supplies and equipment that have been diverted to the black market is expensive. Mission performance can be greatly reduced when needed supplies and equipment are not available to commanders.

9-3. Illegal trafficking of legally purchased PX items or issued supply items is a constant problem. This type of trafficking can be harder to stop than cargo diversions. Legal or authorized access to these items by host countries and third-country nationals (TCN) makes it difficult to stop black marketing. Cases involving negotiable-dollar instruments on the black market need special attention because these instruments can cause a direct-dollar loss to the US.

9-4. Black market activities contribute to the commission of other crimes. Black marketing can promote fraud, the stealing and selling of government property, counterfeiting, forgery, and drug offenses. It may also promote the violation of foreign, currency exchange laws, import laws, tax laws, and SOFAs.

9-5. Black market operations may be the result of an organized group or an individual's efforts. They are often the result of a combination of the two.

BLACK MARKET RING

9-6. One type of black market organization and operation is the black market ring. While the type of contacts may change from one ring to another, the basic organization and its operation stay much the same. The leader and suppliers are the most essential members in a ring. The supplier discovers a supply source, which may be an Army facility or activity that stores, handles, or uses the item. The supplier gets the item through purchase or theft. He may be a military person, a civilian employee, or an acquaintance of someone who will get the item for him.

9-7. After the supplier gets the item, he takes it to the operator of a temporary storage point or uses a transporter to deliver it. The transporter may or may not be a regular member of the ring. The item is then passed, on demand, to the retailer. Sometimes a wholesaler may act for the leader and have a transporter get the item from the storage point and take it to the retailer, or the retailer may get it directly from the warehouse. The retailer then sells the item to the consumer.

9-8. The leader may have any number of suppliers, transporters, warehousemen, wholesalers, and retailers. If he has more than one wholesaler, each will normally handle only one type of item. Most often, the only individuals in the ring who have direct contact with or know the leader are the suppliers and the wholesalers.

BLACK MARKET RING BREAKUP

9-9. A means to breaking up a black market ring is to trace commodities found on the black market. Examining these items may help an investigator find the supply source. When an Army facility is among the supply sources, a check of the facility records can show if items are being removed illegally. Enlisting the aid of the commander of this facility will expedite this action. Surveillance may be used to detect the supplier or the transporter. If either one is seen, he may be placed under surveillance to find other members of the ring. Investigators should also check on personnel or activities at the military depot or warehouse to identify ring members or individuals being exploited as suppliers.

9-10. Transporters may be spotted when items in bulk storage are being removed. Investigators may buy information from transporters, especially if they are not members of the ring. The pay for their work may be small, and they may be willing to tell what they know for a small fee or other consideration. Information may also be gained in the operation of the ring by planting would-be transporters. Individuals like taxi drivers work well as plants in black market areas. A retailer may approach a plant for a onetime job. A retailer, wholesaler, or warehouse representative may be found in this manner. Other ways of making contacts with retailers is by purchasing items or stationing personnel in black marketing locations to watch for transactions. When a retailer is spotted, maintaining surveillance may reveal other members of the ring.

9-11. Large-scale diversions and inventory shortages are major signs of organized rings of black marketers. Good security controls over black marketable items in supply channels can reduce these rings. Screening reports of supply shortages or thefts may yield important information. They consist of police reports, reports of survey (ROSs), inventory adjustment reports, and so forth. Because federal employment requires frequent relocation of personnel, these reports must be screened promptly. Any leads should be checked out as soon as possible. Experience will show which missing supplies are likely to go on the black market.

OBTAIN INFORMATION

9-12. People working at commissaries and PXs may detect irregularities on the part of fellow employees. They may spot the sale of black marketable items to certain individuals. Merchants often know about commodities that are being procured for the black market. The black market is competition for the merchants' businesses so they may provide the names of individuals who have competing commodities for sale. Gate guards and taxi drivers are often good sources for black marketing information. Gate guards can answer questions regarding who is taking a lot of controlled items off an installation and can identify the items in demand on the local black market. Taxi drivers, by virtue of the nature of their work, often come in contact with many people. They too, may know what items are in demand. Personnel assigned to a ration control office can be good sources of information because they can provide information in the line of duty (LOD).

9-13. Men who have acquired money illegally may keep or patronize prostitutes. These women, by association with such men, may pick up information. If prostitutes can be persuaded to talk, they may give tips on men who have money in excess of normal amounts. They may also give tips on actual black marketers and their operations.

9-14. Civil affairs personnel deal with the economy of a HN area. They may be able to help trace black market transactions. For example, if the problem is big enough to affect the country's economy, they may be able to pinpoint where this activity exists. Such pinpointing can help investigators concentrate their efforts in these areas.

LOCATE A SUPPLY SOURCE

9-15. If an investigator suspects that an item on the black market is coming from a certain source, other like items at that source may be marked for future tracing. One of several inks or powders may be used for this. They are invisible unless developed by specific reagents or exposed to infrared or ultraviolet (UV) light. With commissary or PX items, you may mark the price on the items in special ink or like substance.

9-16. With petroleum products, an identification reagent (an additive) can be added at POL storage and supply facilities. The reagent can be detected by chemical testing after the seizure of suspected petroleum products. This is normally done under the supervision of USACIDC personnel. USACIDC is responsible for local administration and control of operations where the reagent will be added. They also set the type and quantity of fuel to be identified. Normally, the additive is blended in with the POL products as the storage tanks are being filled. The proper ratios for blending are discussed in the appropriate Army materiel command technical bulletins. If the additive is blended as prescribed, there should be no interference in the operation of a motor. These approved additive reagents can be requested on an as-needed basis. One field-expedient reagent is 2 2/3 ounces of phenolphthalein in one pint of alcohol added per 1,000 gallons of gasoline. This should be coordinated with POL technical personnel.

9-17. If there is reason to believe a certain place has black market items, an investigator may wish to have it raided. Contact the office of the SJA to ascertain what authorities have jurisdiction in a given location. Civilian police may have to conduct the raid. Jurisdiction, for this purpose and with respect to the apprehension and search of individuals found there, is affected by applicable treaties, laws, or other directives.

IDENTIFY SUSPECTS

9-18. Certain acts or conditions may show black market involvement. A person may be a suspect if he meets certain criteria. Such criteria includes—

- Having more money than would normally be expected for someone of that rank or position.
- Spending more money than legally received or spending large amounts of money on friends.
- Purchasing unusually large quantities of items.
- Purchasing items not normally used, such as a light smoker who buys large quantities of cigarettes or a male soldier who frequently buys quantities of perfume or lipstick.
- Carrying a lot of goods off the compound on a regular basis.

9-19. Special consideration should be given to known narcotic addicts. They may engage in black marketing for money in order to buy drugs or to obtain drugs in exchange for their services.

9-20. Once enough information about a suspect is gathered, an investigation should be initiated. If it is believed that a suspect has unusual amounts of money, make inquiries with the post and finance offices. In response to an official request, postal officials may provide information on the purchase of money orders. The dollar amount purchased may show that a buyer had more money than he or she might be expected to have. This information may provide grounds for further investigation. The finance office may be able to disclose information regarding military personnel who have exchanged large amounts of money at its location. Travel agencies may also be a source of information for potential black marketing activities. They keep records of the trips they arrange for people, which may show frequency of travel and consistency in the destination of individuals.

SUSPECT'S ASSOCIATES

9-21. Investigators must make contact with a suspect's associates. The contacts may provide pertinent information on the suspect and may disclose other members of the group. An investigator should study a suspect's habits and customs to learn about his character. This may help determine whether the person would or would not engage in unlawful activities. Other checks an investigator must make of a suspect are—

- His personnel records for anything of value.
- His bank account. It may show if he has deposited more money than he is known to have received legally. For military personnel, this check should include the soldier's deposits and other authorized investments.
- His private property. It may show income that exceeds what he is known to have obtained legally.

AGENCY CHECKS

9-22. Agencies such as the FBI, US Citizenship and Immigration Services (USCIS), and US Customs and Border Protection (CBP) may be checked in the investigation of a suspect. Local agencies often keep records that may be of value to the investigation. These records may show that a suspect had money in excess of what he should have legally. Civilian police records may provide information leading to a suspect, especially if he is a local civilian or has been residing in the area for a while.

SUSPECT SURVEILLANCE

9-23. A suspect may be put under surveillance to complete the investigation. The investigator may choose to use that suspect in gaining new information by letting him lead the investigator to others engaged in black marketing. A surveillant may be placed at banks, finance offices, or other places that convert money instruments into dollars to watch for suspects who make monetary transactions.

Chapter 10

Burglary, Housebreaking, and Larceny Investigations

Property crimes are among the most common crimes reported and investigated by the law enforcement community. However, many of these crimes are never reported due to a lack of public interest or moral outrage.

OVERVIEW

10-1. Property crimes require extensive knowledge and expertise to solve because quite often, there are no witnesses, victims cannot identify the stolen property, or the property is never recovered. Physical evidence in property crimes is often similar to that found in crimes against people including fingerprints, foot and tire impressions, toolmarks, trace and serological evidence, and objects left at the crime scene.

LEGAL DESCRIPTION OF BURGLARY

10-2. Burglary is the breaking and entering of a dwelling house of another during the hours of darkness with the intent to commit a crime, whether the intent is carried out or not. Burglary is primarily an offense against the security of habitation. The break-in may be by physical force or by trickery, such as posing as a utility worker. Entering by removing or opening any part of a dwelling, such as a screen, window, or door meets the requirement of "breaking." Breaking and entering must occur between sunset and sunrise when there is not enough light to discern an individual's face. It must be done with the intent to commit murder, manslaughter, rape, sodomy, carnal knowledge, larceny, wrongful appropriation, robbery, forgery, arson, extortion, maiming, or assault *(UCMJ, Articles 118* through *128, except Article 123a).* Any person (subject to the *UCMJ)* who, with the intent to commit these offenses, breaks and enters the dwelling house of another in the hours of darkness is guilty of burglary *(UCMJ, Article 129).* As soon as any part of the body is inserted into the dwelling, the requirement of "entry" is met. Inserting an object into the dwelling to extract property also qualifies as "entry." It is immaterial whether the intended offense at the time of breaking and entering was committed or even attempted, but the intent to commit the act is essential to the proof of burglary.

LEGAL DESCRIPTION OF HOUSEBREAKING

10-3. Housebreaking *(UCMJ, Article 130)* is like burglary in that the intruder enters a structure unlawfully with the intent to commit a criminal offense within the structure. However, the offense that the intruder intends to commit does not need to be covered under *UCMJ, Articles 118* through *128.* The intent need only be to commit some criminal offense. Housebreaking also differs from burglary in that the place entered does not need to be a dwelling, does not need to be occupied, and does not require the opening of any part of a dwelling. The intent to commit some criminal offense is an essential element of housebreaking and must be alleged and proven in order to support a conviction of this offense.

LEGAL DESCRIPTION OF UNLAWFUL ENTRY

10-4. Unlawful entry *(UCMJ [Article 134])* is closely related to housebreaking. It is an entry on land or into structures on that land, effected peacefully without force, accomplished by means of fraud or some other willful wrong. Unlike housebreaking, the intent to commit an offense within the place entered is not necessary to constitute this offense. The element of proof is that the entry was accomplished by means of fraud or some other willful wrong.

LEGAL DESCRIPTION OF LARCENY AND WRONGFUL APPROPRIATION

10-5. Larceny is when any person (subject to the *UCMJ)* wrongfully takes, obtains, or withholds, by any means, from the possession of the owner or of any other person any money, personal property, or article of value of any kind with the intent to permanently deprive or defraud another person of the use and benefit of the property or to appropriate the same to his own use or the use of any person other than the owner and steals that property *(UCMJ [Article 121(a)(1)])*. Wrongful appropriation is the same as larceny except that the intent is to temporarily, rather than permanently, deprive or defraud the owner or any other person of the property *(UCMJ [Article 121(a)(2)])*.

10-6. The crime of larceny includes common-law larceny, false pretenses (fraud), and embezzlement. A wrongful acquisition, assumption, or exercise of dominion over the property of another is what each of these crimes have in common. Each contains the proof of intent of the accused to permanently deprive the owner of the property.

- **Common-law larceny.** This offense requires the taking (by trespassing) and felonious carrying away of property belonging to another with the intent to deprive him of that property permanently. This includes shoplifting and pilferage.
- **False pretenses (fraud).** This offense contains all the elements of larceny, but the property is obtained by a misrepresentation of an existing fact or condition that the victim relied on.
- **Embezzlement.** This offense occurs when an individual who lawfully receives the property of another through a position of trust intentionally and unlawfully withholds it. For example, if a bank teller receives money to pay customers, retains part of the money, and then alters his records to cover up the amount he has taken.

10-7. As previously stated, the crime of larceny often results from burglary or housebreaking. However, other forms of larceny may include frauds against soldiers, larcenies involving safes, larceny of motor vehicles, and shoplifting.

Use the same investigative techniques to investigate these crimes as you would other crimes against property.

10-8. The element of proof for larceny is that the accused wrongfully took, obtained, or withheld the owner's or other individual's possession of the item in question. Generally, moving the property or having dominion over it with the intent to permanently deprive the owner of the item without his consent meets the element of proof. Simply receiving, buying, or concealing stolen property or being an accessory after the fact are not included.

10-9. It must also be shown that the property is of some value and belonged to a known owner. The value of stolen property, other than items procured or issued from US government sources and listed by value in an official publication, is generally determined by its legitimate market value.

10-10. The lesser crime of wrongful appropriation has the same elements of proof as those for larceny. However, the intent is to temporarily, rather than permanently, deprive the owner of the property.

INVESTIGATIVE PROCEDURES

10-11. Investigating a burglary or housebreaking usually leads to other possible crimes, such as larceny. The investigative techniques for larceny often apply to both burglary and housebreaking. The investigator's goal is to identify and apprehend the offenders and recover stolen property. Crimes against property are generally investigated using the same procedures described in *Chapter 5*. Those procedures include—

- Measuring, photographing, videotaping, and sketching the scene.
- Searching for evidence.
- Identifying, collecting, examining, and marking physical evidence.
- Questioning victims, witnesses, and suspects.
- Documenting all statements and observations in notes.

INVESTIGATOR'S TIPS

10-12. The following tips are provided to assist an investigator when conducting an investigation of a burglary or housebreaking:

- Establish who had the means, opportunity, and motive to commit the offense.
- Eliminate the individuals who had access to the stolen property.
- Use the crime scene examination to determine how access was gained, how many people were involved, and what was the time frame of the occurrence. The presence of excessive damage is indicative of juvenile involvement.
- Establish what the suspect and victim did before and after the offense. This will aid in eliminating them as suspects or in establishing their ability or opportunity to participate in the crime.
- Canvass, thoroughly, the people who actually live, work, and spend time in the area where the crime was committed and inquire as to who might know the victim or suspect.

- Chase the property not the people. Investigators must make timely entries of stolen property into the NCIC files. They must expeditiously make checks of pawnshops, flea markets, and other sources used for quick resale of property. Internet sale sites should also be considered.
- Believe nothing without first verifying it. Use polygraph early.
- Send the evidence to the lab (always).

CRIME SCENE PROCESSING

10-13. The initial notification begins the investigative process. Obtain as much detail as possible during this phase of the investigation. Fully identify the caller and get detailed information about how the caller discovered the crime and what action the caller took at the crime scene. Inform the caller to avoid rearranging items and prevent others from entering the area. In many cases, the caller may have been notified of the crime by a third party. Identify the third party for later questioning. Use this information to develop a preliminary plan.

10-14. Proceed to the crime scene in a safe manner. On the way to the crime scene, watch for individuals fleeing the area, suspicious looking individuals still at the crime scene, and automobiles that appear to be out of place. Note the time of arrival, and record weather conditions. Verify that it is the correct location of the reported incident. If necessary, assist injured personnel. Set up crime scene security or if security has already been established, ensure that it is adequate. Initiate a personnel roster to document all personnel who enter the crime scene. Record this information in your notes.

10-15. Look for evidence in areas outside the immediate crime scene, such as parking lots, sidewalks, or footpaths. Consider having the military police record the license plate numbers of vehicles parked in the area. Observe the general condition of the crime scene.

10-16. During the initial observation, determine if a crime was committed and, if so, what type it was. Larceny is the offense most often involved during the investigation of a housebreaking or burglary. Get a detailed description of the missing property and record the exact location from which the item was taken. Get the location of all parties initially thought to be involved in the crime and the details concerning how the property was secured. Find out where the owners or occupants were at the time of the crime. Observe the mannerisms of the victims during the initial interview. If there is more than one victim, interview them separately. Take note of inconsistencies in their account of the incident.

10-17. Proof of ownership must be established early in the investigation. Obtain the names and addresses of individuals who can verify ownership or possession of the item. The owner may be able to provide documents or photographs that can help establish ownership, possession, and the value of the stolen property. If the stolen property was insured, get the amount, policy number, name and address of the insuring company, and the names of any beneficiaries of the policy. Insurance companies record detailed information on items that are insured.

10-18. Check for holes sawed or hacked through walls, floors, partitions, or ceilings. The size, shape, and location of openings are important clues in the investigation. Note the height of the openings from the ground or from where the offender stood. If possible, determine if entry was made bodily or with an instrument. Investigate the possibility that someone from inside the building assisted in the crime. Look carefully to see if any evidence was destroyed. In an attempt to mask a crime, offenders often wipe off fingerprints, wear gloves, deface toolmarks, or try to obscure footprints and tire tracks.

EVIDENCE SEARCH

10-19. The circumstances of a case must always guide the actions of the investigator when processing a crime scene. However, experience has shown that systematically searching for and collecting evidence is helpful in preventing errors.

10-20. Determine the initial points of entry and exit. Generally, evidence can be found at these points. Check locks or fasteners for evidence of tampering. Ask the victim to assist in the evidence search. The victim can identify items missing from the scene or moved during the crime.

10-21. Once the points of entry and exit have been determined, inspect the ground surrounding the entry and exit points for signs of evidence, such as footprints or other impressions. Examine broken windows for clothing fibers, blood, other trace evidence, and the method used to break or force the window open. Ask the victim when the dwelling or structure was last known to be secure. Look for cut locks in and around the crime scene. Check dumpsters for discarded tools or evidence.

10-22. Look for fingerprints and other evidence at points where the offender was likely to be. Such evidence may suggest how much time the offender spent on the premises, his skill, or his familiarity with the location. Determine if the offender's search was systematic, thorough, selective, or haphazard. The manner of search may reveal the crime to be the work of a professional or an amateur.

10-23. Examine garbage containers and nearby dumpsters for food items identified by the victim as missing from the refrigerator. These items may have fingerprints or bite marks from the offender. Canvass the neighborhood to identify possible witnesses.

EVIDENCE

10-24. Physical evidence must be protected and processed according to *AR 195-5*. Fragile evidence must be protected from the elements, animals, and people. All evidence, if possible, must be photographed, sketched, preserved, collected, and recorded. Place your initials and the time and date of the recovery on each item of evidence for positive identification at a later date.

10-25. Evidence gained through the questioning of people or by other means must be verified. A confession does not negate the need for evidence. The elements of the confession must be supported by evidence. Look for evidence (such as fingerprints, palm prints, footprints, or identifiable tire tracks) that places the suspect or his vehicle at the scene of the crime. Match soil or rock samples from the suspect's clothing or vehicle to the samples taken from the crime scene to help place the suspect at the scene.

10-26. Evidence may lead to finding stolen property in the possession of a suspect or in a place under his control. Finding the item or evidence of the item in the possession or control of the suspect is not, by itself, enough to convict him of theft. The investigator must show that the suspect knowingly and illegally deprived another of possession of the item.

10-27. Evidence usually common to breaking and entering includes—

- Glass.
- Cigarette butts.
- Fingerprints.
- Fabrics.
- Blood.
- Hair and fiber.
- Dust prints.
- Paint.
- Salvia.
- Shoe prints.
- Small objects.
- Soil.
- Tape.
- Tire tracks.
- Toolmarks or impressions.

CRITICAL INTERVIEWS

10-28. During the investigation, identify potential witnesses. Speak with neighbors, individuals who frequented the area, and coworkers of the victim. Determine who had the motive, means, and opportunity to commit the offense. Determine who had the desire to commit the act and who would benefit the most from the act. Keep in mind that motive is, perhaps, the least important factor to consider in determining who committed the crime. However, the means is very important in determining who committed the offense. Determine if the suspect had the means and if he was mentally and physically capable of committing the offense. Determine if the suspect can be placed at the scene of the crime and if his alibi can be proven false.

10-29. Interview the victim and obtain a sworn statement as soon as possible. Get full details concerning possible suspects and a full description of the missing property. Study the sworn statement to determine if it matches the crime scene.

10-30. Once a potential suspect has been identified, question him and obtain a sworn statement as soon as possible. Have the suspect describe his actions before, during, and after the crime. Develop elements of the crime during the interview. Obtain proper legal authority, and search the suspect's house, vehicle, and storage areas.

10-31. To develop leads, inquire at local pawn shops in the area. Provide the shop owner with a full description of the stolen items. If such items were pawned, coordinate with local police agencies to have the items seized or a "police hold" placed on them.

Chapter 11

Computer Crimes

Computer crimes are among the fastest growing crimes in our society today. Electronic devices can be used to commit a crime, can contain the evidence of the crime, and can even be the targets of crime. Understanding the role and nature of electronic evidence that may be found and how to process the crime scene containing potential electronic evidence are crucial issues facing all law enforcement personnel. Due to a rapid increase and the serious nature of these crimes, several federal agencies have organized special units that investigate computer crimes. The Computer Crime Investigation Unit (CCIU) and USACIL are the primary supporting agencies for military law enforcement investigators with regard to electronic evidence.

OVERVIEW

11-1. Technology is advancing at such a rapid rate that the procedures described in this chapter must be examined through the prism of current technology and the practices adjusted, as appropriate. Electronic evidence can be found in many new electronic devices available to today's consumers. *Appendix C* describes a wide variety of the types of electronic devices commonly encountered in crime scenes. It provides a general description of each type of device, describes its common uses, and presents the potential evidence that may be found in each type of equipment.

ELECTRONIC EVIDENCE

11-2. Electronic evidence is information and data of investigative value that is stored on or transmitted by an electronic device (see *Appendix D*). As such, electronic evidence is latent just as DNA and fingerprints are latent. In its natural state, we cannot see the evidence that the physical object holds. Equipment and software are required to make the evidence visible. Testimony may be required to explain the examination process and any process limitations.

11-3. Electronic evidence is, by its very nature, fragile. It can be altered, damaged, or destroyed by improper handling or improper examination. For this reason, special precautions should be taken to document, collect, preserve, and examine this type of evidence. Failure to do so may render it unusable or lead to an inaccurate conclusion. This chapter suggests methods that will help preserve the integrity of such evidence.

11-4. When dealing with electronic evidence, general forensic and procedural principles should be applied to include the following:

- Actions taken to secure and collect electronic evidence should not change that evidence.
- Persons conducting the examination of electronic evidence should be trained for that purpose.
- Activity relating to the seizure, examination, storage, or transfer of electronic evidence should be fully documented, preserved, and available for review.

11-5. This evidence is acquired when data or physical items are collected and stored for examination purposes. The following are characteristics of electronic evidence:

- It is often latent just as fingerprints and DNA are latent.
- It can transcend borders with ease and speed.
- It is fragile and can be easily altered, damaged, or destroyed.
- It is sometimes time sensitive.

11-6. Precautions must be taken in the collection, preservation, and examination of electronic evidence. Handling electronic evidence at the crime scene normally consists of—

- Recognition and identification of the evidence.
- Documentation of the crime scene.
- Collection and preservation of the evidence.
- Packaging and transportation of the evidence.

INVESTIGATIVE TOOLS AND EQUIPMENT

11-7. Special tools and equipment may be required to collect electronic evidence. Experience has shown that advances in technology may dictate changes in the tools and equipment required. Preparations should be made to get the equipment required to collect electronic evidence. Investigative agencies should have general crime scene processing equipment, such as cameras, notepads, sketch pads, evidence forms, crime scene tape, and markers. Each aspect of the process (documentation, collection, packaging, and transportation) dictates tools and equipment. The following are additional items that may be useful to have in a tool kit at an electronic crime scene:

- Documentation tools such as—
  - Cable tags.
  - Indelible felt-tip markers.
  - Stick-on labels.
- Disassembly and removal tools in a variety of nonmagnetic sizes and types that include—
  - Flat-blade and cross-tip screwdrivers.
  - Hex-nut and secure-bit drivers.
  - Star-type nut drivers.
  - Needle-nose and standard pliers.

- Small tweezers.
- Specialized screwdrivers (manufacturer specific).
- Wire cutters.
- Packaging and transporting supplies such as—
  - Antistatic bags and bubble wrap.
  - Cable ties.
  - Evidence bags.
  - Evidence and packing tape.
  - Packing materials (avoid materials that can produce static electricity, such as foam peanuts).
  - Sturdy boxes of various sizes.
- Other items such as—
  - Evidence tags.
  - Evidence tape.
  - Forms (keystroke or mouse click log, photograph log, *DA Forms 4137 [Evidence/Property Custody Document]*, and *DA Forms 4002)*.
  - Gloves.
  - A hand truck.
  - Large rubber bands.
  - A list of contact telephone numbers for assistance.
  - A magnifying glass.
  - Printer paper.
  - A seizure disk.
  - A small flashlight.
  - Fully-formatted floppy diskettes (3 inch and 5 1/4 inch).

CRIME SCENE SECURITY AND EVALUATION

11-8. The investigator should take steps to ensure the safety of all persons at the crime scene and protect the integrity of all evidence, both traditional and electronic. All activities should be in compliance with Army policy and federal, state, and local laws.

11-9. After securing the scene and all persons on the scene, the investigator should visually identify potential evidence (both physical and electronic) and determine if perishable evidence exists. He should then evaluate the scene and formulate a search plan.

SECURE AND EVALUATE THE CRIME SCENE

11-10. The investigator should secure and evaluate the crime scene by—

- Following jurisdictional policy for securing the crime scene. This would include ensuring that all persons are removed from the immediate area where evidence is to be collected. At this point in the investigation, do not alter the condition of any electronic devices. If it is off, leave it off. If it is on, leave it on.

- Protecting perishable data (physical and electronic). Perishable data may be found on pagers, caller identification (ID) boxes, electronic organizers, cell phones, and other similar devices. The first responder should always keep in mind that any device containing perishable data should be immediately secured, documented, and/or photographed.
- Identifying telephone lines attached to devices such as modems and caller ID boxes. Document, disconnect, and label each telephone line from the wall rather than the device, when possible. There may also be other communications lines present for local area network (LAN), wide area network (WAN), or other network technologies. Consult the appropriate personnel or agency in these cases.
- Preserving the computer mouse, keyboard, diskettes, compact disks (CDs), or other components that may have latent fingerprints or other physical evidence. Chemicals used in processing latent fingerprints can damage equipment and data. Therefore, latent prints should be collected after the completion of electronic evidence recovery.

CONDUCT PRELIMINARY INTERVIEWS

11-11. The investigator should conduct preliminary interviews by—

- Separating and identifying all individuals (witnesses, subjects, or others) at the scene and recording their location at the time of entry.
- Being consistent with departmental policy and applicable laws in obtaining information from these individuals, such as—
  - Passwords and user names of owners and/or users of electronic devices found at the crime scene and the Internet service provider (ISP). Obtain any passwords required to access the system, software, or data. An individual may have multiple passwords, such as basic input-output system (BIOS), system login, network ISP, application files, encryption pass phrase, e-mail, access token, scheduler, or contact list.
  - The purpose of the system.
  - Any unique security schemes or destructive devices.
  - Any off-site data storage.
  - Any documentation explaining the hardware or software installed on the system.

CRIME SCENE DOCUMENTATION

11-12. Documentation of the crime scene creates a permanent historical record of the crime scene. Documentation is an ongoing process throughout the investigation. It is important to accurately record the location and condition of computers, storage media, other electronic devices, and conventional evidence. Moving of a computer system while the system is running may cause changes to system data. Therefore, the system should not be moved until it has been safely powered down. The initial documentation of the physical crime scene should include—

- Observing and documenting the physical crime scene, such as the position of the mouse and the location of components relative to each other (a mouse on the left side of the computer may indicate a left-handed user).
- Documenting the condition and location of the computer system, including the power status of the computer (on, off, or in sleep mode). Most computers have status lights to indicate that the computer is on. Likewise, if fan noise is heard, the system is probably on. Furthermore, if the computer system is warm, it may also indicate that it is on or was recently turned off.
- Identifying and documenting related electronic components that will not be collected.
- Photographing the entire scene to create a visual record as noted by the first responder. The complete room should be recorded with 360° coverage, when possible.
- Photographing the front of the computer, monitor screen, and other components. Take written notes on what appears on the monitor screen. Active programs may require videotaping or more extensive documentation of monitor screen activity.
- Performing additional documentation of the system during the collection phase.

AUTHORIZATION TO SEIZE ELECTRONIC EVIDENCE

11-13. Search authorization may be obtained from a US magistrate, a civilian judge at the state or federal level, or the property owner is required. However, in almost all cases, courts have held a relatively high standard with regard to the specificity of computer-related search authorizations. Investigative personnel seeking search authorization must be able to articulate specific and recent information pertaining to the individual items cited on the affidavit and authorization in order to establish probable cause. In many instances, information that is several months old cannot in and of itself be used to generate probable cause. More recent information, gained through "pretext" phone calls or online undercover operations, may be required to develop current and reliable information (see *Appendix E* for a sample affidavit and authorization). Additionally, if during the conduct of a search for one offense, evidence of an unrelated or different type of offense is identified, the scope of the search authorization must be expanded accordingly. If probable cause cannot be developed, consideration should be given to requesting a consent search. However, this may make the suspect aware of law enforcement interest and cause investigators to lose potential evidence.

11-14. When information is required from private business computers, investigators must take the appropriate measures to avoid any financial liability to the government by determining if the seizure of these computers will cause the business to lose money in any way. If it is determined that the seizure will create liabilities to the government, coordination must be made with CCIU or USACIL to have an expert image the appropriate drive information on the site. Additionally, investigators must take reasonable steps to determine if copyrighted information is contained on a computer to be seized. If a determination is made that such data is stored on a target computer, consult with CCIU or USACIL, as appropriate, to seize the required information without collecting the copyrighted information. Failure to comply with this requirement will likely result in the suppression of all collected evidence from courtroom presentation and could possibly leave the Army legally liable.

EVIDENCE COLLECTION

11-15. Computer evidence, like all other evidence, must be handled carefully and in a manner that preserves its evidentiary value. This relates not just to the physical integrity of an item or device, but also to the electronic data it contains. Certain types of computer evidence, therefore, require special collection, packaging, and transportation. Consideration should be given to protect data that may be susceptible to damage or alteration from electromagnetic fields, such as those generated by static electricity, magnets, radio transmitters, and other devices.

11-16. Electronic evidence should be collected according to departmental guidelines. In the absence of departmental procedures for electronic evidence collection, use the procedures outlined below.

NONELECTRONIC EVIDENCE COLLECTION

11-17. Recovery of nonelectronic evidence can be crucial in the investigation of electronic crimes. Take proper care to ensure that such evidence is recovered and preserved. Items relevant to subsequent examination of electronic evidence may exist in other forms (written passwords and other handwritten notes, blank pads of paper with indented writing, hardware and software manuals, calendars, literature, text or graphical computer printouts, and photographs) and should be secured and preserved for future analysis. These items are frequently in close proximity to the computer or related hardware items. All evidence should be identified, secured, and preserved in compliance with departmental procedures.

STAND-ALONE AND LAPTOP COMPUTER EVIDENCE COLLECTION

11-18. Multiple computers may indicate a computer network. Likewise, computers located at businesses are often networked. In these situations, specialized knowledge about the system is required to effectively recover evidence and reduce your potential for civil liability. When a computer network is encountered, contact the forensic computer expert in your department or an outside consultant identified by your department for assistance. Computer systems in a complex environment are addressed later in this chapter.

11-19. A stand-alone personal computer (PC) is a computer that is not connected to a network or another computer. Stand-alones may be desktop machines or laptops.

11-20. Laptops incorporate a computer, monitor, keyboard, and mouse into a single portable unit. Laptops differ from other computers in that they can be powered by electricity or a battery source. Therefore, they require the removal of the battery in addition to stand-alone, power-down procedures.

11-21. If the computer is on, document existing conditions and call your expert or consultant. If an expert or consultant is not available, document all actions taken and any changes observed in the monitor, computer, printer, or other peripherals that result from actions taken. Observe the monitor and determine if it is on, off, or in sleep mode. Then decide which of the following situations applies and follow the steps for that situation.

- **Situation 1:** The monitor is on and the work product and/or desktop are visible.

  Step 1. Photograph the screen and record the information displayed.

  Step 2. Proceed to situation 3, step 3.

- **Situation 2:** The monitor is on and the screen is blank (sleep mode) or the screensaver (picture) is visible.

  Step 1. Move the mouse slightly (without pushing buttons). The screen should change and show the work product or request a password.

  Step 2. Do not perform any other keystrokes or mouse operations if mouse movement does not cause a change in the screen.

  Step 3. Photograph the screen and record the information displayed.

  Step 4. Proceed to situation 3, step 3.

- **Situation 3:** The monitor is off.

  Step 1. Make a note of the "off" status.

  Step 2. Turn the monitor on, then determine if the monitor status is as described in either situation 1 or 2 above and follow those steps.

  Step 3. Regardless of the power state of the computer (on, off, or sleep mode), remove the power source cable from the computer, not from the wall outlet. If dealing with a laptop, in addition to removing the power cord, remove the battery pack. The battery is removed to prevent any power to the system. Some laptops have a second battery in the multipurpose bay instead of a floppy drive or CD drive. Check for this possibility and remove that battery as well.

  Step 4. Check for outside connectivity (telephone modem, cable, integrated services digital network [ISDN], and digital subscriber line [DSL]). If a telephone connection is present, attempt to identify the telephone number.

  Step 5. Avoid damage to potential evidence by removing any floppy disks that are present, packaging the disk separately, and labeling the package. If available, insert either a seizure disk or a blank floppy disk. Do not remove CDs or touch the CD drive.

  Step 6. Place tape over all the drive slots and over the power connector.

  Step 7. Record the make, model, and serial numbers.

  Step 8. Photograph and diagram the connections of the computer and the corresponding cables.

Step 9. Label all connectors and cable ends (including connections to peripheral devices) to allow for exact reassembly at a later time. Label unused connection ports as "unused." Identify laptop computer docking stations in an effort to identify other storage media.

Step 10. Record or log evidence according to departmental procedures.

Step 11. Package any components as fragile, if transport is required.

COMPUTERS IN A COMPLEX ENVIRONMENT

11-22. Business environments frequently have multiple computers connected to each other, to a central server, or both. Securing and processing a crime scene where the computer systems are networked poses special problems, because an improper shutdown may destroy data. This can result in loss of evidence and potential severe civil liability. When investigating criminal activity in a known business environment, the presence of a computer network should be planned for in advance, if possible, and the appropriate expert obtained. It should be noted that computer networks can also be found in a home environment and the same concerns exist.

11-23. The possibility of various operating systems and complex hardware configurations requiring different shutdown procedures make the processing of a network crime scene beyond the scope of this chapter. However, it is important that computer networks be recognized and identified, so that an expert can be obtained if one is encountered.

11-24. Indications that a computer network may be present include

- Multiple computer systems.
- Cables and connectors running between computers or central devices, such as hubs.
- Any information provided by informants or individuals at the scene.
- Network components.

EXAMINATION REQUEST

11-25. At the point when the USACIL laboratory request is prepared, the initial report and all documentation related to the efforts undertaken by the investigator should be forwarded to USACIL with the items of evidence. It is important that all actions, including every keystroke and mouse click, be recorded by the investigator to ensure that the laboratory examiner can fully appreciate what impact the investigator's actions could have had on the operating system. Additionally, special consideration should be undertaken when shipping computer-related items to USACIL; they should be containerized separate from fungible evidence that will require special treatment, such as refrigeration.

ON-SITE SEARCHES WITHOUT SEIZURE AUTHORIZATION

11-26. In some instances, investigative personnel will find themselves in a situation where there is an authorization to search a computer, but the agent lacks the authorization to seize it. Typically, this occurs in one of two situations. The first is when consent to search is authorized by a suspect who agrees to allow the investigator to search the computer; however, he does not agree to allow the investigator to seize or ship it to the laboratory for examination. The second situation frequently results when a commander suspects that criminal activity has been committed using a government computer, but he has no means to verify it and does not want to deprive the organization of the use of the computer while it is awaiting examination. Investigators in the field should never conduct a search of a computer or open electronic files under any circumstance other than when it is impossible or impractical to seize the device for laboratory examination.

11-27. If an investigator conducts a consent type search of a computer, based on the scenario indicated above, it is essential for him to terminate the search as soon as the first file containing evidence of a crime is identified. At this point, probable cause has been met, and a formal search authorization should be obtained from a competent authority. The subsequent seizure of the computer is not based on the consent to search, but rather the evidence identified during the search and the authority of a formal search authorization or warrant. The investigator must then document all of his activities including every keystroke and mouse click that led to the discovery of the criminal material. It is important that additional files not be accessed, because the date-time group of the accessed file are modified and will likely result in the suppression of them in the prosecutorial process.

EVIDENCE PACKAGING, TRANSPORTING, AND STORING

11-28. Computers are fragile electronic instruments that are sensitive to temperature, humidity, physical shock, static electricity, and magnetic sources. Therefore, special precautions should be taken when packaging, transporting, and storing electronic evidence. To maintain the chain of custody of electronic evidence, document its packaging, transporting, and storing.

PACKAGING

11-29. If multiple computer systems are collected, label each system so that it can be reassembled as found (system A: mouse, keyboard, monitor, and main base unit; system B: mouse, keyboard, monitor, and main base unit).

11-30. When packaging evidence at a crime scene—

- Ensure that all collected electronic evidence is properly documented, labeled, and inventoried before packing.
- Pay special attention to latent or trace evidence and take action to preserve it.
- Pack magnetic media in antistatic packaging (paper or antistatic plastic bags). Avoid using materials that can produce static electricity, such as standard plastic bags.
- Avoid folding, bending, or scratching computer media, such as a diskette, compact disk-read only memory (CD-ROM), or tape.
- Ensure that all containers used to hold evidence are properly labeled.

TRANSPORTING

11-31. Ensure that computers and other components that are not packaged in containers are secured in the vehicle to avoid shock and excessive vibrations. For example, computers may be placed on the vehicle floor and monitors placed on the seat with the screen down and secured by a seat belt. When transporting evidence—

- Keep all electronic evidence away from magnetic sources. Radio transmitters, speaker magnets, and heated seats are examples of items that can damage electronic evidence.
- Avoid storing electronic evidence in vehicles for prolonged periods of time. Conditions of excessive heat, cold, or humidity can damage electronic evidence.
- Maintain the chain of custody on all evidence transported.

STORING

11-32. Ensure that evidence is inventoried according to *AR 195-5*. Store evidence in a secure area away from temperature and humidity extremes. Protect it from magnetic sources, moisture, dust, and other harmful particles or contaminants.

11-33. Be aware that potential evidence, such as dates, times, and system configurations may be lost as a result of prolonged storage. Since batteries have a limited life, data could be lost if they fail. Therefore, appropriate personnel (such as the evidence custodian, laboratory chief, and forensic examiner) should be informed that a device powered by batteries is in need of immediate attention.

Chapter 12

Death Scene Investigations

Many agencies are responsible to the commander for investigating suspicious deaths. Close liaison must be made within commands between investigative, medical, and related forensic personnel to have an effective death investigation. Matters of mutual interest include jurisdiction, investigative responsibilities, local agreements with the civil authorities, SOFAs, and legal issues to be considered by military police, USACIDC, medical personnel, and pathologists.

RESPONSIBILITIES

12-1. Many people have inherent responsibilities to a death scene investigation. The interest of the crime goes beyond the involvement of law enforcement personnel and requires the special qualifications of certain individuals, such as a medical officer, a pathologist, and even an intelligence officer from the decedent's organization.

PROVOST MARSHAL

12-2. The PM and USACIDC are responsible for obtaining all facts pertinent to deaths occurring under suspicious circumstances. An investigator must determine the manner of death to be a homicide, a suicide, or an accidental or natural death. Regardless of the initial manner of death, all circumstances concerning that death must be developed to ensure that the facts support the initial finding of the manner of death.

MEDICAL OFFICER

12-3. The medical officer is responsible for conducting a medicolegal (forensic) autopsy of the decedent. It is important to understand that there are several components of the autopsy. These include a complete review of the medical history, a complete crime scene examination, and the prosecution of the defendant. This autopsy is a joint venture in which the investigator and pathologist must work together. The purpose of the medicolegal autopsy is to determine—

- The cause of death.
- The manner of death.
- The identification of the decedent.
- The location of the decedent. Determine if the decedent was moved.
- The identification or presence of physical evidence. If such is present, it should be collected.

CRIMINAL INVESTIGATOR

12-4. Investigators are encouraged to set up a liaison with the pathologist who does the autopsy. Investigators must tell the pathologist the known facts of the death and the initial investigative findings before the autopsy. This enables the pathologist to select the proper ways to determine the cause of death and to give an opinion about the manner of death. The medical officer must complete certain military records and official certificates of death. The investigator works with the medical officer to collect and preserve evidence. The investigator's direct responsibilities include photographing and obtaining full record fingerprints of the decedent. He should obtain two complete sets.

12-5. *Appendix F* contains checklists for various violent crimes. They will assist the investigator in keeping track of the minute details concerning even the most complicated investigation.

LINE-OF-DUTY AND SAFETY INVESTIGATING OFFICER

12-6. The LOD investigating officer determines the duty status and personal conduct of the deceased. The LOD officer has no jurisdiction with the criminal investigation. The safety officer determines any safety factors, or lack of them, in an accident. His interest in accidents is limited to safety.

INTELLIGENCE LIAISON

12-7. Close intelligence liaison is needed and directed by *AR 195-2*. A report of death due to a homicide, an accident, or a suicide must be relayed at once to the nearest intelligence agency. If the victim had access to classified material, the intelligence officer determines if any of the material is missing. The intelligence officer is responsible for security measures. His main concern is to ensure that classified material is not compromised. This is very important if the death is a suicide. Ensure that intelligence officials are kept fully advised until no further security interest exists.

12-8. In some instances, the post commander or higher authority may call a board of inquiry to find out the facts connected with a death. Such a board has broad powers and may check into all areas of the matter.

INVESTIGATIVE PROCEDURES

12-9. The basic aim of the investigation is to determine if the death was natural, an accident, a suicide, or a homicide and, in the case of a homicide, to collect evidence leading to the conviction of the guilty party. The technique of investigating any violent death is basically the same. The circumstances, conditions, and events leading to and following the death are investigated. Learning and tracing the events and actions involving the victim before his death can show the likelihood of an accident or a clear intent for suicide or homicide.

12-10. Accidental deaths often occur under suspicious conditions. Many of these deaths will look violent but will lack criminal likelihood. Circumstances may show, for example, a logical reason for the presence of a weapon and the likelihood of it being an accident. On the other hand, circumstances may strongly suggest a death by suicide. Prior suicide attempts, earlier written or oral statements of intent, and suicide notes at the scene are strong evidence of suicide.

12-11. Homicide is often shown by conditions and events leading to and following the death. A disturbed scene, wounds to nonvital areas, punctured clothing, no weapon at the scene, no signs of suicidal intent or hazardous conditions, lack of hesitation wounds, signs of a fight, and signs of flight or surprise are all factors pointing to homicide.

12-12. The lack of a visible weapon at the scene most often suggests that the death was homicidal, but a suicide victim may live long enough to dispose of a weapon, or he may arrange a contraption to cause the weapon to disappear after being fired. Relatives fearing social disgrace or having an interest in the life insurance of a suicide victim may try to hide the deceased's suicidal intent and circumstances. Similarly, a murderer may try to make things look like a suicide or an accident. When distinguishing between a suicide and a homicide, it is very important to learn motive. Opportunity is also a factor to be considered where there are signs that an apparent or alleged suicide may be a homicide.

ARRIVE AT THE SCENE

12-13. Upon arrival at the scene and before entering the scene, the investigator should—

- Identify the lead investigator at the scene and present his identification.
- Identify other essential officials at the scene (such as law enforcement, fire, EMS, or social or child protective services) and answer any questions about his activities at the crime scene.
- Identify and document the identity of the first essential official (the first "professional" arriving at the scene for investigative follow-up) to the scene to ascertain if any artifacts or contamination may have been introduced to the death scene.
- Determine the safety of the scene (before entry).

DETERMINE SCENE SAFETY

12-14. Determining scene safety for all investigative personnel is essential to the investigative process. The risk of environmental and physical injury must be removed before initiating a scene investigation. Risks can include hostile crowds, collapsing structures, traffic, and environmental and chemical threats. The investigator must attempt to establish scene safety before entering the scene to prevent injury or loss of life to include contacting appropriate agencies for assistance with other scene safety issues.

12-15. Upon arrival at the scene, the investigator should—

- Secure his vehicle and park as safely as possible.
- Obtain a clearance or authorization to enter the crime scene from the individual responsible for scene safety, such as the fire marshal or disaster coordinator.

- Protect the integrity of the scene and the evidence (to the extent possible) from contamination or loss by people, animals, and elements while exercising scene safety.
- Assess and/or establish physical boundaries.
- Identify the incident command.
- Use personal protective safety devices (physical and biochemical).
- Arrange for the removal of any animals or secure them (if they are present and if it is possible).
- Have the body removed before the scene investigation continues because of potential scene hazards, such as crowd control, collapsing structures, poisonous gases, or traffic.

CONFIRM OR PRONOUNCE DEATH

12-16. Appropriate personnel must make a determination of death before the initiation of a death investigation. The confirmation or pronouncement of death determines jurisdictional responsibilities.

12-17. Upon arrival at the scene, the investigator should—

- Locate and view the body.
- Check for a pulse, respiration, and reflexes, as appropriate.
- Identify and document the name of the individual who made the official determination of death (include the date and time of the determination).
- Ensure that the death is pronounced, as required.

12-18. Once death has been determined, rescue and/or resuscitative efforts cease and medicolegal jurisdiction can be established. It is vital that this occur before the medical examiner or coroner assumes any responsibilities.

PARTICIPATE IN THE SCENE BRIEFING

12-19. Scene investigators must recognize the varying jurisdictional and statutory responsibilities that apply to individual agency representatives, such as law enforcement, fire, EMS, and judicial or legal. Determining the investigative responsibility of each agency at the scene is essential in planning the scope and depth of each scene investigation and the release of information to the public.

12-20. The investigator should identify specific responsibilities, share appropriate preliminary information, and establish the investigative goals of each agency present at the scene.

12-21. When participating in a scene briefing, the investigator should—

- Locate the staging area (such as the entry point to the scene or the command post).
- Document the scene location (such as the address, mile marker, and building name) consistent with other agencies.

- Determine the nature and scope of the investigation by obtaining preliminary investigative details (such as suspicious versus nonsuspicious death).
- Ensure that initial accounts of the incident are obtained from the first witnesses.

12-22. A scene briefing allows for initial and factual information exchange. This includes scene location, time factors, initial witness information, agency responsibilities, and investigative strategy.

CONDUCT A SCENE WALK-THROUGH

12-23. Conducting a scene walk-through provides the investigator with an overview of the entire scene. The walk-through provides the investigator with the first opportunity to locate and view the body, identify valuable and/or fragile evidence, and determine initial investigative procedures providing for a systematic examination and documentation of the scene and the body. The investigator should conduct a scene walk-through to establish pertinent scene parameters.

12-24. Upon arrival at the scene, the investigator should—

- Reassess scene boundaries and adjust, as appropriate.
- Establish a path of entry and exit.
- Identify visible, physical, and fragile evidence.
- Document and photograph fragile evidence immediately and collect, if appropriate.
- Locate and view the decedent.

12-25. The initial scene walk-through is essential to minimize scene disturbance. It will also prevent the loss and/or contamination of physical and fragile evidence.

ESTABLISH A CHAIN OF CUSTODY

12-26. Ensuring the integrity of the evidence by establishing and maintaining a chain of custody is vital to an investigation. This will safeguard against subsequent allegations of tampering, theft, planting, and contaminating evidence. Before the removal of any evidence, the evidence custodian should be designated and should generate and maintain a chain of custody for all evidence collected.

12-27. Throughout the investigation, those responsible for preserving the chain of custody should—

- Document the scene location and the arrival time of the death investigator at the scene.
- Determine the evidence custodian, the agency responsible for the collection of specific types of evidence, and the evidence collection priority for fragile or fleeting evidence.
- Identify, secure, and preserve the evidence with the proper containers, labels, and preservatives.

- Document the collection of the evidence by recording its location at the scene.
- Develop the personnel and witness lists and document the arrival and departure times of personnel.

FOLLOW LAWS FOR THE COLLECTION OF EVIDENCE

12-28. The investigator must follow *AR 195-5* and state and federal laws for the collection of evidence to ensure its admissibility. The investigator must work with law enforcement, the office of the SJA, and other legal authorities to determine laws regarding the collection of evidence.

12-29. The investigator, before or upon arrival at the death scene, should work with other agencies to—

- Determine the need for a search warrant and discuss this with the appropriate agencies.
- Identify local, state, federal, and international laws and discuss them with the appropriate agencies.
- Identify medical examiner and coroner statutes and/or office SOPs and discuss them with the appropriate agencies.

SCENE DOCUMENTATION AND EVALUATION

12-30. The process of documenting and evaluating the scene must be meticulous and timely. The process includes photographing, writing descriptive documentation, and establishing the probable location of the incurred injury or illness of the deceased. It also includes the collecting, inventorying, and safeguarding of property and evidence and documenting interviews or comments of witnesses at the scene.

OBTAIN PHOTOGRAPHIC DOCUMENTATION

12-31. The photographic documentation of the scene creates a permanent historical record. Photographs provide detailed corroborating evidence that constructs a system of redundancy should questions arise concerning the report, witness statements, or position of the evidence at the scene. The investigator should obtain detailed photographic documentation of the scene that provides both instant and permanent high-quality (35-millimeter) images. *Chapter 6* provides technical guidance on crime scene photography.

12-32. Upon arrival at the scene and before moving the body or evidence, the investigator should—

- Remove all nonessential personnel from the scene.
- Obtain an overall (wide-angle) view of the scene to spatially locate the specific scene to the surrounding area.
- Photograph specific areas of the scene to provide more detailed views of specific areas within the larger scene.
- Photograph the scene from different angles to provide various perspectives that may uncover additional evidence.

- Obtain some photographs with scales to document specific evidence.
- Obtain photographs even if the body or other evidence has been moved or removed.

12-33. If evidence has been moved before photography, it should be noted in the report. The body or other evidence should not be reintroduced into the scene in order to take photographs.

DEVELOP DESCRIPTIVE DOCUMENTATION OF THE SCENE

12-34. Written documentation of the scene provides a permanent record that may be used to correlate with and enhance photographic documentation, refresh recollections, and record observations. Investigators should provide written scene documentation.

12-35. After photographic documentation of the scene and before removal of the body or other evidence, the investigator should—

- Diagram or describe (in writing) items of evidence and their relationship to the body with any necessary measurements.
- Describe and document (with any necessary measurements) blood and body fluid evidence including volume, patterns, spatters, and other characteristics.
- Describe scene environments including odors, lights, temperatures, and other fragile evidence.

12-36. If physical evidence is moved or tampered with, this also needs to be documented. The investigator should document the original state of the evidence as reported by whomever moved it, and the state it was in when observed by the investigator.

ESTABLISH PROBABLE LOCATION OF INJURY OR ILLNESS

12-37. The location where the decedent is found may not be the actual location where the injury or illness that contributed to the death occurred. It is imperative that the investigator attempts to determine the locations of any and all injuries or illnesses that may have contributed to the death. Physical evidence at any and all locations may be pertinent in establishing the cause, manner, and circumstances of a death.

12-38. The investigator should obtain detailed information regarding any and all probable locations associated with the individual's death. The investigator should—

- Document the location where the death was confirmed.
- Determine the location from which the decedent was transported and how the body was transported to the scene.
- Identify and record any discrepancies in rigor mortis, livor mortis, and body temperature.
- Check the body, clothing, and scene for consistencies or inconsistencies of trace evidence and indicate the locations where any artifacts were found.
- Check for any drag marks on the body, clothes, and ground.

- Establish postinjury activity.
- Obtain any dispatch records, such as police and ambulance.
- Interview family members and associates, as needed.

COLLECT, INVENTORY, AND SAFEGUARD PROPERTY AND EVIDENCE

12-39. The decedent's valuables or property must be safeguarded to ensure proper processing and eventual return to the next of kin. Evidence on or near the body must be safeguarded to ensure its availability for further evaluation. The investigator should ensure that all property and evidence is collected, inventoried, safeguarded, and released as required by law.

12-40. After personal property and evidence have been identified at the scene, the investigator (with a witness) must inventory, collect, and safeguard the following at the scene and/or the office:

- Illicit drugs and paraphernalia.
- Prescription medication.
- Over-the-counter medications.
- Money.
- Personal valuables or property.

INTERVIEW WITNESSES AT THE SCENE

12-41. The documented comments of witnesses at the scene allow the investigator to obtain primary source data regarding the discovery of the body, witness corroboration, and terminal history. The documented interview provides essential information for the investigative process. The investigator's report should include the source of the information, including specific statements and information provided by the witness.

12-42. Upon arriving at the scene, the investigator should—

- Identify possible witnesses.
- Collect all available identifying data on witnesses, such as full name, address, date of birth (DOB), and work and home telephone numbers.
- Establish the relationship and/or association of the witness to the deceased.
- Establish the basis of the knowledge of the witness. (For example, how does the witness have knowledge of the death?)
- Obtain information from each witness.
- Note discrepancies from the scene briefing (challenge, explain, and verify statements).
- Tape statements when equipment is available and retain the tapes.

BODY DOCUMENTATION AND EVALUATION

12-43. The process of documenting and evaluating the body is intricate and demands the attention of the investigator to ensure that even the most minute detail is documented. The following paragraphs describe the requirements to document and evaluate the body.

PHOTOGRAPH THE BODY

12-44. The photographic documentation of the body at the scene creates a permanent record that preserves essential details of the position, appearance, identity, and final movements of the body. Photographs allow the sharing of information with other agencies investigating the death. The investigator should obtain detailed photographic documentation of the body that provides both instant and permanent high-quality (35-millimeter) images.

CONDUCT AN EXTERNAL BODY EXAMINATION (SUPERFICIAL)

12-45. Conducting the external body examination provides the investigator with objective data regarding the single most important piece of evidence at the scene, the body. This documentation provides detailed information regarding the decedent's physical attributes; his relationship to the scene; and the possible cause, manner, and circumstances of his death.

12-46. The investigator should properly and thoroughly document the body before it is moved. Documentation includes photographing and making detailed notes concerning the condition and location.

12-47. After arriving at the scene and before moving the decedent, the investigator should do the following without removing the decedent's clothing:

- Photograph the scene. Include the decedent as initially found and the surface beneath the body after the body has been removed.
- Photograph the decedent with and without measurements, as appropriate. Include a photograph of the decedent's face.
- Document the decedent's position with and without measurements, as appropriate.
- Document the decedent's physical characteristics.
- Document the presence or absence of clothing and personal effects.
- Document the presence or absence of any items or objects that may be relevant.
- Document the presence or absence of marks, scars, and tattoos.
- Document the presence or absence of petechiae, injury, or trauma, and so forth.
- Document the presence of any treatment or resuscitative efforts.
- Determine the need for further evaluation or assistance of forensic specialists (such as pathologists and/or odontologists) based on the findings.

PRESERVE ANY EVIDENCE ON THE BODY

12-48. The photographic and written documentation of evidence on the body allows the investigator to obtain a permanent historical record of that evidence. To maintain a chain of custody, evidence must be collected, preserved, and transported properly. In addition to all of the physical evidence visible on the body, blood and other body fluids that are present must be photographed and documented before collection and transport. Fragile evidence (that which can be easily contaminated, lost, or altered) must also be collected and/or preserved to maintain the chain of custody and to assist in the determination of the cause, manner, and circumstances of death.

12-49. Once evidence on the body is recognized, the investigator should—

- Photograph the evidence.
- Document any blood or body fluids (such as froth, purge, or substances from orifices) found on the body. Also, document the location and pattern before transporting the body.
- Place the decedent's hands and/or feet in unused paper bags (determined by the scene).
- Collect trace evidence before transporting the body (such as blood, hair, and fibers).
- Arrange for the collection and transport of evidence at the scene, when necessary.
- Ensure the proper collection of blood and body fluids for subsequent analysis if the body is to be released from the scene to an outside agency without an autopsy.

ESTABLISH THE IDENTIFICATION OF THE DECEDENT

12-50. The establishment or confirmation of the decedent's identity is paramount to the death investigation. Proper identification allows notification of the next of kin, settlement of estates, resolution of criminal and civil litigation, and proper completion of the death certificate. The investigator should engage in a diligent effort to establish or confirm the decedent's identity.

12-51. To establish the identity of the decedent, the investigator should document and/or use the following methods:

- Direct visual or photographic identification of the decedent, if visually recognizable.
- Scientific methods, such as fingerprints, dental, radiographic, and DNA comparisons.
- Circumstantial methods, such as (but not restricted to) personal effects, circumstances, physical characteristics, tattoos, and anthropologic data.

DOCUMENT POSTMORTEM CHANGES

12-52. Documenting postmortem changes to the body assists the investigator in explaining body appearance in the interval following death. Inconsistencies between postmortem changes and the body's location may indicate that the body was moved and can validate or invalidate witness statements. In addition, postmortem changes to the body, when correlated with circumstantial information, can assist the investigators in estimating the approximate time of death. The investigator should document all postmortem changes relative to the decedent and the environment.

12-53. Upon arriving at the scene and before moving the body, the investigator should note the presence of each of the following in his report:

- Livor that is consistent or inconsistent with the position of the body (include the color, location, and blanchability, and if Tardieu spots are present).
- Rigor (include the stage and/or intensity and location on the body if it is broken or inconsistent with the scene).
- Decomposition (include the degree, such as putrefaction, adipocere, mummification, or skeletonization).
- Insect and animal activity.
- Scene temperature (document the method used and time estimated).
- Description of the body temperature (such as warm, cold, or frozen) and the measurement of body temperature (document the method used and time of the measurement).

PARTICIPATE IN THE SCENE DEBRIEFING

12-54. The scene debriefing helps investigators from all participating agencies to establish postscene responsibilities by sharing data regarding particular scene findings. The scene debriefing provides each agency the information, special examinations, and requests requiring interagency communication, cooperation, and education.

12-55. The investigator should participate in or initiate an interagency scene debriefing to verify specific postscene responsibilities. When participating in a scene debriefing, the investigator should—

- Determine postscene responsibilities (identification, notification, press relations, and evidence transportation).
- Determine and/or identify the need for a specialist (such as crime laboratory technicians, social services, entomologists, and Occupational Safety and Health Administration [OSHA]).
- Communicate with the pathologist about responding to the scene or to the autopsy schedule, as needed.
- Share investigative data, as required, to further the investigation.
- Communicate special requests to appropriate agencies being mindful of the necessity for confidentiality.

DETERMINE NOTIFICATION PROCEDURES (NEXT OF KIN OR CHAIN OF COMMAND)

12-56. Every reasonable effort should be made to notify the appropriate persons as soon as possible. Notification initiates closure for the family and disposition of the remains and facilitates the collection of additional information relative to the case. The investigator should ensure that the appropriate individuals are notified of the death and that all failed and successful attempts at notification are documented.

ENSURE THE SECURITY OF THE REMAINS

12-57. Ensuring the security of the body requires the investigator to supervise the labeling, packaging, and removal of the remains. An appropriate identification tag is placed on the body to preclude misidentification upon receipt at the examining agency. This function also includes safeguarding all potential physical evidence and/or property and clothing that remain on the body. The investigator should supervise and ensure the proper identification, inventory, and security of evidence or property and its packaging and removal from the scene.

12-58. Before leaving the scene, the investigator should—

- Ensure that the body is protected from any further trauma or contamination (if not, document it). Document any unauthorized removal of therapeutic and resuscitative equipment.
- Inventory and secure any property, clothing, and personal effects that are on the body. Remove them in a controlled environment with a witness present.
- Identify property and clothing to be retained as evidence (in a controlled environment).
- Recover blood and/or vitreous samples before the release of the remains.
- Place identification on the body and body bag.
- Ensure and/or supervise the placement of the body into the bag.
- Ensure and/or supervise the removal of the body from the scene.
- Secure the transportation.

DECEDENT PROFILE INFORMATION ESTABLISHMENT AND RECORD

12-59. Establishing and recording the decedent's profile information is important to a death scene investigation. This information will help determine if the death occurred by natural means or if criminal activity was involved.

DOCUMENT THE DISCOVERY HISTORY

12-60. Establishing a decedent profile includes documenting the discovery history and circumstances surrounding the discovery. The basic profile will dictate subsequent levels of investigation, jurisdiction, and authority. The focus (breadth and depth) of further investigation is dependent on this information. The investigator should document the discovery history, available witnesses, and apparent circumstances leading to the death.

12-61. For an investigator to correctly document the discovery history, he should—

- Establish and then record the name of the person or persons who discovered the body and when.
- Document the circumstances surrounding the discovery (who, what, where, when, and how).

DETERMINE THE TERMINAL-EPISODE HISTORY

12-62. Preterminal circumstances play a significant role in determining the cause and manner of death. Documentation of medical intervention and/or procurement of antemortem specimens help to establish the decedent's condition before death. The investigator should document known circumstances and medical intervention preceding death. In order for the investigator to determine the terminal-episode history, he should—

- Document when, where, how, and by whom the decedent was last known to be alive.
- Document the incidents before the death.
- Document complaints and/or symptoms before the death.
- Document and review the complete EMS records including the initial electrocardiogram.
- Obtain copies of relevant medical records.
- Obtain relevant antemortem specimens.

DOCUMENT THE DECEDENT'S MEDICAL HISTORY

12-63. The majority of deaths that are referred to the medical examiner or coroner are natural deaths. Establishing the decedent's medical history helps to focus the investigation. Documenting the decedent's medical signs or symptoms before death determines the need for subsequent examinations. The relationship between disease and injury may play a role in the cause, manner, and circumstances of death. The investigator should obtain the decedent's past medical history.

12-64. Through interviews and review of the written records, the investigator should document the decedents—

- Medical history including medications taken, alcohol and drug use, and family medical history from family members and witnesses.
- Information from the treating physicians and/or hospitals to confirm history and treatment.
- Physical characteristics and traits (such as left or right handed, missing appendages, or tattoos).

DOCUMENT THE DECEDENT'S MENTAL HEALTH HISTORY

12-65. The decedent's mental health history can provide insight into the behavior or state of mind of the individual. This insight may produce clues that will aid in establishing the cause, manner, and circumstances of the death. The investigator should obtain information pertaining to the decedent's mental health history from sources familiar with the decedent. The investigator should document—

- The decedent's mental health history. Include hospitalizations and medications.
- Any history of suicidal ideations, gestures, and/or attempts.
- The names of the decedent's mental health professionals (such as psychiatrists, psychologists, and counselors).
- Any family mental health history.

DOCUMENT SOCIAL HISTORY

12-66. Social history includes marital, family, sexual, educational, employment, and financial information. Daily routines, habits, and activities, and friends and associates of the decedent help in developing the decedent's profile. This information will aid in establishing the cause, manner, and circumstances of death. The investigator should obtain social history information from sources familiar with the decedent.

12-67. When collecting relevant social history information, the investigator should document the decedent's—

- Marital or domestic history.
- Family history (similar deaths or significant dates).
- Sexual history.
- Employment history.
- Financial history.
- Daily routines, habits, and activities.
- Relationships, friends, and associates.
- Religious, ethnic, or other pertinent information (such as religious objection to an autopsy).
- Educational background.
- Criminal history.

INVESTIGATION COMPLETION

12-68. Completing the investigation of a death scene includes measures necessary for maintaining jurisdiction over and releasing the body, performing exit procedures, and assisting the family. These measures, like the previous, require accurate documentation.

MAINTAIN JURISDICTION OVER THE BODY

12-69. Maintaining jurisdiction over the body allows the investigator to protect the chain of custody when the body is transported from the scene for autopsy, specimen collection, or storage. To maintain jurisdiction over the body, the investigator should use a secure conveyance to transport the body.

12-70. When maintaining jurisdiction over the body, the investigator should—

- Arrange for and then document secure transportation of the body to a medical or autopsy facility for further examination or storage.
- Coordinate and document procedures to be performed when the body is received at the facility.

RELEASE JURISDICTION OF THE BODY

12-71. Before releasing jurisdiction of the body to an authorized receiving agent or funeral director, it is necessary to determine the person responsible for certification of the death. Information to complete the death certificate includes demographic information and the date, time, and location of death.

The investigator should obtain sufficient data to enable the completion of the death certificate and the release of jurisdiction over the body.

12-72. When releasing jurisdiction over the body, the investigator should—

- Determine who will sign the death certificate (such as the individual's name or the agency).
- Confirm the date, time, and location of the death.
- Collect (when appropriate) blood, vitreous fluid, and other evidence before the release of the body from the scene.
- Document and arrange with the authorized receiving agent to reconcile all death certificate information.
- Release the body to a funeral director or other authorized receiving agent.

PERFORM EXIT PROCEDURES

12-73. Bringing closure to the scene investigation ensures that important evidence has been collected and the scene has been processed. In addition, a systematic review of the scene ensures that artifacts or equipment is not inadvertently left behind (such as used disposable gloves, paramedical debris, or film wrappers.) and any dangerous materials or conditions have been reported. At the conclusion of the scene investigation, the investigator should conduct a postinvestigative walk-through and ensure that the scene investigation is complete.

12-74. When performing exit procedures, the investigator should—

- Identify, inventory, and remove all evidence collected at the scene.
- Remove all personal equipment and materials from the scene.
- Report and document any dangerous materials or conditions.

ASSIST THE FAMILY

12-75. The investigator provides the family with a timetable so they can arrange for the final disposition. The investigator should offer the decedent's family information regarding available community and professional resources that may assist the family.

12-76. When the investigator is assisting the family, it is important to—

- Inform the family if an autopsy is required.
- Inform the family of available support services (such as victim assistance, police, or social services).
- Inform the family of appropriate agencies to contact with questions (such as medical examiner's or coroner's office, law enforcement, or sudden infant death syndrome [SIDS] support groups).
- Ensure that the family is not left alone with the body (if circumstances warrant).
- Inform the family of the approximate body release timetable.
- Inform the family of the information release timetable, such as toxicology or autopsy results (as required).
- Inform the family of available reports. Include any costs.

TYPES OF DEATH

12-77. This section describes methods of death through criminal means. In the type of investigations in which footwear, tire wear, or other distinguishable marks are identified, the investigator must carefully document the impressions, photograph the impressions both with and without a measuring device, and collect a cast impression (if appropriate). The photographs and impressions can be used to make forensic identification at the crime laboratory between the suspect's objects and the collected impressions.

FIREARMS

12-78. Homicides and suicides occur most commonly as a result of the discharge of a firearm. Accidental death from the discharge of a firearm is also common. These violent deaths are often not witnessed. Unlike other forms of violent or unnatural death, deaths from firearms will often have trace evidence left by the weapon in or near the victim's body. This evidence can be scientifically compared with the suspect's weapon. It can often provide information about the circumstances surrounding the death.

12-79. In a medicolegal investigation of death by firearms, scientific evidence is very important. Deciding the manner of the death and solving the homicide, if there is one, often hinges on that evidence. Thus, the investigator must take every precaution to ensure that such evidence is not lost. For example, gunshot residue (GSR) must always be collected from the victim's hands at the scene, if possible. The residue is easily lost when a body is moved. If the residue cannot be collected at the scene, personnel transporting the suspect must be told to touch and move the hands as little as possible. This includes preventing the suspect from moving his hands freely. (See *Chapter 25* for GSR.)

12-80. A study of gunshot wounds in a body can tell a lot about the type of gun involved. It can identify the ammunition, the range of fire, and the direction and angle of fire. Sometimes it can tell the number of shots that hit the body. It can also give an idea of the fatal or disabling effects of an injury.

SHOOTINGS

12-81. For self-inflicted gunshot wounds, unless some special contraption is arranged, the victim generally must hold the gun close to his or her body. Rifles and shotguns are sometimes fired by using a stick or string hooked to the trigger guard or by pushing the trigger with a toe or a device. Riggings made to pull the trigger or the removal of a shoe, strongly suggest suicide.

12-82. Suicide wounds are usually single, close-range, and/or contact wounds on a part of the body that is easily reached. Sometimes an individual who commits suicide will shoot himself more than once before being disabled or dying. The presence of misfired rounds in or ejected from a weapon may also suggest suicide. An individual who commits suicide will sometimes fire shots to check the weapon while working up the nerve to complete the act.

12-83. An individual who commits suicide will often expose the part of his body that he plans to shoot. For example, an individual who commits suicide will tend to open his shirt before placing the muzzle against his chest. The chest and abdomen is often the target when a rifle or shotgun is used. The temple, the mouth, and the chest (over the heart) are common sites for suicidal attacks with a handgun. Many individuals who commit suicide with a handgun will shoot at the head area just in front of or over the ear.

12-84. An individual who commits suicide may guide the gun by holding the barrel with his nonfiring hand. In which case, the nonfiring hand of a suicide victim will have burns caused by the flame from the muzzle and breach. His hand may also have singed hairs and powder residue. Finding GSR on a victim's hands is not in itself conclusive proof of a suicide. It must be considered in light of other facts in the case. Residue can be present on a victim's hands because they were close to the muzzle blast of a shot fired by someone else.

12-85. The condition of the weapon can suggest the manner of death. The gun may be defective. The gun's safety may be defective. The gun may not have a safety catch or perhaps dropping it can discharge the gun. Evidence may show that the trigger caught on something, discharging it accidentally. Finding a serviceable weapon that needs normal force to pull the trigger and has a good safety device may help rule out an accidental shooting.

12-86. Most accidental shootings occur because of a victim's careless handling of or unfamiliarity with a gun. Perhaps the victim was on a hunting trip or was cleaning, loading, or otherwise working on the weapon. Evidence may show that the victim was handling the weapon unsafely, showing how another person killed himself, or that he was playing "quick-draw." Children and young people often become accident victims by playing with guns.

12-87. Accidental deaths are often witnessed and reported. Often, the person who fired the gun makes a report of the shooting. Accidental gunshot deaths that are not witnessed may look like a suicide. But the known attitude and lifestyle of the victim (in most cases), plus the lack of a clear case for suicide or homicide, are strong signs of an accident.

12-88. Most deaths that are due to multiple gunshot wounds have proven to be homicides. The murderer is usually related to or closely acquainted with the victim and fires a gun in a fit of rage, panic, or other strong emotions.

12-89. The location and number of empty shell cases at the scene may indicate the number of shots fired and the relative positions of the gun and victim. Lining up the final resting point of the bullet, the position of the victim, and the entry and exit holes on the victim can help tell the position from which a gun was fired. A gun may have been fired close to a surface or while resting on a surface. If so, it will have left powder residue. This also may show the position from which the gun was fired. All feasible surfaces of weapons, shells, and magazines must be checked for fingerprints.

12-90. When bullets are recovered at a crime scene, the investigator must record the exact details. These details include—

- The location and condition of the bullet.
- The type of material the bullet pierced and its depth of penetration.

- Irregularities in the size and shape of the bullet and the approximate angle of impact.
- Any other information that may help the lab examiner.
- The crime scene sketch documenting the point at which each discharged bullet or fired cartridge case was found.

12-91. Markings may be placed on a bullet by the bore of the weapon. Other marks may be placed on the cartridge case by the firing pin, breach block, chamber extractor, or ejector. Also, a lead bullet impacting on cloth may receive a patterned impression of the weave of the fabric. This may be useful in proving that a particular bullet passed through the victim.

12-92. At the laboratory, powder residue on evidence is tested chemically and microscopically. Bloodstains, hairs, fibers, and similar trace evidence is identified and compared. The lab may be able to tell from the GSR or burns on the clothing the approximate range from which the bullet was fired.

BULLET WOUNDS

12-93. A bullet passing through a body makes a wound with traits that can be recognized. All wounds must be medically confirmed during an autopsy. An investigator can usually tell entry wounds from exit wounds, however, sometimes the distinction is hard to make. External determination is hard if a body has started to decompose or has been mangled. The uneven surface and tumbling action of ricocheting bullets may make ragged punctures. Bullets passing across a body can cut gashes that may look like knife wounds. The energy of a high-speed bullet destroys tissue as the shock waves of its impact radiate away from the bullet. This makes a track of permanently disrupted tissue much wider than the bullet.

12-94. In tough cases, inspecting marks and effects on clothing may be the best way to tell the direction of the flight of the bullet. An autopsy examination of the track of the bullet may show the path of travel by pieces of cloth, metal, and bone fragments carried forward by the bullet. Metal debris is scientifically detectable by spectrography and X-ray methods. The nature of bone damage may show the travel path of the bullet. Determining which wounds are exit wounds and eliminating them from consideration helps locate entrance wounds. The wounds associated with bullets are—

- Entrance wounds.
- Contact wounds.
- Intermediate-range wounds.
- Distant wounds.
- Exit wounds.

Entrance Wounds

12-95. Entrance wounds are commonly round holes showing minor bleeding. Often, skin resistance is stretched by the impact of the bullet. This makes the hole somewhat smaller than the bullet. Sometimes a narrow ring around the entrance shows grayish soiling from carbon and oils on the bullet and a reddish-brown abrasion collar caused by the impact of the bullet. Some bullet entry wounds are inconspicuous or hidden. Small caliber bullets often cause such wounds. They may be hidden under clothing, in hair, in body folds or openings, or behind closed eyelids.

12-96. Identifying entrance wounds does not always tell the number of shots fired into the body. A single bullet can sometimes account for a number of entry wounds by piercing the body more than once. For example, a bullet may go through an arm before entering the torso. One of the aims of a pathologist tracing the path of the bullet is to try to match multiple wounds to the same bullet. A bullet striking a bony surface at an angle may split into two or more projectiles. The multiple projectiles can cause many exit and reentry holes. Ricocheting bullet pieces may also cause several wounds from a single bullet. On the other hand, more than one bullet may go through the same entrance wound. For example, in one rare suicide case, a defective round failed to exit the barrel and a second round pushed the first in tandem through a single entrance.

12-97. Bullets and other products from the discharge of a weapon have characteristic effects on skin and clothing. These effects can indicate the distance from which the gun was fired.

Contact Wounds

12-98. Contact wounds are wounds that are incomplete or have partial contact with the victim's body. Contact wounds, especially the ones on bony surfaces, are likely to be large, ragged stellate wounds. The explosive force of gases from the discharge often tears skin and tissue around the bullet hole, producing ragged lacerations radiating from the hole called stellate tearing. A contact wound made when exploding gases are received and expended by a large body cavity like the chest may not be large or irregular. On the other hand, a contact wound to the head made by a high-powered rifle may show massive bursting fractures of the skull from the explosive effect of gas forced into the skull where it has no chance to expand. Contact wounds from small caliber guns like a .25- or .22-caliber pistol tend to be smaller and less devastating than wounds from larger caliber weapons. This is because the discharge from small caliber weapons may not be forceful enough to disrupt the surrounding tissues.

12-99. Contact wounds are categorized as either hard contact (complete contact) that normally leaves a pattern impression in the shape of the weapon muzzle or loose contact (partial contact) that leaves a more irregular shape. Contact wounds leave an abrasion collar that is normally obscured. The edges of the contact wound and the bullet track are burned. If the gun is fired through clothing, the surrounding fabric is also burned. The flame and smoke may cause a sooty, grimy halo around the wound. When the muzzle of the gun is held tightly against the skin, the bullet hole is not "tattooed" with powder grains embedded in the surrounding skin like it is in intermediate-range wounds. This is because most of the unburned powder and other explosive products are blown right into the bullet track. The contact wound may also show a bruise pattern from swelling gases blowing the skin back against the muzzle of the gun. It may be shaped like the muzzle end of the gun, sights, or extractor spring rod.

Intermediate-Range Wounds

12-100. Intermediate-range wounds are made when the muzzle is held between 1 and 48 inches from the victim. The wounds are often round, but their edges may show minor splitting. They differ from contact or longer-range wounds by having burns and powder tattooing in the skin around the bullet hole. Powder residues and other discharge products are projected onto the victim in ample amounts when a gun is fired within 2 feet of the target. Recognizing powder marks and residues can help distinguish entrance wounds from exit wounds. Their pattern and composition help determine the range of fire and the kind of ammunition used. Precise range of fire tests can be made by laboratory test firing the same weapon and ammunition. The type of powder used in the ammunition will affect the distance it travels in the air. Types of powder residues can also be distinguished by chemical, photographic, radiographic, and spectrographic tests.

12-101. The burned and tattooed area is roughly circular. It becomes larger and more diffused as the distance between the weapon and the victim grows. The area has three zones. The first zone, known as the flame zone, is the zone of burned skin around and in the bullet hole. The second zone surrounds this zone and is where most tattooing powder grains and combustion products burn and stick to the skin. The last zone is under the skin where sparsely scattered powder grains and residues are embedded in the dermis. Washing will not remove powder grains in the dermis.

12-102. If the burned, tattooed, and abraded areas form a concentric-symmetric margin around the entry wound, the bullet probably struck the body at a right angle. A bullet striking at a shallower angle shows marginal bruising and abrasions at the point where the bullet first meets the surface. A bullet striking at an extreme angle may cause a shallow-furrowed wound.

12-103. If a gun is fired at close range and at an angle to the body, powder residues will seem to spread away from the bullet hole in an uneven V-shape. The point of the V will point toward the weapon. The size of the ammunition and the type of powder also affect the nature and extent of powder residues. At a distance of 3 to 4 feet, powder residues may not be present on a victim shot with a handgun.

Distant Wounds

12-104. Distant wounds are made by muzzles held more than 48 inches from the victim. The wounds are generally round holes with circular abrasion collars. There are no burns or powder tattooing. Small caliber contact wounds and other contact wounds over soft-tissue areas may look like long-range wounds. They can be distinguished deep in the tissues in the bullet track by the powder residues.

Exit Wounds

12-105. Exit wounds often show more damage than entrance wounds. Exit wounds are ragged and rough in shape. They are often larger than the bullet. Tissues compressed in front of the bullet burst when the bullet breaks through and exits the body. The bullet is often fragmented, deformed, and tumbled by impact. It is, therefore, more destructive. Thus, exit wounds may bleed more than entrance wounds, and pieces of internal tissue may protrude from the wound.

12-106. Because a bullet loses momentum as it passes through the body and its tough, elastic skin, it sometimes uses up its remaining energy at the point of exit. For that reason, a bullet may be found protruding from the skin or loose in a victim's clothing. It may also be found under the skin, where it has caused swelling or bruising. If a bullet is lodged in a body, advise the surgeon of its potential value as evidence. Request that he not probe for the bullet except as a last resort. If probing is needed, request that rubber-tipped forceps be used to remove the bullet.

SHOTGUN PELLET WOUNDS

12-107. Shotgun wounds are very different from wounds caused by other firearms. The destructive force of a shotgun blast at close range is great. If the wound is to the head, the shape of the head may be greatly changed. Large sections of the head or face may be destroyed. Close-range wounds of the trunk and abdomen may cause loops of intestine or other organs to hang out of the body, or it may remove a large portion of the victim's body.

12-108. When a shotgun is fired from a distance of about 10 feet or less, the shot strikes as a fairly compact mass. It leaves a large central, circular hole with very ragged edges from the many single and overlapping punctures made by the shotgun pellets. This is known as the cookie-cutter effect. Scattered around the large central hole are smaller holes made by individual shot beginning to disperse in flight. When a shotgun is fired at close range, the wounds are grossly burned and tattooed. As the distance increases between the weapon and the victim, the wound shows less tattooing and no burning. Beyond 10 feet, the shot spreads in flight and strikes the body in a more scattered grouping so that no central hole occurs.

12-109. The length of the shotgun barrel and the type of ammunition also influence the dispersion of the shot and the scattered pattern of the wound. A sawed-off barrel allows quicker spreading. The dispersion may be cut if the shotgun is choke-bored. The slightly narrowed muzzle focuses the shot and delays its dispersion. Bird shot, even when fired at close range, usually does not go through the trunk or abdomen of an adult. But when the shot load goes through a thinner portion of the body like the neck, limb, or shoulder, it makes large lacerated exit wounds. Sometimes small, ragged exit wounds are made when only some of the bird shot exits the body. At close range, buckshot (having a greater weight and energy) causes wounds similar to those made by large bullets. A general rule of thumb in shotgun spread is that for every 1 yard of distance, the wound will grow by 1 inch in diameter. This can only be verified through specific firearms testing at the crime lab.

12-110. Shotgun pellets cannot be linked to a certain gun by ballistics markings as rifled bullets can. However, the size of shot may be learned from printed material on the top wad or by marks left in the wadding. It can also be learned from printed information in the shot column. One may learn the gauge of a gun by comparing the diameter of the wad with other wads. If the wadding has not struck an intervening zone, it can be found within 50 feet of where the gun was fired. If the gun is fired within 10 feet of the victim, the wadding is often carried into the body with the shot. Investigators must be aware that, during the crime scene search, the shot cup wadding may contain evidence of the lands and grooves of weapons that maintain a rifled barrel.

ASPHYXIA

12-111. When the body or any vital part of it is deprived of oxygen, asphyxia occurs. Death from asphyxia alone is most often due to natural or accidental causes. Many diseases and infections can hinder airways. Foreign bodies like meat or bone can become trapped in the throat, causing asphyxia. Food particles are often the cause of accidental choking deaths in adults. Choking deaths of children are common from food and small plastic or metal toys. Pressure on the outside of the chest that restricts breathing can also cause asphyxia. This pressure can occur in cave-ins, building collapses, or traffic accidents.

12-112. Inhaling chemicals like ammonia, chloroform, carbon monoxide, and carbon dioxide also may cause asphyxia. Sometimes these chemicals are the cause of suicidal or homicidal deaths. Homicide and suicide by asphyxia alone are rare. In learning the reasons for death by asphyxia, anything suspicious must be pursued through a background investigation and an autopsy. The death may be ruled accidental or natural only after a thorough investigation.

Strangulations

12-113. Strangulation is asphyxiation from compression of either the jugular vein or the carotid artery in the neck. Many times, when strangulation occurs, the trachea is not obstructed. Strangulation can be done manually or with a ligature like a binder, a rope, or a necktie. Hard blows to the neck may also cause strangulation. Open-handed strikes to the throat may cause damage to the larynx, followed by suffocation.

12-114. Manual strangulation is normally a homicide. A person cannot kill himself with his hands because when he loses consciousness his hands relax and his breathing resumes. In manual strangulation, the attacker's fingernails often make small telltale bruises or marks on the neck. Marks on the neck will not normally show the direction from which the victim was attacked. Another sign of manual strangulation is hemorrhaging in the throat area. This can be seen in an autopsy. Sometimes a fracture of the hyoid bone, a U-shaped bone at the base of the tongue, is also found.

12-115. Strangulation by ligature may be homicidal, suicidal, or accidental. Strangulation is a fairly common form of suicide but it is a rare form of homicide. The ligature is often made from something handy at the scene. Pajamas, neckties, belts, electrical cords, ladies' stockings, and other items can be used. Strangulation by a garrote of rope or wire sometimes is used in homicidal strangulation but it is not seen very often. Close inspection of the marks left on the skin may show the type of garrote used. If possible, leave the ligature in place for a pathologist to remove during the examination. If the ligature is removed from the body, leave the knots in place and cut the ligature on the opposite side of the knot.

12-116. When investigating strangulation, search the scene and the victim for signs of struggle. Check the body for signs of defense wounds that may suggest homicide. Also look for the presence of hesitation marks, which hint at attempted suicide by other means before ruling it a homicide.

Hangings

12-117. Hanging is asphyxiation by strangulation using a line of rope, cord, or similar material to work against the hanging weight of the body. Hanging is most often suicidal. But can sometimes be accidental. It is seldom homicidal, except in a lynching. A person does not have to be fully suspended to die from hanging. Hanging may occur if a victim jumps or is pushed from a height while tied by a line to a rafter or a tree limb. If the height is more than just a few feet, the victim's neck may break. The neck is seldom broken in suicidal or accidental hangings.

12-118. At the scene, an investigator—

- Checks the beam or rafter over which the line is laid for marks showing the direction of travel of the line. This must be checked.
- Inspects the scene for signs of a fight and signs of defensive marks or rope burns. Bears in mind that an unconscious victim may convulse, knocking over items in the immediate area.
- Does not untie the knots when taking the body down. The type of knot may provide an investigative lead.
- Removes the hanging line from the victim's neck by cutting the line on the side opposite to the knot.
- Makes a careful inspection of the groove around the neck. A close look at the edges of the groove will often show black and blue marks from minute bleeding. Ruptured blood vessels in the skin mean that the victim was alive at the time of the hanging. The lack of these marks does not necessarily mean the victim was dead at the time of hanging. Combined with other conditions, however, it could raise suspicions.
- Notes the position of the groove as it relates to the location of the knot. The mark of the ligature should agree with the location of the knot. For example, if the knot is in front of the face, the deepest part of the groove should be on the nape of the neck. Anything different suggests homicide.

12-119. When a fixed knot is used in a hanging, the groove will form an inverted V on the side of the knot. The bruise on the skin in the groove is greatest opposite the knot. It tapers off as it reaches the knot. If a slipknot is used, the groove may be uniform around the neck; however, it will still be up in directionality.

12-120. An accidental hanging from sexual activity may be suspected when a victim is nude and is suspended before a mirror or is suspended in an unusual manner. Accidental deaths may occur from autoerotic sexual acts using restraints like ropes, cords, chains, and handcuffs. The victim, trying to reach sexual contentment, uses these items to restrain his or her hands, arms, legs, and neck. When strain on the neck causes unconsciousness or when the victim loses balance during the act, an accident results. The victim is unable to release himself because of the binding on his hands, arms, and legs. Sometimes, when a victim uses binding material or plastic bags on his face, he suffocates. A notable feature of this type of death is the presence of female attire and articles on or near a male body. Erotic material and evidence of masturbation is often present. In the past, these deaths were often incorrectly labeled suicides. But they are accidental and they must be listed as such.

12-121. Other accidental hangings differ from autoerotic deaths in the lack of female attire, erotic material, or constrained hands or feet. Accidental hangings often involve infants and young children. Infants can get caught in restraining devices, or they can get their clothing caught on things. They may also get their heads caught between crib or fence slots. If they are unable to get themselves free, they may strangle. For no known reason, young children (especially boys) will put nooses around their necks. They too may strangle to death.

Drownings

12-122. Drowning is asphyxiation from water or liquid being inhaled into the airways, blocking the passage of air to the lungs. Water inhaled into the windpipe causes violent choking. The choking irritates the mucous membranes of the airways causing a large amount of sticky mucus to form. The mucus, mixed with the water and agitated by violent attempts to breathe, turns into thick, sticky foam that fills the windpipe.

12-123. Most drownings occur when the victim submerges in a body of water. A small number of "drowning" deaths among swimmers are actually caused by their hearts stopping from the shock of submersion. Most commonly, a drowning victim has a violent spasm of the neck, throat, and chest muscles. This prevents breathing. The victim submerges, inhaling water. The victim may stay submerged the first time he goes under, or he may go under and surface many times until he can no longer struggle to the surface. The loss of consciousness often occurs quickly. Because the human body is heavier than water, when unconsciousness occurs, the victim sinks and tends to lay at the bottom with his head down. Breathing may continue briefly with varying amounts of water inhaled. The heart may beat briefly after breathing stops. Death by asphyxia occurs within a few minutes. Banning strong currents, a body sinks fast. It often comes to rest at a point close to where it was last seen on the surface.

12-124. Rigor mortis may start earlier than in other deaths because of violent muscular struggle. Postmortem lividity occurs but is often light red in color and is noted mostly in the head and upper body. This is because of the tendency of the body to sink head down. The foam that forms in the airway may exude from the mouth and nose. Often, the victim's hands will be grasping gravel, mud, or grass. The hands and fingertips may be scratched from violent grasping efforts. The victim's palms may have cuts caused by his fingernails during the violent clenching motion of the hands. These factors are good circumstantial signs that the victim was alive when he entered the water.

12-125. After a few hours, depending on the temperature and movement of the water, postmortem changes peculiar to submersion begin to occur. The skin, especially on the hands and feet, becomes bleached and waterlogged. The palms develop a very wrinkled condition called "washerwoman" hands. The constant churning of water currents or long periods of submersion may cause maceration. This is the wearing away of skin and flesh, especially of the hands and feet. Mutilation may occur from propellers of boats. This causes the appearance of postmortem dismemberment. Parts of the body, notably the face, may be eaten by marine life. As bacteria mounts in the body, putrefaction begins. As putrefaction progresses, gases build up in the tissues, organs, and body cavities. The body becomes distended with gas. This makes the white foam in the airway come out of the nose and mouth. As the gases build up, the body becomes buoyant. Warm water speeds putrefaction and cold water slows it. In warm water, buoyancy may occur in a couple of days. In winter, the action may be slowed for weeks or until spring. As putrefaction advances, the skin loosens from the tissues. Sections of skin, especially hands, feet, and scalp may fall from the body.

12-126. Unless a body is heavily weighted down or firmly caught on underwater debris, buoyancy will eventually cause it to rise to the top and float. If a body is prevented from rising, the gases eventually escape. Then buoyancy leaves and a body may stay down forever. When a "floater" rises and is exposed to the air, decomposition proceeds at a much faster rate.

12-127. Prolonged submersion and decaying may dim or destroy the external signs of asphyxia. Signs of violence or another cause of death may also be lost. Prolonged submergence makes death by drowning medically difficult to diagnose. Medical evidence may show signs of asphyxia like foam in the airways and an enlarged heart. It may also show changes in the blood from water absorbed during drowning. Algae and other substances from water may be found in the stomach or airways. Chemical tests conducted during an autopsy can show if the person was alive when he entered the water. However, chemical tests are nonspecific and none are diagnostic of drowning.

12-128. Suicidal drownings in places like bathtubs are hard to distinguish from accidents unless a reason is suggested or some other means of suicide was also attempted. Check for marks that may show suicidal intent. A weighted body strongly suggests homicide. Persons who commit suicide may weight their bodies to speed drowning and stop recovery. Weighted bodies should be inspected carefully for injuries that suggest homicide. It should be determined if the binding and weighting method was used by the victim. An assessment for self-inflicted injuries, such as cut wrists or other sign of suicide, should also be made.

12-129. Homicidal drownings are rare. Unless accompanied by signs of homicidal violence or other such conditions, the autopsy shows only signs of asphyxia by drowning. There have been times when submerged bodies have shown no signs of violence but, after the body dried out, bruise marks and small abrasions appeared that could not be seen when wet.

Electric Shocks

12-130. Death by asphyxiation can occur as a result of electric shock. The shock stops the action of the heart, and the brain (deprived of oxygen) ceases to function. The effect of electric shock on a person depends on many things. It depends on the individual's health, his location, and how wet or dry it is. Other factors include the amount of voltage the person received, how long he was in contact with the voltage, and the aftereffects of the shock.

12-131. Electric shocks often leave marks, although it is possible for a body not to show outer or inner damage. Usually, electrical shocks leave entrance and exit wounds on the body. The wounds have a gray or white puckered look. Lightning deaths leave a characteristic mark that resembles a fern leaf. High-voltage shocks may leave marks where metal objects have melted on the person. There may also be extensive fractures of the bones. Distinguishing between the entry and exit of electric wounds cannot be made from looking at the injury alone.

12-132. When investigating a death by electric shock, an investigator must check the weather conditions, electric appliances that the victim may have been using, and the victim's location and activity at the time of death to determine if the death was accidental. Deaths from electric shock are most often accidental. Homicide by electrocution is rare, but it is possible.

SHARP FORCE INJURIES

12-133. The vital functions of the body can be fatally impaired by injuries from sharp-edged instruments. Deaths or injuries from stabbing, cutting, and chopping are hard to evaluate without extensive experience. The type of wound and the victim's personal history can help decide if death was an accident, a suicide, or a homicide.

Stabbings

12-134. Any object with a fairly sharp point may make stab wounds. Knives, scissors, ice picks, triangular files, or hatpins can all make stab wounds. Sometimes stab wounds look like other kinds of wounds. A wound made with a stiletto or ice pick may look like a bullet wound. If examination fails to show a sure sign of stabbing, the wound may have been made in some other way. X-rays may help to locate an unsuspected bullet or piece of a weapon, such as a knife or stiletto that may be inside the body. Most stab wounds involve some cutting. This occurs as the weapon is pushed in or drawn out.

12-135. The shape of the wound depends on the direction from which the weapon penetrates and the shape of the weapon. It also depends on the movement of the weapon while in the wound. If a blade pierces a body at less than a right angle, it makes a beveled wound. If a blade moves around in the wound, an uneven-shaped scrimmage wound is made. The weapon is often turned slightly as it is withdrawn. This causes a wound that has a notch in one side.

12-136. The depth and shape of a fatal stab wound, fixed during an autopsy, may give a clue to the type of weapon used. The track of a weapon may be very clear in fleshy areas. However, when a weapon penetrates inner organs, its track may not be accurate. Inner organs change in shape and position after death and when a body is moved. Also, a strong stabbing force against a soft area like the stomach can depress the area, making the wound deeper than the true length of the weapon. Likewise, a longer blade may not penetrate its full length, making the path of the wound shorter than the blade.

12-137. A homicidal stab wound often penetrates a victim's clothing. For this reason, investigators must exercise special care when removing and checking the victim's clothing.

12-138. Pierced bony surfaces like the skull, sternum, or spine often show the shape of the part of the weapon that passed through the bone. Sometimes weapons break off or are left in the bone. The blade or portions of it may project from the inner part of the bone. If a blade is broken in a fatal stab wound, the part of the bone with the blade in it may be removed at the autopsy. It can be used to support the main elements of the crime, especially if the matching part of the weapon could be located and preserved as evidence.

12-139. It is difficult to determine if a wound was made before or after death. A good inspection of the wound made before the body is moved is very important. If the wound was made before death, there should be evidence of blood clotting, swelling, wound healing, or infection.

12-140. Accidental stab wound deaths are rare. When they do occur, the victim's falling through glass doors or windows often causes them. The victim is stabbed by the larger pieces of broken glass. Other stabbing accidents may occur when victims fall on sharp pointed surfaces of tools or equipment. Sometimes large splinters, vehicle surfaces, or tusks or horns of animals pierce victims.

12-141. Most fatal stab wounds are homicidal. Often, there is only one wound that pierces a vital organ or nerve center causing death from shock, hemorrhaging, or the interruption of a vital function. Homicidal stab wounds often appear on the back, neck, and upper chest. When many wounds are present on different parts of the body, homicide is strongly indicated. Wounds of the same depth, wounds of nonvital areas, scrimmage wounds, and multiple wounds of a vital area strongly support homicide. Several stab wounds to the breasts and genitals are suggestive of a sex-related homicide. Defense type wounds on the hands and arms and wounds to the back or other areas not easily reached by the victim also hint of homicide.

12-142. Many stabbings are not instantly fatal. The victim may live for days and then die from acute infection or other medical problems. Stabbings are usually not immediately disabling. Unless the victim is unconscious or otherwise helpless, the scene is likely to show signs of a struggle. Signs of flight and traces of blood are likely to be scattered over a large area.

12-143. Suicidal persons most frequently stab themselves in the chest over the heart. Suicidal stab wounds may be made on any area of the body that can easily be reached. Like suicidal shootings, the victim will often open up clothing or uncover the selected stab area. Often, the knife is left sticking in the wound. In some cases, the suicidal person may jab the weapon into his chest a number of times. In such cases, the wounds often vary in depth. Many of them may barely penetrate the chest. These hesitation wounds are made as the victim works up nerve to force the weapon through. Suicidal persons sometimes stab themselves repeatedly in different directions, through the same wound, without completely withdrawing the weapon. This causes more stab tracks than outer wounds. Hesitation cuts on the wrist or thighs are good signs of the suicidal intent of a victim. Investigators must look at the societal history of the victim to support suicide as a manner of death.

Cuttings

12-144. A cut is an incised wound made by a sharp edged object. The sharp edge is pressed to and drawn over the surface of the body to inflict a cut. Knives and razors account for almost all incised wounds. Cutting wounds can cause fatal hemorrhaging and infection. They can also be fatal if the victim inhales blood from a cut airway. Cuts are often made on exposed surfaces like the head, neck, and arms. Where many cuts are involved, those on the palms of the hands and the outer surfaces of the forearms of the victim are often defensive cuts. They may indicate a homicidal attack. Heavy maiming and dismemberment may accompany sex murders.

12-145. Homicidal cuttings are usually deep, clean cuts without hesitation marks. The wounds may be on various parts of the body. Most often, they involve the head and neck. Homicidal slash wounds may be present. Such a wound may be a single deep cut on the side of the face and neck, or it may be one of many deep slashes crisscrossing each other. Sometimes, when a victim tries to dodge slashes, there are small shallow cuts near larger wounds. These defensive wounds may be confused with suicidal hesitation cuts.

12-146. Suicidal cuts are often many, parallel, overlapping incisions of varying length and depth. Many times they have a lot of smaller shallow hesitation cuts on the lead edge of the injured area. Fatal suicidal cutting wounds are often on the throat. When a throat is cut, fatal bleeding sometimes results from a fairly shallow cut that severs a large vessel.

12-147. Other self-induced cuts may be made to the groin, thighs, ankles, knees, and the inside of the forearm at the elbow. Suicidal cuts on the limbs are often not fatal but are frequently found on individuals who have killed themselves some other way. They support a judgment of death by suicide.

12-148. Accidental incised wounds are rarely fatal. They occur most often from broken glass or contact with moving machinery or sharp tools. Most of the time, the situation clearly shows the accidental nature of the injury.

Choppings

12-149. A chopping wound is a mangling, tearing cut. The wound is usually made with a heavy instrument like a cleaver, a machete, a hatchet, or an ax. Death from chopping wounds may come from shock, hemorrhaging, or the interruption of vital functions.

12-150. Most chop wounds are homicidal. They are usually made on the head, neck, shoulders, and arms. Injuries may be multiple. Injuries received by the victim in an attempt to defend himself may include total or partial loss of fingers, hands, or arms. Fatal accidental chop wounds sometimes occur from propeller blades of fans, boats, or planes. Suicidal chop wounds are rare.

12-151. Because the shape and size of chopping wounds often resembles the shape and size of the weapon that made them, autopsies may provide medical evidence. A pathologist may be able to determine the type of weapon that was used by examining the depth, width, and appearance of the wound and the amount of tissue damage. He may be able to link the injuries to a suspected weapon. It may even be possible to take toolmark impressions of the weapon from bone or cartilage.

BLUNT FORCE INJURIES

12-152. Blunt force damages the body by direct physical violence. Generalized blunt force affects the whole body or a large part of it. Deaths caused by such force may happen in vehicle accidents, explosions, or falls from a height. Localized blunt force impacts a limited area. Death involving localized blunt force can be caused by contact with a fist, weapon, or foot.

12-153. Blunt force injuries of the skin and tissues under the skin are of three general types: abrasions, bruises, and lacerations. Abrasions are surface injuries to the outer layer of the skin at the point of impact. An abrasion may duplicate the surface appearance of the impacting object. It may look like the grill pattern of an automobile or the rough edges of a file. It may look like a threaded pipe or the treads of an automobile tire. Bite and nail marks are considered abrasions, but they actually may be small puncture wounds. Abrasions normally are caused by direct violence from hands, blows of a weapon, or collision with a vehicle. They may also be caused when a body falls and strikes a surface.

12-154. The presence of a contusion on a decedent is a good indicator of blunt force trauma. It is not possible to give a precise estimate of the age of the contusion. The contusion may appear at an area of the body that is not the point of impact. Blood will often travel under the skin for a distance (generally associated with gravity) and then appear. Bruises may also be larger or smaller than the object used to inflict an injury. No conclusion concerning the type of weapon can be drawn.

12-155. The size of a bruise may indicate the degree of violence that caused it, but not always. Females tend to bruise more readily than males. Bruises occur more easily on the very young, old, fat, soft-skinned, poorly conditioned, and sick. A light blow to soft tissues like the eyelid or genitalia may cause gross bruising, while a heavy blow to dense, fixed tissue like the scalp may cause only mild bruising.

12-156. Lacerations are caused by a shearing force or violent depression to the skin that tears or splits tissues. Blows from fists, sticks, or hammers may cause lacerations. They may also occur from the impact of a motor vehicle or as a result of a fall. Lacerations, characteristically, are bruised and ragged. The tissues are unevenly divided and the blood vessels and nerves are crushed and torn. The crushed ends of vessels may show only slight bleeding. Lacerations may contain foreign material like soil or glass from the impacting object. Lacerations of the scalp, face, eyebrow, or skin near bone have a linear splitting effect. These may be hard to tell from cuts.

12-157. Normally, it is not possible to tell lacerations made at the time of death from those made shortly after death. The distinction depends on the presence or absence of vital reactions like bleeding and bruising in the wound.

12-158. Homicidal deaths may occur from either generalized or localized blunt force. Victims may be struck with fists or blunt objects, or they may be thrown from heights, pushed in front of moving vehicles, or crushed with heavy objects. Sometimes homicidal blunt-force deaths involve fatal injuries from negligence. This may occur in highway accidents.

12-159. Suicidal deaths from blunt force usually involve generalized blunt force. The victim jumps from a high place or in front of a moving vehicle. Sometimes a suicidal person may ram his head into a wall or in some other way create enough impact or crushing force to cause fatal injuries. Accidental deaths from blunt force are usually falls.

12-160. Often, investigations of death involving generalized blunt force will show that the injuries have resulted from impact with a vehicle. The investigator must be able to link the vehicle to the victim by trace evidence left at the scene or found on the persons or vehicles involved.

VEHICLE TRAUMA

12-161. If circumstances suggest a hit-and-run accident or a vehicular homicide, the investigator must initiate an immediate search and apprehension plan. Such a plan may include setting up coordinated military police patrol activities and roadblocks. It may include searching and patrolling parking lots, service stations, residential parking areas, motels, taverns, bar rooms, garages, and body repair shops. If a military vehicle could be involved, it must include alerting and checking motor pools. It includes checking known license and registration data.

12-162. At the scene, the investigator—

- Looks for evidence supporting the crime and linking the vehicle and victim.
- Checks for skid marks to learn about the speed of the vehicle, the alertness of the driver, and verification of the accuracy of the driver's and witnesses' statements.
- Photographs all skid marks, with and without a measuring device.

NOTE: Photographs intended to serve for later comparison purposes to suspect's vehicle tires must be taken with a 35-millimeter camera, with and without a measuring device.

- Takes samples of dirt from under the vehicle to link it to the point of impact and to use for future comparison.
- Photographs and casts tire tracks before they are disturbed.
- Records the bumper height measurements to match them to the victim's injuries. This may help determine if the vehicle at the time of impact was braking, maintaining, or increasing its speed.
- Collects material from the scene and from the victim or the victim's vehicle that may have come from the offender's vehicle. Broken glass, vehicle parts, trim, paint chips, or other evidence may identify the vehicle type and may be compared with a suspect's vehicle. The laboratory may be able to recover glass fragments, paint chips, or other material from a victim's (pedestrian's) clothing that originated from the suspect.
- Seizes trace evidence from the victim like blood, body tissue, hairs, and textile fragments.
- Checks vehicles suspected of having been involved for signs of impact and traces of the victim or victim's clothing. Often, in hit-and-run accidents, the victim's clothing leaves patterned rub marks. The

pattern may show in the chassis paint. It may also show on grease and mud on the undercarriage. Likewise, hand, finger, and even lip prints of the victim may be left on the vehicle. Hair, tissue, bone fragments, fabric, fibers, and other trace evidence of the victim may be stuck to the suspect's vehicle. Blood of the victim, often found on the undercarriage, must be typed with a sample from the victim. The sobriety of the victim at the time of death must always be learned.

12-163. A follow-up investigation may include—

- Checking on individuals with a history of speeding and reckless or drunken driving.
- Checking with insurance agents, vehicle sales, and transfers of registration.
- Contacting the press, radio, and television for coverage to help seek out the perpetrator.

12-164. The follow-up investigation must include—

- Conducting a thorough canvass interview.
- Contacting medical facilities to see if anyone has sought medical attention after an accident.
- Checking stolen vehicle reports. A driver of a hit-and-run vehicle may report that the vehicle was stolen. Likewise, a hit-and-run driver may file a false accident report to cover a real accident.

BEATINGS

12-165. Beatings involve localized blunt force. Death from a beating is usually not planned. Beatings leave extensive bruises on a body. Autopsies often show ruptured vital organs and brain hemorrhaging. When a weapon is used in a beating, it often leaves distinct pattern injuries that may disclose the type of weapon used. When a person is kicked or stomped, the shoe often leaves impressions and clear-cut marks on the clothing or body.

12-166. Sometimes, in a beating death, the body is moved and a simulated vehicle accident is staged, or a vehicle is driven over the body to stage a hit-and-run accident. An autopsy may show that the injuries are not like motor vehicle injuries. There may not be a point of impact. A thorough search of the area may show evidence inconsistent with an accident.

EXPLOSIONS

12-167. It is imperative that in any investigation involving explosives, the EOD detachment respond to the crime scene and render the area safe before any investigation. Most deaths caused by explosives encountered on a military installation are related to training accidents. However, with today's changing times, WMD are a concern. It is important to note that an explosion, whether it is large or small, will create similar injuries. The injuries will only vary in magnitude.

12-168. In an explosion, a body may be shattered or hurled against a hard surface, causing blunt-force injuries. It may receive many lacerations and punctures from pieces of the explosive device and nearby objects. It may be burned by a thermal blast, flames, and steam. Compression injuries may occur in the lungs and elsewhere from swelling gases. Foreign material in the body must be examined at the autopsy to help learn the nature of the explosion and the explosive device used.

FALLS

12-169. Deaths from falls are usually accidental. A person may be pushed or thrown from a height, but such events are rare. It is sometimes helpful to consider the blood alcohol content of the victim. An investigator must consider the height from which the victim fell and the distance from the base of the object to the point of impact. Do not overlook the fact that a victim found at the very base of an object may have been knocked out and rolled over the edge. An inspection of the point of departure of the body must get ample attention. The totality of the circumstances is the driving force behind a determination if a person fell from, for example, a cliff while hiking on an elevated trail.

THERMAL DEATHS

12-170. Most deaths by fire are accidental. Connecting the death and the cause of the fire may show that a homicide is involved. The fire may have been the cause of death or it may have been used to try to cover up a crime. The investigator must take steps to investigate for arson. Sometimes a person who commits a homicide with a firearm will try to hide the crime by setting fire to the scene. In cases of death, by burning, request that the remains be x-rayed. This may show the presence of a bullet in the body.

12-171. The two toughest facts to establish in a death by fire are the victim's identity and a connection between the death and the cause of the fire. Investigating a death by fire is difficult. The fire may have mutilated the victim, or the scene of the fire may have been (as is often the case) unavoidably disturbed by firefighting activities. Identifying unknown victims requires the help of pathologists. Pathologists can check skeletal remains for size, race, and sex distinctions. They can compare the remains to dental records and X-rays.

12-172. The investigation of a death from fire depends greatly on the pathologist's report of the cause of death. If the victim was alive at the time of burning, the autopsy will show inhaled smoke particles or carbon monoxide in the blood. The presence of these suggests life at the time the fire started but its absence does not support death before the fire. A body is rarely burned to the point that a meaningful autopsy is not possible. Even if death occurs some time before the fire is brought under control and the body is badly charred, the inner organs are usually preserved well enough. The cremation of a body takes 1 1/2 hours at 1600° to 1800° Fahrenheit. Even then, bone fragments are seen.

12-173. The investigator must rely on the pathologist to identify wounds on a burn victim. There are many types of burn injuries that are misleading at first glance. The body may have a pugilistic attitude. The fists and arms may be drawn up like a boxer's stance from contracting muscles and skin. Bones fracture in an odd, curved way when cooling begins. Skull fractures may be present. The cracks, radiating from a common center, are made by the release of steam pressure rather than from a blunt force. (See *Chapter 7* for information on arson.)

DEATHS INVOLVING TOXIC SUBSTANCES

12-174. Death from toxic substances may occur if substances that are safe only for external use are taken internally. Death from toxic levels of substances safe for internal use may occur if the substances are taken in amounts greater than the body can support. In either case, the death may be an accident, a suicide, or a homicide.

Poisonings

12-175. The term poison is relative when describing a substance. A poison is any agent that, when introduced into a living organism, causes a detrimental or destructive effect.

12-176. Accidental death may result from industrial, home, or food poisoning. Sometimes poisonings result from gross negligence like that occurring in bad liquor or criminal abortion cases. Although homicide by poisoning is fairly rare, it must not be ruled out without a thorough investigation. Murder by poison can often be made to look like suicide. For example, the scene of a murder by poisonous gas may be fixed to look like that of a suicide or an accident.

12-177. Investigation of the crime scene is of special importance in the case of poisoning because postmortem detection of poison may be difficult if its presence is not suspected. The presence of any one poison may be hard to identify unless medical personnel have some idea of the type of poison they are looking for. The crime scene search for such poisons is most important.

12-178. When death is suspected to be the result of poison, it is important to give the pathologist performing the autopsy as much information as possible about the circumstances of the death, the on-the-scene investigation, and the type of poison suspected. If this information is provided before the postmortem examination, it allows the pathologist to use the right autopsy methods and to keep good specimens for toxicological tests.

12-179. The military treatment facility must conduct a medical inquiry to learn the immediate cause of a death by suspected poisoning. Results of the inquiry are recorded in the postmortem report. The report is a full record of what the medical authorities know about the person who has died. It includes a record of—

- The clinical treatment given to the victim.
- The statements, accusations, or an utterance made by the victim before death.
- The known facts pertaining to the death.
- The immediate cause of death.
- The autopsy.
- The pathological and toxicological examinations conducted to support the autopsy.
- The medical examinations of items of physical evidence.

12-180. The autopsy may identify the specific poison that caused the death, its concentration in the body, and the period of time the poison was in contact with the soft tissue before and after the death. In some cases, the specific poison may be unidentifiable because the dose was too small to detect or the materials in the compound were the same as natural body products.

12-181. The pathologist must be asked to obtain specimens of the victim's blood, bile, gastric contents, and urine. All toxicology specimens should be sent to the Division of Forensic Toxicology, Office of the Armed Forces Medical Examiner, Armed Forces Institute of Pathology, Washington, DC. Do not send toxicological evidence to USACIL for examination. Remember that body fluids found on a floor are likely to be contaminated. They are of little use in toxicological tests for poisons.

12-182. An investigator must—

- Take samples of food, medicines, beverages, narcotics, fuels, and chemicals that the victim may have consumed.
- Check sinks, pipes, drain traps, garbage cans, cupboards, and refrigerators for possible evidence of the poison. Poison can also be easily hidden in spices, sugar, flour, baking soda, and so forth.
- Collect soiled linen or clothing that may contain traces of poison in stains from food, liquid, vomit, urine, or other matter. Submit these samples, despite an admission or confession, in any case that may involve criminal charges.
- Collect containers that could have held a substance consumed by the victim. Include cups, glasses, and utensils that may have been used to prepare or serve food or drink.
- Check medicine containers for prescription numbers and the name of the dispensing pharmacy. In difficult cases, it may be necessary to take the contents of the medicine chest to search for materials that might have been taken in amounts large enough to cause toxic effects.
- Seize any items like hypodermic needles and syringes that could introduce poison into a victim's body.

12-183. Identification and analysis of the poison may help locate its source. Few laymen know enough about poison in its pure form to purchase or obtain any but the most well-known types. Many common retail products, not often thought of as poison, are toxic under some conditions. It is these materials that will be easily accessible to the poisoner. Their ready availability may make them easy to overlook. Household sprays, paint and paint solvents, pesticides, liquid fuels, patent medicines, many antiseptics, and some cosmetics contain poison.

12-184. To learn the source of a poison, consider its availability and who would have the easiest access to it. A poisoner usually uses a poison he knows. His familiarity with a substance can come from his occupation, hobbies, or past experience. Hospitals, dispensaries, laboratories, pharmacies, and illicit narcotic channels can be sources of medicines and drugs to be used as poisons. Offices, homes, and grocery stores contain cleaning substances, rodent and insect poisons, and medicines that may be toxic. Depots, warehouses, storage areas, farms, and similar places may be sources of rodent and insect poisons.

Motor pools, fuel depots, and other places containing fuels with alcoholic bases and cleaning and solvent compounds may also be sources.

12-185. Locating the source of the poison and determining its availability may suggest the mode of poisoning. Knowing that a poison was contained in a food or beverage may help you ascertain where the victim ate the food or drank the beverage. The place where a poison takes effect is not always the place where the victim consumed the poison.

12-186. There are rarely, if ever, witnesses to an act of poisoning. Consequently, you must gather as much concrete evidence as possible to find out if a crime was committed and, if so, who committed it. Such evidence is not limited to the poisonous substance.

12-187. A background check on the victim and his activities must be completed to learn key information about the poisoning. Conduct an interview of the individuals who may have—

- Witnessed the act of poisoning.
- Known about a suspect's utterances or actions that could establish a motive for the crime.
- Known what the victim ate or drank within the time he probably received the poison.
- Sold drugs or medicines to the victim or suspects.
- Known of the victim's movements before he was stricken.
- Been familiar with the victim's eating and drinking habits, use of drugs or medicines, and attempts at self-medication or treatment from sources outside military medical channels.
- Been familiar with the victim's financial status, family background, social life, or business associates.

Overdose

12-188. A preliminary inquiry into a death may suggest that a victim died from an overdose of drugs. General observations of the crime scene, the victim, and the victim's clothing or conclusions about the victim's lifestyle may suggest this. Note the quality and quantity of food and liquor supplies. These clues may clarify the circumstances of the death or at least give explicit information concerning the victim and the lifestyle he may have led.

12-189. Suspicion of intravenous drug abuse should be aroused when long-sleeved garments are worn and the weather does not justify it. A sleeve that is severely wrinkled in contrast to the other shirtsleeve may have been used as a makeshift tourniquet. Frequently, in cases where a drug death is acute and related to intravenous drug abuse, the abuser will not have time to conceal his drug cache or paraphernalia before his collapse. Thus, cellophane envelopes, balloons or paper packets, syringes, needles, bottle caps or other devices used as cookers, cotton balls, matches, and cigarette lighters may be present. Sometimes a tourniquet or other constrictive device may be dropped after a victim collapses. Syringes are commonly still at the injection site or grasped in the hand.

12-190. Check the body for needle marks and scars. Most intravenous drug injections are made with very small needles that are designed for intradermal injection. If there have been only a few relatively recent injections not associated with puncture hemorrhage, a magnifying glass may be required to detect the punctures. In most chronic addicts, there is no difficulty in detecting the tracks. In addition to the linear scars of intravenous drug use, flat-ovoid or circular scars from lesions caused by unsterile injections given immediately under the skin can sometimes be seen. Chronic addicts may conceal punctures by injecting at unusual anatomic sites. They often inject in and around the genitalia, the nipples, the tongue, the mouth in general, and the scalp. Some addicts, who apparently do not care whether or not puncture sites are seen, may use the jugular vein in the neck to inject. Check the body for signs of nervous tension like the short, irregular edges of fingernails (characteristic of nail biting) or the yellow staining of the fingers (characteristic of excessive smoking). Make detailed notations of papillary diameter, even though this is not a reliable postmortem sign of drug abuse.

12-191. If the cause of death appears to be accidental and there are no signs of criminal acts or negligence, record any evidence supporting that judgment. Sometimes accidental death from drugs does not lend itself to early, clear resolution. Rule out all aspects of other-than-natural causes. Ensure that there was no motive for murder found and that no threats could be learned. See that individuals who may have had a chance to cause ingestion of the lethal dose, either by force or trick, have been searched for leads and that there is no credible sign that the death was other than accidental.

DEATHS INVOLVING CHILDREN

12-192. Investigations of deaths of infants and children are particularly complex. Great caution must be exercised. The investigation must be fully coordinated with medical personnel, social welfare agencies, and the office of the SJA. Suspicious deaths that involve infants and children can be grouped into three types: SIDS, infanticide, and battered child syndrome.

Sudden Infant Death Syndrome

12-193. SIDS is the sudden, unexpected, or unexplained death of a child under the age of one. This diagnosis cannot be made without a thorough investigation of the death. This investigation must include a complete crime scene examination, thorough interviews and canvasses, a review of the medical history of the child, and a complete autopsy. The diagnosis, if made without all of these factors, will likely not answer all questions that must be answered. SIDS is a medical finding where no criminality is involved. As such, it is imperative that the investigator conduct a complete investigation.

12-194. Many cases of accidental or intentional suffocation by parents have historically been misdiagnosed as SIDS cases as investigators rely upon medical personnel and not the crime scene examination. A SIDS investigation at the crime scene must be viewed as any other suspicious death, and all investigative avenues must be pursued. SIDS can only be concluded by the elimination of all other potential causes of death. As such, a complete toxicological examination must be conducted to exclude poisoning, an autopsy must be conducted to look for signs of smothering; and a crime scene examination must be conducted to exclude possible accidental smothering by way of sheets, pillows, or stuffed animals.

Infanticide

12-195. Infanticide is the criminal death of an infant by neglect or deliberation. Sometimes newborns are left to die of neglect in garbage cans, furnaces, restrooms, secluded places, and public dumps. Sometimes they are simply allowed to die at home or in a car in the expectation that they will be disposed of later. The cause of death in cases like these is usually a combination of acute congestion of the respiratory system, dehydration, and lack of basic life-sustaining care. Sometimes parents actively kill their infants. They may choke the baby with the umbilical cord, cup a hand over its mouth and nose, drown it in a bathtub, or drop it into a river or sewer. Sometimes, however, infants are stillborn or die soon after unattended births. Here, the criminal intent may only be to avoid reporting the birth and to illegally dispose of the body.

12-196. Determining that a death is a case of infanticide is often difficult. Most of these deaths are due to asphyxia that can also occur from natural and accidental conditions. But when death occurs from strangulation or other forms of direct violence or when the circumstances show criminal abandonment or disposal with criminal intent, infanticide is strongly suggested. The following three questions must be resolved in a suspected infanticide:

- Did the infant breathe after birth?
- Would the infant have lived if given proper care?
- What was the cause of death?

12-197. At an autopsy, the pathologist checks the infant's lungs to learn if the infant was breathing after birth. Usually the lungs of a stillborn and a live birth appear quite different. But sometimes the signs are not distinct. Then the pathologist must make vast microscopic and hydrostatic tests to find out if lung tissues have been aerated. Even then, there is a chance that breathing may have occurred only inside the uterus or birth canal; and the infant was later strangled by the umbilical cord during birth or was suffocated by extruded membranes, blood, or the mother's weight and position. Attempts at artificial respiration may also account for the air in a stillborn's lungs. Even when signs of asphyxia are present, the death may be wholly natural or accidental. Finding food in the stomach is the only positive proof of life after birth.

12-198. The pathologist medically assesses the completeness of the infant's prenatal development. He also checks for certain vital changes that occur immediately after birth. He considers the apparent general health of the infant and evaluates any congenital defects and injuries received at birth. From his findings, he decides whether or not the infant could have lived if given minimal care. The military standard for a fetus to be considered viable is a gestation of 28 weeks.

12-199. Identifying the victim may be impossible without finding the mother. The body of an abandoned infant usually has no identity. Identifying the mother is not easy because she probably hid the pregnancy and birth. However, a suspect may be found if she seeks medical attention after the birth. She can be medically identified as the mother of the victim if her physical condition is compatible with the birth of the dead infant. Blood tests can show close blood grouping. At all autopsies of abandoned infants, blood samples are taken and analyzed for future comparison.

12-200. If a baby has died from injuries, the child's medical record must be checked to see if the injuries were treated or hidden. An investigator must try to learn if the mother showed signs of mental depression after the birth of the child. In such a case, she would be capable of seriously or fatally injuring the child or even herself. Make a review of the mental history of the father. Medical personnel, neighbors, and friends of the parents can provide information about the temperament of the family. Military or civil police will have records of any complaints or past investigations of the parents.

Battered Child Syndrome

12-201. Battered child syndrome occurs in cases of child abuse. It accounts for a number of deaths of young children under violent conditions. Assigning criminal liability for deaths due to child abuse is often difficult. The victims are most often small children under three years old. If they are still alive, they are usually unable or unwilling to describe what happened.

12-202. A major step in looking into the death of a battered child is to be able to spot signs of battering. The victim of abuse is commonly an infant, most often under 3 years of age. One child in the family is usually the main target of abuse. This child may be the product of an unwanted pregnancy or a premarital pregnancy, or the child may be unwanted for other reasons. The home may be basically clean and the remainder of the children in it well cared for, fed, and clothed. It should be noted that sometimes the family is financially set, well educated, and socially oriented.

12-203. Many times battering parents were targets of abuse in their childhood. A statement such as, "If you think he is mistreated, you should have seen the way my old man kicked me around," shows a trend of child abuse from generation to generation. Parents raise their children the way their own parents raised them, because they know no other way. A battering parent often shows signs of emotional immaturity and mental and environmental stress.

12-204. Another factor that must be recognized is the presence of an extreme sense of competition between the parents. This competition can cause resentment that is taken out on the child. In most abusive families, there is a constant stress of one kind or another.

12-205. Emotional outbursts from aggravation or frustration are responsible for many abusive deaths. Most parents feel some degree of guilt even though their children have been injured accidentally. They make statements like—

- "I shouldn't have bought him such a big bike."
- "Why didn't I watch him more closely?"
- "Why did I let his bath water run so hot?"

12-206. The battering parent, on the other hand, often shows anger and a hostile, argumentative outlook. They may cry harassment on the part of an investigator.

12-207. An investigator must assess the parents to try to detect undue frustration, belligerence, or nervousness when child abuse is suspected. Do not overlook the chance that a brother or sister beat the child. A small child, 18 months and older, may think and feel that his position in the family has been invaded by the arrival of a new baby. Parents may unthinkingly talk of the new baby in a way that the older child will resent. A child has many toys and objects at hand that can cause battering injuries.

12-208. The nonfatally battered child is hard to identify. This child may appear at a medical facility with a broken arm, a cut lip, a black eye, or extensive bruises. Parents easily explain these injuries. A fall or toys thrown by an older child are often the excuses used by battering parents. Only repetitive injuries of this type can alert the doctor to a battered child. To avoid discovery, the parent will take the child to a different doctor or hospital each time.

12-209. Some battered children show no outer signs of injuries. Others show extensive injuries. There may be deep bruises of the face and arms. Deep lacerations are rare. They are probably only seen when a blunt object is used to strike a child on the head or face. Lacerations on the inside of the mouth are more common, caused by the child biting himself when hit.

12-210. Almost all children have one or two scars from falls, but multiple scars on a small child show a pattern lending evidence to abuse. Small, round burn scars may indicate cigarette or cigar burns. Sometimes a parent will bite a child. The bite often leaves a pattern of human teeth marks on the child. An investigator must remember to document bite impressions in his notes and report. Additionally, the investigator must photograph the impressions using a 35-millimeter camera with and without a measuring device. The investigator should also collect a cast of the impression. The photographs and impressions can be used for elimination and suspect identification purposes.

12-211. An investigator's main tool is his eyesight. Look the child over, paying attention to any signs that the child was abused. Look at parts of the child's body that are normally covered by clothing, like the armpits and the inside of the upper thighs. Check the soles of the child's feet for burns. Look at the child's nutritional state and his general cleanliness to check the parent's care of the child.

12-212. Most of a battered child's internal injuries occur in the head or the stomach. The face and scalp may not show outer signs of abuse. But heavy hemorrhaging may be present under the skull. Subdural hematomas, common among battered children, are caused by severe force. They can occur when a child is beaten repeatedly on the head, or when a child is being held by the ankles and swung against a wall. They may also occur from a child being dropped down a staircase. Blunt-force injury to the stomach often causes a lacerated, torn, or ruptured spleen spilling into the peritoneal cavity. The small and large bowels may be perforated, causing the feces to enter into the cavity. Pancreatic substances or bile may be sent to the stomach by injuries to the liver or pancreas. All of these injuries will cause much pain, crying, listlessness, shock, and finally a coma. Because the lining of the stomach is soft, these injuries may not be apparent. One clue to an intra-abdominal injury in the absence of obvious skin injury is a swollen stomach. A careful investigation will usually result in findings that injuries were nonaccidental.

CHILD ABUSE INVESTIGATION

12-213. When investigating the death of a child, the first step is to get a brief background from the person finding the child. Where and in what position was the child found? When was the child last fed? Find out if the child had been ill or irritable on the day of, or two days before, his death. The child's medical background, if known, can be of great benefit. While investigating an infant's death, find out if there were any problems during pregnancy and the infant's birth weight. Find out if the infant had routine visits to the doctor or well-baby clinics. Ask about the infant's history of shots, illnesses, and hospital admissions. Find out the ages of the parents and the number of children in the family and if there is illness among family members. Check the size of the body for consistency with the infant's age. Consider the child's state of nutrition, sickness, dehydration, and cleanliness. Look for old scars and new or old bruises, lacerations, and abrasions. Examine the child's body, bed, and anything else relevant to the child. Include reports and interviews from neighbors, babysitters, and other children in the family.

12-214. Determine how the child was cared for and who was responsible for the care. In most cases, there is one main person responsible for the care of the child. Get information about the family structure and number of relatives or persons frequenting the household. If the child has injuries, one of these parties may be responsible for the injuries. Include information about anyone who may feel competitive toward the child, like a mother's boyfriend. Information may be available from the local welfare agency and hospital and doctor records. Question the child's brothers, sisters, parents, neighbors, and babysitters. Many times a babysitter becomes the confidant of abused children but from fear or disbelief, she may not report the abuse that the children have related.

12-215. Cases of battered children often surface by conflicting statements of what the parents said happened and what the autopsy shows. You must listen for any conflicting statements, no matter how small. In many cases, the parents of a dead child have rehearsed their alibis.

12-216. The pathologist must have as much background about the child as possible before the autopsy. If there is no traumatic injury, the cause of death may be ruled as natural disease or crib death. When trauma from mechanical force is present, the distinction must be made between accidental injury and homicide. Grabbing, gripping, and shaking the child by one or more extremities and physical blows to the child may cause injuries to bones. Blunt-force injury is the major cause of death of a battered child.

12-217. X-rays are crucial and vital. An X-ray of new injuries will show the type and fracture, whether it is transverse or spiral from twisting forces. A radiologist can also find out the age of the injury. Some injuries to the head and stomach when used with X-ray evidence and autopsy findings of old injuries show repeated abuse and develop a pattern of injuries. Other injuries are of such a profound nature that accidental cause is hard to believe.

12-218. A thorough crime scene investigation, as detailed in *Chapter 5*, will aid the investigator in accomplishing his mission. Although the investigation of cases involving the "littlest victims" is emotionally trying upon the investigator, staying on track and being thorough is the only way to get to the bottom of this type of investigation. Close coordination with social workers and legal and medical professionals is crucial. The successful investigation of child abuse allegations, whether fatal or nonfatal, is a multidisciplinary approach in which information is freely shared between agencies. In about 50 percent of fatal child abuse cases, more than one agency was aware of prior allegations of abuse. Information sharing cannot be understated.

DEATHS FROM SEXUAL ASSAULT

12-219. A death will sometimes occur during a sexual assault. The method of killing the victim can be from accidental or intentional asphyxia, sometimes associated with bondage play or strangulation from anger. Determining the difference is dependant upon a thorough investigation. It is important to consider a sexual assault in any death involving a female.

DEATH INVESTIGATION TOOLS AND EQUIPMENT

12-220. The investigation of a death scene normally requires an extensive list of tools and equipment. Army law enforcement agencies should maintain and store required items to be readily available for investigators and to prevent delays in the investigation of serious crimes. Investigative tools and equipment should include:

- Gloves (universal precautions).
- Writing implements (pens, pencils, and markers).
- Body bags.
- Communication equipment (cell phone, pager, and radio).
- A flashlight.
- Body identification tags.
- A camera (35-millimeter with extra batteries, film, and so forth).
- An investigative notebook (for taking scene notes and so forth).
- Measurement instruments (tape measure, ruler, a measuring wheel, and so forth).
- Official ID (for yourself).
- A watch.
- Paper bags (for hands, feet, and so forth).
- Specimen containers (for evidence items and toxicology specimens).
- A disinfectant (universal precautions).
- Departmental scene forms.
- An instant camera (with extra film).
- Blood collection tubes (syringes and needles).
- Inventory lists (clothes, drugs, and so forth).

- Paper envelopes.
- Clean, white linen sheet (stored in a plastic bag).
- Evidence tape.
- Business cards or office cards with telephone numbers.
- Foul-weather gear (raincoat, umbrella, and so forth).
- A medical equipment kit (scissors, forceps, tweezers, exposure suit, scalpel handle, blades, disposable syringes, large-gauge needles, cotton-tipped swabs, and so forth).
- A telephone listing (of important phone numbers).
- Tape or rubber bands.
- Disposable (paper) jumpsuits, hair covers, face shield, and so forth.
- An evidence seal (use with body bags and/or locks).
- A pocketknife.
- Shoe covers.
- A trace evidence kit (such as tape).
- Waterless hand wash.
- A thermometer.
- Crime scene tape.
- A first aid kit.
- A latent print kit.
- Local maps.
- Plastic trash bags.
- GSR analysis kit (such as a scanning electron microscopy/energy dispersive [X-ray] spectroscopy [SEM/EDS]).
- Photo placards (signage to identify the case in the photo).
- Boots (for wet conditions, construction sites, and so forth).
- Hand lens (magnifying glass).
- Portable, electric area lighting.
- Barrier sheeting (to shield the body and/or the area from public view).
- A purification mask (disposable).
- A reflective vest.
- A tape recorder.
- Basic hand tools (bolt cutter, screwdrivers, hammer, shovel, trowel, paintbrushes, and so forth).
- Body bag locks (to secure the body inside the bag).
- A video camera (with an extra battery).
- Personal comfort supplies (insect spray, sun screen, hat, and so forth).
- A presumptive blood test kit.

Chapter 13

Drug Types and Identification

Military law enforcement personnel investigate criminal drug activities, including their use, sale, manufacture, and distribution. Investigators employ surveillance, use informants, and operate undercover to identify those involved in illegal drug activity. They execute search warrants to seize illegal drugs, confiscate money derived from the sale of drugs, and apprehend those involved. Investigators also collect and analyze CRIMINT gathered during drug investigations. Investigators help reduce the criminal threat to US and friendly forces by sharing CRIMINT with other federal, state, local, and HN police agencies.

LEGAL CONSIDERATIONS

13-1. The *Controlled Substances Act (CSA)* is the legal foundation of the government's fight against the abuse of drugs and other substances. This law is a consolidation of numerous laws regulating the manufacture and distribution of narcotics, stimulants, depressants, hallucinogens, marijuana, anabolic steroids, and chemicals used in the illicit production of controlled substances.

13-2. The *CSA* is included in *Part B, Subchapter I, Chapter 13, Section 812, Title 21, USC (21 USC 812).* It regulates the manufacture and distribution of all legal and illegal drugs and places all substances regulated under existing federal law into one of five schedules. This placement is based upon the substance's medicinal value, harmfulness, and potential for abuse or addiction. Schedule I is reserved for the most dangerous drugs that have no recognized medical use, while Schedule V is the classification used for the least dangerous drugs. The act also provides a mechanism for substances to be controlled, added to a schedule, decontrolled, removed from control, rescheduled, or transferred from one schedule to another.

13-3. *UCMJ, Article 112a,* interprets the *CSA* in terms of military justice. Schedules I, II, and III have maximum punishments of 5 years for possession and 15 years for distribution. Schedules IV and V have maximum punishments of 2 years for possession and 10 years for distribution. Marijuana has specific penalties. Possession of 30 grams or less has a maximum punishment of 2 years. Possession of 30 grams or more has a maximum punishment of 5 years. Possession of any amount to distribute has a maximum punishment of 15 years.

PRELIMINARY IDENTIFICATION OF DRUGS

13-4. A major challenge to law enforcement personnel is the ability to identify drugs found at a crime scene or in a suspect's possession. Because of the numerous drug types, colors, sizes, and street names, the aid of a pharmaceutical reference book is often necessary. An excellent reference source for drug identification is the *Physician's Desk Reference (PDR)*. When the *PDR* or another means (an investigator purchasing a particular type of drug through undercover operations) cannot assist an investigator in the identification of a drug, he can conduct a preliminary identification of the drug by using a chemical-reagent field test kit. There are several drug identification kits available. Each kit contains vials of reagent grade chemicals that a qualified chemist uses to prepare the same test when testing substances in a laboratory environment. The kit shows a color reaction. Although the color reactions are valuable, they have the following limitations:

- The tests are qualitative and not quantitative (the drug percentage will not be known).
- The tests are only precursory and presumptive (a chemist's testimony as to his scientific analysis is the most valid evidence).
- A negative test does not necessarily preclude the presence of a drug, nor does a positive test guarantee the presence of a drug.

TESTS OF SUSPECTED DRUG SUBSTANCES

13-5. The first rule an investigator must adhere to when testing suspected drug substances is to never taste or try the substance in any fashion. Secondly, the investigator must remember to use only the amount of the suspected substance needed to conduct the field test and to preserve enough of the substance for laboratory testing. Instructions are provided with all test kits and should be followed.

13-6. Investigators will have to test drugs in powder, solid, or liquid form. Drugs found in powder form can be mixed with other substances and contained in capsules. If suspected drugs are found in capsules, an investigator can open one capsule to pour some of the powder into a test vial to determine its identity. When encountering solid forms of suspected drugs, such as tablets, the investigator can either crush the tablet into powder form or break the tablet in half and scrape some powder from the inside of the tablet into the test vial. For vegetable material, such as marijuana or hashish, a few flakes or the crumbled substance can be put into the test vial. When suspected liquid drugs are found, only add the necessary amount to the vial to receive a reading from the test kit.

13-7. Upon completion of the test, the investigator must use the neutralizing agent that comes with the kit. These kits contain reagent grade chemicals and are considered hazardous waste, which requires proper disposal. Every investigator should identify where HAZMAT can be discarded on the installation. These kits are not to be disposed of in office trash receptacles.

SAFETY

13-8. Safety is an important consideration for the investigator who conducts preliminary drug identification tests. The chemicals are sometimes caustic and can cause burns or irritate the body. Some commercial tests are manufactured in glass vials and when opened, small fragments of glass may scatter about. Rubber gloves should be worn when available and, when not available, the investigator should wash his hands as soon as possible after conducting the test. The investigator should always wash his hands before eating, drinking, or smoking if he has been handling suspected drugs. In addition to wearing gloves, the investigator should wear protective goggles and use the field test kits in a well-ventilated area.

DOCUMENTATION OF FIELD TESTING

13-9. Remember that field tests are only presumptive and must be confirmed by laboratory analysis. In the event that there is not enough product to be field tested and sent to the laboratory, the investigator must decide which is more imperative for the investigation.

13-10. When an investigator conducts a field test, he must document the results using *CID Form 36 (Field Test Analysis on Non-Narcotic Substances). CID Form 36* will allow the investigator to document what the suspected drug was, what type of field test kit was used, and the results of the test, which is an important aspect of the investigation. *CID Form 36* must accompany *DA Form 4137*; both forms establish and show the chain of custody of the suspected drugs. *CID Form 36* is used to document that a portion of the suspected drug was consumed for testing and will be recorded on *DA Form 4137* in the chain of custody portion.

OPIUM AND OPIUM DERIVATIVES

13-11. Initially, opium and its derivatives were considered a cure-all for many ailments but very little was known of their pharmacological affect or toxicity. Certain individuals began to glamorize the stupefying effects of the drugs and, shortly thereafter, large numbers of people began to abuse the drugs. Through continuous promiscuous use, the numbers of addicts began to swell in countries throughout Europe. The following addresses raw opium, morphine, and heroin:

- **Raw opium.** To fully understand the physical, chemical, and pharmacological aspects of opium and its derivatives, the papaver somniferum (opium poppy) must be examined first. The word "papaver" is a Greek word meaning poppy. "Somniferum" is a Latin word meaning to dream or induce sleep. The opium poppy is indigenous to many climates, growing from the southernmost tip of Africa to as far north as Moscow, with the largest quantities coming from the following three main areas of the world:
  - The area known as the Golden Triangle (Laos, Burma, and Thailand).
  - The area known as the Golden Crescent (Afghanistan, Pakistan, and Iran).
  - Mexico.

NOTE: The opium poppy provides opium in its raw form and is generally smoked.

- **Morphine.** Raw opium contains approximately 10 percent morphine, 1/2 percent codeine, 1/5 percent of thebaine, and 1 percent of papaverine, plus more than 35 additional alkaloids in smaller amounts. To produce 1 kilogram of morphine, approximately 10 kilograms of raw opium are required. The process of extracting morphine from opium involves heating water to the proper temperature and then thoroughly mixing the raw opium with the water. The remaining opium gum is removed and remixed 5 or 6 times. All of the water is combined and condensed. At this point, a chemical fertilizer is added, leaving the morphine suspended in a chalky white substance near the surface of the water. The chemist pours that solution into another container, filtering it through cheesecloth. This leaves chunky white kernels of morphine on the cloth. This free-base morphine is then dried and packaged for shipment. It is quite common to encounter morphine in blocks that are approximately 3 x 4 x 1 inch and weigh from 10 to 12 ounces. The blocks will sometimes bear trademarks or brands, such as "999" or "AAA." The morphine is then transported to a much more elaborate laboratory setup where the chemist now begins the conversion from morphine to heroin on a 1-to-1 basis with the addition of other chemicals and processing. Morphine is most generally injected or swallowed by illicit users or by individuals who are prescribed the drug. Legally produced morphine and codeine are available and necessary drugs that are extensively used throughout the medical world. They are used to reduce severe pain, quiet nervous individuals, and arrest heart disease, among many other additional uses.

- **Heroin.** Heroin (diacetylmorphine) is the most commonly abused narcotic. It is synthesized from morphine (a derivative of opium poppy). In general, there is a long chain of intermediaries who will handle the heroin on its journey from the laboratory to the addict. Each handler will attempt to make a profit by reselling the heroin or charging for its transportation. To maximize his profits, a dealer will usually "cut" the heroin. Once a dealer has a general idea of the purity of the heroin, he will mix a small quantity of the heroin and the cutting agent in the desired proportions and retest the heroin as a final check. When the dealer is sure of the ratio he wishes to use, he proceeds to dilute the remaining heroin. When a dealer cuts the heroin once, he will take an equal portion of the heroin and the cutting agent and mix them together. The cutting procedure varies a great deal from area to area. The most common cutting agent or dilutant is lactose (milk sugar). The heroin is repeatedly cut during the journey along the distribution chain until it finally reaches a strength and purity commonly ranging from 1 to 3 percent. Sometimes the cutting procedure involves tremendous danger to the potential user because of the other substances, which may have been substituted for milk sugar. This is especially true of the street level dealer who does not take the time to go to the store to buy a cutting agent. This individual may simply go into a kitchen cabinet and remove any substance that looks similar and can be assimilated into the system. Such substances include baking soda, powdered sugar, powdered milk, baby laxatives, starch, cake flour, coffee, tea, and cocoa. These common kitchen substances may be in addition to other adulterants that might have already been used at a higher level of cutting, such as barbiturates, caffeine, methamphetamineadone, amphetamine, and quinine. The color and texture of heroin is not necessarily an indicator of its quality. Heroin naturally comes in various colors and textures. Cutting agents only add to those variations. The heroin that comes from the Far East is usually a white or off-white color, ranging from a tan or light tan shade. Southwest Asian heroin is generally light gray, but on occasion may be pure white. Mexican brown heroin will commonly range in color from off-white to a dark brown or black. The consistency of heroin will range from a fine powder to a heavy granular or chunky substance. Heroin can be used in almost every conceivable manner. It can be taken orally, inhaled, injected, or smoked (referred to as "chasing the dragon").

COCAINE

13-12. Cocaine (methybenzoylecgonine or C17H21NO4) is a white crystalline alkaloid (any of a class of nitrogenous organic bases, especially one of vegetable origin having a physiological effect on animals and man) that is chemically synthesized using the leaves of the coca bush (erythroxylon coca) and acts as a stimulant to the central nervous system.

13-13. The coca plant is an evergreen that is native to South America, particularly the countries of Peru, Bolivia, Brazil, Chile and Columbia, and should not be confused with the cocoa plant from which chocolate is made. Although the coca plant is natural to South America, it has been successfully cultivated in Java, West Indies, India, and Australia.

Manufacturing Cocaine

13-14. The following are three basic processes of extracting cocaine from the coca leaf:

- In the first process, the dried coca leaf is treated using a chemical process with an acid solution (such as sulfuric acid) that produces raw cocaine or coca paste. The coca paste (which contains approximately 70 percent cocaine) is put through another chemical process with hydrochloric acid, creating a hydrochloric salt or cocaine hydrochloride (which is soluble in water). This process is very time-consuming and can take from 1 to 2 weeks to complete. Both the legitimate and illicit manufacturers of cocaine use this process.
- In the second process, the dry coca leaf is treated using a chemical process with a basic solution (such as sodium carbonate) that produces raw cocaine. The raw cocaine is then put through another chemical process with hydrochloric acid, creating a hydrochloric salt or cocaine hydrochloride. This process is less time-consuming than the first process and is probably the one that is preferred by illicit manufacturers.
- The third process is more advanced and technical than the other two processes. The basic advantage of this process is that it provides a greater yield. The dried coca leaf with its various alkaloids, including cocaine, is broken down into ecgonine, which is the chemical base or core of the cocaine molecule. The ecgonine is then treated with

methamphetamineyl iodide and benzoic anhydride in a chemical process that creates pure cocaine.

13-15. The Peruvian coca leaves, because of their richness, are commonly used in the first and second extraction process. When the dried coca leaves have low cocaine content, the third process (ecgonine) is preferred. Normally, it takes approximately 125 pounds of dried leaves to produce 1 pound of cocaine.

Trafficking Cocaine

13-16. Illicitly manufactured cocaine from the various clandestine cocaine labs in South America is smuggled to various countries for black market trafficking and use. The smuggling methods are unlimited and vary with one's imagination. Often, while en route to the United States, cocaine is first smuggled into Mexico rather than directly from South America. Illicit cocaine, basically, comes in the following three forms:

- A hard, tiny rock form that is readily available, especially to the large wholesaler or dealer.
- A flake form that is generally pure cocaine that has been broken down into tiny flakes and considered a delicacy among users of cocaine.
- A powdered form that is usually rock or flaked cocaine diluted with other substances, such as lactose or procaine.

Using Cocaine

13-17. Cocaine can be applied, taken orally, injected or sniffed into the nostrils (snorting or horning). When cocaine is taken orally, it can be mixed with a liquid or semisolid for ingestion, placed in a capsule and ingested, or swallowed in its powdered form. Sometimes users of cocaine will directly apply the cocaine on their gums, the side of their eyelids, their genitalia, or underneath their tongue. Some users inject the cocaine directly into the veins, and the process is similar to that of injecting heroin. The most common method of using cocaine today is by snorting (inhaling the cocaine into the nostrils).

MARIJUANA AND ITS DERIVATIVES

13-18. Cannabis sativa L (from the genus Cannabis and the family Cannabinaceae) is the botanical name for a tall, annual, woody shrub commonly known as marijuana. The Federal law definition, *Part A, Subchapter I, Chapter 13, Section 802, Title 21, USC (21 USC 802),* of "marijuana" is as follows: "The term 'marihuana' means all parts of the plant Cannabis sativa L., whether growing or not; the seeds thereof; the resin extracted from any part of such plant; and every compound, manufacture, salt, derivative, mixture, or preparation of such plant, its seeds or resin. Such term does not include the mature stalks of such plant; fiber produced from such stalks; oil or cake made from the seeds of such plant; any other compound, manufacture, salt, derivative, mixture, or preparation of such mature stalks (except the resin extracted therefrom); fiber, oil, cake, or the sterilized seed of such plant which is incapable of germination."

13-19. The Cannabis plant contains several alkaloids. The principal ones are cannabinol, cannabidiol, and tetrahydrocannabinol (THC). THC is the most active alkaloid and is considered to be the agent responsible for the hallucinogenic effect of marijuana.

13-20. Marijuana and hashish are derivatives of the cannabis plant that has been cultivated for centuries for its fiber, oil, and psychoactive resin. There are two varieties of the cannabis plant. One is resin producing and the other is fiber producing. THC is found most abundantly in the upper leaves, bracts, and flowers of the resin-producing variety.

13-21. The flowering tops, leaves, and small stems are gathered, dried, and usually smoked in a pipe or as a cigarette. Its use in cigarettes is most often the chosen method.

Hashish

13-22. Hashish, or concentrated cannabis as it is legally and medically known, is the concentrated resin that has been extracted from marijuana. It is logical to assume that the potency of THC content within the hashish is directly related to the marijuana from which it was extracted. However, a basic rule is that hashish is 8 to 10 times stronger than commercial grade marijuana on the average. A general range in THC content would be 0.5 to 22 percent.

13-23. Hashish is usually granular or solid and chunky in form, ranging from mustard yellow to dark brown in color, and is usually smoked or eaten. A hash pipe or regular pipe is normally used when it is smoked. It can be eaten as is or used in cooking.

Hashish Oil

13-24. Sometimes called "marijuana oil" or "honey oil," hashish oil is legally considered concentrated cannabis. This substance is an illicitly manufactured form of what was formerly known in the pharmaceutical and medical professions as tincute or extract of cannabis, a lawful product once used for medicinal purposes.

13-25. In general, hashish oil is about 3 to 4 times stronger than hashish and 30 to 40 times stronger then commercial grade marijuana. It appears on the street as a very thick liquid and is many times so thick that it must be heated to allow it to flow. It varies in color, but can generally be found in amber, dark green, brown, or black. Many users smoke hash oil by adding it to a marijuana cigarette or a commercial cigarette. Some users take hash oil by mouth, such as adding it to food preparations or liquids like hot teas.

13-26. The symptoms of marijuana abuse and its derivatives include problems with memory and learning, distorted perception, difficulty in thinking and problem solving, loss of coordination, and increased heart rate, anxiety, and panic attacks.

STIMULANTS

13-27. The two most prevalent stimulants are nicotine (found in tobacco products) and caffeine (the active ingredient of coffee, tea, and some bottled beverages). When used in moderation, these stimulants tend to relieve fatigue and increase alertness. They are an accepted part of our culture.

13-28. There are, however, more potent stimulants that, because of their dependence-producing potential, are under the regulatory control of many countries. These controlled substances are available by prescription for medical purposes. They are also clandestinely manufactured in vast quantities for distribution on the illicit market.

13-29. Stimulants are compounds that affect the central nervous system by accelerating its activities. Stimulants are either natural (such as epinephrine) or synthetic (such as amphetamine or methamphetamine).

13-30. There are many symptoms of stimulant abuse. They include—

- Excessive activity.
- Irritability.
- Nervousness.
- Argumentativeness.
- Excessive sweating.
- Excitability.
- Talkativeness.
- Trembling hands.
- Dry mouth.
- Overreaction to normal stimuli.
- Flushed skin.
- Headache.
- Tremors.
- Euphoria.
- Dilated pupils.
- Increased blood pressure or pulse rate.
- The ability to go long periods without eating or sleeping.

DEPRESSANTS

13-31. Depressants are compounds that affect the central nervous system by decelerating its activities. They may be synthetic or natural. Depressants can also be categorized as hypnotics (producing or inducing sleep), sedatives (producing a relaxed state that can lead to sleep), or tranquilizers (bringing about relief of anxiety, relaxation of muscles, and calming without sleep or drowsiness).

13-32. Since physical dependence results from the abuse of depressants, there is also a withdrawal syndrome. Withdrawal from nonnarcotic depressants can be fatal and should be medically supervised. These nonnarcotic depressants can be divided into three main categories: barbiturates, tranquilizers, and nonbarbituric acid drugs.

13-33. The symptoms of depressant abuse include behavior that is like alcohol intoxication, but without the odor or apparent use of alcohol. These symptoms include—

- Staggering.
- Stumbling.
- Disorientation.
- A general lack of interest in activity.
- A quick temper.
- A quarrelsome disposition.
- Slurred speech.
- Poor coordination.
- Drowsiness.
- Impaired memory and judgment.
- An unrealized loss of motor control.
- Problems with perception.
- Dilated pupils.

HALLUCINOGENS

13-34. The term "hallucinogens" refers to a group of drugs that affect the central nervous system and produce perceptual alterations, intense and varying emotional changes, ego distortions, and thought disruption. Most of these substances have no medical use and are taken simply because of the subjective effect they produce. They are not considered to be addictive, although they can and do produce psychological dependence. Hallucinogens are exotic drugs that have received considerable attention from the media and drug abuse educators. While many of these drugs enjoy periods of fad-like popularity with an accompanying temporary demand for them on the illicit drug market, others (such as phencyclidine [PCP]) have been abused over the last decade.

13-35. Hallucinogens (also called psychedelics) are capable of provoking alterations of time and space perception, illusions, hallucinations, and delusions. The results are variable because the same person may experience a "good trip" or a "bad trip" on different occasions. Many drugs will cause a delirium accompanied by hallucinations and delusions when taken by individuals who are hypersensitive to them. Extraordinarily large amounts of other types of drugs may also produce hallucinations because of their direct action on the brain cells. Because of the potentially adverse effect of hallucinogens on the human body, these drugs pose special dangers to anyone handling them. Moreover, since very small quantities (micrograms in some cases) may have great potency, the police investigator must be extremely careful in how he handles and packages hallucinogens seized as evidence. Under no circumstances should an investigator taste this or any other type of drug or narcotic. Equally important, he must avoid any direct physical contact with the suspected drug. Drugs categorized as hallucinogens include lysergic acid diathylamide (LSD); PCP; psilocybin and psilocyn; and 3,4-methylenedioxymethamphetamine (MDMA).

Lysergic Acid Diathylamide

13-36. The most powerful and possibly the most widely used of the "mind-expanding" drugs is LSD, a semisynthetic alkaloid substance extracted from a fungus that grows on rye, wheat, and other grains. It is an extremely potent drug, requiring only a small amount to induce a "trip." The affects of an average dose (about 100 micrograms) usually last for 6 to 12 hours. One ounce is enough to provide 300,000 doses. LSD is encountered as a liquid or powder. In its original state, it is colorless, odorless, and tasteless. It is often put on or in things, such as sugar cubes, toothpicks, aspirin, crackers, postage stamps, or bread.

Phencyclidine

13-37. PCP, known on the streets as "angel dust" and numerous other exotic names, is popular among youthful drug users. The affects of PCP vary widely. In small doses, it causes sedation like most depressants. In moderate doses, analgesia and anesthesia occur, characterized by sensory disturbances. In large doses, PCP may produce convulsions and a coma leading to death. Most persons using PCP experience a confused state characterized by feelings of weightlessness, unreality, and hallucinations. Reports of difficulty in thinking, poor concentration, and preoccupation with death are frequent. Other effects include nausea, vomiting, profuse sweating, involuntary eye movements (nystagmus), double vision, and restlessness. It has also been reported that PCP users have increased rates of fetal loss, chromosome breakage, and decreased fertility.

Psilocybin and Psilocyn

13-38. Psilocybin occurs naturally in several species of mushrooms. Psilocybin is relatively unstable and upon ingestion is converted to psilocyn by the enzyme, alkaline phosphatase. Therefore, it seems likely that psilocyn is actually responsible for the drug effects accredited to psilocybin.

3,4-Methylenedioxymethamphetamine

13-39. MDMA (ectasy) is a synthetic chemical that can be derived from an essential oil of the sassafras tree. MDMA was first synthesized and patented by Merck Pharmaceuticals in 1914. It was not until the mid-1970s that articles related to its psychoactivity began showing up in scholarly journals. In the late-1970s and early-1980s, MDMA was used as a psychotherapeutic tool and became available on the street. Its growing popularity led to it being made illegal in the United States in 1985, and its popularity has continued to increase since then.

13-40. The symptoms of hallucinogens and dissociative drug abuse vary depending on the particular drug. Drugs with street names like acid, angel dust, and vitamin K distort the way a user perceives time, motion, colors, sounds, and self. These drugs can disrupt an individual's ability to think and communicate rationally or even to recognize reality, sometimes resulting in bizarre or dangerous behavior. Hallucinogens such as LSD cause emotions to swing wildly and real-world sensations to seem unreal, sometimes with frightening aspects. Dissociative drugs like PCP and ketamine may make a user feel disconnected and out of control. Such drugs as MDMA, rohypnol, gammahydroxybutyrate or gamma hydroxybutyric acid (GHB), and ketamine can cause a sharp increase in body temperature (malignant hyperthermia), leading to muscle breakdown and kidney and cardiovascular system failure when taken in high doses.

CLANDESTINE LABORATORIES

13-41. The Drug Enforcement Administration (DEA) defines a clandestine drug laboratory as, "an illicit operation consisting of a sufficient combination of apparatus and chemicals that either has been or could be used in the manufacture or synthesis of controlled substances." Clandestine laboratories are known to cause serious injury or death to the lab operator and to the investigator. Fire and explosion from organic solvents and reactive material remain a serious threat. However, the silent killers are deadly gases and by-products emitted from the process. Red phosphorous labs emit phosphine gas, hydriodic gas, and hydrochloric acid gas. Ammonia labs emit ammonia gas and hydrochloric acid gas. Phenyl-2-propanone (P2P) by means of mercuric chloride labs emits methylamine gas and acid gas. Thionyl chloride labs emit hydrochloric acid gas, sulfuric acid mist, and chloroform.

NOTE: There are many health and safety hazards associated with clandestine laboratories.

13-42. The precursor chemicals used in clandestine drug labs are continually changing and vary from state to state. The local HAZMAT team, fire department, or federal or state law enforcement agency should have the latest information and technologies in dealing with clandestine laboratories for the area. The DEA and the Environmental Protection Agency (EPA) are two federal agencies that may be able to provide assistance in dismantling, removing, and cleaning up a clandestine drug lab.

Chapter 14

Environmental Crimes

Environmental crimes are not new. However, recent national events have caused increased emphasis on law enforcement agencies training to conduct investigations in a hazardous or contaminated environment. Investigating an environmental crime or a crime with an environmental impact is very similar to investigating any other crime. A crime scene examination, evidence collection, and interviews are still required. Environmental crimes do not consist of just an intentional dumping of HAZMAT. They are also criminal investigations with an environmental impact that include things such as a package delivered with suspected anthrax, a fatal traffic accident involving a tanker truck, or a clandestine drug laboratory. These crimes could easily encompass or be part of other crimes, such as fraud, arson, or assault, and could involve the use of computers or other electronic records, such as HAZMAT disposal records. Even the most common investigation could involve environmental concerns.

OVERVIEW

14-1. The EPA and state, local, and even foreign agencies (depending on the applicable SOFA) hold military installations responsible for environmental damage that they cause, even though the majority of this damage is caused by the military's main peacetime mission, which is training. Because of excessive environmental fees (9.9 million dollars between 1992 and 1994) assessed to Army installations, the Army has developed a strict policy to protect the environment *(AR 200-1)* with specific guidelines for vehicle maintenance during field operations. Therefore, investigators are charged to thoroughly investigate the intentional dumping of used motor oil, parts-cleaning solvent, and other HAZMAT.

14-2. Other concerns include conservation areas on military installations. Many installations border or encompass national forest lands, historic sites, landmarks, wildlife refuge areas, or parks. US law requires the preservation of these areas. Although training in these areas may be forbidden or severely restricted, activities (such as noise, pyrotechnics, or dust from heavy vehicle movement) in adjacent areas could have a negative impact on the environment.

14-3. Generally, the major significant difference between a crime with an environmental impact and any other crime is the contamination of the crime scene. Traditionally, fire department personnel have controlled these scenes. This is primarily due to their specialized training on working in protective clothing within a hazardous environment and their responsibility to contain, control, and clean the area to prevent the spread of contamination. As with any arson investigation, the techniques used to contain and control these scenes are very destructive and not conducive to the preservation of physical evidence. Therefore, the investigator should respond to the scene (either with or at the same time as fire and medical personnel) as a member of the hazardous-incident integrated response team.

DEFINITION OF AN ENVIRONMENTAL CRIME

14-4. Before defining an environmental crime, it must be determined what is meant by "environment." An environment is conceptual and encompasses anything (including circumstances, conditions, and other elements) that affects the existence and development of an individual, an organism, or a group. The environment of an organism is made up of everything around it. Sunlight, temperature, air, soil, minerals, water, and other living things are all elements of the environment of an organism. An environmental crime is any crime that has, or possibly could have, a negative impact on any specifically protected element of the environment.

14-5. Not all actions that cause a negative impact on an environment are crimes. A person who cuts down a tree in the woods behind his house is damaging the environment of that tree, other trees, animals, and every organism in the area of the tree, but that would not usually be a crime. If the same individual chose to dump a significant amount of poison into the soil around that same tree and the poison not only killed the tree but also washed over to his neighbor's property and seeped down into the water table, then a criminal act might have taken place. A violation of the law must occur before an incident can be considered an environmental crime.

14-6. US Army installations and activities do not develop environmental laws. They develop regulations and policy guidelines to comply with the statutes. Although all federal laws apply to the US Army, each individual installation must be aware and comply with other laws and regulations in its area. The following are sources of laws that may apply to US installations and activities:

- **Federal law.** The *Federal Facilities Compliance Act (FFCA) of 1992* requires military installations to comply with environmental laws. The *FFCA* allows regulator agencies to impose civil fines on other federal agencies, like the DA, for violations of the *Resource Conservation and Recovery Act (RCRA)*. Congress expanded the concept of federal facility compliance to other US environmental laws, including the *Clean Water Act (CWA)*. Changes to the *FFCA* now require federal agencies (including military installations) to comply with state, regional, county, and local laws; allow for the imposition of fines for violations of these laws; and allow these governmental bodies or even individuals to sue US Army installations.
- **State law.** Each state has its own regulatory agencies charged with developing and implementing environmental regulations. All state regulations should at least parallel federal regulations. Some federal statutes allow states to set standards that are more stringent than the

federal requirements. When the EPA approves a state program, that state has "primacy" for that particular environmental program.

- **Regional, county, and local law.** Local laws and ordinances address the concerns of the local community. Generally, these ordinances will be based on state and federal laws. However, each municipality or community may place restrictions that are more stringent on certain activities, such as noise restrictions during certain hours of the day and restrictions on pollution, which require car pool lanes in major cities. The car pool lanes encourage people to ride to work together and reduce the number of vehicles on the road, which reduces pollution. Individuals driving without additional passengers are restricted from access to the lane, thereby, thinning traffic to allow those who carpool a quicker trip. Although regional and county ordinances could affect installations within their boundaries, it is unlikely that local or municipal ordinances will apply since most installations are not within municipal boundaries. However, the potential for conflict exists when installations are located along a border with cities or towns.
- **Host nation law.** While serving in areas outside the continental US (OCONUS), individuals and installations are required to maintain cooperative relationships with regulatory agencies of the HN and comply with their environmental control standards. SOFAs that permit or require standards other than those of the host country are considered part of the environmental abatement standards that apply to the military in the HN or its jurisdiction.

HAZARDOUS-INCIDENT RESPONSE

14-7. Hazardous incidents are uncontrolled, illegal, or threatened releases of hazardous substances or hazardous by-products of substances. When the *Superfund Amendments and Reauthorization Act (SARA)* was passed in 1986, it regulated the storage, transportation, use, and disposal of HAZMAT into the environment. Within the *SARA*, *Title I* and *Title III* were established to provide specific guidance for responding to hazardous incidents.

14-8. *Title I* of *SARA* mandates that the OSHA and the EPA establish regulations on training, emergency response, safety, and associated HAZMAT activities. Within this title, *Section 120, Part 1910, Title 29, Code of Federal Regulations (29 CFR 1910.120)* for OSHA and *Section 1, Part 311, Title 40, CFR (40 CFR 311.1)* for EPA were established. These federal regulations outlined training standards and mandated written SOPs for HAZMAT incidents.

14-9. *Title III of SARA* sets requirements for industries to report materials used or stored in the workplace. These reports supply emergency responders with an inventory of what chemicals may be found in an industrial setting. Part of the reporting includes a requirement to supply a material safety data sheet (MSDS) for certain substances, based on their particular level of toxicity or danger.

GENERAL CONSIDERATIONS

14-10. Investigators must meet OSHA training and medical requirements before entering potentially hazardous sites. The base bioenvironmental offices and military public health offices should be contacted to determine and arrange for necessary training and medical monitoring. Investigators are not first responders to environmental crime scenes. They should enter potentially hazardous sites only under the guidance and supervision of environmental regulatory or emergency response personnel. Do not risk exposure. Responsible environmental management personnel make safety determinations and specify the use of protective gear. The investigator must follow their instructions. Never direct an investigator to enter potentially hazardous sites or handle potentially HAZMAT unless he is specifically trained and experienced in operating in a hazardous environment.

14-11. Gather as much information about the hazardous substance as possible through interviews and document reviews. The investigator should determine the following:

- The products involved, the types of containers, and the extent of the discharge.
- The potential offenders, the ownership of the materials, and the property involved.
- Any ongoing dangers posed by the discharge.
- The circumstances surrounding the discovery of the incident.

14-12. Once the scene is determined to be safe and released to investigators, decide on a search technique and assign a search team to conduct an area search. The search team should look for any—

- Discoloration of soil, water, and vegetation.
- Distress or absence of vegetation.
- Sheen on the water.
- Dead or sick wildlife.
- Unusual odors.
- Residue on hoses, storm drains, grates, and so forth.
- Drums and other containers.
- Tanks (above and below ground).
- Recent soil movement.
- Tire tracks and footprints.
- Labels on the containers.
- Paperwork associated with the incident.

14-13. Aerial photography can document visible water, soil, and vegetation contamination and changes due to discharges of HAZMAT. Special films can show a variety of conditions (temperature changes due to the presence of chemicals, bacterial growth, and vegetation discoloration).

Chapter 15

Fraud Investigations

Fraud is an intentional deception to cause an individual to give up property or some other lawful right. It differs from theft in that fraud uses deceit rather than stealth to obtain goods illegally. Fraud is committed in many ways, including identity theft, checks and credit card fraud, and forgery.

OVERVIEW

15-1. Frauds against the US government range from intentional submission of claims for travel not performed to collusion in contracting for, or disposing of, government property. Fraud against the US government may be an intentional deception to unlawfully deprive the government of something of value. It may also be an intentional deception to secure from the government a benefit, privilege, allowance, or consideration to which the securer is not entitled.

IDENTITY THEFT

15-2. The DOJ prosecutes cases of identity theft and fraud under a variety of federal statutes. In the fall of 1998, for example, Congress passed *Section 1028, Title 18 USC (Identity Theft and Assumption Deterrence Act) (18 USC 1028 [The Identity Theft and Assumption Deterrence Act]).* This legislation created a new offense of identity theft, which prohibits knowingly transferring or using, without lawful authority, a means of identification of another person with the intent to commit, or to aid or abet, any unlawful activity that constitutes a violation of Federal law or a felony under any applicable state or local law.

15-3. Criminals may steal an individual's identity by co-opting his name, SSN, credit card number, or some other piece of personal information for their own use. In short, identity theft occurs when someone appropriates personal information from another individual without his knowledge to commit fraud or theft. Methods used by identity thieves include—

- Opening a new credit card account using another individual's name, DOB, and SSN.
- Establishing a cell phone service in another individual's name.
- Opening a bank account in another individual's name and writing bad checks on that account.

Generally, techniques used to investigate larceny are used to investigate these crimes as well.

15-4. Identity theft often has a devastating effect on members of the military and their families. Victims must act fast to minimize the damage to their credit and financial status. Army law enforcement personnel assist victims by providing them with information designed to help them take the appropriate action to prevent further loss and to recover from identity theft.

15-5. The following is information that should be provided to the victims of identity theft when there is reason to believe that these conditions exist.

- If your keys were taken—change or rekey the locks that need to be changed for your protection.
- If your checks or credit cards were taken—notify your bank, if you have not already done so. Get "fraud alert" placed on your account so new credit will not be issued without contacting you. Do this by reporting your loss to the following three credit reporting bureaus:
  - Experian—1-888-397-3742 <www.experian.com>

    Experian Fraud Department
    PO Box 9556
    Allen, TX 75013.

    Notice will appear for 90 days. If a victim wants an extension, he must send in a current telephone bill.
  - Trans Union—1-800-680-7289 <www.tuc.com>.

    1-714-870-5565 (Fax 714-447-6034)
    Trans Union Fraud Victim Assistance
    PO Box 6790
    Fullerton, CA 92834

    Notice appears for seven years.
  - Equifax—1-800-525-6285 (Fax 1-770-612-2533) <www.equifax.com>

    Equifax Credit Information Services
    PO Box 740250
    Atlanta, GA 30374-0250

    Notice will appear for two years.
- If your social security card was taken—call the Social Security Administration (SSA) fraud hotline to notify it of the loss and get information on how to get a duplicate card. Contact the SSA fraud hotline at 1-800-269-0271, <www.ssa.gov>.
- If your driver's license was taken—apply for a new driver's license as soon as possible and ask if anyone has applied for a license since yours was stolen. They can refer you to an investigator.
- If new checks or cards have been mailed to a different address—call the local office of the US Postal Inspection Service to inform it of the situation. Follow its guidance.
- If your stolen checks or cards have been used—contact the banks and/or businesses that accepted your checks or cards to notify them of the fraud and offer to sign any affidavits of forgery, as needed. Encourage

the banks and businesses to pursue pressing charges against any suspects identified.

- If someone has stolen your identity to get new credit—call the police department and report the crime. Also, call the Federal Trade Commission (FTC) identity theft hotline (1-877-438-4338) to notify it and get advice on how to proceed. To report fraud to the FTC, other than identity theft, call 1-877-382-4357.

CHECK FRAUD

15-6. A significant amount of check fraud is due to counterfeiting through desktop publishing and copying to create, chemically alter, or duplicate an actual financial document, which consists of removing some or all of the information and manipulating it to the benefit of the criminal. Victims include financial institutions, businesses that accept and issue checks, and the consumer. In most cases, these crimes (including forgery) begin with the theft of a financial document. It can be perpetrated as easily as by someone stealing a blank check from a home or vehicle during a burglary, searching for a canceled or old check in the garbage, or removing a check mailed to pay a bill from a mailbox. Types of check fraud include—

- **Forging.** Forgery typically takes place when an employee issues a check without proper authorization. Criminals will also steal a check, endorse it, and present it for payment at a retail location or at the bank teller window, probably using bogus personal identification.
- **Counterfeiting.** Counterfeiting can either mean wholly fabricating a check by using readily available desktop publishing equipment consisting of a PC, a scanner, sophisticated software, and a high-grade laser printer or by simply duplicating a check with advanced color photocopiers.
- **Altering.** Using chemicals and solvents, such as acetone, brake fluid, and bleach to remove or modify handwriting and information on the check. When performed on specific locations on the check such as the payee's name or amount, it is called "spot alteration." When an attempt to erase information from the entire check is made, it is called "check washing."
- **Paperhanging.** This problem primarily has to do with people purposely writing checks on closed accounts (their own or others).
- **Check kiting.** Check kiting is opening accounts at two or more institutions and using "the float time" of available funds to create fraudulent balances. This fraud has become easier in recent years due to new regulations requiring banks to make funds available sooner, combined with increasingly competitive banking practices.

15-7. There are several signs that may indicate a bad check. One sign on its own does not guarantee a check to be counterfeit; however, the greater the number of signs, the greater the possibility that the check is bad. Signs of bad checks are as follows—

- The check lacks perforations.
- The address of the bank is missing.

- The customer's address is missing.
- The check number is either missing or does not change.
- The check number is low (like 101 up to 300) on personal checks or (like 1001 up to 1500) on business checks.
- The type of font used to print the customer's name looks visibly different from the font used to print the address.
- Any additions to the check, such as phone numbers written by hand.
- Any stains or discolorations on the check possibly caused by erasures or alterations.
- The numbers printed along the bottom of the check (called magnetic ink character recognition [MICR] coding) are shiny. Real magnetic ink is dull and nonglossy in appearance except if applied with a laser printer when it may have a shine or gloss.
- The MICR coding at the bottom of the check does not match the check number.
- The MICR numbers are missing.
- The MICR coding does not match the bank district and the routing symbol in the upper-right corner of the check.
- The name of the payee appears to have been printed by a typewriter. Most payroll, expenses, and dividend checks are printed by using a computer.
- The word "void" appears across the check.
- Notations appear in the memo section listing "load," "payroll," or "dividends." Most legitimate companies have separate accounts for these functions that eliminate a need for such notations.
- The check lacks an authorized signature.

15-8. Tips for detecting counterfeit checks include—

- **Perforation.** Many checks produced by a legitimate printer are perforated and have at least one rough edge. However, many companies are now using in-house laser printers with MICR capabilities to generate their own checks from blank stock. These checks may have a microperforated edge that is difficult to detect.
- **MICR line ink.** Some forgers lack the ability to encode the bank and customer account information on the bottom of a check with magnetic ink. They will often substitute regular toner or ink for magnetic ink. If a counterfeit MICR line is printed or altered with nonmagnetic ink, the sorting equipment at the bank will not be able to read the MICR line, thus causing a reject item. Unfortunately, the bank will normally apply a new magnetic strip and process the check. This works to the forger's advantage because it takes additional time to process the fraudulent check, reducing the time the bank has to return the item. But banks cannot treat every non-MICR check as a fraudulent item because millions of legitimate checks are rejected each day due to unreadable MICR lines.
- **Routing numbers.** The nine-digit number between the brackets on the bottom of a check is the routing number of the bank on which the check is drawn. The first two digits indicate in which of the 12 Federal

Reserve Districts the bank is located. It is important that these digits be compared to the location of the bank because a forger will sometimes change the routing number on the check to an incorrect Federal Reserve Bank to buy more time.

CREDIT CARD FRAUD

15-9. Many laws cover larceny or fraudulent use of credit cards. The elements of the crime of larceny by credit card are possessing a credit card obtained by theft or fraud, using the card to obtain services or goods, and signing the cardholder's name. Investigating credit card fraud normally requires obtaining samples of handwriting from sales slips signed by the suspect. Handwriting samples should be sent to USACIL for examination. Investigators should examine credit cards for alteration of the cardholder's name, signature section, and the numbers on the card. Merchandise listed on the sales receipts found in the suspect's possession could also provide evidence of illegal use. Potential witnesses include sales clerks, food service personnel, airline and hotel employees, and car rental personnel.

15-10. If investigators suspect that the US mail system was used in credit card fraud, they should contact the local office of the US Postal Inspection Service. The US Postal Inspection Service is the law enforcement arm of the US Postal Service that is responsible for investigating cases of identity theft. The US Postal Inspection Service has primary jurisdiction in all matters infringing on the integrity of the US mail.

FRAUD AGAINST THE GOVERNMENT

15-11. The USACIDC and the FBI have concurrent jurisdiction over persons who are subject to the *UCMJ* and commit fraud against the US government. When fraud against the government is committed outside military installations and involves a person subject to the *UCMJ*, it may be investigated by the FBI or USACIDC, unless the DOJ determines otherwise.

15-12. When fraud against the government is committed on a military installation and involves persons subject to the *UCMJ*, it is investigated by USACIDC to determine the nature and extent of the crime. If the fraud is determined to be a minor offense as defined by *AR 27-10*, the investigation may be continued by the military. If the fraud is a serious offense, prompt notification is made to the FBI. While awaiting a response, the military maintains authority to apprehend and detain persons subject to the *UCMJ*, and the investigation is continued until the DOJ notifies the military commander to withdraw from the investigation. Even then, the military commander may make inquiries for administrative action related to the offense as long as no action is taken that would interfere with the FBI investigation and subsequent prosecution of the case.

15-13. USACIDC may conduct or participate in investigations of persons not subject to the *UCMJ* if the military has a substantial interest in the investigation, such as identifying military property or determining facts on which to base security or administrative action. If the appropriate government agency declines to investigate, USACIDC may investigate suspected fraud for the above limited purposes, regardless of who is suspect.

15-14. In occupied territory, USACIDC may investigate fraud against the US. In liberated areas, USACIDC investigates fraud committed against the United States by persons subject to military law. In liberated countries or in countries in which US Armed Forces are present as guests, investigations by USACIDC of fraud committed by nationals of those countries against the United States are conducted according to the agreements between the United States and the country.

15-15. There are five main categories of fraud against the government that may require an investigation. These are frauds involving claims, supply, petroleum distribution, contracting, and property disposal. Investigate claims, supply, and petroleum distribution frauds to determine if an offense has been committed. Then, using standard investigative skills and techniques follow the investigative process to bring your inquiries to a successful conclusion. Investigate contracting and property disposal frauds to determine if an offense has been committed. Once it is determined that an offense has occurred, request assistance from economic crime investigative specialists. The successful resolution of contracting and property disposal frauds generally requires the training and experience of a specialist.

COORDINATING FRAUD OFFENSES INVOLVING CLAIMS

15-16. The crime of defrauding the government by the claims process is illusive in nature. It is strongly recommended that any investigation involving these frauds be closely coordinated with the local office of the SJA. This legal advice can help the investigator avoid many of the pitfalls inherent in establishing the existence of offenses in this highly technical area of criminal law.

MAKING AND PRESENTING FALSE AND FRAUDULENT CLAIMS

15-17. There are two common elements of proof needed to substantiate the offenses of making and presenting false and fraudulent claims. The investigator must show the false or fraudulent nature of the claim and show proof that the accused knew of the dishonest or fictitious character of the claim in question. For example, a false or fraudulent claim is made against the government when a person files a claim for property lost in military service knowing that the articles were not lost. Making a false or fraudulent claim, by its very nature, requires the claimant to personally make a false statement. But presenting a claim for payment when the claimant knows that it has been paid or that he is not authorized to present it does not require him to make a false statement. Someone who submits a legitimate voucher a second time is presenting a false claim, but he or she is not making a false statement.

MAKING OR USING A FALSE WRITING OR OTHER PAPER WITH A CLAIM

15-18. Making or using a false writing or other paper in connection with a claim is fraud against the government. The offense of making a false writing for the purpose of obtaining an allowance, payment, or approval of a claim is complete with the writing of the paper, whether or not the writer attempts to use the paper or to present the claim. If a person makes or uses a false writing in connection with a claim and if the false writing contains statements intended to mislead government officials considering or investigating the claim, he is chargeable.

MAKING A FALSE OATH WITH A CLAIM

15-19. Proof that a fraud against the government has been committed by means of a false oath requires evidence that the accused knowingly made a false oath to a fact or to a writing to obtain an allowance, payment, or approval of a claim. For example, a claimant filing a sworn statement requesting quarters for a person to whom he is not married is making a false oath to support his claim.

FORGING A SIGNATURE WITH A CLAIM

15-20. Under the *UCMJ,* forgery of a signature in connection with a claim constitutes a separate and distinct offense from the crime of forgery. The offense is complete once it can be demonstrated that the accused forged an individual's signature on writing or knowingly used a forged signature for the purpose of obtaining an allowance, payment, or approval of a claim.

CLAIMS FRAUD

15-21. To investigate a fraudulent claim against the government, make a discreet inquiry into the circumstances surrounding the allegation of fraud to determine if an offense has been committed. This should be accomplished without endangering any sources of information or placing suspects on their guard. If it is determined that a fraud has been committed, continue the investigation to learn the extent of the offense and the identity of the individuals involved.

15-22. The investigator should learn the specific transactions by which the fraud was committed. He should identify the roles of the suspects in the alleged fraud and check for jurisdictional problems. Based on the investigator's findings, he should make an estimate of the technical skills needed to establish the offense and the identity of the offenders. Look for probable types and locations of evidence of the fraud. The investigator should carefully question individuals who—

- Prepared or submitted the claim.
- Received and approved the claim at local or intermediate levels of command.
- Witnessed or attested to the circumstance on which the claim was based.
- May have been in collusion with the suspect to prepare or justify the claim.
- Witnessed or knew of any motive, incident, or circumstance that may point toward the fraudulent nature of the claim.
- Witnessed conversations or observed correspondence between the individuals involved in making, justifying, or approving the claim.

15-23. The investigator may need to audit many pieces of documentary evidence to find those that bear on suspected fraud. Claims, applications, travel vouchers, receipts, business and finance reports, audits, bank deposits and withdrawals, and records of monetary conversions and transmittals can all be used to substantiate this form of fraud against the government. Guided by the elements of proof required for the specific offense, the investigator should search for documents to substantiate the allegations of a claims fraud.

15-24. Take action early to secure cooperation from the appropriate commands, and refer undeveloped leads to the appropriate commands. This will expedite the investigation and give other agencies time to comply with requests. If there is a need for more information or additional documents on the fraudulent actions under investigation, coordinate with any other agencies involved in the investigation. Try to do this without disclosing the results of the preliminary investigation. While awaiting replies or action, check every available local source of information. Make careful use of selected sources, and seek out reliable individuals who possess information material to the investigation.

15-25. Arrange the evidence to point directly to the elements of proof of the specific alleged offense. The final case report for a fraud must be specific in its allegations and in its information. When undeveloped leads are to be checked by investigators in other fields of study, provide them with information allowing them time to proceed logically in their work.

SUPPLY FRAUD

15-26. Fraud in the supply system of the US Army, commonly called supply diversion, is the most frequent crime occurring within logistics channels on military installations. Supply diversion ranges from ordering self-service items for personal use or resale to requesting supplies to be shipped by rail and then routing the railcars to areas of low-density traffic to steal their contents.

COMMON SUPPLY FRAUDS

15-27. Common supply frauds include ordering items under the wrong national stock number (NSN) or a false document number and ordering unauthorized items. If a perpetrator puts the wrong NSN of the item in the stock number block on the request form while putting a correct item description in the description block, the automated system issues and ships the NSN item, not the description item. When the perpetrator receives the requested item, he diverts it for his own private gain. To spot the diverter, investigators should use the document number and trace the document from the requestor to the issuing activity and back to the receiver obtaining copies of all requests and receipts.

15-28. If a perpetrator places an order under a false document number, the investigator should trace the audit trail to establish the diversion pattern and find the perpetrator. If a perpetrator is ordering unauthorized items, trace the complete audit trail. Take statements from key witnesses, and then compare a copy of the table of organization and equipment (TOE) or table of distribution and allowances (TDA) against the property book. The authorized allowances are filled out in pencil. Thus, they could be erased. But most of the time, the perpetrator makes new pages because the items are not authorized or are not authorized in the quantity ordered under the TOE or TDA of the unit.

INVESTIGATIVE APPROACH

15-29. The first step in investigating supply fraud is to identify the supply system in which the fraud or theft occurred. Then determine if the system is at the retail (installation or organization) or wholesale (depot or manufacturer) level of the US Army logistical system. Determine if the system is manual or computer-automated. Manual and automated systems use the same forms, but their operational principles differ at the local level. The manual system uses a property book reflecting TOE and TDA equipment on individual property pages. The automated system uses computer listings reflecting all authorized and on-hand equipment on a single printout.

15-30. After determining the system from which a supply item is missing, review the supply transaction register, called a document register, and see which unit or organization requested the item. Obtain the document number of the requisition. Then carefully follow it through the audit trail. Check each level of the supply system furnishing material to the supply activity that has physically issued and shipped the item to the requestor. Obtain a copy of the request at each step of this initial investigative path for backup. Then begin following the issue trail that leads from the supply activity that was the issuer to the requestor or user. The points along the path of issue will reflect at what point the item was taken from US Army control. Obtain copies of all requests for issue, issue documents, shipping reports, or the like of. When the investigator has copies of all these documents, continue the investigation as if investigating a larceny.

15-31. Not all supply frauds occur as diversions from a supply system. Many items are reported stolen from a storage area. To investigate the loss, obtain the supply documents verifying that the items were physically present at the activity reporting the loss. Determine the inventory procedures of the activity. Then establish the time frame extending from the date when the items were last seen at the activity to the date when the loss was noted. If the items were last present at an inventory, apply larceny investigative techniques and procedures to find the perpetrator. If the items were known to be missing before the last inventory and they were carried on the inventory as being on hand, the provisions of *AR 735-5* apply. The property book officer must make a ROS. Be aware that inventory shortages are often reported as supply larcenies. This is done in an attempt to cover poor supply management techniques and to generate a criminal investigation instead of an ROS.

PETROLEUM DISTRIBUTION FRAUD

15-32. Fraud in the petroleum distribution system can be minor pilferage or systematic theft. It can also be falsification of multimillion dollar orders by a purchasing conspiracy among contracting officials and oil companies. A study of *AR 70-12* and *FM 10-67-1* should give the investigator the knowledge of petroleum operations needed to investigate most petroleum fraud.

Investigations of extremely large losses from conspiracies are usually outside the Army law enforcement purview.

15-33. Pilferage may occur in "nickel-and-dime" losses of petroleum in amounts as low as 5 or 10 gallons a day. The methods of pilferage may range from recording the wrong amounts on *DA Form 3643 (Daily Issues of Petroleum Products)* to siphoning gas from a vehicle tank. Investigators can discover these losses by simply monitoring the amount of gas used and then comparing that amount with the amount stated on the form. If pilferage is discovered, use the gasoline theft detection kit and undertake surveillance to catch the offenders.

15-34. Larger, systematic losses are usually from theft by a supplier. Suppliers may use false tanks. They may trap petroleum in buckets inside the delivery vehicle or add air or heat to the delivery line just before it connects to the meter. They may also conspire with a government attendant to leave some of the petroleum in the delivery vehicle. Large-scale theft usually means the government attendant is not making the checks required by *AR 70-12* or is conspiring with the supplier. In the latter case, a fluid-like water is usually mixed with the petroleum to cover the shortage.

15-35. Sometimes paperwork is falsified to cover a loss. It is easy to cover shortages by adding gallons to those a driver signs for on the *DA Form 3643* or by falsifying entries on the form. The driver, for example, may be receiving 10.2 gallons and signing for 11 gallons. At a large issue point, several hundred gallons a week can be lost by this method. Use surveillance and cross-check the logbook against the *DA Forms 3643* to help prove the fraud.

CONTRACTING FRAUD

15-36. Contracts embrace all types of agreements to procure supplies or services. The investigation of crimes like fraud and bribery involving government contractors is within the purview of the FBI. However, under *AR 27-10*, which affects the MOU between the FBI and DOD the investigation may be done by military investigators depending on whether any federal statutes were violated. In cases where it appears that a government employee has violated a departmental regulation involving standards of conduct but has not violated any federal statutes, military investigators normally conduct the inquiries. They investigate to obtain the detailed information on which the commander needs to base his action. An investigation of this nature, while mainly of administrative interest, may be conducted concurrently with a criminal investigation.

15-37. All suspected criminal conduct and noncompetitive practices related to contracting must be reported. Reports of possible fraud or violations of antitrust laws must contain a certified statement of the facts of the dereliction. The reports must include affidavits, depositions, records of action (if applicable), and other relevant data. This reporting may require preliminary investigation of allegations of a criminal nature for referral to the DOJ and the FBI for determination of prosecutive interest. It may include supplying details for consideration of debarring persons or firms from participating in procurement contracting. It may include furnishing information to a commander to help him decide whether or not to take administrative or disciplinary action in connection with procurement.

15-38. Government personnel engaged in contracting may violate statutory prohibitions and administrative regulations by accepting gratuities or conspiring to defraud the government. Their wrongful act and malfeasance in the performance of duty, when established as fact, may be both legally and administratively actionable. Government contracting personnel may perform a lawful act in a manner prohibited by regulations or perform the act in a manner not directed by regulations. Their misfeasance would be administratively actionable. Their actions violate the *UCMJ*. Government contracting personnel who fail to follow procedures required by acquisition regulations are guilty of nonfeasance. Even if the omission is not a part of a scheme to defraud the government, it is nevertheless actionable.

STANDARDS-OF-CONDUCT VIOLATION

15-39. Regulatory standards of conduct and ethics apply to contracting officers and all military or civilian personnel engaged in contracting actions and related processes. In contracting, many decisions are largely a matter of personal judgment. Contracting is necessarily carried on, to a great extent, through personal contacts and relationships. Thus, high ethical standards of conduct are essential to protect the interests of the government. The standards of conduct for government civilians and military personnel are set forth in *Part 1, Title 48, CFR (48 CFR 1)* and *DODD 5500.7.*

15-40. Any act that compromises the DA or that impairs confidence in government relations with industry or individuals must be avoided. Violations of regulatory standards of ethics and conduct may involve such variable factors as judgment, previous experience and relationships, and individual interpretation of ethics. Whatever the circumstances, the ethical standards of all individuals charged with the administration and expenditure of government funds must be above reproach and suspicion in every respect and at all times.

INVESTIGATING

15-41. An investigator should familiarize himself with the contracting process and the laws and regulations that apply before conducting an investigation. Irregularities often occur in the contracting process due to the complexity of statutory provisions, administrative regulations, and departmental or agency procedures. The investigator must be reasonably familiar with these laws, regulations, and procedures to recognize deviations from normal contractual processes.

15-42. Discovering contracting irregularities requires continuous scrutiny of each step of the process from the inception of the contract to its termination. Easy identification of the exact spot where an irregularity has occurred is a rarity. It takes an extensive study of a contract and the regulations pertaining to it before the investigator can expect to successfully undertake a contract investigation. The investigator's familiarity with these matters is a basic tool for exploring the causes of, and contributing factors to, contract irregularities.

15-43. Begin the investigation by methodically and carefully separating pertinent issues and completely reviewing all related records, regulations, and procedural requirements. Approach contractors, government contracting personnel, and others connected with the issues in question on an informed and perservering basis. Appropriate curiosity is essential to a definitive investigation. Take nothing for granted. Check and confirm verification information, statements, time sequences, and observations. Seek corroborative evidence. Exhaust all leads to clear up matters not fully understood or completely clear. Seek to clarify and verify dates at the beginning of the investigation. Delays may permit suspects to develop collusive measures or cover stories to alter or substitute records.

15-44. The most valuable sources of information will be government employees. They have a basic obligation to report suspected wrongdoings. Nurture their confidence and trust. If you receive information with a stipulation of confidence, honor it.

15-45. Former employees are often willing to become involved in an investigation; especially if they feel they have been treated unfairly during their employment or in connection with their separation from government service. Review records of employees separated from government service to find those who may have observed a questioned action during their employment.

15-46. A discreet inquiry among trade groups can often produce revealing information as to whether or not procurement actions involving a particular agency or firm are "clean." Perhaps the most willing, if not the most knowledgeable sources of information, will be disgruntled, unsuccessful bidders.

15-47. Most of the human sources of information are likely to have only a general suspicion or a fragmentary knowledge of an alleged irregularity. However, some may be able to supply enough information to permit a rapid and thorough evaluation of the situation. Use your knowledge of the contracting processes to evaluate and convert their statements into leads.

15-48. The investigator should get full information on any allegations. It may indicate which individuals and processes are suspect. If allegations are in writing, contact the writers to seek more information. Often, they can provide names, dates, or places not initially reported. Check their motives for making the allegations. Anonymous allegations are often unfounded and made for ulterior motives. Investigate all such allegations to confirm or refute them.

CHECKING THE ACTIONS OF GOVERNMENT EMPLOYEES

15-49. There may have been a premature and/or unauthorized release of procurement information. Contractors may have been permitted access to areas or offices where contracting actions were discussed and where prerelease information could have been obtained. Contracting officers could have failed to furnish complete information to boards of award. Boards of award may have failed to consider all relevant factors. This is particularly true if the senior, best informed, or dominating member is in a position to exert undue influence. Contracting officers could have failed to enforce all provisions of a contract. Inspections, delivery of government-furnished property, delivery schedules, or closing of completed contracts are particularly open to fraud. See if government-furnished property was released to a contractor before it was needed, enabling the contractor to use it on other products. Supervisors may have failed to ensure proper use of government-furnished or government-owned property, or they may have failed to exercise adequate controls over, or accountability for, such property particularly upon completion of a contract.

CHECKING INSPECTION PROCEDURES

15-50. The preaward survey inspections may have been inadequate. The reports of inspection of the contractor's facilities may be false or misleading. Inspectors may have failed to inspect contractor products. They may have permitted the contractor to use inferior materials. They may have allowed the contractors to meet weight specifications by adding unauthorized materials. They may have allowed contractors to deviate from weight or density specifications. See if the contract administrator failed to document actions in the contract file that could result in savings or that could be detrimental to the government.

15-51. Check the actions of contractors. Learn if gratuities were given to a government employee. See if frequent visits or telephone calls that could have gained information resulting in a more favorable position for the contractor were made to government employees. Check for substitution of rejected or substandard items with acceptable items in shipments, with or without the inspector's knowledge.

15-52. See if the contractor could have presented false data or incorrect information before the award of a contract. Also, check specifications and sole-source procurements.

15-53. Specifications can be slanted to favor the product of a particular manufacturer. Sole-source contracts must be checked to ensure that individuals in engineering, supply, maintenance, or the like of have not inserted specifications for their own interest.

Chapter 16

Robbery

Robbery is a serious offense that may be carried out by a variety of means. Robbery occurs when an item of value is deliberately taken from a victim. The item could be taken from his presence and against his will by the use of force or violence or by instilling fear. The victim's fear can be of immediate or future threat of injury to him, his property, or the person or property of a family member or anyone in the victim's company at the time of the robbery. Because larceny or attempted larceny is an important element of robbery, it is essential for the investigator to be familiar with the elements of larceny and related offenses.

OVERVIEW

16-1. Serious crimes on military installations often come under the jurisdiction of federal law enforcement agencies, due to exclusive federal jurisdiction. In some cases, bank robberies, car thefts, and thefts of US property are investigated by the FBI. Federal agencies outside the military who have sole or concurrent jurisdiction are called immediately. Local authorities are contacted as a matter of routine police coordination.

16-2. The most common commercial robberies involve shoppettes, gas stations, liquor stores, cab drivers, and other businesses or activities that operate during the evening hours and involve cash transactions. Other robberies include—

- Street robberies.
- Vehicular robberies.
- Residential robberies (home invasions).
- Bank robberies.
- Robberies in schools.

16-3. The types of robbery investigated most frequently by military investigators are muggings and planned robberies of post facilities. The principles and techniques used to investigate these robberies also apply to residential, vehicular, bank, and other robberies that are encountered less often. The elements of proof for the offense of robbery remain the same regardless of the type of robbery being investigated.

16-4. Normally, robberies are reported soon after they are committed and law enforcement response is quick. The likelihood of locating the offender of a robbery depends on how long it takes to obtain a description of the assailant and broadcast it to responding patrols, which is the beginning of the investigative process. Frequently, mugging type robberies are committed by people who live on the street or who are involved in drug-related activities.

Consequently, patrols responding to a robbery scene may drive past a suspect who fled the immediate area, then continued walking away from the area to avoid raising suspicion. Rapid dissemination of suspect descriptions can aid responding patrols in detaining potential suspects who may not otherwise be identified.

ELEMENTS OF A ROBBERY

16-5. There are three basic elements of a robbery. These elements state that—

- There must have been a larceny committed with all of the elements of larceny present.
- An item must have been taken from a victim or from his presence.
- An item must have been taken from a victim by actual or threatened force or violence.

16-6. If the threat of force is enough to cause a victim to fear that force will be used and keeps him from resisting, it is robbery. Holding a victim at gunpoint is sufficient threat to support the offense of robbery.

16-7. For example, an intruder enters a house and points a gun at the owner. The owner, who is afraid of the gun, tells the intruder where his valuables are hidden in the house. The intruder then ties the owner up and goes into the room where the valuables are hidden. The intruder commits the offense of robbery, as well as housebreaking or burglary (as appropriate), when he takes the valuables.

COMBINATION OF LESSER-INCLUDED OFFENSES

16-8. The offense of robbery combines the offenses of assault *(Chapter 8)* and larceny *(Chapter 10)*. Thus, if the elements of proof do not support a charge of robbery, they may support a charge for either assault or larceny (lesser offenses). However, the elements of proof can only support both offenses if the larceny occurred before the presentation or use of force. For instance, someone steals the property of another and is subsequently observed doing so by the owner. The owner of the property then confronts the suspect. Pursuant to the confrontation, the suspect assaults the owner and flees the scene. This would warrant both larceny and assault charges but would not support the charge of robbery. If there is not enough evidence to show requisite force or engendered fear, a charge of larceny may result. If evidence fails to support a charge of larceny and the element of force is present, a charge of assault may result.

16-9. More than one robbery may occur at one time. If a group of people is threatened and property is taken from each person, there is more than one offense. There are as many robberies committed as there are victims who were deprived of their property. Each instance of taking is considered a separate offense. However, when several people are threatened and property is taken from only one victim, there is only one robbery and several assaults.

SPECIAL CONSIDERATIONS FOR ROBBERIES

16-10. When investigating robbery complaints, investigators should remain aware of other illegal activities that may be associated with the offense under investigation. For instance, during the course of an investigation, it may be determined that an alleged victim may not have been robbed. The alleged victim may have reported robbery to claim money from the government or to legitimize his theft of other funds. Additionally, in some instances, a drug dealer or prostitute who is not paid for his merchandise or services may report that he was robbed. If he reports the theft of the illegal substances or services honestly, he may be charged with appropriate criminal charges. If he reports the robbery as the theft of funds, which he calculated based on his respective value, the victim may be charged with false swearing or making a false complaint, as appropriate.

INVESTIGATION OF A ROBBERY

16-11. An investigator who responds to a robbery should observe areas for possible assailants while en route to the scene, such as a person running or driving recklessly away from the scene. Upon arrival at the scene, the investigator follows the basic steps in crime scene processing. Actions at the scene include—

- Interviewing the victim and/or the complainant.
- Identifying and interviewing witnesses.
- Conducting a crime scene investigation.
- Identifying the assailant.

16-12. An investigator determines and documents the events that led up to the robbery. This includes—

- The movements of the victim and others.
- A detailed physical description of the assailant and any accomplices.

NOTE: The investigator should get this information from individuals while it is still fresh in their minds.

- A detailed description of the methods and actions used by the assailant and any accomplices to the crime.
- A description of the weapons and vehicles used in the offense.

INTERVIEWING THE VICTIM AND/OR THE COMPLAINANT

16-13. Interview the victim of a robbery more than once. At a later interview, the victim may recall details not remembered or thought to be unimportant during the initial interview. It will help even if the victim recalls only a few details about the assailant.

16-14. The initial interview of the victim should be restricted to descriptive data that will aid in the detention of potential suspects. A subsequent interview should be conducted to obtain detailed information in the form of a formal statement.

16-15. Question the victim to get a description of the assailant. Ask about the assailant's voice, mannerisms, jewelry, tattoos, scars, distinguishing features, physical appearance, clothing, and so forth. Note what conversation occurred between the assailant and the victim. If the assailant had accomplices, note what they said to each other. The wording of verbal threats or demands uttered by the assailant must be carefully documented. Written threats and demands must be retained and examined.

16-16. Assailants can be difficult to identify because many robberies, especially muggings, occur after dark or under conditions that make the assailant's features hard to see. Even when the assailant directly confronts his victim, the victim is not able to provide an accurate description because of his emotional state.

16-17. Try to ascertain what type of approach the assailant used, and obtain a detailed description of the items taken from the victim and their value. Determine where the victim was immediately before the attack and what he was doing at the time of the attack. When a victim does not have a viable reason for being where the attack took place, the disclosure of additional information pertinent to the crime may result. Determine which direction the assailant went when he left the scene, and determine if he left on foot or by vehicle.

16-18. If the victim is injured, request permission to take photographs of the injuries. Follow the same photographing procedures used with an assault victim described in *Chapter 8*.

CONDUCTING A CRIME SCENE INVESTIGATION

16-19. Verify the crime scene by having the victim show the exact location and actions of the suspect during the robbery to ensure that the crime did occur. Look for items that the suspect may have discarded at or near the scene. Refer to *Chapter 5* for the procedures to follow when conducting a crime scene investigation.

16-20. The collection and preservation of evidence usually common to a robbery includes—

- Bite marks.
- Ligatures.
- Glass.
- Paint.
- Cigarette butts.
- Documents.
- Fingerprints.
- Seminal stains.
- Fabrics.
- Shoe prints.
- Blood.
- Small objects.
- Firearms and ammunition.
- Soil.

- Hairs and fibers.
- Tape.
- GSR.
- Tire tracks.
- Dust prints.
- Tool impressions.

IDENTIFYING AND INTERVIEWING WITNESSES

16-21. Identifying and interviewing witnesses is critical to solving robberies. Military police or other investigators must help get the names of every person in the facility. They must check for witnesses outside the building in the event that someone may have seen the assailant flee. The witnesses should be interviewed separately.

IDENTIFYING THE ASSAILANT

16-22. As leads develop to the robbery, there are many people and places that an investigator can seek out to gain more information. Investigators should check—

- Known or suspected drug addicts. Addicts are often desperate for money to support their habit and are compelled to rob others as a rapid means in which to obtain more drugs.
- Pawnshops periodically for items taken in robberies. Many addicts are careless as to how they exchange stolen items for cash and are quick to make exchanges without hiding their identity.
- The computer for like vehicles when automobiles have been involved. The victim should then be asked to view the photographic identification file of the local USACIDC office and, if possible, the local police department files.

16-23. Because robbery is a recidivistic crime, the techniques and mannerisms of an assailant are clues to his identity. Investigators should—

- Look for the use of same or similar locations. Perhaps there is a pattern of using parking lots, parade grounds, or stairwells.
- Determine what types of weapons were used, if any.
- Compare methods of approach and the number of assailants involved in each attack.
- Ascertain what the assailant said as an opening statement to the victim. If any conversation was held with the victim or among the accomplices, learn what was said.
- Ask about the assailant's peculiarities of accent or pronunciation of certain words.
- Note any violence that was used against the victim. Specifically, note how and where an injury may have been inflicted.
- Recognize that most people who find something that works will, by nature, continue to use it. This action solidifies an MO and will assist investigators in linking crimes that may have been committed by the same individual or group of individuals.

16-24. One of the most effective means in identifying a suspect is to compile a composite sketch, which is the result of several victim and witness accounts. This allows for the integration of key features provided by different individuals, who observed several different aspects of the assailant. If the composite can be developed in sufficient detail, prepare and distribute *CID Form 88 (Wanted Poster)*. Investigators should consider if requesting and offering a reward is warranted and, if so, disseminate copies of the wanted poster to local and resident federal agencies. Post them on the installation and around the area where the offense was committed. A few days after posting them, return to the area where the offense occurred to determine if they have been removed. If the posters are being removed, a surveillance operation may be an effective means to identify who is removing them and may help determine if that person is the suspect. On occasion, the post and local newspapers and local TV stations may be used as other means of publicizing the poster.

16-25. When an assailant is identified, obtain legal authorization to search him. Search him for injuries and items transferred from the victim or scene. Recover his clothing and collect known samples, if appropriate, and consider searching the suspect's home and possessions, such as his car and storage areas for evidence that links him to the victim or scene.

ESTABLISHMENT OF MODUS OPERANDI

16-26. Investigators must note facts pointing to a certain MO. Matching the MO of an unsolved robbery with cases from other police agencies may lead to identifying and apprehending the responsible parties. Investigators must—

- Consider if the target was cased weeks in advance.
- Determine if there was a detailed timetable of operations. Did the suspect appear to be monitoring or calling out time? This may be evidence of a plan, rehearsals, and an MO.
- Note the number of individuals used to commit the robbery and how their tasks were split. Sometimes one person directs the operation and others perform the actual work of the robbery. Tasks may be handled by roles, such as a driver, a gunman, a lookout, and an inside man.
- Check the techniques used and how individuals were positioned during the robbery. Did they use verbal commands, written or verbal demands, and/or visual signals? What did they call the other team members when issuing directions?
- Establish what kind of equipment was used and the types of facial disguises were used.
- Attempt to capture the conversations between the suspects in each witness interview and determine if the conversations provide any insight into the robbery and the suspect's escape plan.

16-27. Both positive and negative actions may have influenced the suspect's plans. Consideration should be given to—

- Whether fewer employees worked on the day of the crime because of lighter patronage. This factor may have entered into the suspect's planning. Habitual movements by employees could also have been used to the suspect's advantage.
- The level of familiarity of the location robbed to the assailants, which may indicate that they received assistance from an employee, or they expended a significant amount of time in learning the routines and weaknesses of security personnel and procedures.
- Reviewing surveillance tapes for several weeks before the robbery to detect suspicious activity of suspects that was not noticed during surveillance activities.

16-28. In the commission of some robberies, especially those of convenience stores and gas stations, the suspects may not have taken the time to extensively plan and observe their target location. In the vast majority of robberies involving banks and other cash facilities, it is almost certain that suspects reconnoiter the target location to determine when and how cash is moved, what the security protocols are, and how they can be defeated. Whenever a robbery occurs at the peak of cash flow periods or while funds are being moved, there is an extremely high likelihood that extensive planning preceded the robbery. The suspects may have even rehearsed their actions at a mock-up facility, which will likely only be determined by disseminating information to the public and requesting public involvement in solving the crime.

STRONG-ARM ROBBERY OR MUGGINGS

16-29. Assailants who mug others are often the least professional of all assailants; they generally use strong-arm tactics or other unsophisticated means of committing their crime, such as using a gun or weapon to overcome potential resistance. Most muggings occur out of necessity and opportunity; therefore, assailants who mug do not generally know their targets, nor can they control other factors, such as passers-by who may witness the attack. The actions of an inexperienced assailant who mugs someone may be based only on a need for money and a sudden chance to victimize a person who is alone. Due to these factors, they are likely to commit one or more careless mistakes or complacently walk into a nearby establishment subsequent to the attack.

16-30. Individuals who take wallets, purses, or other items (which can be linked to the victim) generally do not want to possess them any longer than they have to. They will frequently flee the immediate area and hide where they believe they will not be discovered. Once in hiding, perhaps behind a shrub or garbage dumpster, they will sort through the items taken and discard the items they do not want. Conduct canvass interviews and a crime scene examination that includes a search beyond the scene. Look for discarded items and review surveillance tapes of local area establishments (such as liquor stores, gas stations, automated teller machines, and so forth) to help identify the assailant. Frequently, latent fingerprint impressions will be deposited on discarded items because it is cumbersome to rifle through the contents of a wallet or purse while wearing gloves. The recovery of these items may prove to be valuable evidence.

16-31. Experienced individuals who mug often plan their actions or use the same methods to identify their target. They may enlist the aid of others to serve as watch outs or helpers in the attack. Those who mug normally try to select individuals who are alone or will not likely be able to defend themselves and are believed to have a large sum of money. They select locations that are free of witnesses and that will provide the advantage of surprise. The patterns or MOs associated with these assailants can aid investigators in predicting additional attacks and can be used to proactively thwart these crimes.

ROBBERY OF MONEY-HANDLING FACILITIES

16-32. The fastest way to effect the apprehension of individuals who rob post facilities is to quickly seal off the area and catch the assailant before he can flee the scene and dispose of the stolen property. This is accomplished through planning and conducting mock exercises between post facilities and the military police. Employees of a facility notify the military police desk of a robbery and the desk dispatches patrols to the training site and orders the closure of all access control points. This exercise allows military police to respond in a preplanned, rehearsed manner, which leads to the successful blocking of avenues of escape in an actual robbery.

16-33. In most post facilities where high-dollar amounts of money are handled, there are video-recording devices inside and outside. Investigators should obtain these vital pieces of evidence and analyze them for clues to identify the assailants. The video should be viewed away from the victims of the robbery and then compared to the interviews of the victims to prove or disprove the account of the robbery.

Chapter 17

Sex Offenses

Investigators may be required to investigate the alleged commission of sex offenses contained in the MCM. They may also investigate alleged sexual activity of individuals subject to the *UCMJ* if these activities conflict with current Army policy, bring discredit upon the military, or involve security matters. The investigation may result in an administrative action under the provisions of ARs or it may result in a court-martial action under the provisions of the MCM. In all cases, an investigator must conduct a complete, thorough, and impartial investigation.

OVERVIEW

17-1. The investigation of alleged sex offenses must be tactful and discreet, regardless of gender and age. Detailed information about a sex offense may become public knowledge at the time of the trial, but the investigative process must not start or add to rumors that often circulate after the discovery of a sex offense. The provisions of *AR 195-2* govern the dissemination of publicity related to such incidents.

17-2. The reporting of an alleged commission of a sex offense can create public pressure to identify and apprehend a suspect and to prevent future offenses. This pressure may hurt a suspect's right to a complete and fair investigation of the charges. Because of the nature of sex offenses, an investigator must work quickly during the preliminary investigation. Investigators must be aware of hasty or rash conclusions during this phase of the investigation because they can cause innocent individuals to be falsely branded as sex offenders.

17-3. Every sexual offense is different. When arriving at the scene, the investigator must determine the psychological and physical state of the victim. This will aid in developing a course of action. Rape is perhaps the most serious crime, excluding homicide, that an investigator will investigate. The trauma of rape can be a long-lasting one. It is essential that investigators who are assigned rape cases have a special knowledge and understanding of rape victims and suspects.

17-4. During the conduct of a sex offense investigation, the investigator—

- Directs the main effort of the investigation toward finding out if an offense did occur, the specific nature of the offense, and who committed the offense.
- Collects evidence to prove or disprove the fact of the offense.
- Maintains records so the chain of custody can be shown at a trial by court-martial or can support administrative action.

- Does not analyze the mental condition of the subject in the investigator's report.
- Avoids recording a personal evaluation of the suspect.
- Directs the investigation toward apprehending the suspect.
- Allows legal and medical authorities to determine the disposition of the suspect according to their professional analyses.

TYPES OF OFFENSES AND ACTIVITIES

17-5. Adultery, prostitution, solicitation, and pandering violate the *UCMJ*. When a person solicits or advises (with wrongful intent) someone to commit a sexual offense, it is a violation of the *UCMJ, Article 134*. It does not matter if the solicitation or advice is acted upon. The solicitation may be by means other than verbal or in writing. The solicitor may act alone or through other individuals to commit the offense. An attempt or conspiracy to commit any sex offense may be charged under the *UCMJ*. This may be important when a sex offense was planned or tried but not completed to the point that the needed elements of the offense can be shown.

ADULTERY

17-6. Adultery is sexual intercourse between a male and female when one of the two is lawfully married to a third person at the time of intercourse. Adultery violates the *UCMJ, Article 134*.

PROSTITUTION

17-7. Prostitution is the engaging of sexual intercourse for pay or reward. Prostitution by members of the US Armed Forces is punishable under the *UCMJ, Article 134*. When prostitution of a person not subject to the *UCMJ* occurs on a military base, it is investigated by military police and referred to the proper authorities.

PANDERING

17-8. Pandering is the wrongful or unlawful compelling, inducing, enticing, or procuring of an individual to engage in acts of prostitution with individuals directed to the prostitute by the panderer. It is also the arranging by the panderer of sexual intercourse or sodomy between two people.

VOYEURISM

17-9. Voyeurism invades an individual's privacy. It is the trespassing of an individual's property to gaze through an opening at seminude or nude individuals without their consent. Voyeurism is often (not always) done for autoerotic purposes. Generally, the voyeur must deviate from normal activity or trespass on an individual's property for the express purpose of invading an individual's privacy. If done from the voyeur's abode or from a public vantage point, it may not be a violation. This offense is punishable under the *UCMJ* as it may bring discredit upon the Armed Forces.

INDECENT EXPOSURE

17-10. Indecent exposure is the willful, wrongful exposing to public view the private portions of one's anatomy in an indecent manner. It applies to both male and female exhibitionists. Indecent exposure and any oral or written communication between individuals that contains indecent, insulting, or obscene language violates the *UCMJ*.

INDECENT ACTS

17-11. The taking of any immoral, improper, or indecent liberties with or the commission of any lewd or lascivious act upon or with the body of a child under age 16 is a violation of the *UCMJ*. The intent is to arouse or gratify lust, passion, or sexual desire. The desire can be that of the person committing the act, the child, or both. Actual touching is not required.

ADULT-CONSENSUAL SEXUAL MISCONDUCT

17-12. Indecent, lewd, and lascivious acts like mutual masturbation or indecent fondling of another are also violations of the *UCMJ*. Either or both participants may be prosecuted. Whether or not a participant is prosecuted depends on the individual's ability to intend to commit or to cooperate in such an act. Similarly, consensual homosexual activity defined as two individuals of the same sex engaging in mutual sexual activities including but not limited to sodomy, mutual masturbation, or anal intercourse is a violation of the *UCMJ*. Consensual homosexual activity is different from nonconsensual homosexual activity where a person of the same sex victimizes an individual sexually. In those instances, a criminal investigation must be conducted.

17-13. Current DA and DOD policy mandate that law enforcement will not conduct an investigation of adult-consensual sexual misconduct when it is the only allegation involved. Such allegations are investigated only if approved by the Commander, USACIDC. As this is an ever-changing policy, investigative personnel must remain vigilant in their review of *DODI 5505.8* and *AR 600-20*.

CARNAL KNOWLEDGE AND SODOMY

17-14. Carnal knowledge and sodomy are violations of the *UCMJ*. Carnal knowledge is an act of sexual intercourse between a male and a female who is under 16 years of age and to whom the male is not married. Any penetration is enough to complete the offense.

17-15. Sodomy is an act of unnatural carnal copulation with a person of either sex or with an animal. Any penetration is enough to complete the offense. Emission is not necessary. If the act is done with a child who is under 16 years of age, the penalty is more severe.

SEXUAL ASSAULT

17-16. Indecent assault, assault with the intent to commit sodomy or rape, and rape are all in violation of the *UCMJ*. Indecent assault occurs when a suspect takes indecent, lewd, or lascivious liberties with a person to whom the suspect is not married. The liberties must be without the victim's consent and against the victim's will. The intent is to gratify the suspect's lust or sexual desire. The offense applies to both males and females.

17-17. Assault with intent to commit sodomy is made on a victim without the victim's consent and against the victim's will. This offense includes attempts to commit cunnilingus, fellatio, or anal intercourse. Assault with intent to commit rape occurs when a man intends to have sexual intercourse with a woman by force and without her consent. It is enough that he intends to overcome by force any resistance to his penetrating the woman's person. Actual touching is not needed.

RAPE

17-18. Rape is an act of sexual intercourse committed by any person by force and without consent. Any penetration is enough to complete the offense. Among the offenses that may be included in a charge of rape are assault, assault and battery, and assault with intent to commit rape. Indecent assault and taking indecent, lewd, and lascivious liberties with a female may also be included.

INVESTIGATION OF SEX OFFENSES

17-19. When called to investigate a sex offense, the investigator—

- Makes a note of the time, date, and person making the notification.
- Records the weather conditions and any other information that may help when prosecuting the offender.
- Notes the time he arrives on the scene.
- Gets as many details as possible from those who report sex offenses. The who, what, when, where, why, and how should be answered quickly and clearly to establish jurisdiction.
- Coordinates with required agencies and those that may assist in the investigation.

17-20. The investigator's first contact with the victim is of great importance. The investigator must not assume that the victim is old enough or mature enough to cope psychologically with the offense. An investigator's interest in the victim and a concern for the victim's welfare are factors in the victim's future cooperation. If the victim is a child, appoint an individual who is comfortable around children to stay with the child until the parents, other family members, or a family friend arrives.

17-21. At the crime scene, quick, thorough security must be established. A lot of physical evidence at a sexual-assault crime scene is fragile and special care must be taken to protect it. One investigator, at a minimum, should be assigned the responsibility of conducting a crime scene examination. Additional details concerning methodologies for initiating a crime scene search can be found in *Chapter 5.*

17-22. Medical personnel must examine all victims as soon as possible because the value of serological evidence is reduced by delay. Detailed questioning of the victim can be done later to get leads and information related to the offense. Find out by whom, in what setting and manner, and to what extent the victim has been questioned about the offense. The more victims are interviewed, the more reluctant they may be to talk. Do not allow other interested investigators to question a victim. Because of the nature of sexual assaults, one investigator should be assigned exclusively to the victim. This will enable that investigator to develop a rapport with the victim, which will hopefully enhance later cooperation.

17-23. During the initial contact with the victim, the investigator must instill confidence in the victim that he is qualified to investigate the offense. Be sincere and display true concern for the victim's situation. Explain to the victim what is being done and why. When there is enough information to enable other personnel to begin crime scene processing, the victim must be taken quickly to the nearest medical facility for a thorough examination. Explain that the clothing worn during the attack must be examined for evidence, and make arrangements for the victim to obtain a change of clothing.

17-24. Sometimes, the suspect is still at the scene. In these cases, a third investigator is required to apprehend and process the suspect. In such cases, the suspect must also be examined. Ensure that the victim and suspect are not transported in the same vehicle or interviewed in the same office. Otherwise, trace evidence may be transferred after the crime.

17-25. Victims and suspects who are subject to the *UCMJ* are examined by medical officers at the nearest military or civilian medical facility. Before the examination, the investigator should provide the examining physician with details about the victim, the subject, and the crime scene. This will facilitate a more comprehensive examination where the required items of evidence are obtained for a successful resolution and prosecution of the investigation.

17-26. The examination must be done in a reasonable way for both the victim and the suspect. If possible, obtain the consent of both parties. If the victim or the suspect will not consent to the examination and collection, a legal search authorization must be obtained. Such searches are allowed if they are not unreasonable or morally reprehensible. A search of any part of the body not normally open to public view may be made without an individual's consent if it is incident to a lawful apprehension. Use only the degree of force needed to do the search and ensure that a complete chain of custody is kept for all collected evidence.

17-27. Individuals not subject to the *UCMJ* may choose to be examined by either a civilian doctor of their choice or a medical officer. They cannot be forced to submit to an examination by a medical officer. The fact that the suspect is military or that the military is investigating the offense does not alter a civilian's right to choose an examining doctor or to refuse a military examination.

17-28. Obtain a parent's or guardian's written permission before a child is examined or treated by a medical officer. A parent or guardian should be with the child and present during the examination. Explain very tactfully that an examination is needed for the investigation. Advise them that it should be shown by medical opinion that the offense did take place.

17-29. The examining physician may not be aware of the evidence-seeking objective of the examination unless he is told. His main concern is for the welfare of his patient. For this reason, the investigator must advise the physician of the areas of interest to the case and the evidence samples needed for the investigation.

17-30. USACIL can examine most of the evidence needed for sex offense investigations. Examinations of parts of a human or animal body and of materials from a human or animal body require other services.

17-31. Vaginal, anal, and oral swabs and blood samples collected by the doctor who examines a victim or a suspect should be processed at USACIL. A vaginal smear should also be taken and sent to the hospital to be tested for the presence of motile sperm to verify a fresh complaint. Motile sperm presence would not be able to be determined by the time the specimen would reach the crime lab. Separate the samples, and provide one vaginal swab for motile sperm and the remainder of the swabs for the crime lab. The samples must meet the requirements of the sexual assault kit, NSN 6640-01-046-2693. Submit all of the swabs used to make smears on glass slides. They are best for crime lab purposes.

17-32. If the victim indicates that he is sexually active, hair and blood standards should be collected from the victim's sexual partners for elimination purposes. This should also be done for roommates or other persons who live in the immediate area of a crime scene, because these persons may have contributed hair or other DNA material that is recovered during crime scene activities. Send all physical evidence, such as hair, blood, and foreign materials taken from the body of the victim or suspect and comparison samples to the laboratories immediately.

EVIDENCE OBTAINED FROM VICTIMS

17-33. The victim's body is the crime scene and it may conceal evidence of a crime. From this evidence, the examining physician can give expert testimony. Wounds, bruises, cuts, abrasions, and irritations may help to show penetration, violence, or resistance. These should be described in the doctor's notes, reports, and testimony. This evidence may provide leads to the type of suspect and the weapons used. Photographic records are quite helpful to the prosecution.

ALTERNATE LIGHT SOURCE

17-34. Frequently, there is evidence present on the victim, whether the victim is living or dead that is not visible to the naked eye. The victim should be examined using an alternate light source. Bite marks, contusions, and fiber evidence, which may not be evident, become readily available when viewed under a light source. To conduct the examination, first ask the victim (if he is able to communicate) where to look for evidence. Other victims will require a greater area of search. To conduct the examination, activate the light source, put on the appropriate goggles, and slowly examine the area. Different substances will fluoresce at different light wavelengths and may be readily visible with the goggles. Any potential evidence discovered should be photographed and documented. If visible evidence is present, collect it according to *Chapter 5*.

SWAB SPECIMENS

17-35. Whether the victim is living or dead, the genitalia, anus and nearby areas may show traces of the victim's or suspect's blood and semen. The following exams and measures will be conducted by the medical officer when the victim is living and by the forensic pathologist when the victim is deceased:

- Swab specimens of material from the vagina, breasts, mouth, and anus for examination.
- Swab specimens from bite marks. Swab the bites with water and air dry the swabs before sending them to the crime lab.
- Collect foreign material from the pubic and anal areas, the legs, and the stomach for examination of semen, blood, or other evidence. Often, dried semen may be present in the pubic area. This can be cut off and placed into a coin envelope for later analysis.
- View and document wounds, abrasions, skin damage, feces, mucous materials, and lubricants found in the pubic, anal, and oral areas.
- Make the same checks of both individuals in an alleged sodomy case when the active or passive roles of the individuals are unknown.

17-36. All of the approved sexual assault determination kits contain the required materials for the examination. If an approved kit is not available, standard medical treatment facilities will usually have adequate supplies.

HAIR AND FIBER SAMPLES

17-37. Foreign hairs and fibers on the body must be collected as evidence. A method most often used on homicide victims is combing with a new fine-tooth pocket comb. It is the best way to find loose hairs and fibers. Pack the teeth of the comb with absorbent cotton along the comb's base where the teeth meet the spine. Use separate combs for the head and pubic areas. Use transparent tape to collect fibers from other areas.

17-38. Medical personnel take samples of hair from the head and pubic area by pulling the hairs. Entire strands of hair should be taken. Care should be taken not to dislodge foreign materials or damage the hair ends. Each sample should consist of 15 to 20 hairs.

17-39. Each hair sample must be packed in a white, totally sealable envelope or plastic bag. It must be labeled carefully and identified by the initials of the doctor who took it. The samples should be sent to the appropriate crime lab for examination.

SALIVA, URINE, AND BLOOD SAMPLES

17-40. Saliva, urine, and blood samples should be taken for typing; for testing for venereal diseases; and for examining for alcohol, narcotics, and poisons. If there is any suspicion of the use of a "date rape" drug, a urine specimen must be expeditiously obtained. This evidence quickly metabolizes and will vanish. Drug analysis is accomplished by way of a urine specimen obtained and forwarded to the Division of Forensic Toxicology at the Office of the Armed Forces Medical Examiner's Office, Armed Forces Institute of Pathology (AFIP), Walter Reed Army Medical Center, Washington, DC. Evidence should be submitted on *AFIP Form 1323 (AFIP/Division of Forensic Toxicology—Toxicological Request Form).* This form is not only the laboratory request, but the chain of custody for that specimen. If a blood alcohol test is requested, blood should be obtained in a "gray-top" (sodium chloride) tube and submitted in the same manner as the urine specimen.

17-41. If the victim was menstruating at the time of the offense, any tampons found should be submitted (after they are dried). If a sanitary napkin was worn at the time of the crime, it should also be taken to compare with evidence stains. An early morning, first-time urine specimen should also be collected for a pregnancy test. Sufficient blood samples should be sent to the crime lab to compare with evidence stains. All evidence collected by the doctor must be put in separate containers and marked.

FINGERNAIL CLIPPINGS

17-42. The doctor should take fingernail clippings from each finger. Right-hand clippings must be kept separate from left-hand clippings. The clippings, properly packed and marked, should be sent to the crime lab. This is especially critical if the victim or subject has been scratched. Hands and fingers can then be swabbed using a sterile swab in cases involving digital penetration.

CLOTHING

17-43. The clothing the victim wore during the crime is needed as evidence. The investigator must legally and promptly obtain the victim's clothing. The victim's family may destroy valuable evidence in an effort to clean or to dispose of the garments. Even if a garment has been cleaned, it should be secured.

17-44. Mark each item for identification. Clothing must be packaged in separate paper containers to be sent to the laboratory. If a garment is wet or has damp blood or seminal stains, dry it at room temperature without a fan or artificial heat. The garment must not become contaminated while it is being dried or stored. Do not allow it to come in contact with any other clothing. Clothing should be packaged and sealed as soon as possible in order to preserve any evidence present. These items should be delivered to the evidence custodian in a sealed package and remain that way until the evidence is examined at the laboratory.

BED LINENS

17-45. If the offense is committed in a bed, all bed linens should be marked for identification. Packaging requirements are the same as clothing.

VICTIM PHOTOGRAPHS

17-46. Take photographs to preserve graphic evidence of the appearance of the victim. The photos can show wounds, bruises, and lacerations that may heal or disappear by the time a suspect is brought to trial. Use color film, as it will enhance the evidentiary value of the photographs. The photographs and negatives should be taken, processed, and handled in a way that preserves their integrity. This will ensure their admissibility in court and prevent them from falling into the hands of unauthorized individuals. The photographs should be used only for the purpose of investigation and prosecution.

17-47. Photographs of living victims taken by law enforcement personnel should be limited to those parts of the body normally visible when the victim is clothed. Photographs may be needed of the genetalia to substantiate and illustrate medical testimony. Photographs may not be taken of these parts of the victim's body except with his express written consent. If the victim is a minor, consent must be received from the victim's parents or guardians before taking the photographs. The photographs should only be made under the supervision of the examining physician.

17-48. A female must be present when a female is photographed. Military police may photograph the bodies of deceased victims without permission of the next of kin. Bruises should be photographed four times with 24 hours of separation (initial, 24 hours, 48 hours, and 72 hours later). Ensure that the photographs exposed do not unnecessarily embarrass the victim or his family members.

17-49. If the victim of an alleged sodomy is an animal, have it examined by a veterinarian. It should be searched for wounds, bruises, abrasions, human semen, blood, and hair and clothing fibers. Hair and blood samples should be taken and swab samples should be taken from body openings. If the animal is to be destroyed, a picture of it (showing evidence of the assault) should be taken first. If the animal is dead, the veterinarian should do a complete autopsy. The veterinarian can give expert testimony in court about the examination and findings.

EVIDENCE OBTAINED FROM SUSPECTS

17-50. The suspect is examined for wounds, bruises, cuts, or abrasions. These may have been caused by the victim's struggle or from an act of forced sex. The entire body, particularly the genitals and pubic area, should be searched for blood, semen, hair, feces, vaginal debris, or other matter from the victim. For sodomy suspects, the oral and anal areas should be examined as well. Evidence is collected to compare with materials from the victim's body or the crime scene. Hair, blood, saliva samples, and fingernail scrapings are taken. The methodology and amounts of each of these samples are similar and in the same amounts. All evidence must be properly sealed, marked, and sent to the crime laboratory.

17-51. Suspects do not have the legal right to refuse to have their photograph taken. Wounds on the suspect should be exposed and photographed. As with victims, bruises should be photographed four times with 24 hours of separation (initial, 24 hours, 48 hours, and 72 hours later).

SCENE PROCESSING

17-52. While the victim is receiving medical attention, the crime scene should be processed. Investigators should—

- Process the crime scene according to *Chapter 5*.
- Make photographs and sketches of the scene.
- Gather and process items of evidence.
- Collect the clothing worn by both the victim and the suspect.
- Search for seminal stains on clothing and bedding.
- Search for pubic and head hair of the suspect.
- Look for blood at the scene if the sex offense involved an assault.
- Check entrance and exit points when the incident occurred indoors, and look for items left by the assailant.
- Make a thorough search of the area that the victim declared in the initial interview to be the exact point of the attack.
- Search the bathroom if the attack was made in a residence. Many sex offenders will use the bathroom after committing a sex offense.
- Check the kitchen, such as the refrigerator and trash cans for any evidence that the assailant may have used to discard any evidence.

CONSENT DETERMINATION

17-53. Whatever the age of the victim, if consent to a sex offense is material to the case, find out if the victim encouraged, resisted, or consented to the act. Intimidation, coercion, threats, or fraud may be given as reason for not resisting. If such a reason is given, make note of the acts, threats, or statements that were made. Obtain a detailed description of any weapons used.

17-54. Do not assume from a victim's occupation, associates, habits, or appearance that the victim is promiscuous and most likely consented. Do not infer this from the fact that the victim has associated with the accused. The courts decide whether or not consent was given. The investigator collects evidence that may show consent or the lack of it.

17-55. Testimony on the character of the victim of a sex offense may be of value in showing whether consent was given. If there is conflicting character testimony, polygraph tests may be in order. An investigator may also secure testimony from unbiased individuals who are familiar with the victim. Try to learn if the victim was mentally incapable of legally consenting to sexual intercourse at the time of the act because of drugs, alcohol, age, disease, injury, or psychiatric condition.

17-56. Check to see if the victim has ever made false sexual allegations, and determine if the victim has a motive for a false complaint. The complaint may be a means of concealing an indiscretion. Consider the following questions:

- Did the victim fail to get the expected pay for an act of prostitution?
- Is the victim conscience-stricken because of a seduction?
- Does the victim want revenge against the suspect or want to force a marriage?

- Does a female victim think that she is pregnant and hopes to remove the stigma of indiscretion?
- Is the victim hoping for a cash settlement from the suspect or his family?
- Does the victim believe the US government will pay damages to someone assaulted by a service member?

17-57. Statements and accusations that the victim is of lewd repute, habits, or associations or has engaged in specific sex acts with the suspect or other individuals may be admissible evidence. These points should be checked in detail. Although these points may be true, such individuals can be bona fide victims of sex offenses. Investigators must be alert for attempts by an accused to make up untrue stories of past sex acts with the victim. A suspect may produce other individuals who claim such experiences. On the other hand, check on efforts by relatives or friends to provide the victim with a good reputation that is not deserved. It is important to remain unbiased, gather all of the facts, and remain vigilant in the pursuit of the truth.

SUSPECT IDENTIFICATION

17-58. As with any other crime, sex crimes require that your investigation be a marriage of physical and testimonial evidence. The objective of a criminal investigation is to solve the crime by placing the suspect and the victim at the crime scene at a particular point in time. This, coupled with your ability to answer who, what, where, when, why, and how will result in a successful investigation.

17-59. When checking leads, keep in mind that sex crimes do not just have one type of a suspect. Anyone can commit a sex offense. Think of the unusual. Whoever had the chance to commit the crime can be suspect. This is in spite of an excellent reputation, law abiding past, or high station in life.

17-60. The first steps in apprehending a suspect often relate to whether the victim or witnesses know or can identify the person. If the suspect is known or if a description is available, send this promptly to military police patrols to make the apprehension. All resources must be used to identify, search, and apprehend the suspect. Descriptions of individuals recently apprehended by military police may be matched with the description of the wanted suspect.

17-61. Think about what the victim or witness said the suspect wore. Make a detailed visual examination of all the suspect's clothing. It should be examined for evidence of the offense, physical contact with the victim, and presence at the crime scene. Clothing should be packaged and sealed as soon as possible in order to preserve any evidence present. These items should be delivered to the evidence custodian in a sealed package and remain that way until the evidence is examined at the laboratory. An investigator must be particularly attuned to all types of physical evidence. Trace evidence is minute and easily overlooked, it is frequently not seen at all with the unaided eye.

17-62. Sometimes items belonging to friends or associates of a suspect need to be examined. This is done if there is evidence that the suspect has loaned, borrowed, or exchanged clothing. Places where the suspect may have disposed of incriminating evidence should be searched. Many sex offenders take items of clothing from their victims. Physical objects symbolic of sex and obscene literature and photographs are often found in the possession of suspects. Because some suspects will not want these items left around their homes, the investigator should get permission to search a suspect's place of business. Other items that might give a clue to the identity of a sex offender are tape recordings of previous sexual acts or letters from friends discussing his shared participation in unusual sexual activity.

17-63. To obtain suspect leads if the victim is dead, trace the victim's recent movements. Identify individuals with whom the victim was last seen. Seek witnesses who saw any individuals near the crime scene or the place where the victim was last seen alive. Photographs of the victim may help in locating witnesses. Check the victim's associates and places he frequented for leads. The only lead may be the name of the person with whom the victim was supposed to be with. It could also be the place at which the victim was believed to be just before or at the time of the offense.

17-64. Medical personnel can help identify and locate the suspect. They may be asked to identify persons recently seeking treatment for injuries that may be associated with a sexual assault. They may also be able to account for the activities and recent whereabouts of patients who may be suspect. There are special considerations with regard to privileged medical information. Normally, there is no right to privacy in military treatment facilities.

17-65. If leads do not develop elsewhere, check MO offenses or sex offender files. Try to identify individuals who have committed similar crimes or other sex crimes or who have used similar criminal methods. Accurate and detailed records of unrelated sex crimes can lead to early detection of a sex offender. A check of these individuals and their recent activities may result in leads. Check assault records. Assaults may sometimes be sex-related. Remember that rape, although a sex crime, is not about having sexual relations but rather about control and domination. Often, an assailant with latent-sadistic sexual overtones makes unprovoked assaults on women and children. Check on the larcenies of or mutilation of women's garments. These offenses are often treated as juvenile pranks. But experience has shown that these offenses may be the first step toward deviate sexual offenses.

Chapter 18

War Crimes

Today, the lawmaking treaties and customary laws that first determined what constituted a war crime are firmly intact. However, with the judgments of recent tribunals, decisions have been made to declare certain misconduct inclusive of war crimes. The following pages describe the expected behavior of any person (regardless of whether he is friend or foe or military or civilian) toward another person during times of war as established by the provisions of treaties or conventions and internationally recognized customs of nations. When the behaviors of individuals do not comply with these provisions, they may constitute war crimes and require investigation.

OVERVIEW

18-1. The law of land warfare, which is both written and unwritten, exists to regulate the conduct of armed hostilities. It is inspired by the desire to diminish the evils of war by—

- Protecting combatants and noncombatants from unnecessary suffering.
- Safeguarding certain fundamental human rights of individuals who fall into the hands of the enemy, particularly prisoners of war (POWs), the wounded and sick, and civilians.
- Facilitating the restoration of peace.

18-2. The laws of war are derived from the following two principal sources:

- International treaties or conventions like the Hague Conventions and the Geneva Conventions.
- The body of unwritten laws that have been established by custom and are recognized by authorities on international law.

LAWMAKING TREATIES AND CUSTOMARY LAWS

18-3. The laws of war treaties have a force equal to that of laws enacted by Congress. They must be observed by all individuals in the United States. Customary laws are binding upon the United States, its citizens, and other individuals serving it.

18-4. The *Geneva Conventions* spell out the customary laws of war. In the case of armed conflict (not on an international level) in the territory of one of the High Contracting Parties, each party is bound to apply some basic provisions. Individuals who do not take an active part in hostilities shall be treated humanely. This includes members of armed forces who have laid down their arms and those removed from the conflict by sickness, wounds, detention, or other cause. No distinction in treatment will be made by race, color, religion, sex, birth, wealth, or any other similar criteria. The wounded and sick will be collected and cared for. Certain acts are prohibited, at any time and place, with respect to these nonparticipants. There will be no—

- Violence to life and person, murder of any kind, mutilation, torture, or cruel treatment.
- Hostages taken.
- Outrages upon personal dignity. Humiliating and degrading treatment are expressly forbidden.
- Sentences passed or executions carried out without prior judgment by a legitimate court affording all the judicial guarantees viewed as essential by civilized individuals.

18-5. See *FM 27-10* and *DA Pam 27-1* for war crimes, lawmaking treaties, and customary laws.

DEFINITIONS

18-6. To better understand war crimes, the following terms are defined:

- **War crimes.** The technical expression for violations of the laws of war by any individual or individuals (military or civilian). Every violation of the laws of war is a war crime.
- **Rape.** Sexual intercourse by a man or woman without his or her consent and by force or deception.
- **Genocide.** The deliberate and systematic destruction of a racial, political, or cultural group. (Genocide is punishable under international law regardless of whether the offense was committed in time of war or peace.)

COORDINATION OF SUPPORT

18-7. War crime investigations often have high-level government interest. They generate national and international news media coverage. War crime investigations must be coordinated directly with the Army theater commander and his staff. They can coordinate directly with the US Embassy and senior officials of an occupied country and the major combat units within the area.

ESTABLISHING LIAISON WITH HIGHER HEADQUARTERS

18-8. When an investigator is called to investigate an alleged war crime, his initial contact is with the Army theater PM. Access to the US Embassy and the Army theater commander should be coordinated through USACIDC elements within the theater of operations (TO).

18-9. The theater HQ can provide various types of personnel and logistical support, such as office space, communications, billeting spaces, and field equipment. Support should be established through an MOU or a formal tasking for support to investigative operations. MPIs should coordinate with USACIDC elements within the theater to determine what resources are available.

REPORTING TO USACIDC

18-10. There must be a means for reporting to USACIDC in Washington, D.C. Established satellite communications in conjunction with secured cell phone lines should be used. They offer real-time, secured conversations and a quick means of reporting information. These lines also allow for three-way conversations to be established in the event that legal or other advice is necessary during the reporting process. Military police and DOD investigators coordinate appropriate investigative information and findings with USACIDC, within a deployed environment, based upon the reporting criteria established by the USACIDC field element.

TAKING SPECIALIZED EQUIPMENT

18-11. The USACIDC team must be well supplied with proper equipment and able to operate under adverse conditions. The USACIDC team must bring the specialized equipment it will need. Computers and their component parts, writing materials, forms (even if they are programmed on the computer), and portable files are a necessity. Maps, aerial photographs, tape recorders, and cameras will be useful. A camera and tape recorder are useful for referral to statements and individuals. Digital, 35-millimeter, and/or instant cameras should be on-hand to photograph interviewees at the end of each interview. The photographs are valuable as an identification tool and may be useful in future trials.

DETERMINING THE ATTIRE

18-12. In some areas, the combat uniform is needed. In others, civilian clothing is preferred. The determining factor for attire, in addition to climate and war conditions, is the established policies within the TO. If civilian clothing is authorized and determined to be more advantageous to a successful interview, then that clothing should be worn.

IDENTIFICATION OF WAR CRIMES

18-13. Violations of the laws of war may include conspiracy and attempts to commit war crimes. An individual may commit war crimes in which the individual and the commander or only the commander may be held accountable, or war crimes may be committed by an organization to include that of a given government. In all cases, the war crime has to be identified and investigated.

DETERMINING WHAT CONSTITUTES A WAR CRIME

18-14. Violations of the *Geneva Conventions* are specific crimes of the laws of war and include—

- The willful killing of noncombatants.
- The torture or inhumane treatment of combatants and noncombatants, to include performing biological experiments that willfully cause great suffering or serious injury to the body or health.
- The unlawful deportation, transfer, or confinement of protected persons, such as a clergyman.

- The forcing of protected persons to serve in the forces of a hostile power.
- The taking of hostages and excessive destruction and appropriation of property not justified by military necessity and carried out indiscriminately.

18-15. In addition to the grave breaches of the *Geneva Conventions*, the following are some of the representative acts identified in *FM 27-10* that constitute war crimes:

- Making use of poisoned or otherwise forbidden arms or ammunition.
- Requesting quarters by using treacherous means.
- Maltreating dead bodies.
- Misusing the Red Cross emblem or its equivalent.
- Abusing or firing on the flag of truce.
- Pillaging or purposelessly causing destruction of protected places.
- Violating the terms of surrender.
- Firing on hospital zones.

18-16. In 1998, the Tribunal for the Former Yugoslavia became the first international court to determine that an individual accountable for rape was also accountable for war crimes. The Tribunal determined that rape is now inclusive of war crimes when committed under the laws of land warfare. In 2001, the Tribunal set another precedence by convicting former Bosnian Serb soldiers of systematically raping and torturing Muslim females. This same court established sexual enslavement as a war crime.

DETERMINING WHAT IS NOT A WAR CRIME

18-17. The occurrence of criminal acts taking place before the declaration of war or after the termination of a war by agreement or unilateral declaration of one of the parties, by complete subjugation of an enemy before termination of war, or by termination of war or armed conflict by simple cessation of hostilities with exception *(see FM 27-10)* are not war crimes. These criminal acts will be investigated as are crimes that are committed in time of peace.

WAR CRIME INVESTIGATIONS

18-18. Investigations of war crimes are conducted when committed by an enemy against US personnel and when committed by US personnel against an enemy. One of the most important actions an investigator can take during the investigation of a war crime is to maintain steady communications with his chain of command. These communications keep the command informed and allow the command the precise moment of knowing if and when to pass the investigation on to another investigative agency. An example of such an investigation is that of a suspected series of murders that turns into a suspected genocide. The investigation is required to move out of the hands of investigators and into the investigatory purviews of the United Nations (UN), although investigators may continue to support the UN in the investigation.

18-19. When investigating war crimes, investigators work closely with the office of the SJA, which answers to the commander for the administration of war crime matters. Investigators also collect and report pertinent police intelligence to military intelligence, counterintelligence, and other investigative agencies of the United States and the HN.

INITIATING THE INVESTIGATION

18-20. When a CID investigator receives a report of a war crime, he takes the same steps that he does in the notification of any criminal activity. He initiates *CID Form 28 (Agent Activity Summary [AAS])* and annotates the date, time, and details of the notification. Each time the investigator makes a related action to the case, he annotates it in detail on *CID Form 28.* When the investigator makes or observes significant achievements during the investigation, he annotates these results on *CID Form 94 (Agent Investigation Report [AIR]).* This form is compiled by the agent and is attached to the final CID report that is submitted to a suspect's commander. In addition to these two forms, the investigator maintains a daily journal in which he records, in abbreviated style, the significant operational decisions and developments of the day. (*See CID Regulation 195-1* for prescribed formats.)

18-21. War crime investigations are conducted using standard investigative techniques that include—

- Protecting and processing the crime scene using standard procedures documented by the investigator's notes, photographs, and sketches.
- Collecting, marking, packaging, and shipping evidence.
- Interviewing witnesses.
- Identifying suspects.

18-22. At war crime scenes, investigators must be aware of potential environmental hazards, such as areas devastated by war that may have unexploded munitions present. Investigators must exercise due caution in moving in and around the scene and ensure that onlookers are carefully removed from the scene. The onlookers could be potential witnesses and subsequent claimants and must be evaluated as such.

18-23. Investigators will not discuss claims with potential claimants. It is advisable to have SJA personnel available to answer any claims questions from victims and their families that arise against the US government for injuries sustained by war crimes.

OVERCOMING INTERVIEW OBSTACLES

18-24. Standard interview techniques must be modified with war crime survivors and witnesses. A very prominent problem that must be overcome in these interviews is the language barrier. The investigative team must have experienced, reliable, and competent interpreters. Interpreters must be able to convey the attitude and personality of the investigators. It is best if part of the investigative team is fluent in languages of those being interviewed. Investigators will be able to convey their own ideas and thoughts much more clearly to the interviewees. US interpreters from a military intelligence unit or the supporting unit may be used; however, these US interpreters lack USACIDC background, investigative experience, and the ability to reflect the investigator's personality.

18-25. A less effective alternative is to use a local national. Even with good language skills, he may hurt the investigation. Local national interpreters are often indifferent to the outcome of an investigation. They may have no patience with very old, young, or confused witnesses. Local nationals are often unreliable or do not associate themselves with the USACIDC mission. Like US interpreters, they lack USACIDC background and investigative experience and are often unable to reflect the personality of the investigators for whom they interpret. If United States or local national interpreters must be added to the USACIDC investigation team, provisions for payment, billeting, and messing must be made. Only interpreters of the highest caliber should be selected for the mission, and an extensive background check should be considered.

18-26. A second problem bearing on interviews involves cultural differences. Interviewees may be fearful and apprehensive, and they may be illiterate and lack sophistication. In some parts of the world, every day points of reference are nonexistent, such as standard units of measure, western calendars, and directional points of the compass. Consideration must be made to overcome these differences.

18-27. A third problem stemming from investigators is the human tendency to be less disciplined and systematic in questioning when conducting a long series of interviews on the same topic. The chief investigator can help eliminate this issue by soliciting the entire investigative team's help in preparing a comprehensive list of key questions to ask during the interviews. This list will elicit the most complete statements from the interviewees and will ensure uniformity of coverage from one interviewee to the next.

CHOOSING AN INTERVIEW SITE

18-28. Choosing an interview site is important to the results of the interview and can help offset some of the above problems. Interviews are best done in an atmosphere near the witnesses' homes, which allows many more witnesses including the very young and the very old to be questioned. USACIDC funds can be used to provide cigarettes, gum, and like items for interviewees and will also be used to supply the investigative team with the national currency. With this currency, investigators can ensure interviewees that they will be reimbursed for out-of-pocket expenses incurred incidental to the interviews.

PART FOUR

Evidence

Part Four lays the foundation for evidence management and control and describes the important role of the evidence custodian. *Chapter 19* establishes proper procedures for maintaining the physical integrity of evidence and lawfully seized items for temporary and permanent custody. The remainder of the chapters in Part Four describe the collection, preservation, and forensic analysis of evidence. Additionally, these chapters are intended to educate the reader on the many innovative forensic and technical services provided by USACIL.

Chapter 19

Evidence Management and Control

This chapter describes the evidence custodian's responsibilities for managing and controlling evidence. In addition to the guidance provided by *AR 195-5,* law enforcement managers are encouraged to develop local SOPs that serve as quick references for evidence management and control. Patrol and security personnel, desk sergeants, and investigators should be familiar with the procedures for properly receiving, handling, accounting for, and storing evidence.

OVERVIEW

19-1. Evidence is the legal data that conclusions or judgments may be based on. It is the documentary or verbal statements and material objects admissible as testimony in a court of law. There is an increased demand for scientific evidence since it is objective by nature and not subject to confusion, memory lapse, and perjury like human witnesses. To admit lawfully seized items into evidence at a trial, they must be authenticated. Critical to authentication of evidence is a well-documented chain of custody. Often, the person obtaining the evidence, such as the military police desk sergeant and the evidence custodian all share this responsibility.

19-2. Military police patrols and security personnel obtaining evidence are required to accurately list all evidence on a *DA Form 4137* and properly mark and tag evidence. The desk sergeant ensures that a correctly completed voucher, tag, and MPR accompanies all evidence received at the military police desk. He safeguards the evidence until it is properly released to an investigator or the evidence custodian.

PRIMARY AND ALTERNATE EVIDENCE CUSTODIANS

19-3. For both military police and USACIDC activities, the evidence custodian is appointed in writing and should meet the requirements of *AR 195-5*. Appointment orders will be kept in the evidence room under file number 310-2c. They will remain on file until the primary and alternate custodian no longer retain their positions. The evidence custodian ensures the integrity and physical characteristics of property and evidence items for temporary and permanent custody. He maintains a continuous permanent record showing the chain-of-custody signature log of all items under the control of the evidence room. He is responsible for all necessary maintenance of the evidence until authorization is granted for proper disposal. The evidence custodian also manages all temporary releases of evidence and, when required, the submission of evidence to USACIL for laboratory examination.

REPOSITORY GUIDELINES

19-4. The evidence custodian ensures that a *DA Form 4137* is correctly completed before accepting the actual evidence. Physical evidence will be released to the evidence custodian no later than the first working day after it is acquired. If the evidence room is geographically in a separate location, it will be released physically or through registered mail no more than two working days after it is acquired.

19-5. If the evidence is suspected drugs, the evidence custodian should check to see if a completed *CID Form 36* is attached and that the field test is recorded properly (if conducted). *Figures 19-1* and *19-2* is a sample of a completed *DA Form 4137*.

19-6. Once the evidence custodian receives a *DA Form 4137*, a document number is assigned to it (not the item). The evidence custody document number will consist of the document number "001", which indicates that it is the first evidence custody document for the investigating office followed by the calendar year, such as "03". Subsequent *DA Forms 4137* are given document numbers in increasing order, such as 001-03, 002-03, and 003-03.

19-7. Evidence is then assigned a location within the evidence room. This location is written in pencil at the bottom of *DA Form 4137*. This is done in case the evidence is moved to a new location within the evidence room.

19-8. The evidence custodian will retain the original and the first copy of *DA Form 4137*. The second copy will be given to the investigator or military police to be placed in the proper case file.

19-9. The original and first copy of the evidence vouchers will be placed in a file no thicker than 3/4 inch and will contain no more than 50 vouchers. The number and year of the documents in the folder will be shown on the outside, such as 01-02 through 50-02. When a new folder is needed, it should be marked, such as 51-02 through 100-02. As many folders as required should be made, but should not exceed 50 vouchers per folder.

19-10. The original *DA Form 4137* will accompany the evidence when it is temporarily released from the evidence room. When evidence must be released to the USACIL or another authorized agency, the first copy of *DA Form 4137*

EVIDENCE/PROPERTY CUSTODY DOCUMENT For use of this form see AR 190-45 and AR 195-5; the proponent agency is US Army Criminal Investigation Command	MPR/CID SEQUENCE NUMBER 0005-85-CID063 CRD REPORT/CID ROI NUMBER 0005-85-CID063-46846
RECEIVING ACTIVITY Fort McClellan FO, Third Region, USACIDC	LOCATION Fort McClellan, AL 36205-5000
NAME, GRADE AND TITLE OF PERSON FROM WHOM RECEIVED ☐ OWNER ☒ OTHER Crime Scene	ADDRESS *(Include Zip Code)* N/A
LOCATION FROM WHERE OBTAINED 3673-B Church Road (Living Room) Fort McClellan, AL 36205	REASON OBTAINED: Evidence TIME/DATE OBTAINED: 0900 to 1130, 2 Jan 85

ITEM NO.	QUANTITY	DESCRIPTION OF ARTICLES *(Include model, serial number, condition and unusual marks or scratches)*
1	1	Baseball, approximately 2 1/2" in diameter, brand name Wilson, white and red in color, leather like construction, scuffed, marked under the brand name Wilson, PGP, 2 Jan 85, 0900. (Right top desk drawer)
2	1	Drinking glass, about 4 inches in height, clear in color, glass construction; marked on bottom of glass PGP, 2 Jan 85, 0904. (Right middle desk drawer)
3	1	Bottle labeled Jim Beam, quart size, about 10 inches in height, no visible contents, clear in color, glass construction, marked on bottom of bottle PGP, 2 Jan 85, 0905. (On top of coffee table)
4	1	Cigarette butt, about one inch long, white in color, partially burnt and flattened condition, labeled Kent, placed in vial, both vial and seal marked PGP, 2 Jan 85, 0915 0005-85-CID063. (In ash tray on coffee table)
5	1	Bag, about four by five inches, brown paper construction, approximately half full, containing suspected marihuana, placed in a clean heat seal bag, and heat sealed, marked on bag PGP, 2 Jan 85, 1130 0005-85-CID063. (Left top dresser drawer)
XXXXXXXXXXXX	XXXXXXXXXXXXX	XXXXXXXXXXXXXLAST ITEMXXXXXXXXXXXXXXXXXXXXXXXXXXXXXXXX

CHAIN OF CUSTODY

ITEM NO.	DATE	RELEASED BY	RECEIVED BY	PURPOSE OF CHANGE OF CUSTODY
	2 Jan 85	SIGNATURE: NA NAME, GRADE OR TITLE: NA	SIGNATURE: Peter G. Paul NAME, GRADE OR TITLE: Peter G. PAUL, SA	Evaluation as Evidence
	2 Jan 85	SIGNATURE: Peter G. Paul NAME, GRADE OR TITLE: Peter G. PAUL, SA	SIGNATURE: Roger R. List NAME, GRADE OR TITLE: Roger R. LIST, SA	Rel to Evidence Custodian "SCRCNI"
	3 Jan 85	SIGNATURE: Roger R. List NAME, GRADE OR TITLE: Roger R. LIST, SA	SIGNATURE: Reg Mail NAME, GRADE OR TITLE: #1234	Fwd to USACIL-☐ for exam
	6 Jan 85	SIGNATURE: Reg Mail NAME, GRADE OR TITLE: #1234	SIGNATURE: Ralph E. Davis NAME, GRADE OR TITLE: Ralph E. DAVIS, CW3	Rec'd at lab for exam
	20 Jan 85	SIGNATURE: Ralph E. Davis NAME, GRADE OR TITLE: Ralph E. DAVIS, CW3	SIGNATURE: Reg Mail NAME, GRADE OR TITLE: #5678	Ret to Submitter

DA FORM 4137, 1 JUL 76 Replaces DA FORM 4137, 1 Aug 74 and DA FORM 4137-R, Privacy Act Statement 28 Sep 78 Which are Obsolete LOCATION Bin 5 DOCUMENT NUMBER 7-85 USAPPC V1.00

Figure 19-1. Sample DA Form 4137, Page 1

CHAIN OF CUSTODY *(Continued)*

ITEM NO.	DATE	RELEASED BY	RECEIVED BY	PURPOSE OF CHANGE OF CUSTODY
	24 Jan 85	SIGNATURE: Reg Mail NAME, GRADE OR TITLE: #5678	SIGNATURE: Roger R. List NAME, GRADE OR TITLE: Roger R. LIST, SA	Rec'd by Evidence Custodian
	14 Feb 85	SIGNATURE: Roger R. List NAME, GRADE OR TITLE: Roger R. LIST, SA	SIGNATURE: Peter J. Kane NAME, GRADE OR TITLE: Peter J. KANE, MAJ, JAGC	Rel to TC for Court
	15 Feb 85	SIGNATURE: Peter J. Kane NAME, GRADE OR TITLE: Peter J. KANE, JAGC	SIGNATURE: Roger R. List NAME, GRADE OR TITLE: Roger R. LIST, SA	Ret to Evidence Custodian
	8 Mar 85	SIGNATURE: Roger R. List NAME, GRADE OR TITLE: Roger R. LIST, SA	SIGNATURE: Paul J. Kelley NAME, GRADE OR TITLE: Paul J. KELLEY, CW3, USA	Ret to owner Final Disposition
	8 Mar 85	SIGNATURE: Roger R. List NAME, GRADE OR TITLE: Roger R. LIST, SA	SIGNATURE: Item 4,5 Burned/Item 3 Crushed NAME, GRADE OR TITLE: DESTROYED	Final Disposition
		SIGNATURE NAME, GRADE OR TITLE	SIGNATURE NAME, GRADE OR TITLE	
		SIGNATURE NAME, GRADE OR TITLE	SIGNATURE NAME, GRADE OR TITLE	
		SIGNATURE NAME, GRADE OR TITLE	SIGNATURE NAME, GRADE OR TITLE	

Sample

FINAL DISPOSAL ACTION

RELEASE TO OWNER OR OTHER *(Name/Unit)* Items 1 and 2, CW3 Paul J. KELLEY, Co B, HQ Comd, Ft. McClellan

DESTROY Items 3, 4, and 5

OTHER *(Specify)*

FINAL DISPOSAL AUTHORITY

ITEM(S) 1 thru 5 ON THIS DOCUMENT, PERTAINING TO THE INVESTIGATION INVOLVING PFC *(Grade)*

John S. DOE *(Name)* Co A, 1st Bn, 5th Tng Bde, Ft. McClellan, AL *(Organization)* (IS) (ARE) NO LONGER

REQUIRED AS EVIDENCE AND MAY BE DISPOSED OF AS INDICATED ABOVE. *(If article(s) must be retained, do not sign, but explain in separate correspondence.)*

Hugh H. JOYCE, CPT, JAGC *(Typed/Printed Name, Grade, Title)* *(Signature)* 7 Mar 85 *(Date)*

WITNESS TO DESTRUCTION OF EVIDENCE

THE ARTICLE(S) LISTED AT ITEM NUMBER(S) 3, 4, and 5 (WAS) (WERE) DESTROYED BY THE EVIDENCE CUSTODIAN, IN MY PRESENCE, ON THE DATE INDICATED ABOVE.

SA Hubert L. HARRISON, Ft. McClellan Field Office *(Typed/Printed Name, Organization)* Hubert A. Harrison *(Signature)*

USAPPC V1.00

Figure 19-2. Sample DA Form 4137, Page 2

will be separated from the original and retained in the appropriate suspense folder. The original *DA Form 4137* will accompany the evidence being released.

SUSPENSE FOLDERS

19-11. The following three suspense folders are required in the evidence room:

- **USACIL.** The first copy of the evidence voucher that is sent to the crime laboratory will be placed in this folder.
- **Adjudication.** This folder is for evidence that is temporarily released to Article 32 investigating officers, courts, SJA officers, or other legal proceedings.
- **Pending disposition approval.** The first copy of the evidence voucher is stored in this folder when the original is sent to the SJA for approval of disposition.

19-12. Other folders can be used if needed as management tools. The evidence custodian should establish a suspense system to maintain the status of evidence at all times and assist in the timely disposition of evidence.

19-13. After all of the items of evidence listed on *DA Form 4137* have been properly disposed of, the original evidence custody document will be placed in an inactive evidence custody document file under file number 195-5a. These files are established for the original evidence custody documents closed during each calendar year, rather than the numerical sequence method of controlling active evidence custody documents.

19-14. The closed evidence custody document files will be held in the evidence room or the administrative files area in the office, and disposed of as specified in *AR 25-400-2*. Once the original evidence custody document has been placed in the inactive evidence custody document file, any remaining copies may be destroyed.

SENDING EVIDENCE

19-15. When evidence has been identified for USACIL examination, the case agent will provide the evidence custodian with the original and three copies of a completed *DA Form 3655 (Crime Lab Examination Request)*. Refer to the sample *DA Form 3655* (*Figures 19-3* and *19-4*, page 19-6). The original and one copy of the lab request is sent to USACIL with the evidence. One copy is retained in the USACIL suspense folder with the first copy of *DA Form 4137*. One copy (annotated with the registered mail number and the date the evidence was forwarded to USACIL) is returned to the case agent to be placed in the case file.

19-16. Upon mailing the evidence, the evidence ledger should be updated to reflect the submission of the items of evidence to USACIL. This entry should be recorded in pencil to allow erasure upon the return of the item from USACIL.

CRIME LAB EXAMINATION REQUEST
For use of this form, see AR 195-6; the proponent agency is the United States Army Criminal Investigation Command.

TO: *(Include Zip Code)*	FROM: *(Include Zip Code)*	LAB USE ONLY
Laboratory Director USACIL 4553 N 2nd St Forest Park, GA 30050-5122 ATTN: Latent Print Div. Firearms Div.	Special Agent in Charge Fort Leonard Wood RA (CID) 6th Military Police Group CID USACIDC Fort Leonard Wood, MO 65473 Fax: 856-6388 e-mail: wim5396@cid037	REFERRAL NUMBER RECEIVED: REGIS MAIL / RY EXP / HAND / DATE RETURNED: REGIS MAIL / RY EXP / HAND / DATE RECEIVED BY EVIDENCE RECEIPT: RECEIVED / INITIATED

1. CONTRIBUTOR CASE NUMBER	2. INVESTIGATOR'S NAME	3. AUTOVON AND PHONE NUMBER
0123-02-CID053-xxxxxx	SA William J. MAC	DSN: 856-3308/3932

4. SUSPECT(S) *(Last, first and middle name(s))*
Unknown

5. VICTIM(S) *(Last, first and middle name(s))*
WILLIAMS, Janet A.

6. TYPE OF OFFENSE	7. ONE COPY OF EVIDENCE RECEIPT ENCLOSED WITH EVIDENCE	8. OTHER EVIDENCE PREVIOUSLY SUBMITTED ON THIS CASE
Undetermined Death	☒ YES ☐ NO	☐ YES ☒ NO

9. IF "YES" IN ITEM 8, LIST OTHER SUSPECT(S), DATE SUBMITTED, UNIT CASE AND LABORATORY REFERRAL NUMBER(S)

Sample

10 EVIDENCE SUBMITTED

a. EXHIBIT	b. DESCRIPTION OF EXHIBIT
1	Drinking Glass (Item 4, Document# 014-02)
2	Pistol, SN# 854435 (Item 2, Document # 014-02)

DA FORM 1 AUG 74 3655 REPLACES DA FORM 3655-R, 1 NOV 70, WHICH IS OBSOLETE.

Figure 19-3. Sample DA Form 3655

10. EVIDENCE SUBMITTED (Continued)	
a. EXHIBIT	b. DESCRIPTION OF EXHIBIT

Sample

11. EXAMINATION(S) REQUESTED (*Briefly furnish any information or instructions that might assist the laboratory in examining, evaluating or returning evidence and/or report*).

Please See Attached Report for Additional Information

Latent Print Division: Examine exhibit 1 to determine if there are any identifiable latent prints, if so, please retain on file for future reference.

Firearms Division: Examine exhibit 2 to determine if it is in working order and has been fired since the last cleaning.

Conduct any additional examination you deem as appropriate.

Evidence above has not been examined by another expert in the same scientific field.

TYPED/PRINTED NAME OF REQUESTOR	SIGNATURE	DATE
SA William J. MAC	William J. Mac	2 Jan 02

Figure 19-4. Sample DA Form 3655, Reverse Side

RETURNING EVIDENCE

19-17. Upon the return of the evidence from USACIL, the evidence custodian must verify that the items returned from USACIL match the items on the evidence custody documents included with the evidence. Mistakes by USACIL in returning the proper items of evidence are rare, but they can occur. Complete the chain of custody on both the original and first copy (suspense copy) of the evidence custody documents. Return the evidence to the proper bin or file, and refile the evidence custody document. The evidence custodian must then update the evidence ledger to show the return of the evidence and its location in the evidence room (if recorded in the "remarks" column).

19-18. When either a portion of a controlled substance or an entire item of evidence is consumed during analysis, the evidence custodian will ensure that a copy of the laboratory examination report documenting the amount of evidence consumed and returned is attached to the evidence custody document.

19-19. When a controlled substance (evidence) is returned to the evidence custodian after temporary release for other than laboratory examination, any apparent changes in the substance will be annotated in the "Purpose of Change of Custody" column on *DA Form 4137*. A memorandum for record (MFR) explaining the apparent changes will be prepared and attached to the evidence custody document by the evidence custodian. If the change is due to the conduct of a field test, a copy of the field test form will be attached and no MFR is required.

19-20. Whenever documents, such as laboratory reports, field test results, and memorandums are attached to an evidence custody document, they become a permanent part of the chain of custody for that evidence. All documents must be maintained with the evidence custody document, even after the final disposition has been completed and the inactive evidence custody document placed in the closed evidence custody document file.

19-21. MFRs and/or evidence custody documents generated by USACIL examiners are not required for items otherwise exempt from evidence custody documents. They are also not required for evidential material standards or controls recovered or removed from items properly accounted for on existing evidence custody documents (such as hair, fibers, debris, pieces of questioned fabric with stains, fabric standards, soil, paint, and glass fragments). Such evidence will be properly packaged, preserved, and returned to the original container. If practical, this evidence should be physically attached to the item from which it was removed or recovered. For example, hair collections from a sheet should be placed in a heat-sealed bag and stapled to the sheet. In this way, the item is returned essentially as it was received with each container holding the original item and all appropriate samples collected or derived, as appropriate.

WRAPPING, PACKING, AND TRANSMITTING EVIDENCE TO THE LAB

19-22. Each item of evidence within the shipping container should be in a separate, sealed container and marked appropriately with an individual's initials and the date and time. Failure to adhere to this procedure may result in the loss and/or contamination of evidence and problems with the chain of custody. Each item should be packed in a way that will minimize friction and prevent the item from shifting, breaking, leaking, or contacting other evidence. (See *Figure 19-5, a.*) Evidence should be packed in cotton or other soft paper items (not susceptible to being broken, marred, or damaged). Each packaged and sealed item should be labeled using a *DA Form 4002* that corresponds with the entry on *DA Form 4137*. The individually sealed items of evidence should be placed in a shipping container (envelope or box). The envelope or box should be sealed with paper tape. The individual's initials or signature should be placed across the paper tape used to seal the shipping container and then covered with transparent tape. (See *Figure 19-5, b.*)

19-23. Supporting documentation, such as photographs, sketches, copies of the initial report, and victim and subject statements (when available) are often very useful to the laboratory examiner (particularly in violent crimes) and should be forwarded to the laboratory with a *DA Form 3655*. These documents frequently give the examiners vital information that they would not necessarily get from the laboratory request. Supporting documentation should be attached or packaged with *DA Form 3655* and placed on the outside of the shipping container. (See *Figure 19-5, c* and *d.)* These documents should not be placed in the shipping container because the documentation is reviewed at the division level before the case is assigned to an examiner and before the evidence is picked up and inventoried.

19-24. The original and one copy of *DA Form 3655* and the original *DA Form 4137* obtained from the evidence custodian should be placed in an envelope. The envelope should be sealed and addressed to the laboratory with an attention line to the specific division, such as the document, fingerprint, or firearm division. This sealed envelope should be taped securely to the box or envelope containing the evidence. (See *Figure 19-5, e.*) The box should be wrapped in heavy paper or the envelope should be sealed inside another envelope. Address the package to USACIL. (See *Figure 19-5, f.*)

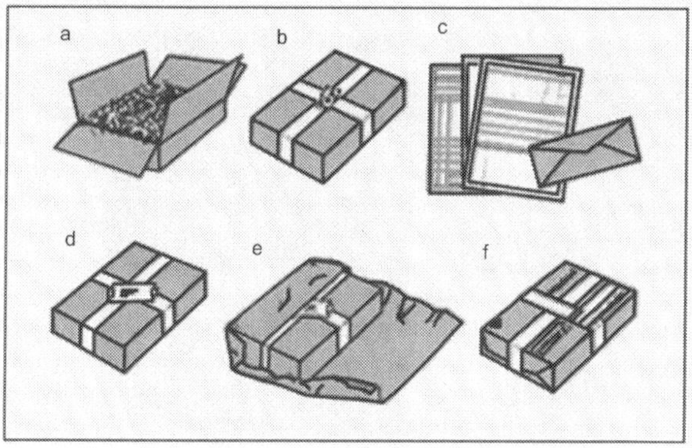

Figure 19-5. Wrapping and Packaging Evidence

19-25. Packages containing items of evidence that require careful or selective handling while in transit should be labeled: "corrosive," "fragile," "keep away from fire," or "keep cool" (as appropriate). Evidence needing refrigeration can be damaged or destroyed if left unattended in a post office over a weekend.

19-26. The method for transporting evidence to the crime lab depends on the type of evidence and the urgency of need for the results. Evidence is normally sent by registered mail but may be hand-carried or transported by government carrier. Federal law prohibits transmitting certain types of merchandise through postal channels. The nearest postmaster should be consulted if there is any question on mailing. Coordination should also be made with the USACIL before these items are forwarded.

19-27. Chemicals, gases, unexploded bombs, detonators, fuses, blasting caps, and other explosive or inflammable materials cannot be sent by mail. Transmittal of these items of evidence must conform to the provisions of *US Postal Service (USPS) Publication 52*, CFR, OSHA guidelines, interstate commerce regulations, and appropriate state and municipal ordinances. Before these items are forwarded, the laboratory should be notified about the planned shipment. The notification should include instructions on packing the materials. This will reduce the danger involved in unpacking these items at the laboratory.

EVIDENCE LEDGER

19-28. All evidence logged into the evidence room is recorded in an evidence ledger. The evidence ledger is the bound record book (federal serial number [FSN] 7530-00-286-8363) kept by the evidence custodian. The evidence ledger provides double accountability through cross-reference with *DA Form 4137*. The evidence ledger accounts for the document number assigned to *DA Form 4137* and *DA Form 4002*. The evidence ledger follows the disposition schedule set by *AR 25-400-2*. The ledger is destroyed 3 years after a final disposition is made of all items entered in it.

19-29. Evidence ledger books can be used for more than a single calendar year. Smaller offices that do not receive large quantities of evidence use ledger books for several years. A single book is used for consolidated evidence rooms, recording evidence from different offices in the same numerical sequence that it was received in the evidence room.

19-30. The cover of the evidence ledger book is labeled and marked with Modern Army Record-Keeping System (MARKS) file number 195-5a, the organization or activity responsible for the evidence room, and the dates spanned by the ledger book entries.

19-31. The evidence ledger is prepared with six columns that span two facing pages when the book is opened. Each page does not have to show the column headings. As a minimum, the first page of the ledger and the first page beginning a new calendar year should show the column headings. A commonly used practice is to mark the column headings on the inside covers of the book, then trim off the top of the ledger pages so the headings are visible when the book is opened. Ledger headings include—

- **Document number and date received.** This column contains the document number assigned to the evidence custody document. The date *DA Form 4137* was received in the evidence room is entered below this number.
- **Criminal Investigation Division control number or military police report number.** The office investigative file number assigned to the investigation to which the evidence pertains is entered in this column. A USACIDC evidence custodian will also enter the MPR number in the remarks column when the evidence pertains to both a CID and military police investigation.
- **Description of the evidence.** A brief description of the evidence is entered in this column, such as ".38 cal pistol, SN# 000000," or "25 handwriting exemplars of SGT Smith." A single line in the ledger is used to record each single item of evidence. The same item number listed on *DA Form 4137* will identify each item of evidence in the ledger. Items that are considered fungible or other evidence that is sealed in a container will be described briefly with information taken from the data on *DA Form 4137*. This entry does not imply that the evidence custodian has inventoried these items.
- **Date of final disposition.** The date the evidence was disposed, as shown in the "Chain of Custody" section of *DA Form 4137,* is entered in this column. When *DA Form 4137* contains several items that are not disposed of on the same date, the date of disposition for each item should be recorded on the same line. When all the items in an entry are disposed of on the same day, only one date is required to be entered, followed by the words "all items" (such as 12 Jan XX - all items).
- **Final disposition.** The method of final disposition for each item of evidence is entered in this column on the same line. When all items in the entry have been disposed of in the same manner, the means of disposal may be listed once, preceded by the words "all items" (such as "all items burned"). Both the date and method of final disposition must match what is shown in the chain-of-custody portion of the evidence custody document for each item of evidence.
- **Remarks.** The evidence custodian may use this column to record necessary information. When fungible or other evidence is received in a sealed container and cannot be inventoried, the notation "SCRCNI" is required in this column. If .0015 funds are retained as evidence, ".0015 funds" must also be noted in this column. The other types of entries are optional and commonly made in pencil to facilitate changing. Some examples include the names of the case agent, owners, subjects, victims, location of evidence in the evidence room, and the dates to and/or from USACIL for examination.

19-32. Both vertical and horizontal lines are used to separate entries. Black ink should be used to make the entries. The lines separating the entries may be in a different color. No blank pages or lines will be left between ledger pages or ledger entries. Erroneous entries will be lined through with a single line, so they may still be read and initialed by the evidence custodian. Any spaces between ledger entries or at the bottom of a ledger page will be lined through and the word "VOID" written in capital letters in the space and initialed by the custodian. No white correction fluid will be used in the evidence ledger to conceal mistakes.

EVIDENCE DISPOSITION

19-33. Evidence will be expeditiously disposed of as soon as possible after it has served its purpose or has no further evidentiary value. Methods of disposition are detailed in *AR 195-5*.

19-34. Evidence returned to the owner for final disposition requires a receipt for registered mail. The original *DA Form 4137* will not be sent with the evidence. When the receipt for registered mail is returned, it will be permanently attached to the original *DA Form 4137*.

19-35. The final disposition of evidence also has particular requirements when sending it to another office. When evidence is permanently transferred from one evidence room to another, the original and a duplicate copy of the evidence custody document (with properly annotated chains of custody) will be sent with the evidence. The transferring office will make a legible copy of the original evidence custody document, which will be placed in the closed evidence custody document file in place of the original copy of the evidence custody document. The evidence custodian who receives this evidence will enter the next document number of the receiving evidence room on both copies of *DA Form 4137*. The previous document number will be lined through so that the number is still legible.

19-36. Sometimes evidence is made a permanent record of trial. Evidence released to the TC for judicial proceedings will be returned as soon as possible to the custodian for final disposition. If an item of evidence is made part of the record of trial, the TC will immediately notify the custodian so *DA Form 4137* can be annotated properly. This will be considered the final disposition of the evidence.

19-37. Evidence obtained by investigators or military police at a crime scene or during an investigation may be released to another law enforcement agency without SJA approval when that agency assumes full investigative jurisdiction and responsibility or jurisdiction and responsibility for just that portion of the investigation to which the evidence pertains. The evidence does not need to be processed into the evidence room before release unless the release cannot be made before close of business on the first working day after the evidence is acquired. Investigators and military police will ensure that the evidence is properly recorded on a *DA Form 4137* and that the chain-of-custody section is complete upon release of the evidence.

19-38. The USACIDC commander, SAC, or PM (in the case of military police investigations) will review and give approval of the release of the evidence by completing the final disposition authority section of the evidence custody document. Since the final disposition approving authority may not be available to give prior authorization for the release of evidence, this approval may be given after the fact with the final disposition authority annotated on the final copy of the evidence custody document. A copy of the evidence custody document will be maintained in the case file and release of the evidence will be documented in the report. Consideration should be given to photographing the evidence before its release.

19-39. Sometimes evidence is of no value after laboratory analysis. These items (contained in the evidence room) may be disposed of after consulting with the proper SJA or civilian prosecutor if a subject has been identified or upon approval of the USACIDC commander or SAC in an investigation with no subject identified.

19-40. Controlled substances received by the evidence custodian that do not apply to an investigation may be immediately disposed of after approval has been received from the USACIDC commander, SAC, or PM (as appropriate). A copy of *DA Form 4137* will be filed with the appropriate police report. Disposition may be made immediately after determining that the substance cannot be linked to a suspect.

19-41. Items of potential evidence that are determined to have no value as evidence by the USACIDC special agent or military police before they are released to the evidence custodian may be disposed of by the agent or military police. This does not include controlled substances. The evidence custody document chain of custody will be completed to show the final disposition made for the item and the evidence custody document filed in the case file. The method of disposal will be according to *AR 195-5* (released to the unit or activity, returned to the rightful owner, or destroyed, as appropriate).

19-42. When it is not practical or desirable to keep items of evidence, such as vehicles, serial-numbered items, items required for use by the owner, undelivered mail, large amounts of money, and perishable or unstable items, disposal action may be taken immediately. If such items can be immediately disposed of, it will not be necessary to enter them into the evidence room. This should be coordinated with the office of the SJA. The office of the SJA will complete the final disposal authority portion of *DA Form 4137*. When it is not possible to get written approval from the office of the SJA before the disposal of the evidence, oral permission should be obtained followed by written approval. Photographing the evidence before release should be considered and discussed with the office of the SJA.

19-43. When final action has been taken in known subject cases, the original evidence custody document will be taken by the evidence custodian or sent to the SJA officer appointed as TC for the commander of the subject or the civilian prosecutor if the subject is a civilian. An SJA or civilian prosecutor will complete the final disposal authority part of *DA Form 4137* if the evidence is no longer needed. When evidence must be retained, this part of the form will not be completed; a brief statement giving the reason for keeping the evidence will be furnished to the evidence custodian using separate correspondence. In unusual cases, where there is a high risk of losing the original *DA Form 4137* (such as isolated units that must mail a *DA Form 4137* to the servicing SJA officer for disposition approval), a memorandum may be substituted for disposition approval. When this is used, enough information will be furnished to allow the SJA or civilian prosecutor to make a decision.

The return correspondence from the SJA or civilian prosecutor giving disposition approval will be attached to the original *DA Form 4137*.

19-44. Evidence in an investigation for which no subject has been identified may be disposed of 3 months after the completion of the investigation without the approval of the office of the SJA. Evidence may be disposed of earlier with approval of the office of the SJA. However, care must be taken with serious crimes when it is possible that a subject may later be identified. In such cases, it may be advisable to keep the evidence longer than 3 months. When the subject is not known (3 months after completion of the investigation), USACIDC evidence custodians will obtain final disposition approval of the commander or SAC. Military police evidence custodians will obtain disposition approval from the supported PM. Approval will be given by completing the final disposal authority section of the original *DA Form 4137*.

19-45. In some instances, such as unsolved rapes, homicides, or solved cases which will require a protracted appeals process resulting from a general courts-martial (GCM), evidence needs to be retained longer than usually required after the completion of a final report. In these situations, *AR 195-5* provides guidance on the packaging of evidence pertaining to one investigation in consolidated sealed containers. The evidence should be packaged in cardboard boxes in the presence of a disinterested witness. A certificate should be completed listing the evidence documents in the box (by document number) and the date the carton was inventoried and sealed. Then the evidence custodian and disinterested witness should sign the certificate. A copy of the certificate should be attached to the container. The evidence custodian and disinterested witness should sign the container seal according to *AR 195-5*. A copy of the certificate should be attached to each evidence custody document for evidence sealed in the container. The original certificate is attached to the original *DA Form 4137* of the first evidence custody document listed on the certificate. No firearms can be placed in the container. The evidence custody documents continue to be maintained in the open evidence custody document files. The container will remain sealed during inventories unless a breach of the seal or tampering is detected.

INSPECTIONS AND INVENTORIES

19-46. The evidence room is inspected once a month. The results of this inspection are recorded in the evidence ledger and signed by the inspector. This inspection will be made in lieu of a physical security inspection according to *AR 190-13* and is not required monthly if no evidence has been received, maintained, disposed of, or otherwise accounted for in the evidence room since the last monthly inspection. *Appendix C* provides guidelines and a checklist for evidence room inspections or inventories.

MONTHLY INSPECTION

19-47. The monthly inspection is intended to be an inspection of the evidence room operation, not an inventory. The inspector will determine if the evidence room is orderly and clean; structural and security requirements of *AR 195-5* are being met; and evidence is being received, processed, safeguarded, and disposed of according to *AR 195-5*. The inspector also checks to see if evidence temporarily released for laboratory examination or presentation at a judicial proceeding has been released for an excessive period of time. If so, an annotation is made in the evidence ledger and the custodian contacts the appropriate agency to inquire about the status of the evidence.

INVENTORIES

19-48. An inventory will be conducted of all the evidence stored in the evidence room and all temporary evidence facilities once a quarter, upon change of the primary evidence custodian, upon loss of evidence stored in the evidence room, or upon breach of security of the evidence room.

19-49. The evidence custodian and a disinterested officer will conduct the quarterly inventory jointly. The disinterested officer will not be a current member of USACIDC or an officer assigned to military police activities or units on the installation. The guidelines for disinterested officers conducting a joint inventory are established in *AR 195-5* and should be read by the inspecting officer before conducting the joint inventory. The disinterested officer will make a physical count of evidence to verify that evidence in the evidence room corresponds with that shown on *DA Form 4137*. The disinterested officer will cross-reference all *DA Forms 4137* (including those in suspense files) with entries in the evidence ledger to ensure the accountability of all evidence and proper annotation of the copies of *DA Forms 4137* in the suspense file.

19-50. When a primary evidence custodian changes, the incoming and outgoing primary evidence custodians will make a joint physical inventory of all evidence in the evidence room. An entry must be made in the ledger as detailed in *AR 195-5*. Upon satisfactory completion of the change of custodian inventory, all *DA Forms 4137* in the open evidence custody document files will be noted and signed to show the change of custody. When some items of evidence have been retained in the evidence room while other evidence was recorded on a voucher and sent to USACIL (or other external agency), the incoming primary custodian will sign the first copy of the evidence custody document retained in the USACIL suspense file to assume custody of the retained evidence. This situation creates another type of "legal gap" in the chain of custody, which is covered by the change of custodian entry recorded in the evidence ledger.

SECURITY STANDARDS

19-51. The outer door of the evidence room will be secured with an appropriate high-security key-operated padlock, never a combination lock. Combination locks of an approved design may be used inside the evidence room or on the inner door to the evidence room for security. Only the primary and alternate evidence custodians will know the combinations of locks used in an evidence room. Each key-operated padlock will have two keys. One key to each lock will be in the constant possession of the primary custodian. The duplicate key will be put in a separate sealed envelope and secured in the safe of the appropriate supervisor.

19-52. Lock combinations will be changed when the primary or alternate evidence custodian changes. All combinations and key-operated padlocks will be changed upon any possible compromise. Keys will be transferred from the primary custodian to the alternate only if the primary custodian is to be absent for more than one duty day or three nonduty days. Master- or set-keyed padlocks will never be used, in any capacity, in an evidence room. There will be no classified materials or property stored in an evidence depository that is not evidence. *AR 195-5* establishes the physical construction standards for evidence facilities.

19-53. The evidence room is required by the provisions of *AR 195-5* to contain certain items and other specific equipment to facilitate the proper handling of evidence. This equipment includes the following:

- At least one securable container for high-value items and large quantities of narcotics (more than one ounce of cocaine or heroin or more than one kilogram of marijuana).
- Bins, file drawers, or shelves for smaller amounts of evidence.
- An approved safe for the storage of weapons and ammunition.
- A refrigerator (required to be a permanent fixture in the evidence room) to store items such as blood and DNA samples.
- A work table and/or desk to process and package evidence.
- A separate General Services Administration (GSA)-approved safe for the storage of controlled-substance training aids (optional).
- Separate filing cabinets for the evidence custody document files (open and closed) and associated evidence room files.

TEMPORARY EVIDENCE CONTAINERS

19-54. A temporary evidence facility may be needed when investigators are operating in remote locations or other unusual situations. Normally, a safe or secure filing cabinet will be used for temporary storage of evidence. Access to the safe or filing cabinet will be restricted to the person securing the container. A key-opened padlock should be used instead of a combination padlock. One key should be secured in a separate envelope in the safe that contains the extra keys to padlocks to the evidence room. Master- or set-keyed padlocks are prohibited.

19-55. There should be one container per duty agent, depending on the volume of evidence handled, so that each duty agent has a container for use during the absence of the evidence custodian. This will limit the change in custody of evidence when duty personnel change and the custodian is not available.

19-56. Temporary containers will be secured to the structure or fastened together as specified for high-security containers located in the evidence room. A container express (CONEX) or military-owned demountable container (MILVAN) may be converted with specific security reinforcements, outlined in *AR 195-5*, to serve as temporary evidence storage facilities. A separate building or fenced enclosure may be used as long as access can be restricted and controlled by a single person. This is usually required only for unusually large items or quantities of evidence. Since the acquisition of these items cannot be predicted, evidence custodians should have a facility identified to use for this purpose.

Chapter 20

Fingerprints

Fingerprint evidence remains the most positive means of personal identification in forensics, to date. Though often compared with other modern innovations such as DNA, fingerprint evidence results in positive identifications whereas other evidence does not. Fingerprint evidence can also distinguish between identical twins. Identifications can be effected from fingerprints made in a victim's blood, paint, or other contaminants, which no other form of evidence can accomplish. There should not be competition between fingerprint evidence and other innovations because all should work hand in hand to solve identification questions.

DESCRIBING TYPES OF PRINTS

20-1. Latent prints can be seen or unseen and often require development. The word "latent" means "hidden," but normally the term latent prints refers to those prints left at crime scenes and/or on items of evidence. Another category of latent prints is patent prints. Patent prints are impressions that are visible in some form of contaminant. Plastic prints are those impressions left in materials, such as wax, window putty, or other pliable materials.

20-2. Record prints are the controlled recordings of the friction ridge skin contained on the palms of the hands and each finger, using various methods such as fingerprint cards, printer's ink, or electronic recording by way of "live scan." Though there are many variations on how to obtain record prints, the principles are the same. Traditionally, the term major case prints refers to finger and palm prints. Record prints can also be taken from feet, which also bear friction ridge skin. Record prints must be submitted for all victims, witnesses, suspects, medical and law enforcement personnel, and anyone known or suspected of handling evidence or entering a crime scene. Once legible and complete elimination prints for investigators are on file at USACIL, there is no requirement to resubmit record prints for each investigation conducted. In some cases, it may be necessary to record ear and lip prints for comparison. The laboratory should be contacted for guidance in these cases.

SEARCHING FOR, IDENTIFYING, AND PROCESSING LATENT PRINTS

20-3. Prints deposited on items of evidence are generally divided into the following two basic categories:

- **Porous evidence.** This type of evidence can absorb fingerprint residue into its surface. Porous evidence can be best described as a sponge that absorbs residue; for example, paper, checks, currency, unfinished wood, cardboard, and other similar material. These items do not require treatment by the crime scene investigators. In fact, the investigator should not attempt to process fingerprints on porous items of evidence because laboratory-processing procedures are best for this type of evidence. Clean gloves should be worn at all times when handling porous evidence. Little danger exists of destroying latent prints on porous evidence, but a high possibility does exist of accidentally depositing additional latents. All porous evidence should be placed in a paper envelope, bag, box, or wrapped in paper and sealed. The outside of the container should be marked with unique, identifiable markings. Investigators should be aware that the laboratory cautions against the field use of chemical agents commercially marketed for the development of latent prints on porous materials. Some of these products are of poor quality and can damage or destroy latent prints. Latents developed in the field can fade or totally disappear before laboratory examination. An example is latents that are developed using iodine and ninhydrin, which produce "fugitive" prints or prints that fade within a short period of time after initial development. If an investigator believes that a scene or evidence could best be processed using such chemicals, he should consult with USACIL for advice and guidance.
- **Nonporous evidence.** This type of evidence does not readily absorb water into its surface; for example, plastic bags, painted or sealed woods, metal, glass, some glossy magazine covers, knives, guns, computer equipment, and like materials. All nonporous evidence selected for latent-print examination should be processed as soon as possible. If ridge detail is visible, photographs should be exposed of them before any further processing. All nonporous evidence should then be exposed to superglue fuming. For most evidence, this fuming process could be all the processing necessary before shipment to the laboratory. In other circumstances, fumed latents can be photographed, powdered, and lifted. Do not submit evidence to the laboratory if the investigator has powdered the latent prints and they are capable of being lifted. Send only the photographs taken before lifting and the actual lifts. However, evidence requiring examination by other divisions of the laboratory should never be processed with fingerprint powder because contamination can hinder other examination processes. Evidence requiring additional examination merely needs to be fumed with superglue as soon as possible.

  NOTE: Superglue should not be used on any evidence being submitted for trace evidence examinations.

PRESERVING LATENT PRINTS

20-4. Latent prints on nonporous evidence are often deposited on the surface of an item and are extremely vulnerable. Wearing gloves does not protect the latent prints from being destroyed if they are touched, rubbed, or smeared; they only prevent additional prints from being deposited. When it cannot be determined from appearance whether a drop of water would be absorbed into a surface, the evidence should be handled and processed as nonporous (such as a leather wallet, cigarette cartons, and shiny cardboard boxes). Photography, superglue-fuming, and fingerprint-powdering are techniques used to preserve latent prints.

PHOTOGRAPHY

20-5. The very first step in latent print preservation is photography. Visible latent prints should always be photographed to prevent the loss of evidence. Latent prints deposited in grease, blood, paint, and other visible substances will often not require additional processing before photography. Always use a scale in evidence photography and steady the camera using a tripod. It is best to use a macro lens, filling the entire frame. Do not use digital photographic field-issued equipment; digital photography has not advanced technologically for the recording of latent print evidence. Traditional photography is still required for latent print evidence to be suitable for identification. If there is no other choice but to use digital photography, use maximum resolution (largest photo file size) settings combined with good lighting and a tripod.

20-6. Attempt to keep the back of the camera parallel to the surface bearing the latent print. If it is necessary to photograph the evidence from an angle to catch the light in a manner that increases the contrast of the latent print, additional photographs should also be made of the same area with the camera back parallel to the surface bearing the latent print.

SUPERGLUE-FUMING TECHNIQUE

20-7. Superglue fuming, or cyanoacrylate fuming, remains the most effective way to develop, protect, and preserve latent prints on nonporous evidence. The superglue-fuming process can be accelerated using heat or chemicals. USACIL suggests heat to accelerate the fuming process. After the latent prints are developed, package and ship them to the laboratory. Studies conducted by USACIL have shown that latent prints on evidence that was superglue-fumed in the field are preserved better and have a significantly greater chance of being identified than latent prints not superglue-fumed, but forwarded to the laboratory as found. Superglue fuming preserves latent print evidence making it stable for shipment to the laboratory without any further processing. It can simply be placed in an envelope, bagged, or wrapped in paper without special packaging materials and shipped to the laboratory. Superglue-fuming procedures are as follows:

- Evidence should be placed in a suitably sized, sealed container and in an area that is well-ventilated.
- A test print should be placed in a container where it can be seen or checked. A small piece of a clear plastic bag will work.

- A few drops of liquid superglue should be put on a piece of foil or laboratory tin and placed on a coffee cup warmer, or a similar heat source, inside the container.
- The evidence should be observed—this is critical. When the test print has developed, any latent prints on the evidence will also develop.
- The evidence should be removed from the container and placed in a paper envelope, bag, box, or wrapped in paper to be transported to the laboratory.

20-8. Large items of evidence can be fumed in much the same way. The investigator may have to build a makeshift tent or enclosure to seal in the evidence. Latent prints developed with superglue fuming on large or immovable items of evidence should be dusted with fingerprint powder, photographed, and lifted. Only the lifts should be sent to the laboratory. This effort saves shipping and handling costs of large bulky items.

20-9. Caution should always be used to ensure the safety of investigators who are using this fuming process. Superglue fumes should not be inhaled or exposed to the investigator's eyes, especially if he is wearing contact lenses because these situations can create medical illnesses.

FINGERPRINT-POWDERING TECHNIQUE

20-10. The traditional fingerprint-powdering technique is still a vital piece of the identification and preservation process of fingerprint evidence. The preferred method of recovering latent fingerprints from a crime scene (especially those that are located on large, bulky, or immovable items) is to superglue fume it first and then powder and lift the latents.

20-11. Many latent prints can be developed and preserved using a fingerprint brush and powder. All latent prints developed with a brush and powder must be photographed (with a scale) before lifting. Latent prints found in dust, grease, blood, or other contaminants should not be processed using fingerprint powders. Fingerprint powders are supplied in crime scene kits in several colors. In most instances, the best powders to use are the black or gray general-purpose powders. Always choose a powder that contrasts best with the background of the evidence and the color of the lifter used. Fluorescent powders can be used to develop latent prints on multicolored surfaces. These powders require the use of an alternate light source or UV light to be able to photograph. Effective use of these light sources requires training and experience. They are very costly and can cause health issues. Only long-wave, UV light should be used; short-wave, UV light is harmful to the eyes and skin. Anytime UV light is used to develop latent prints, investigators must wear protective goggles and clothing.

20-12. Many types of fingerprint brushes are used to apply fingerprint powder. Examples of these brushes are fiberglass, animal hair, and feather brushes. For overhead work or in situations where it is critical that the brush elements do not come in contact with the surface, magnetic wands and magnetic powders are used. The procedures for using fingerprint powders are as follows:

- Check the surface first using a test print. Lightly brush an area away from the subject surface and determine if any latent prints are present. If none are present, wipe the surface and apply and process a test print to determine how acceptable the surface is to the fingerprint powder processing. The investigator can make a test print by wiping an ungloved finger on his face or neck to collect skin oils. He should apply his finger to the test surface to deposit a latent fingerprint.
- Pour a very small amount of powder out onto a sheet of paper. Never dip the fingerprint brush into the container, this causes contamination and spoils the working properties of the powder.
- Touch the powder only with the tip of the brush. Shake off any excess powder and brush the surface using only the very tips of the powder-filled brush. The key to proper print development is to use a small amount of powder and a delicate touch. Use a twirling method to ensure that the sides of the bristles are not coming into contact with the surface and destroying latent prints.
- Watch for the latent print to become visible to ensure that it is not overbrushed. Overbrushing can destroy the print.
- Brush following the contour of the ridges and stop when the ridge detail is developed.
- Stop brushing when the ridge detail is complete.
- Discard any unused powder; never return contaminated powder to the container.

20-13. All developed prints should be photographed and then lifted. All lifts and photographs should be submitted to the laboratory for evaluation, examination, and comparison. All latent print photographs should include a scale. Sometimes a second lift of the same area is necessary to achieve the best possible lift. Superglue-fumed prints can be powdered and lifted many times without destroying the print; however, latent prints that have not been fixed using the superglue process can diminish or be destroyed while attempting to lift them.

LIFTING LATENT PRINTS

20-14. The most common means used to lift latent prints are commercially produced lifting devices, such as hinge lifters, lifting tapes, rubber and gel lifters, and various types of liquid lifting mediums. Hinge lifters and transparent lifting tape have the advantage of presenting the lifted latent print in its correct perspective. Latent prints on rubber lifters are in a reversed perspective and must be reversed again using photographic techniques to properly visualize and compare the latent print. However, rubber lifters generally work better than hinge lifters. Transparent lifting tape works better for taking prints from curved or uneven surfaces. Transparent tapes used in office work, such as cellophane tape, are not suitable for lifting fingerprints except in dire circumstances. A lift background that contrasts the color of the powder should always be used. A gel lifter is not as tacky as hinge, tape, and rubber lifters. It can be used on surfaces that are more fragile where paint might be pulled away with a powdered print and is excellent for lifting dust prints. Hinge and rubber lifters and lifting tape store well; gel lifters may require refrigeration.

20-15. A lifter large enough to cover the entire print should always be used. The plastic cover should be removed from the rubber lifter with care in one steady movement. Any pause can result in a crease being left on the lifter surface. The adhesive side of the lifter should be placed to the developed, powdered print. It should be pressed down evenly and smoothed out over the surface. If an air pocket is sealed under the surface of the lifter, an attempt should be made to force it out. Use pressure or a pin to puncture the lifter and release the air by applying pressure to the bubbled area. The lifter should be peeled from the surface in one smooth even motion.

20-16. Transparent lifting tape is applied in much the same way as commercial lifters. One end of the tape should be placed on one side of the latent print and smoothed out across the surface of the print. Air bubbles should be worked out using a pin, if necessary, to expel air trapped under the surface of the tape. The tape should be pulled free with one continuous motion. The tape should be mounted on materials that contrast the fingerprint powder used. A black background should be used for gray or white powders. A white background should be used for black or dark powders. Commercial mounting cards usually offer the best types of mounting surfaces and have contrasting surfaces on each side of the card. Lifting tapes can be used to lift large areas of latent prints by being applied in overlapping strips, and a rubberized roller can be used to work out air bubbles. All of the strips should be pulled free from the surface in one continuous motion with all of the strips connected together. They should be mounted as one connected piece.

20-17. Many types of silicone and liquid lifting materials are available for lifting latent prints from uneven surfaces, such as appliances, computer equipment, and vehicle interiors. Most types work by pouring them over the powdered latent print and removing them after they dry.

CONDUCTING CHEMICAL PROCESSING

20-18. Only trained laboratory personnel should conduct the vast majority of chemical processing of latent print evidence in an approved laboratory facility; however, there are some instances where chemical processing can and should be conducted in the field by trained investigators. USACIL should be consulted when there is doubt about using chemical processing. The premature or improper use of chemical processes in the field can result in the loss and/or damage of latent print evidence. Most chemical processes are fugitive in nature, meaning that once the latent prints are developed with chemicals, they will fade and often disappear before the occurrence of proper photography and comparison of the evidence. One type of processing that may be used is small particle reagent (SPR). SPR may be more of a physical process than a chemical process in that the resulting action is physical in nature.

20-19. SPR is used on wet items of nonporous evidence, such as those covered in moisture or submerged in bodies of reagent. Metallic particles suspended in water, lodge themselves in the fatty and waxy residue of the latent print after moisture has washed everything else away. SPR is simply applied and then rinsed away with water. It also works on metal and masonry type surfaces. It can be photographed and lifted as with powdered prints, after drying. SPR comes in contrasting colors and UV formulas.

OBTAINING RECORD PRINTS OF LIVING INDIVIDUALS

20-20. To classify, analyze, and compare record fingerprints, they must be complete and clear. It takes practice to obtain suitable record fingerprints and could take several attempts to obtain suitable prints from a particular individual.

20-21. Both the person being fingerprinted and the person taking the fingerprints should always sign and date the record fingerprint cards before the printing process, which will lessen the chances of smearing wet ink. It is difficult in court to prove the origin of record prints without both signatures. All blocks on the fingerprint card should be completed before using any ink to avoid smearing the prints after they have been transferred to a fingerprint card. The subject should wash and dry his hands thoroughly to remove any dirt, sweat, or grime. The subject's hands should be examined to ensure that they are absent of intentional disguises, such as coatings and any disfiguring. The following equipment is normally required for printing:

- *FBI Form Federal Document (FD) 249 (Arrest and Institutional Fingerprint Card).*
- Fingerprint card holder.
- Printer's ink.
- Ink roller.
- Ink plate.

TAKING RECORD PRINTS

20-22. Record prints are taken to show the entire friction ridge skin surface of the fingers, thumbs, and palms. Record fingerprints for submission to the laboratory should consist of at least two completed FBI fingerprint cards and a set of fully rolled fingers and fully rolled palm prints to include the web and side areas of the palms. (See *Appendix G.*) To prepare for recording the prints, the fingerprint card should be secured in the holding device. A small dab of ink should be placed on the inking plate and rolled until a thin, even film covers the surface. The consistency of the ink should appear almost opaque.

20-23. The motions for inking the finger and recording the finger are the same. The fingers are rolled from nail edge to nail edge and from approximately 1/8 inch below the crease of the first joint to as far up as possible. This area will allow for the recording of all ridge characteristics required for correct classification of each finger. The finger is rolled through the ink and then rolled in the corresponding block of the fingerprint card. When the investigator takes record fingerprints, he should grasp the top of the subject's hand to ensure that the finger to be printed is extended. The investigator uses his other hand to hold the finger at the base where it meets the palm. He tells the subject to look away, relax, and allow him to do all the rolling. Each finger should be rolled in one continuous and smooth motion.

The fingers and thumbs are rolled from awkward to comfortable, meaning from left nail edge to right nail edge for fingers on the right hand and right nail edge to left nail edge for fingers on the left hand. This allows the investigator to work with the anatomic features of the hands without fighting the natural resistance of the hands. The finger should not be rolled back and forth on the ink or the card since this will cause over inking, distortion, and ink lines to appear on the recordings. The pressure should be firm and even. Pressing too hard causes the furrows (grooves between the ridges) to fill in with ink. It is important that the investigator ensures the correct finger is rolled in the designated block.

20-24. The investigator will have to roll each finger in its entirety for cases being submitted to the laboratory. This means the investigator will have to use the ink roller to ink each finger separately and then roll that entire finger from nail edge to nail edge and from the tip where it connects with the palm of the hand. This will ensure that each joint of each finger is recorded. The tips of the fingers should also be rolled. They should be rolled from side to side just above the corresponding finger on the paper used to record the entire fingers. A separate full-finger card or piece of bond paper must be used.

PLAIN OR SIMULTANEOUS PRINTS

20-25. After all fingerprint blocks have been completed, the plain or simultaneous prints at the bottom of the card should be completed. They verify the order of the rolled record fingerprints and show characteristics that are sometimes distorted in rolled prints. Simultaneous prints are made on the card by pressing (not rolling) the four inked fingers onto the card in the appropriate blocks at a slight angle so they fit the space. The subject should hold his fingers straight and stiff. His hand should be level with his wrist. His wrists should be grasped with one hand and the fingers should be pressed onto the cards with the other hand. Thumbs are recorded by inking each thumb and pressing it on the appropriate thumb impression block.

RECORD PALM PRINTS

20-26. The investigator must obtain record palm prints from a person each time his record fingerprints are obtained for an investigation, especially if that case is being submitted to the laboratory. Ink should be applied to the subject's palms using the ink roller. Using the inking plate would cause ink lines, created by the edge of the plate, to appear in the record palm print. The palm print card or a piece of bond paper should be wrapped around a tubular object. The subject's heel or base of his palm should be placed on the tubular object and the palm rolled in a pulling motion from the heel of his hand to his fingertips. The investigator should ensure that he records the entire center areas of the subject's palms, which will require direct pressure being applied to the back of his hand. The investigator should also record his web area (between his thumb and index finger), thenar edge (the edge of his palm on his thumb side), and knife edge (the side of the palm opposite the thumb side). Several recordings of each palm should be taken to ensure that all areas are recorded properly.

PROBLEM RECORD PRINTS

20-27. Excessive perspiration and dirty hands and equipment may cause problems when recording prints. The investigator should always start with clean equipment and clean fingers. When the person whose fingers are being recorded are wet from perspiration, each finger should be wiped with alcohol, quickly inked, and rolled onto the fingerprint card. This process should be followed with each finger. Some people have dry and/or rough hands. Rubbing them with lanolin, lotions, or creams can often make them soft enough for clear, unsmudged prints. If the ridges are very worn or fine, alternate methods must be used to obtain prints, much like the methods for recording "postmortem records" (see *paragraph 20-30*). When nothing seems to work, USACIL should be consulted for suggestions and guidance.

20-28. If the hands and fingers are deformed, normal printing steps cannot be followed. The ink should be applied directly to the fingers with a spatula or small roller, and then a square piece of paper should be rotated around the finger. When an acceptable print has been made, the square is taped to the proper box of the fingerprint card.

20-29. If there is an extra finger (usually a little finger or a thumb), the innermost five are printed as usual on the card. The extra digit is then printed on the reverse of the card. Webbed fingers should be printed as best as possible in the rolled and plain impressions blocks of the fingerprint card. If a finger or a fingertip has been amputated, it should be noted in the proper box (such as AMP, 1st joint, FEB 1993 or TIP AMP).

OBTAINING RECORD PRINTS OF DECEASED INDIVIDUALS

20-30. Full record finger and palm prints are always obtained from deceased individuals. The record prints are used to identify the deceased and/or eliminate them as the source of the latent print evidence. The process of taking postmortem record finger and palm prints has always been cumbersome, but it is too important to take lightly. The investigator only has one opportunity to obtain postmortem prints before the body is interned. This process must be completed with accuracy and diligence. The key is to prepare for the process.

20-31. The means used to record the prints depend on the condition of the fingers and the investigator's ingenuity. For the recently dead, the process is the same as for live subjects. The process of inking the fingers and using inking spoons and square paper tabs on the fingers might be used if rigor has started. When rigor mortis is present, the investigator may have to massage and straighten the fingers. Breaking rigor requires a certain technique, and massaging the fingers and hands takes time (about 10 minutes per hand). Rigor can be broken using finger spoons or by bending the fingers backward and pressing down on the middle joint of the finger. If the investigator is not having any success using conventional methods, he should process the fingers and palms using equipment and other methods.

EQUIPMENT

20-32. The following items can be used for processing fingerprints and palm prints of deceased individuals:

- Black or aluminum fingerprint powder.
- A fingerprint brush (soft-hair type).
- Transparencies made from fingerprint cards without a textured surface.
- White case file labels (precut to finger block size and full-length size).
- Larger mailing labels for palm prints.
- Blank transparencies or document protectors.
- A permanent marker.
- Extra large ziplock plastic bags.
- Latex gloves.

METHODS

NOTE: The methods for obtaining record prints of deceased individuals include before and during procedures.

20-33. Before beginning, the investigator should think "safety first". He should wear latex gloves when processing deceased individuals. When finished and before removing his gloves, he should put the postmortem prints just taken into a ziplock bag and discard the magic marker used to label the prints.

20-34. It is recommended that the body lie out for about an hour before taking the prints so that the body can adjust to room temperature, lessening the problems of condensation during recording. The easiest position from which to take the record prints is to lay the deceased in a prone position (face down) with the arms stretched out in front of the body.

20-35. The hands of the deceased should be clean and dry. The investigator may have to use some alcohol swabs to ensure that the skin is dry enough to receive a light dusting of powder. It may be necessary to massage the fingers and palms to make them more pliable and receptive to the print-taking process. This massaging will open up the palm area for better record taking. A small worktable should be used for laying out the supplies and equipment. This makes the printing process easier.

20-36. During the printing process, the fingerprint powder should be brushed on the palm side of the right thumb of the deceased. If the investigator always starts with the right hand in the following order: thumb, index, middle, ring, little finger, and then the left hand in the same finger order, it will help him stay organized and keep him from making mistakes with labeling. He should place a precut, white case file label on the tip of the finger and gently smooth out the label, molding it to the finger. He should use the same process until the complete fingerprint card is full. The larger labels should be used to complete the simultaneous prints.

20-37. To obtain full record fingerprints, the entire finger should be powdered from tip to base (where the finger joins the palm) and from nail edge to nail edge. Again, the investigator should gently and steadily peel the label from the finger and attach it to the back of the blank transparency. He should immediately write on the front of the transparency just below the applied label which finger it is so as not to lose track or get the labels out of order. The investigator should remember that, when viewing the ridge detail through the transparency, it is a reversal of the pattern on the actual finger. The following includes methods for obtaining fingertip prints, palm prints, and guidance for special cases:

- **Fingertips.** The tip of the finger can be powdered and a label applied across the tip from side to side. This process should ensure that all of the ridge detail available has been captured. The recorded tip should be placed just above the corresponding finger on the blank transparency. The investigator should remember to keep all the labels for the same finger together on a transparency and label each accordingly. These same steps should be repeated for all ten fingers.
- **Palm prints.** The investigator should use the same method for taking palm prints as he did for taking fingerprints, but this time he should use the larger mailing labels. In most cases, the investigator will have to overlap two labels in order to obtain all the ridge detail on the palms. The investigator should remember to keep those two labels together when removing them from the hand and applying them to the back of the transparency. He should gently mold the labels to cover the center of the palms, the edges of each palm, both the little finger and thumb sides, the area where the wrist connects to the forearm (the wrist bracelet area), and the interdigital area where the palm connects to the fingers.

NOTE: In some cases where fingers and/or palms are too damaged to allow for the powdering of the skin, photography or other methods may have to be applied.

- **Special cases.** The hardest record prints to obtain are those from a body that has started to decompose. It may require techniques beyond the investigator's expertise. When the hands are badly damaged, the investigator may need to coordinate with USACIL for guidance on how to proceed.

Chapter 21

Firearms, Ammunition, and Toolmarks

Solving a crime that involves firearms often depends on the scientific examination of evidence by a qualified examiner at USACIL. Laboratory examination may show that a projectile (the part that exits the weapon) or a cartridge case was fired using a specific firearm. Firearm function testing might show that an accidental discharge was possible. Other tests can show the presence or absence of gunpowder residues in the barrel of a firearm. A projectile or cartridge case may show the caliber and type of firearm that fired it. It may also tell the manufacturer of the ammunition. Tests could show the distance between the muzzle of the firearm and the target and the point of entrance and/or the exit of a bullet in clothing, wood, glass, or metal.

OVERVIEW

21-1. Investigators do not perform firearms identification tests in the field. Firearms examiners do the identification tests at the crime laboratory and give the test results to the investigator in the field. They also give expert testimony in court, when needed. While only a qualified examiner may give expert testimony, the solving of a crime involving firearms depends largely on how the investigator collects and preserves firearm evidence.

RECOVERY AND PRESERVATION OF EVIDENCE

21-2. Any item that may need the services of a firearms examiner must be handled with care to ensure that it is not altered or damaged. For instance, the investigator should have medical personnel cut around bullet holes to leave them intact when the victim's clothes are removed. Garments should be cut a good distance away from bullet holes to leave the holes and area surrounding them intact. The investigator should also ensure that the items do not become contaminated. The investigator should ensure that care is used when clothing and similar items are involved.

21-3. Recovering projectiles at a crime scene is often difficult. The investigator should never probe for, or try to extract, a projectile with other than rubber or heavily taped tools. It is often best to take a small section of the wall, ceiling, or the like of with the projectile still in it. By forwarding it intact to the laboratory, the investigator prevents damage to the projectile.

21-4. Latent print processing is very difficult on the parts of firearms that have a slight oily film; however, it is possible to get usable impressions. Latent print techniques will not normally hinder the examinations of the firearms examiner, and superglue fuming does not have an impact on GSR in the barrel or chamber. At the lab, the fingerprint and firearms examiners will coordinate their efforts.

EVIDENCE MARKING

21-5. Evidence should be marked so it can be readily identified later. Firearms, known to be of evidentiary value, are marked immediately. Those seized or impounded to decide their value are not to be marked, scratched, or defaced in anyway. These items are marked only after it is decided that the firearm has value as evidence. Common sense should be used in marking antique firearms and highly engraved firearms. Their value should be protected. Also, areas on the firearm where latent prints could be found should be avoided and not marked.

21-6. The investigator should place his initials and the time and date of recovery on each item of evidence so he can positively identify the evidence at a later date. When several similar items are found, an identifying number should be added on each item. No two items of evidence in the same case should bear the same identifying numbers. The investigator should put a description of each item and all identifying marks involved in the same case in his notes. The identifying number has no bearing on the numbers of the exhibits in the ROI.

21-7. Marking tools may be used for inscribing identifying markings on firearms evidence. Diamond point pencils or scribers are ideal. Dental picks make excellent marking devices when the curved tip is cut off and the point made needle sharp. Firearms are usually marked on the side of the frame. All parts of the firearm that leave imprints on either the bullet or cartridge case should be removed and marked. For example, a conventional 9-millimeter caliber semiautomatic pistol should be marked in three places (see *Figure 21-1*): the barrel that marks the bullet; the slide that contains the extractor, the breech face, and the firing pin; and the receiver that includes the ejector that marks the cartridge case.

21-8. All parts of a firearm should be marked the same. Marks should be put where they can be seen but do not interfere with existing markings or stampings on the firearm. The magazine should be marked on the floor plate (bottom) or on the exterior portion of the magazine body and then submitted with the suspect firearm.

21-9. Some revolvers have interchangeable cylinders that should be marked on both the cylinder and barrel. Some revolvers have a removable side plate and should be marked on the side of the frame that cannot be removed (see *Figure 21-2*).

21-10. Firearms having removable bolts, such as a semiautomatic, an automatic, and a bolt-action should be marked on the bolt, barrel, and frame. If the barrel of a firearm cannot be removed without tools, it does not need to be marked. But marking the barrel, even under these circumstances, adds certainty. A projectile submitted as an exhibit may be jacketed or lead. No markings should be placed on the projectile. Identification marks may cause the loss of trace evidence or evidence marks. Containers, such as pillboxes and plastic vials with cotton packing material can be used for sealing fired bullets.

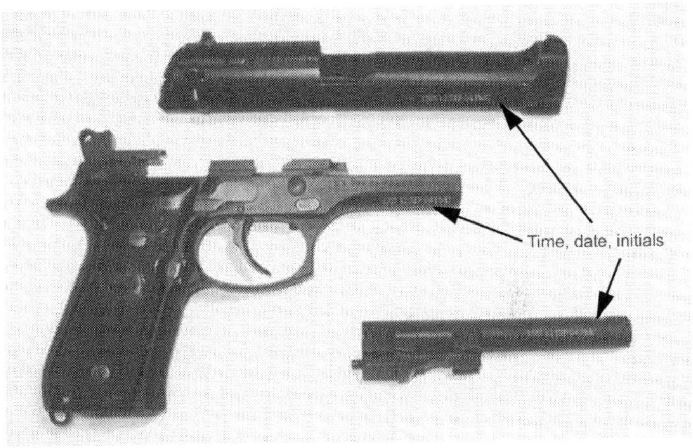

Figure 21-1. Marked 9-Millimeter Pistol

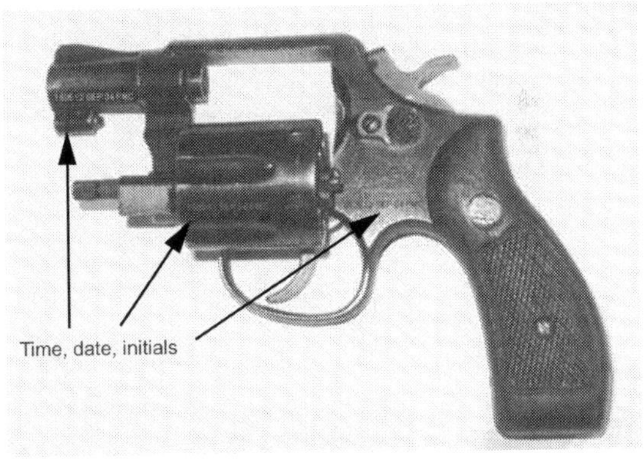

Figure 21-2. Marked Revolver

The containers should be sealed with evidence tape or an equivalent. The containers should then be marked with the date, time, and initials of the individual doing the packing so they cross over the sealed tape. If the package is opened before it arrives at USACIL, it will be apparent. Deformed projectiles and jacket fragments must also be placed in a container and marked as described above.

21-11. Cartridge cases are not marked. They are treated the same as projectiles and then placed in containers. Shotgun shell cases, wads, or shot columns are not marked. Shot pellets, such as bird shot and buckshot known to be from one source can be placed together in a container. Containers should be marked for identification.

EVIDENCE TRANSMITTAL

21-12. Unload the firearms to be examined at the USACIL before preparing them for shipment. If a firearm cannot be unloaded, contact the USACIL for advice and shipping instructions. Firearms may be shipped by US mail as allowed by postal laws and regulations. Live ammunition, propellant powders, primers, or explosives may be sent using a private shipping company, freight, or courier. A local post office can be contacted for more information.

21-13. Firearms should be wrapped in a clean, protective covering. This prevents dust, lint, and other foreign matter from filtering into the mechanisms. They should be packed in suitable shipping containers. Special packaging procedures should be used when the evidence is to be examined for fingerprints. USACIL should be contacted for questions about packing or shipping evidence.

21-14. Firearms should not be cleaned before shipping them to the lab. If there is a lot of moisture in the firearm barrel, is should be removed (as much as possible) to stop rust from forming. A single dry patch should be used. This fact should be recorded in the investigator's notes and on the laboratory request. A collection of rust makes it hard for the laboratory examiner to conduct a comparison test. In special cases, when firearms must be cleaned, the USACIL should be consulted and the cleaning patch should be sent to the laboratory with the firearm.

21-15. All ammunition found in the possession of a suspect or at the scene of a crime is seized and held as evidence. The laboratory may have enough ammunition of the same type to use for test needs. The laboratory firearms division should be contacted to learn if it has the right ammunition. If not, arrangements should be made for the ammunition collected as evidence to be sent to the laboratory with the firearm.

21-16. When revolvers having loaded cartridges or fired cases are obtained, a diagram should be made of the rear face of each cylinder. The position of the loaded cartridges or the fired cases should be shown with respect to one another and the firing pin. An arrow should be scratched on each side or the rear face of the cylinder (under the firing pin) to show the position that the chamber was in when the revolver was found. This should be done on the revolver and also on the diagram. The diagram, complete with legend, lets the laboratory examiner relate the fired cartridges to the chamber of the cylinder in which they were fired (see *Figure 21-3*).

21-17. Clothing items sent to the firearms division for proximity tests should be packed so the area around the entrance hole in the garment does not

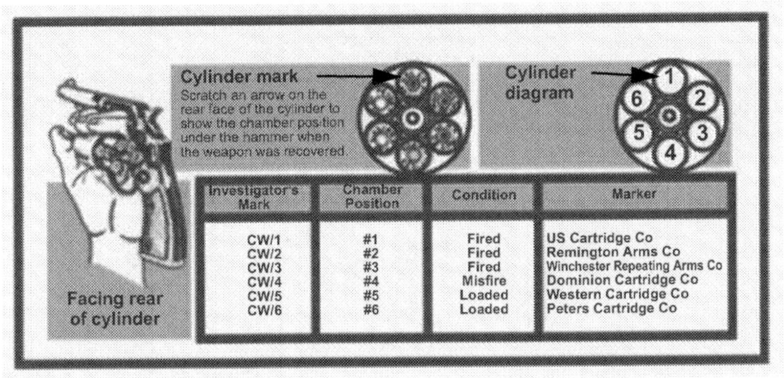

Figure 21-3. Notation of the Position of the Cartridges in a Recovered Revolver

become contaminated. This is done by sandwiching the part of the garment containing the GSR between sheets of cardboard or brown paper.

LABORATORY TESTING

21-18. Testing by laboratory examiners can provide the investigator with information he is not able to determine by field examination. For instance, in gunshot wounds, powder residues may be deposited either on the skin or clothing. Only pathologists or other qualified medical individuals can give an expert opinion on gunshot flesh wounds and their powder patterns. Only laboratory examiners can give an expert opinion on powder residue in clothing. By firing a suspect firearm and using ammunition of the type that left the residue, a laboratory examiner can do tests to learn the approximate distance from the muzzle to the target. These proximity tests are based on the dispersion of gunpowder residues. They are, of course, subject to limitations. A scaled photograph of the wound may be helpful to a firearms examiner who is examining the clothing worn by the victim. Normally, with a muzzle to target distance in excess of 2 1/2 feet, no discernible gunpowder residue pattern will be present.

21-19. Sometimes a firearm has had a serial number or other die-stamped lettering removed. Showing ownership or otherwise identifying the item may depend on discerning the serial number. Restoration of serial numbers or other identification data is performed at USACIL and should not be attempted by the investigator.

21-20. Often, the laboratory can examine a projectile or even a cartridge case alone to learn facts of the class characteristics of the involved firearm. The laboratory can tell the caliber and type of firearm (pistol, revolver, or rifle) from which the projectile was fired. The number and width of lands and grooves in the rifling and the direction of twist may also be provided (see *Figure 21-4*).

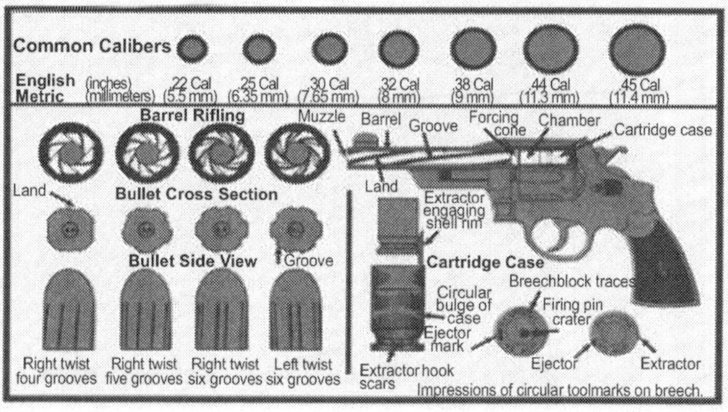

Figure 21-4. Class Characteristics of Firearms

TEST FIRING

21-21. If a firearm is sent to the laboratory with projectiles, cartridge cases, or both, tests can be done to see if these components were fired from that firearm. If the class characteristics of these components are consistent with the characteristics of components fired from a firearm like the exhibit firearm, test firing will be done. The test projectiles and cartridge cases will be microscopically compared with the exhibit items. If there are numerous firearms that are suspect in a case, it may not be wise to ship all the firearms. In such cases, the supporting laboratory should be contacted for advice.

21-22. All firearms uncovered during the investigation of homicides, suicides, assaults, and robberies should be submitted for function testing. The value of learning if a firearm will function and if it functions safely is often overlooked. It might be that a firearm could not have discharged accidentally as stated by a suspect or that a particular firearm could not be of fired at all.

TOOLMARKS

21-23. A toolmark is an impression, cut, scratch, gouge, or abrasion made by a tool in contact with an object. A tool can make a negative impression, an abrasion or friction mark, or a combination of the two.

21-24. A negative impression is made when a tool is pressed against or into a receptive surface. The mark made by a crowbar used to pry open a door or a window is a negative impression. An abrasion or friction mark is made when a tool cuts into or slides across a surface. A pair of pliers, a bolt cutter, knife, ax, or saw can make this type of mark. A drill, a plane, or a die in manufacturing could also make it. A combination mark is made, for example, when a crowbar is forced into the space between a door and its facing and pressure is then applied to the handle of the tool to force the door open. The forced insertion of the crowbar makes an abrasion or friction mark. The levering action produces a negative impression. The visible result is a combination of the two.

21-25. No two tools are alike in every detail. Thus, they will not leave identical impressions. Tools may have obvious differences in size, width, thickness, or shape. They also have minute differences that are only seen when the tools are examined under a microscope. These minute differences can be caused by manufacturing, grinding and finishing, uneven wear, and unusual use or misuse. They may also be caused by accidents, sharpening, and alterations or modifications made by the users of the tools. From these minute differences, it may be possible to identify the tool that made a given impression.

21-26. When a toolmark is discovered, it should be photographed with a measuring device as soon as possible. It should always be photographed before it is moved, disturbed, or altered in any way. Photographs provide a permanent record of the evidence in its original state and location. They match original evidence with any casts or molds made. They also satisfy the legal need for records of original evidence.

21-27. The toolmark should be visually examined to note its gross appearance. This can tell you what type of tool or shape of tool to look for. The gross appearance of a tool impression may not be complete or well-defined. For example, a hammer impression on a steel safe may not include the edges of the hammerhead. Thus, the shape of the head cannot be shown. When this occurs, all suspect tools that could have made the mark must be sent to the laboratory to be examined.

21-28. The surface that bears the impression of a toolmark may have been painted. If so, a careful check may show that flakes or chips of paint were removed. The paint flakes may be sticking to the tool that made the impression and can be compared to the paint from the impression surface to determine possible common origin. If a tool is found with paint like that of the painted surface and the flake patterns look alike, the paint pattern formation should be photographed since some of the paint flakes may become dislodged during transmittal to USACIL. Additionally, flakes of paint could have been removed and transferred from a painted tool to the surface bearing the impression. A trace-evidence examiner can examine paint flakes to determine their origin.

21-29. A tool should never be placed into a toolmark to see if it fits or if it could have made the impression. This could prevent any evidence on the tool and its marks or the paint on the tool and the object bearing the toolmark from being admitted in court.

21-30. Laboratory examination of toolmarks is based on the same principles and techniques used for fingerprint and firearm identification. Tools leave unique characteristic traces that cannot be exactly reproduced by any other tool. In the lab, test marks are made with suspect tools on materials like those on which the toolmarks are present. The test marks are then compared with the suspect toolmarks under a comparison microscope.

21-31. Often, the laboratory will find that the suspect tool made the toolmark in question. Such findings, however, are not always possible. Sometimes the material on which the toolmarks are found does not record minor tool imperfections. These imperfections are needed for positive identification of the evidence mark. In such cases, the examination may yield other valuable information that can be used as a lead for further investigation.

21-32. An examination of toolmarks without a suspect tool can also be of value. Comparing the toolmarks found at each scene may link a series of burglaries. A match of the lengthwise markings on two pieces of wire may show that both were manufactured at the same time, having been drawn through the same die during production. A suspect's possession of a piece of wire that matches a piece found at a crime scene would show the possession was more than accidental. Wood shavings from a drill, a plane, or another tool that is able to produce wood chips may be matched with the tool producing them.

21-33. The toolmarks present on a doorjamb, door, or safe should be checked to ensure that they are sufficient to actually open the door. Sometimes, the suspect will open the door with a key or combination and then close the door on the tool to make it appear as if someone had broken in.

21-34. Each piece of evidence to be sent to the laboratory must be marked for identification and wrapped separately. Evidence samples should not share the same package unless all danger of mixing has been removed. Toolmark evidence should be wrapped and packaged so that the toolmark and the tool will not be damaged and trace particles will not be lost. The cutting blades or tips of the tool should be covered to prevent damage. If the item to be examined has to be removed by cutting, always mark the end that is cut and the questioned end to be examined.

21-35. Original evidence is less subject to attack in court than reproductions. Often, photographs and casts do not show the evidence well enough for identification purposes at the lab. Some authorities recommend that casting or other means of taking impressions of a toolmark should only be used as a last resort. A casting is never as good as an original impression. This is especially true of toolmarks made in soft materials like wood, putty, and paint. Many of the casting media suited to these materials will not reproduce the fine details needed for identification. An impression or a cast cannot reproduce scratches in paint from minute irregularities in the edge of a tool. A cast or mold should only be made from a toolmark when there is a good reason for not removing the original evidence.

21-36. The investigator should decide if the original evidence that bears the toolmark should be removed to send to the lab. This is a judgment call. Sending original evidence to the laboratory is highly desired. But wholesale removal of property or parts of valuable structures is not desired or needed. The decision should be based on the importance of the case and the value of the toolmark compared with other evidence at hand. The distance of the crime scene from the laboratory should be considered. The tool-marked object should be checked to see if it belongs to the US government or is civilian property.

21-37. If the evidence is civilian property, the investigator must contact the owner to make arrangements to return, replace, or pay for the items removed.

He should ensure that competent witnesses are present during the removal of the evidence. This prevents later claims against the US government. It also verifies the original condition of the evidence.

21-38. If a toolmark can be removed to send to the lab, a piece of the object large enough to keep the toolmark from splintering, bending, twisting, or abrading should be removed. If the marked part of a door, window sash, windowsill, or doorsill is removed, that portion of the window or doorframe that is adjacent to the marked area should also be removed. Any window latch, door latch, bolt, hasp, or lock that has been cut, broken, or forced for entry should be obtained. Any tools found at the crime scene must also be sent to the laboratory with the toolmarks.

21-39. Each item of evidence should be clearly marked with the case number, initials, and the date and time of removal. The evidence should also be marked to show the inside, outside, top, and bottom surfaces and the area bearing the toolmark. If the surface bearing the toolmark is painted, samples of the paint should be sent to the lab. Even though paint could not be seen on the tool, enough particles may be present for analysis and comparison at the lab. Since some tools are also painted, there may be paint from the tool on the toolmark surface. This can also be compared for possible common origin.

21-40. The angle at which the tool was held when it made the mark could be known. If it is, all information that can be provided on the various angles formed by the tool when it was used will greatly help the examiner. The details of the incident should be given to include the measurements of the toolmarks from the floor. If a window was involved, the details should include its location (such as in the basement, first floor, or second floor).

21-41. A toolmark may be on metal and not removable. Samples of the metal should be taken and sent to the lab. Particles of metal may adhere to the tool in addition to the paint. The metal particles can be analyzed and identified by the laboratory examiner.

21-42. If cut pieces of wire are to be sent, the suspect end of the wire should be clearly marked. When cutting the wire to send to the lab, it should not be cut with the suspect tool. Matching the cut ends of wire can help identify related items. For instance, if the owner cannot positively identify a stolen automobile radio, matching a radio wire to the wire attached to a car can show it to be originally from that vehicle. Toolmarks also appear on rubber tires. These marks can be successfully matched the same as any other toolmarks.

SERIAL NUMBERS

21-43. Serial numbers are placed on many manufactured objects to distinguish one item or model from another. Serial numbers may consist of numerals, letters, symbols, or a combination of the three. Serial numbers are often the only way to show ownership. Items with serial numbers can often be traced from the manufacturer to the wholesaler and on to the jobber, the retailer, and (finally) the purchaser.

21-44. Owners of items lacking manufacturer's serial numbers often place their own marks or serial numbers on the items. This helps identify the item if it is stolen.

21-45. Serial numbers or private marks may be stamped, molded, etched, or engraved. Some items, such as automobiles, firearms, and watches, bear serial numbers on several parts. If an object is found and the serial numbers seem to have been removed, the object should be searched for other numbers. Such numbers are often found in hard-to-find places.

21-46. Military services buy in large quantities. Often, they do not initially record individual serial numbers. Lot numbers and shipping and receiving documents account for the shipments. Sometimes other means are used to speed the movement of supplies. Often, the manufacturer of a serially numbered item can give the lot number. The manufacturer may also have data of other recorded items bought by military services. As the bulk shipment is broken down for issue to units, the serial numbers are often used for records and identification. Lot numbers or shipping document numbers often narrow the search to the unit of ownership.

21-47. There is no easy way to know if a serial number that has been removed can be restored, and there is no method that can be used in the field to find out. The laboratory can determine whether or not a serial number can be restored. All items that seem to have had the serial numbers removed should be sent to the lab.

21-48. Neither the material from which an item is made nor the method used to affix the serial number automatically preclude restoration. Serial numbers have been restored under the most adverse conditions. Conversely, restoration attempts have failed when conditions seemed most favorable.

JEWELER'S MARKS

21-49. While jewelers' marks are not serial numbers, their use in tracing stolen property can be of value. When an item is given to a jeweler for repair, it is common practice for the jeweler to place a small, identifying mark in a hard-to-see place on the item. This mark is often inscribed with a very fine engraving tool. The mark is engraved under magnification. Therefore, the mark is often visible only when viewed under equal magnification.

21-50. Jewelers in the same location often know each other's markings. When a mark is found, an attempt should be made to locate the jeweler who inscribed it. The jeweler may be able to identify the person who brought the item to the jewelry shop.

Chapter 22

Impressions and Casts

Footwear and tires are highly complex and precisely engineered. Features that are incorporated into the manufacture of footwear and tires can become very useful in impression examination. Logos, lettering, numbers, or other markings may be clues to the brand, style, or manufacturer. Expert examiners take advantage of these unique aspects of footwear and tires to place or eliminate individuals at crime scenes. It is what occurs to the footwear and tires after they are put into use that makes them individual and highly identifiable. It is the individual wear and damage to the friction surfaces, which makes every footwear and tire unique.

FOOTWEAR AND TIRE TRACK IMPRESSIONS

22-1. One of the most overlooked types of evidence at the scene of a crime is footwear and tire track evidence. Due to the lack of awareness and training in the collection and preservation of footwear and tire track impressions, most crime scene investigators do not give proper consideration to this type of evidence. Fortunately, criminals also forget the importance of this evidence. Criminals have long recognized the need to protect themselves from leaving fingerprints at the scene of a crime by wearing gloves, but often forget about footwear and tire track impressions. Generally, perpetrators do not intentionally destroy these types of impressions.

22-2. Crime scenes generally contain numerous footwear and tire track impressions from witnesses, police, fire, and medical responders. Neither the volume of impressions present nor the trampling of the crime scene preclude the perpetrator's footwear or tire track impressions from being found and/or identified. Successful identification is often made solely from partial impressions. Many times footwear impressions are not visible in existing ambient light conditions. Proper lighting and search techniques can assist in locating valuable impression evidence. Most surfaces are conducive to footwear impressions. Surfaces such as rough, uneven carpeting or even masonry should never be discounted without proper examination. Although poor weather conditions can destroy some impressions, this should not be accepted as fact in all cases. Excellent footwear impressions can still be found in mud puddles, under fallen snow, below the overhang of a house, under shrubs, and in numerous other hiding places.

IMPRESSION SEARCHES

22-3. When searching for impressions, a system should be used to ensure that the entire crime scene area is covered to include a search beyond the scene. One small area at a time should be searched in a systematic way, such as using a grid search or concentric circle search. Also, looking at the crime scene from an overall perspective or even from overhead should be considered. Sometimes this could mean using resources such as helicopters to view a field or large area where tire tracks are suspected or entry and/or exit routes are not obvious. Getting above the scene can give a different perspective of the area. Impression evidence normally requires the investigator to view things from a particular angle with particular lighting before it is revealed or can be seen from the best vantage point.

LIGHTING TECHNIQUES

22-4. Existing lighting conditions should be used first to detect obvious footwear impressions. Especially those made in some form of contaminant, such as blood, grease, mud, and other visible residues. Each impression should be marked when it is located using a system of placards, signs, or labels with some form of alphanumeric designator that is easily seen in a photograph and identifiable. The area should be darkened, if possible, and searched using a bright, intense light at an oblique, side-lit angle. The light source should be held just above and parallel to the surface being searched. Raising and lowering the light will cause the shadows to fall differently in the impression and allow for adjustments for the best possible visualization of the impression. The rule is the deeper the impression, the higher the angle required for best visualization. A portable floodlight is most effective; however, a bright flashlight will also work well. Remember, an investigator is normally searching for residue or dust impressions, which are not visible in normal room light. A search using oblique lighting is an excellent technique for detecting impressions on smooth surfaces, such as flooring and furniture. Impressions on carpeting or other rough or porous surfaces may still be invisible even with the oblique light. An electrostatic lifting device may be required to search these types of areas for dry-residue impressions.

CHEMICAL SEARCHES

22-5. Chemical searches are often necessary to detect and develop latent footwear impressions at crime scenes, such as faint blood impressions. In this case, the floor or substrate needs to be removed where impressions are suspected and then sent to the laboratory for processing. In some instances, on-scene assistance from laboratory personnel may be necessary. Each case offers its own unique set of circumstances.

FOOTWEAR AND TIRE TRACK SEARCH AREAS

22-6. When searching for footwear or tire track impressions, there are specific areas that an investigator should search. They are identified as follows:

- Areas to search for footwear impressions include—
  - The point of entry (both interior and exterior).
  - The path traveled from the point of entry to the area where the crime occurred.
  - The immediate area where the crime took place.
  - The path traveled from the area where the crime occurred to the point of exit.

- Other areas where the perpetrator may have walked to include beyond the scene.
- The point of exit (both interior and exterior).
- Areas to search for tire track impressions include—
  - The route traveled to the location where the crime occurred.
  - The area where the crime took place.
  - The route traveled from the location where the crime occurred.
  - Any other areas where the suspect vehicle might have traveled or parked.

NOTE: Searches beyond the scene can sometimes reveal areas where a vehicle has paused waiting for an accomplice or where a vehicle has been parked to hide it until it was needed.

IMPRESSION COLLECTION AND PRESERVATION

22-7. Collecting and preserving footwear and tire track impression requires special attention and care. Impressions can be easily damaged or destroyed.

CRIME SCENE IMPRESSIONS

22-8. Footwear and tire track impressions are collected and preserved using the same methods and techniques. Footwear and tire track impressions are extremely fragile in nature. Environmental elements, improper safeguarding of the scene, time, and improper processing and collection techniques can often destroy this type of evidence. Impressions must be collected and preserved as close to their original state as possible to be useful. The evidence should be protected from destruction by natural elements or accidental damage by covering it with a trash can lid, a cardboard box, or another suitable object. Large areas may have to be roped off, and guards may need to be posted. A suspect's footwear should never be placed in or near a crime scene impression because it could contaminate the crime scene impression and jeopardize its integrity, making future findings invalid.

PHOTOGRAPHY

22-9. Photography establishes the integrity of the evidence and is one of the best techniques for capturing and preserving impressions. There are two types of photography necessary to record crime scene impressions: general crime scene photography and examination-quality photography.

- **General crime scene photography.** This includes long-range photography, which means shooting overall photographs of the entire scene where the impression is located. Footwear or tire track impressions at the crime scene are marked and a general overall photograph is exposed depicting the location of the impression in relationship to the remainder of the scene. General photography also includes medium-range or establishing photography (a photograph of the impression as it relates to a specific area of the scene or other pieces of evidence). General photography is used to help show the direction of travel and position in relation to other impressions. Finally, a close-up is taken of the overall impression showing its appearance at the scene. Digital photography may be used for general photography.

- **Examination-quality photography.** This includes close-up photography of footwear and tire track impressions that will be used by an evidence examiner for evaluation, comparison, identification, and verification. All examination-quality photographs must be taken with a film camera. Never use digital photography for close-up or examination-quality photographs. At present, digital cameras do not produce the resolution needed for examination-quality photographs. For this type of photography, a 35-millimeter film camera with an off-camera flash is required. It is recommended that a 35-millimeter camera be used with ISO 100 black and white film. If black and white film is not available, use ISO 100 or 200 color film. Never take a photograph of footwear or tire track impressions without a tripod, a scale, and a flash. The camera should be focused on the impression; the impression should fill the entire frame. This allows for the greatest possible resolution when enlarging photographs. Photography considerations include the following:
  - **Proper camera position.** A tripod is absolutely necessary for examination-quality photography. Many investigators mistakenly assume that they can handhold a camera steady enough for examination-quality photographs. When the image is enlarged, the distortion caused by any slight movement is exaggerated and becomes obvious. The camera should be placed on a tripod directly over the impression. This sometimes requires inverting the center rail of the tripod so that the camera can be mounted underneath. The back of the camera or the film plane should be parallel to the impression to avoid distortion and allow for the greatest possible resolution when enlarging photographs.
  - **Scales and labels.** A measurement scale or ruler must always appear in evidence photographs. A scale or ruler allows for the photograph to be enlarged to the exact scale of the impression for a one-to-one comparison. Ink pens, coins, or business cards, for example, should not be used. The scale must be placed on the same level and as close as possible to the same plane as the impression. An L-shaped measuring device that measures the length and width will produce the most useful results. The same label or marker used in the overall photographs should be placed in the picture with the scale to identify which impression is being photographed.
  - **Lighting.** Oblique lighting should always be used because it provides the greatest amount of contrast by casting shadows in the impression, thus capturing better details. In order to obtain this shadowing, the flash should be detached from the camera and used through the flash extension cord. Using a flash mounted on the camera does not produce detailed images. To obtain the oblique lighting required, the flash should be held at about a 45° angle and about 3 to 5 feet to one side of the impression. Depending on the depth of the impression, this angle may vary (the deeper the impression the greater the flash angle).

Photographs should be taken with the flash held at four different directions from the impression. North, south, east, and west directions will provide sufficient shadow variances to yield the best details. A shadow indicator (such as a thumbtack, an ink pen, or a golf tee) should be positioned in the shot to allow the examiner to determine the direction the light is coming from. This aids the examiner in his comparison work. If suitable conditions do not exist in the ambient light of the scene (such as too much direct sunlight), a shadow should be cast over the impression with a large piece of cardboard or other suitable material and then illuminated by using an electronic flash.

SPECIAL CONSIDERATIONS FOR TIRE TRACK IMPRESSIONS

22-10. Investigators should be familiar with the information printed on the sidewalls of tires. This information can be crucial in identifying a particular tire. See *Appendix H* for a complete discussion on how to read a tire sidewall.

22-11. Tire track impressions are dealt with in much the same manner as footwear impressions. If tire track impressions are short, the entire impression should be cast. Longer impressions should be cast in sections that are no longer than 3 feet to prevent breakage during shipment to the laboratory. One good 3-foot cast is often sufficient to identify the impression of a specific tire. The likelihood of identifying a tire track to a tire increases with the amount of cast evidence submitted. Casts should be made of each found tire track found.

NOTE: Testimony regarding combinations of designs taken from a set of four tire track impressions at the crime scene corresponding to the designs of a set of four tires from a suspect's vehicle is of obvious value.

22-12. Suspect tires should be sent to the lab with the casts. The rims should be left on the tires and the tires should be inflated at the same level as when they were collected. The lab may need to make test prints with the tires and additional information will be necessary. Each tire sent should be identified as to the wheel position and the type of vehicle from which the tires were collected. Sketches, photographs, and other notes should be used to show the tire positions. Often, digital photographs of the tire track impressions and the tires on the suspect vehicle can be emailed to the laboratory in advance of the laboratory request to determine the evidential value of the suspect tires. This screening process saves time, effort, and shipping costs.

TWO-DIMENSIONAL IMPRESSION LIFTING

22-13. Two-dimensional footwear or tire track impressions should only be lifted when the item bearing the impression cannot be readily removed from the crime scene: for example, footwear impressions on doors or vehicles; footwear impressions on immovable or heavy objects; a dusty tire track impression on a garage or warehouse floor; or any impression on a heavy, bulky, valuable, or sensitive surface or item. Impressions should also be made when it is not cost effective to forward the item to the laboratory. Coordination should be made with USACIL before the collection and shipment of this type of evidence.

22-14. There are various methods available for lifting two-dimensional impression evidence. The following are the preferred methods:

- **Electrostatic dust print lifter.** An electrostatic dust print lifter is a device used to make electrostatic lifts of dry substances, such as dust, powder, and other lightweight dry residues and debris deposited in impression evidence. It works on the principles of static electrical charges. A piece of statically charged Mylar® film attracts the dust particles to the film, collecting the impression. It is a highly effective method of collection and renders excellent detail suitable for identification. An electrostatic dust print lifter can be used on porous or nonporous evidence. Again, the surface must be dry. Electrostatic lifting is considered to be nondestructive and may be used for searching as well as for collecting. Blind searches can be made by laying out the film over an area suspected to contain latent dust impressions, charging the film, and examining the lift. This works well on carpeted entrance and exit ways. Lifts can be made of vertical surfaces as well, such as on doors and walls, by simply taping the film to the surface and lifting as normal. Electrostatic lifts are very fragile in the sense that the film holds much of its charge even after the collection process and continues to attract dust and particles, which can obscure the impression. The film must be placed in a suitable box and sealed as soon as possible to prevent damage. A flat, sturdy box (such as a photographic paper box, shirt box, or clean unused pizza box) is recommended for collection of this type of impression. The box must be wiped out before use to clean away dust or paper residue. Place the film silver-side down and the impression side up into the bottom of the box. Secure the film to the bottom of the box by taping down its four corners. Close the box and tape shut all the edges of the boxes, making it nearly airtight.
- **Gelatin lifters.** Gelatin lifters, also called gel lifters, should only be used after photography and an electrostatic dust print lifter has been considered. Gelatin lifters differ from typical fingerprint lifters and are excellent for footwear impressions. They are made with reduced adhesive properties and can be used successfully on fragile surfaces. They are soft and pliable and can lift good detail from rough surfaces. In fact, they can lift an impression from newsprint without tearing or sticking to the paper. An investigator must be aware of the contrast issues when choosing the appropriate color of lifter. If the footwear impression is in dust, the black gel lifters are the best choice. If the impression is in a darker substance, use the white.
- **Standard lifters.** Most standard lifters are not very useful in lifting footwear impressions because they are normally too small to accomplish the job. There are some products that are specifically cut to footwear size but are too tacky and the backing is too flimsy causing too many air bubbles and distorting the impression. Rubber footwear lifters are sufficient for lifting footwear impressions and come in a variety of colors and sizes. An image it always in the reversed position

after using opaque lifters to lift it and must be reversed at the laboratory using photographic techniques.

- **Chemical processing.** An arsenal of chemical processing techniques is available to develop impressions. However, trained laboratory personnel must perform all chemical processing of impressions in an approved laboratory facility. Laboratory examiners and technicians are trained to use many types of chemical processes in a safe and efficient manner. The premature or improper use of chemical processes in the field will result in the loss and/or damage of footwear impressions. Most chemical processes are fugitive in nature, meaning that once the prints are developed with chemicals in the field, they will fade and often disappear before proper photography and comparison of the evidence.

THREE-DIMENSIONAL IMPRESSION CASTING

22-15. When casting a three-dimensional impression, there are several steps to consider. The first step consists of preparing the impression without contaminating or destroying it while the last step consists of annotating identifying data.

PREPARING IMPRESSIONS

22-16. To obtain a good cast, it is sometimes necessary to prepare the impression. The impression should always be photographed first. When the impression is located outdoors, it should be determined whether any debris might have blown into it. Debris should be removed carefully using a pocketknife or tweezers. Do not attempt to remove debris that is part of the impression or was present when the impression was made. When impressions are made over rocks, sticks, or other debris, excellent reproduction or detail can be adjacent to these items. If a loose leaf or twig has managed to fall into the impression after it was made, it can be carefully removed.

22-17. In some cases, an investigator may want to make a practice cast of his own impression before trying to make the evidence cast to determine the strength of the ground. It may be necessary to strengthen the soil in which the impression is found by spraying it with a plastic spray or lacquer. Hair spray and spray paint may also be used. Spraying directly on the print may damage individual characteristics used to effect identification. Instead, the spray should be directed against cardboard or other material so that a fine mist settles gently into the print. It should then be dried and sprayed again. The number of coats required can be determined by examining the test print. Three to ten coats are not unusual in sandy soil. Thin coats are better than heavy coats, which can damage the impression.

CASTING DENTAL STONE OR DENTAL COMPOUND

22-18. Since 1986, dental stone, dye stone, or other stone-based products have been the recommended material for making casts of footwear and tire track impressions. Materials with the word "plaster" in the title should not be used to cast impressions as they are too soft, are much harder to work with, and dry very slowly. Dental stone is much more durable and stronger than other casting materials; it dries faster, captures greater detail, does not require reinforcement materials, and does not require frames. Dental stone is readily available from local dental supply companies or military dental facilities in emergency situations.

MIXING DENTAL STONE

22-19. Dental stone can be premeasured and stored in large ziplock bags. It takes 6 ounces of water to every 1 pound of dental stone. A footwear impression requires about 2 to 3 pounds of dental stone and 12 to 15 ounces of water. A good measuring device for the proper amounts of water is the use of a beverage can, which typically holds 12 ounces. Large batches are recommended for casting tire tracks. Dental stone is very forgiving. If the mixture is too dry, water is simply added. If the mixture is too wet, more dental stone is added. The water is simply poured into the ziplock bag of material and the two are mixed inside the zip-lock bag and kneaded until the mixture is about the same consistency as pancake batter. Some water may need to be added if the mixture is too thick. A little too thin is better than too thick. This technique should be practiced before actual impressions are collected.

POURING DENTAL STONE INTO THE IMPRESSION

22-20. When using dental stone, forms are seldom required as long as the surface is somewhat level. A partial form can be used to control the flow of castings on the lower side if the impression is on a slope. Normally, the material can be poured and shaped by directing the flow. The dental stone will often act like pancake batter confining itself by its volume. If the dental stone is mixed in a plastic bag, cut or tear off the bottom corner and allow the mixture to flow out the hole. The mixture should not be poured directly into the impression. The mixture should be poured by holding the opening at ground level next to the edge of the impression on the high side and allowing the mixture to flow onto the ground and into the impression. Pieces of cardboard or other stiff materials can be used as a deflection panel to slow the flow of liquid from damaging the impression when pouring the cast. Once the entire surface of the impression is covered, it should be overfilled gently with any excess material to build the thickness of the cast for strength. If necessary, a second batch can be poured directly over the first.

COMPLETING THE CAST

22-21. After pouring, but before it dries, the cast should be marked for identification. The data can be written into the surface of the cast using a paper clip, toothpick, or similar item to make a permanent marking. The minimum data should be the investigator's initials and the time and date. If known, the case number should also be inscribed. An arrow indicating north may be inscribed to help determine the relationship of this evidence to other evidence, and the arrow could help to prove the direction of travel. If more than one impression is being collected and they were marked for photography, that identifying number or letter used should also be annotated on the cast. A typical footwear cast can dry enough to be collected in approximately 30 to 35 minutes. After collection, the cast should be allowed to dry another 72 hours before packaging and shipping to the laboratory. The cast should not be cleaned; only bulky clods of dirt should be removed. The dirt aids in protecting the cast during handling and shipment. Furthermore, cleaning techniques and equipment can damage and scratch the surface of the cast rendering it useless and destroying its individual characteristics.

SNOW CASTING

22-22. There are many ways to collect impressions in snow. One of the best methods (also the easiest and quickest) is to spray the impression with a spray wax made specifically for impression recovery before photographing. After fixing the impression with the spray wax, the required photographs should be taken. The spray wax will allow for better contrast in the photograph. A single can of spray wax is normally enough for about three impressions. The spray wax should be cured for approximately 15 minutes before the dental stone mixture is poured into the impression. Water that is as cold as possible should be used to mix with the dental stone for the best results. One trick is to set a bucket of water in the snow while working and add snow to the water until the snow stops melting. This is a good indication that the water is cold enough not to melt the snow in the impression. The dental stone should be poured into the impression and not off to the side as in other impressions. If it is poured off to the side, especially if the snow is deep, the impression could cave in, allowing the material to find the path of least resistance into the snow. The cast should be dried and then removed from the snow. The snow should not be removed from the cast; it should be allowed to melt away and then air-dried indoors for 72 hours before packaging for shipment to the laboratory. If spray wax is not available, any contrasting spray paint can be used to highlight the print for photography. The spray paint helps to seal the impression for casting the same as the spray wax.

WATER CASTING

22-23. It is very possible to cast an impression that is submerged beneath the water, especially in puddles of water and at the edges or banks of a lake or stream. Sometimes it is possible to dam around the impression and siphon off most of the water. Experience shows it is not a good idea to siphon away all of the water because the water sometimes acts as a stabilizer for the impression. If possible, a form should be placed around the impression extending above the waterline. The form should not be placed so close to the impression that it will disturb or destroy the impression. After removing as much water as possible, dry dental stone should slowly be sprinkled over the impression, allowing it to precipitate down through the water and into the impression. A flour sifter or large saltshaker can help evenly disperse the dental stone. This process should continue until the cast begins to build up. Approximately one inch of dental stone should be allowed to settle into the impression. A separate mixture of enough dental stone material to cover the entire framed area should be prepared. The mixture should be poured into the framed area of the impression, displacing the water from the impression. The casting material should have two hours of drying time before it is removed. The cast should be fully dried before shipping to the laboratory for examination. The cast should not be cleaned.

CRIME LABORATORY SUBMISSION

22-24. Casts must be carefully packaged since they are fragile evidence. One of the most important considerations in sending a cast to the laboratory is ensuring that the cast is completely dry. Casts that are not dry may develop fungus and subsequently deteriorate. Casts should not be packaged in plastic wrappings, just dry paper. Plastic promotes moisture and moisture promotes decay. Other packaging considerations are the same as for preparing evidence to be sent to the laboratory. However, the outer packaging should state, "Do Not Refrigerate" as refrigeration damages the cast. Submit all shoes, tires, and casts on a chain of custody. The impression film does not have to be on a chain of custody.

Chapter 23

Questioned Documents

For many investigations, a document or a document-related item becomes evidence of the crime or about the person who committed the crime. Often, a document is the instrument of the crime.

OVERVIEW

23-1. The Questioned Document Division, USACIL, conducts forensic examinations of document evidence. Examinations commonly conducted include handwriting and handprinting comparison, alteration and obliteration examinations, typewriting examinations, photocopy examinations, and other nonchemical examinations relating to document evidence.

EVIDENCE COLLECTION

23-2. The investigator should take notes about the process when collecting evidence involving a questioned document. The notes will later help refresh the investigator's memory if he is called to testify. The investigator should note the place, time, and date he collected the document. He should also note the name of the person he received the document from and how it was marked. Information about the history and contents of the document should be included. Later the investigator may add notes about the handling and disposition of the document. All of this information may be of value later in the investigation or in court.

23-3. The questioned document must be identified so it will not be confused with other documentary evidence. The document should be marked to identify it at a later date. The investigator should examine it to find a good place to put his initials and the time and date. He should use care in choosing this location. The identification data should be as inconspicuous as possible. It should not, in any way, interfere with any writings or impressions on the document. A corner on the back of a document is most commonly used. The investigator should remember to note in his records how and where it was marked.

23-4. Questioned documents must be protected from damage. A questioned document should never be folded, crumpled, or carried unprotected in a pocket. It should be placed in some sort of protective cover. It is best to use a paper envelope in which the document easily fits. The container should be made to fit the document. The document should not be folded to fit the container. Its evidence tag should be attached to the outside of the envelope. There should be no writing on the envelope with the document inside (see *paragraph 23-41*). When shipping to the lab, enough heavy wrapping material should be placed around it to stop it from being bent, torn, or folded in transit. With torn documents, the pieces should be placed in a protective covering and placed in the most obvious and logical positions. Transparent plastic document protectors are not suitable for use with some document materials. They should be used with caution, if at all. Typewriting made with a carbon ribbon and the toner on some photocopies may stick to the plastic and be lifted off the paper.

23-5. Questioned documents often represent valuable transactions. Sometimes they can be used as evidence for a victim in a civil suit to recover losses suffered because of a fraudulent transaction. The victim will need assurance that the document will be returned after the case is complete. The investigator should give a property receipt to the person who gives him the document. He should ensure that the receipt describes the document in enough detail to permit future identification. It should not have statements as to the value of the document. The description should be limited to the physical aspects of the document. Similar receipts should be given for any other items like pens, pencils, or paper that the investigator collects.

23-6. The document may need to be examined for fingerprints. The investigator should handle it with tweezers or cloth gloves so his fingerprints are not added. The document should not be subjected to any strong light for prolonged periods. But it may be viewed with a UV light for a short time to compare or contrast its fluorescence or reflectance with other similar documents or possible paper sources. Documents should be handled so that any indented markings are not destroyed or added.

23-7. The investigator should make copies of the questioned document for use during the investigation. The original must then be placed in the evidence depository until the lab requires it for examination. Photocopying and photographing are acceptable methods for making copies. The investigator should avoid feeding the document through any kind of sheet-feeding mechanism. This could result in damage to the document.

INTERVIEWS

23-8. The investigator should question all persons affected by the document. For example, in a case concerning a forged check, the investigator should question the cashier or the teller who accepted the check, the person whose signature had been allegedly forged, and a representative of the bank on which the check was drawn. Any bank, business, or other organization that will be affected by the questioned document should be contacted. Information about past dealings with the person whose signature was allegedly forged may give helpful clues. Other incidents in which the same forms or method of operation were used may be discovered.

23-9. If the document was prepared or signed in the presence of a witness, the witness should be questioned about the method of preparation. These questions should include the following:

- Was it written with the right or left hand?
- Was it written quickly or slowly?
- Was it written on top of other papers or on a hard surface?

- Was the writer nervous or intoxicated at the time?
- Was the writer physically impaired as a result of injury or illness?

23-10. If the questioned document is written on a special form, the investigator should talk to the individuals who normally use such forms. This will allow the investigator to examine the place where they are kept and find out who could have had access to them.

23-11. The investigator should encourage all victims and witnesses to name possible suspects. He should ascertain the reasons for their suspicions. This list should be used to check on victims and suspects. It may be of help to look into their financial status and business practices to check for motives. Checking the emotional stability of the victims and suspects may also be of value. The investigator can then try to reduce the number of suspects from which he needs to obtain dictated and collected known writings.

23-12. The investigator should try to learn how and when the document was made or used. He should get a description of the suspect's appearance, actions, conversation, and any identification he may have used. The number of suspects and the number of individuals present when the document was offered or found are also of value. The investigator should find out how the document was determined to be false or why it is suspected of being false.

23-13. If the signature is that of a known person, that person should be interviewed to verify that he or she denies writing or signing the document. In some cases, the questioned document must be shown to the victim. If possible, the investigator should avoid showing the document to the victim until after he has obtained dictated known writings or he should ensure that there is a time lag between the two actions so the format of the questioned document is not fresh in the victim's mind. It is best if the investigator does not let the victim handle the questioned document as it may negate a latent fingerprint examination. It is possible that the victim of a case may have actually made the questioned writing. Known writings of the victim will also assist the lab forensic document examiner in determining whether a questioned writing involves a simulation of the victim's writing style. For these reasons, the investigator should always obtain and submit known writings of the victim for lab examination.

HANDWRITING AND HANDWRITING COMPARISONS

23-14. After the investigator has conducted a preliminary investigation and collected the evidence, he may need to request a forensic document examination in order to attempt to identify the writer of a questioned handwriting or handprinting.

23-15. A forensic document examiner conducts these examinations by comparing the writing on the questioned document to known writings already submitted. The investigator should obtain known writings of victims, suspects, or anyone else he thinks may have written the questioned entries. (See *paragraph 23-23*, page 23-5.)

23-16. Handwriting and handprinting identification is based on the many individual characteristics that distinguish each individual's writing from that of others. These characteristics include size, slant, letterforms, proportions, height relationships, beginning and ending strokes, connecting strokes, i dots, t crossings, spacing, baseline habit, arrangement, and many others. In natural writing, these characteristics are made by habit, and the writer is not usually consciously aware of all of them. In handwriting and handprinting examinations, these characteristics are compared to determine if there are enough matching characteristics or different characteristics to support the identification or elimination of a writer.

LINE QUALITY

23-17. Line quality is perhaps the single most important characteristic evaluated in the comparison of handwriting for identification. The success of a handwriting comparison is largely dependent on the naturalness of the writings involved, both the questioned and the known. Anything other than natural writing is, to some degree, artwork. Artwork is not identifiable as to authorship because it does not contain the habitual, unconscious writing habits that make handwriting identifiable. Line quality is the tool the document examiner uses to gauge the naturalness of the writing submitted for comparison.

23-18. The natural writing of a skilled writer flows smoothly. The beginning and ending strokes are tapered because the pen is moving when it touches the paper and when it is removed from the paper. Long curving strokes are smooth in their curving movements and free of tremor or signs of hesitation. Vertical up and down strokes display natural variation in pen pressure by changes in the width and darkness of the ink line. Connecting strokes between small internal letters are regular direction changes, short smooth curves, and small well-formed loops. There is an absence of false starts and retouching. It appears to have been rapidly and reflexively written, without conscious thought about the writing process. Handwriting with poor line quality lacks one or more of these features.

23-19. Poor line quality sometimes appears in genuine writing. Illness or injury may affect the quality of the written product. Fear or stress may influence the skill displayed by the writer. Handwriting ability may be affected by the ingestion of drugs and alcohol. The conditions under which a person writes may also detract from the quality of the written line. It is very difficult to write fluently while riding in the back seat of a moving vehicle or when the paper is resting on the rough surface of a well-used field table.

23-20. More importantly to the investigator, poor line quality may also be the result of an attempt by the writer to effect a forgery by tracing or simulating the handwriting habits of someone else. Signs of tracings or simulations include blunt beginning and ending strokes; a tremulous writing line indicative of slow, careful drawing; curved lines which lack smoothness; corrected mistakes; and misinterpretations of letterforms. Unskilled forgers are also prone to patch, touch up, or try to improve a completed forgery.

23-21. Poor line quality can also be indicative of disguised writing. To hamper handwriting identification, the suspect may disguise both questioned and known writings. Poor line quality usually results from an attempt to consciously control the writing process. Other indicators of disguise are inconsistent letterforms, bizarre letterforms, unnaturally large or small writing, extreme angularity, and excessively elaborate writing.

23-22. An awareness of the difference between good and poor line quality in writing can help an investigator spot possible forgeries when screening records during an investigation and enable him to recognize disguised, dictated known writings when they are being created by a suspect.

KNOWN WRITINGS

23-23. Because handwriting and handprinting examinations are done by comparison, the known writings must be comparable in kind to the questioned writing. The known writing must contain the same words or, at least, the same letters and letter combinations as the questioned writings. Cursive handwriting generally must be compared to cursive handwriting, not handprinting. Handprinting generally must be compared to handprinting, not handwriting. Capital letters must be compared to capital letters. Lowercase letters must be compared to lowercase letters.

23-24. There are two types of known writings, each with advantages and disadvantages. Both types should be obtained, if possible.

COLLECTED KNOWN WRITINGS

23-25. These are writings collected from various sources that the writer prepared for purposes usually unrelated to the investigation. For example, they may include military records, other government documents, employment documents, financial records, personal correspondence, or negotiated personal checks. The advantages of this type of known writing are that they were usually written naturally with no intent to disguise the appearance and they show the individual's writing over a period of time. The disadvantages are that they may not be fully comparable to the questioned writing, and the number available is likely to be limited.

DICTATED KNOWN WRITINGS

23-26. These are writings prepared under the supervision of the investigator. The advantage of this type of writing is that the investigator can control the form used, the content of the writing, the type of writing, the number prepared, and the manner in which the writings are prepared. The main disadvantages are that the writer may attempt to disguise the writing and the writings only show the individual's writing as it appears on a single occasion.

23-27. When obtaining dictated known writings the investigator should—

- Obtain collected known writings first.
- Be familiar with the appearance of the individual's natural writing.
- Obtain paper or forms similar to the questioned document.
- Use a writing instrument similar to that used for the questioned document (if it is not an ordinary ballpoint pen, he should also get some writings with a black ballpoint pen).
- Use a copy of the questioned document to dictate the questioned text and ensure comparability of the known writings.

- Dictate the questioned writing to the writer.
- Have the writer positioned in a similar position during the dictated writing as that of the questioned document (such as standing or sitting)

23-28. The dictated known writings should be prepared on blank forms (such as government forms or checks) or in a format (such as paper size or arrangement) similar to the questioned document. When using copies of forms or checks, the investigator should ensure that they are clean copies with white backgrounds (not gray or dark), such as those copied from designer checks so that the writing is readable.

23-29. The investigator should include all of the entries that the person may have written. As each sample is completed, the investigator should remove it from the writer's sight. Each sample should be numbered and marked with the investigator's initials and the time and date. It is also necessary to obtain several examples without giving the writer any instructions. If instructions are necessary to ensure comparability (such as cursive versus printing or capital versus lowercase letters), the investigator should note the instructions given on the back of the first sample to which it applies. If the writer is suspected of trying to disguise the writing, additional examples should be obtained.

23-30. The writer should not be shown the questioned writing and multiple samples, such as signatures, should not be put on the same sheet of paper. Writings should be obtained on one side of the paper, only. If the questioned document has writing on both sides, the back should be duplicated separately.

23-31. For questioned signatures, personal checks, and similar brief writings, the investigator should get about 25 repetitions. For documents about the length of a short note, the investigator should get 10 to 15. For questioned text that is one or more full pages in length, he should get at least one complete repetition of the text and then get additional repetitions of important parts, such as admissions or text that is the essence of the crime (such as a threat or obscenity). The investigator should consider getting additional repetitions of the first and last paragraphs if there are no particularly important parts.

23-32. If the investigator has a number of questioned documents with many repeated entries, such as personal checks, the number of known writings to duplicate each document can be reduced. The investigator should ensure that he gets several samples of all of the questioned entries, especially those that only appear on one questioned document.

23-33. One to three samples written with the individual's hand that he does not normally use for writing should be obtained. These are usually enough, but if the person shows real ability with this hand or if the writing looks like the questioned document, the investigator should get more.

23-34. The investigator may have a document that contains obscene or classified words or phrases. Dictated known writings may be made without the objectionable words or classified information if the document is long enough. But the elimination of such words must not leave the dictated known writings incomplete for comparison. Dictated known writings of short documents of this type must normally be produced in full.

23-35. If the person objects to obscene words in the dictated known writings, the investigator should use nonobscene words with the same letters and letter combinations as the obscene words. The investigator should ensure that the dictated known writing (words) he chooses includes the beginning and ending letters of the obscene word. The substitute words should be in the exact same position in the sentence. A similar method can be used for classified documents. If omitting the classified portion can produce the dictated known writing, the investigator should do so.

TRACINGS AND SIMULATIONS

23-36. The author of a tracing or simulation usually cannot be identified by handwriting comparison because these are not the author's natural handwriting. However, a simulation is sometimes so poorly done that identifying characteristics may remain. It may be possible to associate a tracing or simulation with the specific genuine writing used as a model to produce it.

TRACING

23-37. A tracing is a duplicate of another individual's writing, typically a signature, made using the writing line of a genuine writing as a model or guide. Several methods may be used. The most common involve viewing the model through the paper onto which the tracing will be placed, with or without the aid of backlighting, or transferring a guideline from the model signature to another piece of paper using carbon paper or other means. A tracing may show signs of having been slowly drawn. The investigator should be alert to the possibility of recognizing and seizing the genuine writing that was used as a model.

SIMULATION

23-38. A simulation is a freehand imitation of another individual's writing, typically a signature. It may show signs of having been slowly drawn. It commonly involves the use of a genuine writing as a model. The investigator should be alert to the possibility of recognizing and seizing the genuine writing that was used as a model.

WRITINGS ON WALLS AND SIMILAR SURFACES

23-39. Some questioned writings are written on walls, doors, and similar surfaces. If necessary, the investigator should consider removing the surface to secure the writing as evidence. Whether the surface is removed or not, the writing should be photographed. Ideally, photographs should be taken using a normal focal length lens (approximately 50 millimeters) from a position directly in front of the writing using available light and including all of the writing in the photograph, along with a scale indicating size. Because of the lack of space, it may be necessary for the investigator to use overlapping photographs (photographs taken from a variety of angles or a wide-angle lens). It may also be necessary for the investigator to use a flash or other artificial light. A flash pointed directly at a flat surface usually produces a glare in the center of the photograph that obliterates the writing. The investigator should consider using a bounced, diffused, or low-angle flash. He should try a variety of approaches to ensure usable results.

23-40. Dictated known writings obtained to compare with such questioned writings should mostly be obtained in the usual manner, at a table using normal-sized paper. But large sheets of paper taped to a wall or a large tablet on an easel may be used to simulate the vertical surface and large writing size of the questioned writing.

WRITING INDENTATIONS

23-41. Writing indentations are produced when the pressure of writing on a sheet of paper is transmitted to the sheet or sheets of paper beneath it. This often occurs with writing on tablets of paper. It can also occur on loose sheets and any other paper beneath a sheet being written on. Writing indentations can be important in many types of investigations, but they can be especially helpful in cases involving anonymous notes. The note may bear indentations of writing that can lead the investigator to the writer.

23-42. It may be possible to read writing indentations in the field with the help of a light held at a low angle to the page. The investigator should not attempt any other method (such as a pencil or fingerprint powder) to enhance these indentations.The Questioned Document Division, USACIL, is equipped to develop and preserve writing indentations (even those too faint to see).

23-43. The investigator should ensure that no new indentations are added to documents that will be examined for writing indentations. He should protect the document with cardboard or place it in a rigid container. The investigator should ensure that it is not placed in an envelope or beneath other documents that will be written on.

ALTERATIONS

23-44. An alteration occurs when someone tries to change a document or obliterate part of the text on a document. Such documents can be submitted to the Questioned Document Division, USACIL, with a request to attempt to determine if an alteration has occurred or to attempt to decipher the original entry that was altered or obliterated.

TYPEWRITTEN DOCUMENTS

23-45. Sometimes a typewritten questioned document can be linked to the typewriter used to type it. For this to be possible, the typewriter or the typing element (if present) must have developed individual characteristics, usually in the form of damage or other mechanical defects, which appear on documents typed on that machine. The individual characteristics may include damaged letters, alignment problems, or other things. Typewritten text from a typewriter with no individual characteristics may not be distinguishable from that of another typewriter of the same make and model in good condition.

23-46. On typewriters of the older, typebar design, the individual characteristics belong to the typewriter. On newer, single-element (such as daisy wheel or ball element) typewriters, the individual characteristics are likely to be on a removable typing element. It is important to locate the typing element used to type the questioned document.

23-47. It may be possible to reduce the number of suspect typewriters or typing elements by doing a field comparison of the type style on the questioned document with the type style of the typewriters or typing elements.

23-48. First, the investigator should look for obvious differences. A very different type size or type style will tell him that a different typing element (in the case of removable element machines) or typewriter (in the case of typebar machines) is involved. Different letter spacing, such as 10 characters per inch versus 12 characters per inch, can show a difference between older machines; but on newer typewriters, this setting can be changed.

23-49. Next, the investigator should examine the typed characters. He should check the upper and lowercase letters of the M and W first, as they are often the most distinctive in style. Their differences may be easily recognized. The bottom of the staffs of the lowercase may or may not have serifs (cross strokes) at the bottom. The two outside staffs may have serifs, and the center staffs none. The center V-like formation of the capital M may descend to the baseline or stop varying distances above it. If it descends to the baseline, it may or may not have a serif. The inverted V of the center formation of the W may or may not extend to the top of the line formed by the outer portions of the letter and may or may not have a serif at the top. Other characters with designs that help distinguish between typestyles are the letters g, t, a, r, y, i, f, and the numerals. If the letters and numerals are not distinguishable with ease, the investigator should submit typewriter exemplars to the lab.

23-50. Before obtaining typewriter exemplars, the investigator should check the ribbon of the suspect typewriter. If the ribbon is a carbon film ribbon that passes through the typewriter only once and bears transparent images of the letters typed, the investigator must remove the ribbon cartridge. He should not take exemplars on this ribbon. He should seize it as evidence and preserve it for a possible typewriter ribbon examination. The investigator should use another ribbon known to be unconnected with the investigation to obtain the exemplars.

23-51. When obtaining typewriter exemplars, the investigator should duplicate the content and formatting of the questioned document. It is desirable (but not essential) for the document examiner to be able to overlay the questioned and known writings. The investigator should pay particular attention to the letter case (upper or lower), margins, tabs, spacing between letters and words, and line spacing.

23-52. If the questioned document consists of about a one-half page, it should be reproduced in its entirety. If the document is lengthy, the first 20 to 30 lines should be reproduced. The remainder of the questioned document should then be examined. Any words, numerals, or symbols not appearing in the first 20 to 30 lines should be added to the sample. The words proceeding and following the material to be added should be included and typed as it appears in the questioned document.

23-53. The investigator should type or note the make, model, and serial number of the typewriter on the exemplars. He should mark them as evidence in the usual way and account for them according to *AR 195-5*. The investigator should find out when the ribbon on the machine was last changed. He should learn the nature and date of the latest repair work done on the typewriter.

23-54. If possible, the investigator should collect known typewriting produced on the suspect typewriter from office files or wherever they might be found. Typewriter characteristics can change with use, maintenance, or repair, and it may be important to locate documents typed on about the same date as the questioned document. Sometimes changes in the condition of a typewriter can be used, together with dated collected typewriter standards, to determine the approximate date a questioned document was typed.

23-55. Consider seizing the suspect typewriter and typing element as evidence if present. Coordinate with USACIL before shipping a typewriter for examination. However, it may be possible to complete the examination using only the typewriter exemplars. It may be helpful to submit separate typing elements to USACIL with the exemplars and questioned document.

TYPEWRITER RIBBONS AND CORRECTION TAPES

23-56. Some typewriters and computer printers have carbon film ribbons that pass through the typewriter once and bear transparent images of the typed characters. The Questioned Document Division can read these ribbons to attempt to locate a questioned text or to determine what was typed on the typewriter from which they were taken. Fabric ribbons or multistrike carbon film ribbons cannot be read.

23-57. Sometimes it is possible to link a carbon film ribbon to a document by comparison of irregularities in the carbon transfer, paper fiber impressions on the ribbon, or other characteristics. If a typewriter has a correction tape that is used to strike over or lift off typographical errors, the tape should be seized. It may be possible to match characters on the tape to corrected errors on a questioned document.

COMPUTER PRINTER DOCUMENTS

23-58. Most types of modern computer printers are simple and reliable devices that are less likely than typewriters to place individual characteristics on the documents they produce. However, individual characteristics maybe present, so consider obtaining known documents from suspect printers and submitting them for examination. Coordinate with the Questioned Document Division before submitting the printer for examination.

PHOTOCOPIED DOCUMENTS

23-59. A copy produced on a photocopier or similar device can sometimes be linked to the copier that produced it. This is done by matching individual markings placed on the photocopy by the copier with those on known photocopies from the suspect photocopier. Such markings may result from trash particles or marks on the glass platen of the machine, from damage to the copying drum, from images of parts of the machine included in the copy, or from other sources.

23-60. It is also possible to eliminate a particular photocopier or similar device from having produced a questioned copy. This is usually done by comparing class characteristics of a questioned copy with those of known copies made by the suspect machine, but it may also be done by other means.

23-61. When investigating cases in which the evidence is a photocopy, it is important to locate and seize the particular copy that is the evidence or instrument of the crime. Additional copies of such documents are commonly made after the offense for administrative purposes, and these subsequent copies are much less useful, both as evidence and for the purpose of forensic examinations. The copy seized is evidence and must be accounted for according to *AR 195-5*.

PHOTOCOPIER EXEMPLARS

23-62. A clean, blank sheet of paper the same size as the questioned copy should be submitted to USACIL. It should be without paper impurities or other marks that can be copied by the copier. This sheet of paper should be saved as evidence as an item separate from the exemplars and submitted to USACIL. The investigator should—

- Place the sheet of paper on the glass platen of the copier and make 10 copies on paper the same size as the questioned copy.
- Remove the sheet of paper, close the copier lid, and make 10 additional copies.
- Open the copier lid and, with no paper on the platen, make 10 copies.
- Feed the blank sheet through the feeder, if the copier uses a sheet feeder, and make 10 copies.
- Keep the blank sheet and each group of copies separate and note how they were obtained.

23-63. Since the individual characteristics of photocopiers can be changed or eliminated by cleaning and maintenance, the investigator should attempt to collect existing photocopies known to have been made on the suspect machine. Ideally, these existing photocopies should have been made around the same time as when the questioned document was produced. Similarly, collected standards can sometimes be used to determine the approximate date a questioned copy was produced.

PRINTED DOCUMENTS

23-64. Documents produced by one of the various types of printing processes (such as offset, letterpress, or flexography) may become evidence in a criminal investigation. There are several forensic examinations that may be requested for these documents, depending on the circumstances. These include—

- Determining what method was used to print the document.
- Determining whether the document is genuine or counterfeit.
- Determining which printing job the document was printed on.
- Determining the approximate date on which the document was printed.

23-65. Other examinations may be possible. The information and standards needed by USACIL will vary depending on the issues and circumstances. The Questioned Document Division should be contacted for guidance.

MECHANICAL IMPRESSIONS

23-66. Occasionally, investigators will encounter documents with mechanical impressions made by a device, such as a check protector or an embossed seal. A forensic examination may be needed to determine whether the impression is genuine or whether the impression can be associated with the device alleged to have produced it.

23-67. In such cases, the investigator should obtain the suspect device and submit it to USACIL for comparison with the questioned document. If seizing the device is not possible, he should prepare about 20 exemplars with the device and submit them. If the device has data that can be changed, such as on a check protector, the investigator should change it to duplicate the information on the questioned document. Also, he should try to collect standards prepared with the device about the same date as the questioned document.

23-68. If the investigator believes that a mechanical impression on a document is fraudulent, he should obtain specimens of genuine impressions and submit them for comparison. Coordination with USACIL is recommended.

RUBBER STAMPS

23-69. Questioned rubber stamp images on documents can be compared to suspect rubber stamps or to documents bearing rubber stamp images from a known source. Although rubber stamps can be mass produced, they may acquire individual features, such as manufacturing defects or damage and wear resulting from use.

23-70. It is best to seize suspect rubber stamps and submit them to USACIL for comparison with the questioned stamp impression. If the stamp cannot be seized, the investigator should prepare rubber stamp exemplars. They should be prepared using different amounts of ink and with different amounts of pressure. The investigator should ensure that the entire surface of the stamp is reproduced and make impressions from different angles, if necessary.

23-71. Whether the investigator submits the rubber stamp or exemplars, it is important that he attempts to locate and obtain any existing documents on which the stamp was used. The individual characteristics of rubber stamps may change with use and cleaning. It may be necessary to have existing documents on which the suspect rubber stamp was used on around the same time as the questioned stamp image. With such documents, it may be possible to establish the approximate date the questioned rubber stamp image was made.

INK EXAMINATIONS

23-72. The Questioned Document Division is equipped to do nondestructive examinations of inks including infrared, infrared luminescence, UV, and other nonchemical examinations.These examinations are usually performed for the purpose of detecting alterations, deciphering obliterations, or determining that entries were made in different inks.

23-73. All of these examinations attempt to detect a difference between inks. Nondestructive examinations are limited to the following potential results: the inks are different, or no difference was detected between the two inks (a difference may exist though one was not detected). Destructive ink examinations are currently conducted by the US Secret Service (USSS). The optimum potential finding by destructive examination is that no differences exist between two inks.

PAPER EXAMINATIONS

23-74. Paper can be examined to determine its physical characteristics. Pieces of paper can be compared to determine whether they are different or whether no difference can be found. Some characteristics found in some papers, such as watermarks, may help determine the source of the paper or even when it was produced.

23-75. A paper examination, along with other evidence, may be useful if you suspect that the paper used for a document is the wrong kind or a page has been added or substituted in a multipage document. It may also show that the paper of a questioned document is of a type available to a particular suspect or from a particular source.

TORN, CUT, AND SHREDDED DOCUMENTS

23-76. Examining torn, cut, or shredded documents may serve one of two purposes: matching a paper fragment to another piece of paper from which it was separated for the purpose of associating the fragment with a source; and reconstructing torn, cut, or shredded pieces of a document so that the document may be seen whole. In such cases, it is important to recover all fragments and protect them from further damage.

23-77. Sometimes document fragments can be reconstructed in the field without USACIL involvement. In such cases, the investigator should not use tape, glue, or any other permanent adhesive. When a high-security shredder or another good quality shredder has shredded documents, it may be impossible to reconstruct the document.

DOCUMENT DATES

23-78. It is possible to determine the date, or approximate date, that a document was prepared. It is also possible to determine that a document was not prepared on the date alleged.

23-79. The forensic techniques that may allow these findings are varied but may include typewriter, photocopier, ink, and paper examinations. Much of the potential for reaching any results at all will depend on information gathered by the investigator about the date, circumstances, and methods alleged in preparing the questioned document. The collection of various standards may be necessary, especially samples of genuine documents of the same type prepared on or about the alleged date.

23-80. The best approach and potential for a useful result in this type of examination varies considerably with the facts of the case. Coordination with the Questioned Document Division is recommended.

CHARRED DOCUMENTS

23-81. It is often possible to read text on a charred document. A charred document is different from ashes. A charred document has been blackened and made brittle from exposure to high heat without enough oxygen to burn. To be examined, the pieces must be large enough to have legible text. Charred documents are very fragile. Pick them up by sliding a sheet of paper beneath them and, using this sheet as a support, transfer the charred documents to a shallow, cotton-lined box (such as a pie box). Sheet cotton stapled to the top and bottom inner surfaces of the box will prevent movement by the charred document.

23-82. If a single charred document is relatively flat, it may be placed between two panes of glass that you then tape together. If feasible, the charred documents should be sent to the lab by courier. This will preclude unneeded handling and prevent destruction. In some cases, the laboratory examiner should be asked to come to the location of the document. If neither of these two preferred methods is practical, careful packaging is needed to preclude destruction.

COPIES AS EVIDENCE

23-83. Original documents, rather than copies, should always be obtained as evidence when they are available. Originals are the best evidence to present in court, and they are the best for the purpose of forensic examinations. Some forensic examinations can only be performed on an original. Handwriting comparisons using copies typically yield poorer results than could have been obtained with an original.

23-84. Sometimes an original document is not available because it cannot be located or has been destroyed. In other cases, a copying process has been used to fabricate a document that did not exist as an original. In these cases, a copy is obtained as evidence.

23-85. It is important to get the best copy available. If a copy has been used as the instrumentality of a crime, that copy should be obtained, rather than subsequent copies. If a copy is obtained as a substitute for a missing original, the investigator should try to obtain a copy that was made directly from the original. He should try to avoid getting copies of copies.

23-86. If an original document is to be considered as evidence, but is not available, then the investigator must obtain a copy of the original to use as evidence, and it must be accounted for according to *AR 195-5*. The investigator should obtain a copy even if he expects to obtain the original later.

23-87. The copy the investigator submits to USACIL for examination must be the evidence copy, not a case file copy or a copy made especially for laboratory submission. If expert testimony or a laboratory report will be used in court, the same copy examined by the laboratory examiner must also be introduced as evidence in court.

COURT AUTHENTICATION

23-88. In order for expert testimony or a laboratory report to be admitted as evidence in a trial, the evidence examined must have been admitted as evidence. In order to be admitted, the evidence must be authenticated, that is, shown to be what its proponent claims. The presiding judge decides authentication and admissibility.

23-89. It is particularly important to be aware of this requirement with regard to known writings. Dictated known writings may be authenticated by the testimony of the investigator who obtained them. Collected known writings may be authenticated by one of several means. The investigator should review Military Rules of Evidence (specifically rules 901, 902, and 903) or Federal Rules of Evidence (as appropriate) and consult his legal advisor.

LATENT PRINTS ON DOCUMENTS

23-90. Consider requesting a latent print examination on questioned document evidence. Latent prints on paper are relatively permanent, and it is uncommon for additional handling to obliterate them.

23-91. Using gloves or forceps while handling these documents and placing them into an envelope should protect them. The investigator should use a pencil for evidence markings and place the markings in a place less likely to have been handled by the suspect.

23-92. When both questioned document and latent print examinations are to be done, the questioned document examination is done first, and the document is protected for the latent print examination. Latent print examinations on paper normally degrade the document in a manner that would hamper a subsequent questioned document examination.

EVIDENCE SUBMITTED

23-93. All evidence should be submitted at one time. A case cannot be examined until all evidence is received. If evidence or documents are requested from another office, the added material should be obtained before forwarding the referrals to the lab. This precludes the lab having to hold referrals that cannot be examined pending receipt of other evidence. If the examination requires an original document on file at the Defense Finance and Accounting Service (DFAS) and DFAS will not release it to the investigator, request that DFAS send the document directly to USACIL. The investigator should retain any other document evidence (that is to be examined) until notified by DFAS or USACIL that such documents have been sent or received before submitting the other evidence.

ON-SITE ASSISTANCE

23-94. Investigations occur that justify on-site assistance by USACIL Forensic Document Examiners. When large numbers of questioned documents are involved, document examiners can assist investigators in screening the documents for those most likely to be productive for handwriting and other examinations. When large numbers of suspect writers, copy machines, typewriters, and other sources of documents are involved, document examiners can screen the sources (such as personnel records or post locator cards) to identify the source of questioned documents. On-site assistance trips by USACIL document examiners have been extremely successful. The investigative and lab examiner hours saved by these trips have been significant.

Chapter 24

Deoxyribonucleic Acid Evidence

Serological evidence often consists of microscopic particles and is generally not obvious at a crime scene. It may be DNA on glass or blood on a victim's clothing. Serological evidence is similar to trace evidence in that it can easily be overlooked or destroyed. A sound, prosecutable case concerning serological evidence begins the moment that an investigator arrives at a scene. It includes the processes he uses in evaluating, collecting, and preserving serological evidence and the court testimony that he or the evidence examiner provides. A suspect may leave blood, hair, and fibers at a crime scene. Likewise, a suspect may carry this evidence away from the scene.

DEOXYRIBONUCLEIC ACID

24-1. DNA is the basic component of an individual's entire genetic structure. Virtually every cell of the human body contains DNA. The DNA in an individual's cells is the same for each type of cell, such as an individual's saliva, hair, and skin. All cells with the exception of sex cells have the same DNA. An individual's DNA remains the same throughout his life.

24-2. DNA is a powerful tool in an investigation because no two people have the same DNA (with the exception of identical twins). Because of that difference, DNA collected from a crime scene can link or eliminate a suspect to the evidence. It can place a person at a particular location that the person denies having been. DNA from relatives can assist in identifying a victim even when a body does not exist. The individuality of DNA allows the linking of a suspect from one crime scene to another when the evidence of separate crime scenes is compared. This capability may be within a small community, statewide, or even nationwide. The uniqueness of an individual's DNA can positively identify a suspect or exonerate an innocent suspect.

EVIDENCE IDENTIFICATION

24-3. DNA evidence can be collected from basically anywhere. Only a few cells are required to obtain useful DNA information relevant to an investigation. DNA has helped solve numerous investigations when creative investigators collected evidence from unlikely sources. One such investigation involved a murder case being solved by taking DNA from saliva in a dental impression mold and matching it up with DNA swabbed from a bite mark on the victim. Investigators must realize that although they may not be able to see evidence (DNA cells), it does not mean that there is not enough DNA cells present for typing. Typing is a technical process performed by laboratory technicians and, because of this distinction. Typing allows laboratory technicians to identify a specific pattern present in an individual's genetic makeup. *Table 24-1* shows a myriad of locations where DNA evidence can be obtained, what that evidence is, and the biological source of origination of the DNA. DNA testing of urine

Table 24-1. DNA Evidence

Evidence	Possible Location of DNA on the Evidence	Source of the Evidence	Collection
Baseball bat or similar weapon	Handle and end	Skin, blood, and/or tissue	Place in a clean paper bag after allowing the wet fluids to air-dry.
Hat, bandanna, or mask	Inside	Hair and/or dandruff	Place in a clean paper bag.
Eyeglasses	Nosepiece, earpieces, and lens	Skin	Place in a clean paper bag.
Facial tissue or cotton swab	Surface area	Mucus, blood, semen, and/or earwax	Place in a clean paper bag after allowing the wet fluids to air-dry.
Dirty laundry	Surface area	Blood and/or semen	Place in a clean paper bag.
Toothpick	Tips	Saliva	Allow to dry, place in (bond) paper, and complete a pharmacy fold.
Used cigarette	Cigarette butt	Saliva	Allow to dry, place in (bond) paper, and complete a pharmacy fold.
Stamp or envelope	Licked area	Saliva	Allow to dry, place in (bond) paper, and complete a pharmacy fold.
Tape, ligature, or other binding item	Inside and outside surface	Skin	Place in a clean paper bag.
Bottle, can, or glass	Sides and mouthpiece	Saliva	Place in a clean paper bag.
Used condom	Inside and outside surface	Semen and/or vaginal or rectal cells	Allow to dry, place in (bond) paper, and complete a pharmacy fold.
Blanket, pillow, or sheet	Surface area	Hair, semen, urine, and/or saliva	Place in a clean paper bag after allowing the wet fluids to air-dry.
Bite mark	Individual's skin or clothing	Saliva	Swab saliva with a cotton swab and place in a clean paper bag.
Fingernail or partial fingernail	Scrapings	Blood and/or tissue	Place in bond paper and complete a pharmacy fold.

may be conducted to prove or disprove whether an individual is the source of a specimen in which illegal drugs have been identified.

COLLECTION AND PRESERVATION

24-4. Investigators and laboratory examiners work together to determine the most probative pieces of evidence and establish priorities. The collection and preservation of DNA evidence in itself is not defined in this section; however, it is important to note that the initial collection of evidence is a key and vulnerable link in the chain of events leading to successful DNA testing. Investigators should aim to collect and properly preserve potential evidence for DNA testing while minimizing the possibility of contamination. Each type of evidence that may bear potential DNA evidence outlines the procedures required for collecting and processing that particular type of evidence under its individual heading, such as fingernails and blood samples. It is important that personnel handling biological evidence be aware of the potential for the presence of hazardous pathogens, such as the human immunodeficiency virus (HIV) and the hepatitis B virus. Some DNA samples are collected and submitted to the USACIL CODIS laboratory as outlined in this chapter.

CONTAMINATION

24-5. Great attention to contamination issues is necessary when identifying, collecting, and preserving DNA evidence because even minute samples of DNA can be used as evidence. DNA evidence can become easily contaminated when DNA from another source gets mixed with the DNA relevant to an investigation. The initial responders to the scene of a crime, investigators, and laboratory personnel must be aware of this. They must be careful not to cough on potential evidence. They must not run their hands through their hair and then handle evidence. To avoid contaminating possible DNA evidence, the investigator should—

- Wear gloves and change them often.
- Use disposable instruments or clean them thoroughly before and after handling each sample.
- Avoid touching the area where he believes DNA may exist.
- Avoid talking, sneezing, and coughing over evidence.
- Avoid touching his face, nose, and mouth when collecting and packaging evidence.
- Air-dry the evidence thoroughly before packaging.

TRANSPORT AND STORAGE

24-6. It is important to keep evidence that may contain DNA dry and at room temperature during transport and while in storage. While being careful to secure possible DNA evidence in paper bags or envelopes, it is necessary to seal, label, and maintain a proper chain of custody. Proper identification of the evidence and the location from where it was obtained are crucial to the chain of custody. Personnel handling the possible DNA evidence must never place it in plastic bags. This may produce undesirable moisture that would damage the DNA. Just as important, personnel handling possible DNA evidence should not allow it to be exposed to overly hot conditions (such as an investigator's vehicle without air-conditioning) or direct sunlight. Again, these conditions can damage DNA evidence.

ELIMINATION SAMPLES

24-7. The effective use of DNA may require the collection and analysis of elimination samples. The elimination samples may be necessary to determine whether the evidence came from the suspect or from another source. Military police and investigators responding to a scene must think ahead. If a crime occurred in the bathroom of a residence, the investigator should determine who resides at that location that may have DNA present. It is these individuals that elimination DNA samples should be taken from. The DNA (specifically from blood samples) of a deceased victim of a crime is collected from the medical examiner. Collecting elimination samples may be very sensitive in nature. In considering a rape investigation, it may be necessary to collect and analyze the DNA of the victim's recent consensual partners, if any, to eliminate them as potential contributors of DNA suspected to be from the suspect. If this is necessary, the help of a qualified victim advocate should be enlisted. Extreme sensitivity and a full explanation of why the request is being made should be given to the victim.

BODY FLUIDS

24-8. Body fluids, such as blood and saliva, provide the necessary DNA to link or dismiss a suspect to a crime. Body fluids must be collected carefully and forwarded to the laboratory. Even dried body fluids can be collected and forwarded to the laboratory for comparisons.

BLOOD

24-9. In crimes of violence, blood evidence is very valuable if handled properly. It can indicate to an investigator if a victim's body was or was not moved from the location in which the victim was killed. This indication comes from a pool of blood that is in the vicinity of the victim's body or the lack of a pool of blood. The manner in which blood is present, such as splatters or transfers, may also assist the investigator. (See *Chapter 5*.)

24-10. Sometimes liquid blood samples must be sent to the laboratory with other evidence. A medical officer or a trained medical technician should draw blood samples. Medical personnel may take samples of body fluids like blood and urine from soldiers without their consent when authorized to do so by a search warrant or search authorization. Fluid samples may be taken from nonconsenting soldiers without a warrant or authorization if a delay could destroy the evidence. The samples should be taken at a medical facility where proper precautions can be taken to prevent contamination of the samples. Medical facilities have sterile containers available for sending samples to the laboratory.

24-11. The amount of liquid blood needed for laboratory DNA examination is about 5 milliliters or one tube. The tube of blood should be sent with an anticoagulant in a purple-top tube. If there is a delay in sending drawn blood to the laboratory, refrigeration should be used, but the sample should not be frozen. The preferred method is a punch card dried sample.

24-12. Examiners performing a preliminary laboratory examination of an alleged bloodstain use chemical tests to tell if the stain is a bloodstain. If the results are negative, the stain cannot be blood. If the results are positive, further examination and testing are required. The chemical tests may not be conclusive. Other substances, common chemical compounds, and certain body discharges may also give positive results. The inability of the laboratory to provide information on bloodstain evidence is often due to unsuitable samples. Late shipment or contamination of evidence can cause unsuitable samples.

24-13. If testing shows that the stain is a bloodstain, it must then be determined if the blood is human. The evidence value of a bloodstain may be seriously impaired unless the stain is shown conclusively to be human blood. A suspect may claim that the stain is blood from an animal that the suspect has handled in some way.

SEMEN

24-14. In the case of a rape or sexual assault, it may be alleged that the suspect had an emission. If so, the identification of semen is of paramount importance. Semen is a colorless, sticky fluid produced in the male reproductive organs. It is often found in the form of stains on clothing, bedding, or other articles.

24-15. Fresh, undried semen has a characteristic odor. Semen contains thousands of minute organisms, known as spermatozoa, which die as the semen dries. Spermatozoa keep their shape indefinitely if they are not destroyed through handling. In its dried state, semen appears as a grayish-white, sometimes yellowish, stain. It gives a starchy stiffness to the part of the fabric that has been stained. Suspected fluid or stains may be identified as semen by the laboratory even if the attacker has had a vasectomy. Specific tests for semen involve the identification of the spermatozoa and chemical testing of the stain. Items believed to bear seminal stains should be handled with care at all times.

24-16. Inspection of evidence under UV light sometimes helps find the location of semen stains. Semen stains have fluorescent qualities. Laundering may remove traces of seminal stains; the investigator should check for them in any event.

HAIRS

24-17. The value of hairs as evidence in criminal cases has been clearly recognized. Hairs are seldom conclusive evidence but, in conjunction with other details, they have proven to be important and essential aids to investigators. The investigator must capitalize on the importance of this type of evidence during the initial phase of the investigation.

24-18. The origin and texture of hairs found at a crime scene or on the body, clothing, or headgear of a suspect or a victim may be highly important as evidence. This is especially true in homicides and sex crimes. Hairs may be pulled out during the crime and left at the scene or on the victim. Hair transfer may take place during any physical contact between a suspect and a victim. Hair may fall out under conditions that a suspect is not aware of and cannot guard against. Properly handled, hair and fibers may yield excellent investigative leads and add to the evidence facts being assembled.

24-19. Structurally, a hair is composed of the tip end, cuticle, cortex, medulla, and bulb or root. Each of these parts provides the laboratory examiner with definite information. During examination, the laboratory will usually first see if the hair samples are animal or human. If the hairs are from an animal, a general determination of the species may be made to see if the hair came from a cat, a dog, a horse, a cow, or another animal.

24-20. In the case of human hairs, a laboratory determination may yield several findings. It may show the race of a person. It may show where on the body the hair originated, such as the head, face, chest, armpit, limb, or pubic area. A finding may show if the hair was removed naturally or forcibly. The laboratory determination may show if the hair was bleached, dyed, or waved. It may also show whether the hair was cut with a dull or a sharp instrument (if the cutting was recent) and whether it had been crushed or burned. It may also show whether blood grouping and sex can be estimated or determined.

24-21. Laboratory comparisons of hair will generally result in one of the three following conclusions:

- The hairs are dissimilar and did not come from the same individual.
- The hairs match in terms of microscopic characteristics and blood groupings and came from the same person or another person whose hair has the same microscopic characteristics.
- The comparisons show that no conclusion could be reached concerning the origin of the hair.

NOTE: The laboratory can now conduct DNA testing of hair, proving or disproving that hair originated from a particular person.

24-22. Hair and fiber evidence is very susceptible to cross contamination. Evidence gathered from a suspect and a victim must not be intermingled. It must be individually collected, marked, and kept separated during packing for shipment. Detailed examinations of hair and fiber should be left to the laboratory.

24-23. Twenty hair strands are considered a minimum sample. Only a doctor should collect sample hairs from the body of a victim or a suspect. These samples should be obtained from any of the parts of the body that could be involved in the crime. Hair combings and representative samples of pulled hairs should be submitted.

24-24. Hair should be placed on a clean piece of paper; the paper should be folded into a packet and put into a clean container.

24-25. When transparent adhesive tape is used to collect the hairs, the tape should be placed with the adhesive side down on the inside of a document protector, with the paper insert removed, or on the inside of a plastic bag. The document protector or plastic bag should be sealed in another container. Under no circumstances should the tape be affixed to an index card or other paper. Envelopes that are sealable around all edges and plastic or kapok bags should be used as containers for hairs.

FINGERNAIL SCRAPINGS AND BROKEN FINGERNAILS

24-26. Fingernail scrapings should be exploited to the fullest advantage. The cause of abrasions and scratches found on many parts of the body are often from fingernails. The face, neck, arms, thighs, and genitals are the places commonly attacked. These places should be medically examined, carefully. The form, extent, and location of abrasions will depend on the circumstances in each case.

24-27. A victim's resistance to a sexual assault often results in gouges caused by the suspect's or victim's fingernails. Minute particles of fibers, skin, blood, hair, and cosmetics found under the fingernails may help link the suspect and the victim.

24-28. Examination of the fingernails of an unidentified corpse may show that individual's occupation. Fingernails that are trimmed and bear scratches but not regularly manicured may indicate some manual labor. Fingernails that are beveled, brittle, growing tight at the corners, rounded at the ends, and regularly manicured may indicate a lack of manual labor. Fingernail scrapings may also show that a person has handled narcotics, marijuana, or poison.

24-29. The residue under a suspect's fingernails may have traces of substances from the crime scene or from the victim's body or clothing. Scrapings should be taken from all of the suspect's fingers, preferably before the suspect can bathe or clean his nails. Scrapings should be kept separate and placed in appropriate containers. In taking fingernail scrapings from a suspect or a victim, a knife, a file, or any other hard, sharp instrument should not be used. It may cause bleeding and contaminate the nail scrapings. The best item to use is the blunt end of a flat, wooden toothpick. A different toothpick should be used for each finger. As the scrapings from each finger are taken, the toothpick and the scrapings should be placed on a clean piece of paper. The paper should be folded and placed in a proper container. Each container should be marked to show the finger from which the scraping was taken. The packed scrapings are then sent to the laboratory for examination.

24-30. Broken fingernail fragments can also provide DNA evidence that can link a suspect to a crime scene or victim. They should be collected and processed in the same manner as other physical evidence.

COMBINED DEOXYRIBONUCLEIC ACID INDEX SYSTEM

24-31. The CODIS is an FBI program that consists of a database containing DNA profiles. This program allows federal, state, and local crime laboratories to compare DNA profiles and subsequently, when a match is derived, initiate an investigation.

24-32. The CODIS consists of four separate indexes or parts of the database. Each index contains a different type of DNA profile. Searching the indexes against themselves and each other to find matches can generate investigative leads. These indexes include the forensic, the offender, the unidentified human remains, and the relatives of missing persons.

THE FORENSIC INDEX

24-33. This database of DNA profiles is developed from biological material that was left at or carried away from a crime scene. The material is believed to belong to a suspect. Some common sources for the forensic file include—

- Semen from vaginal swabs, panties, bedding, and so forth taken from rape investigations.
- Cigarette butts or drink containers left at burglary scenes.
- Blood from suspects left at crime scenes, such as a burglary, an assault, or a murder.

THE OFFENDER INDEX

24-34. This is a database of DNA profiles developed from individuals convicted of qualifying offenses. The federal government, all states, and the military have laws requiring sample collections from convicted individuals. Under the Military Convicted Offender Program, samples from individuals convicted by a general court-martial or special court-martial of qualifying military offenses are collected and sent to the USACIL CODIS laboratory for processing. The qualifying military offenses include—

- Violent crimes.
- Burglaries.
- Housebreakings.
- A variety of charges under *UCMJ, Article 134*.

THE UNIDENTIFIED HUMAN REMAINS INDEX

24-35. This index contains profiles developed from bodies or body parts found and deduced victim profiles. A deduced victim profile is one developed from DNA thought to belong to the missing victim. Examples include the victim's toothbrush and blood found at the scene believed to belong to the victim.

THE RELATIVES OF MISSING PERSONS INDEX

24-36. This index contains profiles voluntarily contributed by the relatives of missing persons for comparison to unidentified human remains. Search restrictions apply to this index.

COMBINED DEOXYRIBONUCLEIC ACID INDEX SYSTEM PROCEDURES

24-37. The procedure for an investigator to get the profile from his case into CODIS is easy. All the investigator has to do is submit his case to USACIL. The work then rests with the USACIL DNA examiner. The examiner selects the profile from that case that is allowable in CODIS and then submits it to the CODIS laboratory. The CODIS laboratory enters the profile and searches it first against the local database and then against the National DNA Database. The profile continues to be searched routinely.

24-38. Investigators should submit unknown subject cases to USACIL for DNA analysis. These are the types of cases that CODIS can help with the most.

24-39. When an investigator submits evidence from his case to USACIL, if appropriate, the evidence profile is entered into the forensic file regardless of whether there is a match to a suspect. If there is a match, the investigator receives a telephone call followed by a formal report. When a match is seen in the CODIS software, it must go through a confirmation process. After confirmation, the investigator is notified and is given the opportunity to have any questions answered.

Chapter 25

Trace Evidence

Trace evidence can be defined as anything small enough to be easily overlooked by the investigator and transferred from one individual or item to another through contact or other means. Such evidence often consists of microscopic particles. Virtually any type of material can play a potential role as trace evidence. Trace evidence at a crime scene may be as obvious as soil or as inconspicuous as dust particles. Often, trace evidence is easily overlooked, mishandled, and discarded as useless.

OVERVIEW

25-1. Trace evidence is often referred to as contact evidence, contact transfer evidence, or transfer evidence. Its use is based on the belief that every contact leaves a trace; therefore, finding these traces can help establish associations or links. Trace evidence may be left at a crime scene by the suspect or may be carried away by him. The suspect may leave toolmarks, soil, paint chips, and similar traces, or he may carry away items, such as glass fragments, soil, or safe insulation. Similar traces may cling to his person, clothing, or equipment.

25-2. Investigators and other personnel handling evidence must be alert to the effect that poor handling has on trace evidence. Improper handling of trace evidence may negate the value of evidence that would otherwise be admissible in court. For example, if a suspect returns to the scene of a crime before it is completely processed, he could later claim that trace materials found there were left during his return visit. Thoughtless mixing of trace evidence found at different locations of the crime scene can also make the evidence worthless. Personnel collecting, processing, and examining evidence must always observe the cardinal rule (avoid contamination) for its handling.

GUNSHOT RESIDUE ANALYSIS

25-3. GSR may be defined as everything that exits a firearm during its discharge except the intact bullet. These residues originate primarily from the primer mixture, the propellant, the bullet, the bullet jacket, and the cartridge case.

25-4. There are two types of GSR tests conducted at the lab. The first test, referred to as the proximity test, is performed to search for and identify unburnt-powder particles and to measure muzzle-to-target distances using the residue patterns left on the target. The other test, known as the GSR test, is performed to detect primer residue in determining if an individual has handled or fired a weapon.

COLLECTING GUNSHOT RESIDUE

25-5. GSR is extremely fragile. It must be collected as soon as possible, especially on a live subject. As time passes between an incident and the time of collecting GSR, the likelihood that a detectable amount will remain on the subject's hands reduces. When GSR kits are used on live subjects after 12 hours has elapsed, the residue is not normally analyzed since no GSR attributable to the incident under investigation could be reasonably expected to still be present.

25-6. GSR is collected using commercially available kits that are analyzed using SEM/EDS. The person who does the collecting should wear the gloves from the kit to prevent contamination, use the instructions included with the kit, and must not have fired or cleaned a weapon within the past day.

25-7. The determination of whether a subject has recently handled or discharged a firearm is accomplished by analyzing primer particles. Primer particles are formed when components of the primer mixture are vaporized and subsequently cool and condense upon discharge of a firearm. Primer particles are usually microscopically small (1 to 5 micrometers in diameter) spherical particles containing lead, barium, and antimony or some combination of these elements involved in their formation. Spherical particles in this size range (containing all three of these elements) are considered highly specific for GSR.

25-8. The SEM/EDS collection kits allow the USACIL to analyze primer particles. In SEM/EDS analysis, the particles are imaged and the elements present in each individual particle are determined. The collection kits contain everything necessary for proper collection of potential primer particle evidence.

25-9. These GSR kits consist of aluminum stubs covered with an adhesive material (basically double-sided tape). A container and a lid protect the stubs. The lid doubles as a holder for the stub when it is removed from the container. The stub is used to remove potential primer particles by pressing it against a surface suspected of harboring primer particles (such as an individual's hand), lifting the stub, and blotting the area. This is a lot like using a piece of tape to remove lint from clothing. The surface is usually the hands of a suspected shooter; however, suspected primer particles may be collected from other surfaces (such as clothing or weapons) provided the presence of GSR on the object would aid in the investigation.

PREPARING GUNSHOT RESIDUE FOR THE LABORATORY

25-10. When sending shot or powder pattern exhibits to the laboratory for examination, ensure that the following requirements are met:

- Only the clothing that contains the bullet or shot penetration should be sent if the exhibit is on clothing.
- The laboratory should be consulted if the exhibit is on skin, doors, walls, or other surfaces.
- A written description of the garment containing the suspected shot or powder residue (including the location of the suspected shot or

powder) should be sent to the laboratory. The person recovering the exhibit (evidence) must maintain a copy of this description.

- Each article of clothing should be marked by attaching a tag to it to indicate its source of origination. Linings should be marked with ink or an indelible pencil in an area away from the suspected residue.
- Clothing should be wrapped in clean paper after it has been dried and then forwarded to the USACIL laboratory.

FIBERS

25-11. Placing a suspect at the scene of a crime is an important element in a criminal investigation. This can be achieved by locating textile fibers from the victim's clothing or the crime scene on the suspect's clothing or by locating fibers like those in the suspect's clothing at the crime scene.

25-12. Textile fibers can be exchanged between two individuals, between an individual and an object, and between two objects. When fibers are matched with a specific source (such as fabric from the victim, the suspect, and/or the scene), a value is placed on that association. This value is dependent on many factors including the type of fiber found, the color or variation of color in the fiber, the number of fibers found, the location of fibers at the crime scene or on the victim, and the number of different fibers at the crime scene or on the victim that match the clothing of the suspect.

25-13. Whether a fiber is transferred and detected is dependent on the nature and duration of the contact between the suspect and the victim or crime scene. It also depends on the persistence of the fibers after the transfer and the types of fabric involved in the contact.

25-14. Emergency personnel, medical examiners, and investigators must handle the victim's clothing carefully to minimize fiber loss. Fibers transferred onto an assault victim's or suspect's clothing will be lost if the victim and suspect move about and brush or wash the clothing. It is difficult to predict precisely how many fibers might remain on the clothing of a living victim or suspect after a given period of time, but it is important for investigators to retrieve and preserve the clothing of these individuals as soon as possible.

FIBER EVIDENCE (ASSIGNING SIGNIFICANCE)

25-15. Whenever a fiber found on the clothing of a victim matches the known fibers of a suspect's clothing, it can be a significant event. Matching dyed synthetic fibers or dyed natural fibers can be very meaningful; whereas, matching common fibers, such as white cotton or blue denim cotton would be less significant. In some situations, however, the presence of white cotton or blue denim cotton may still have some meaning in resolving the truth of an issue. The discovery of cross transfers and multiple fiber transfers between the suspect's clothing and the victim's clothing dramatically increases the likelihood that these two individuals had physical contact.

25-16. When a fiber examiner matches a questioned fiber to a known item of clothing, there are only two possible explanations: The fiber actually originated from the item of clothing or the fiber did not originate from the item of clothing. In order to say that the fiber originated from the item of clothing, the clothing either had to be the only fabric of its type ever produced or still remaining on earth, or the transfer of fibers from the clothing had to be directly observed. Since neither of these situations is likely to occur or be known, fiber examiners will conclude that the fibers could have originated from the clothing or that the fibers are consistent with originating from the clothing. The only way to say that a fiber did not originate from a particular item of clothing is to know the actual history of the garment or to have actually observed the fiber transfer from another garment.

25-17. It is argued that the large volume of fabric produced reduces the significance of any fiber association discovered in a criminal case. It can never be stated with certainty that a fiber originated from a particular garment because other garments were likely produced using the same fiber type and color. The inability to positively associate a fiber with a particular garment to the exclusion of all other garments, however, does not mean that the fiber association is without value.

NATURAL FIBERS

25-18. Many different natural fibers originating from plants and animals are used in the production of fabric. Cotton fibers are the plant fibers most commonly used in textile material, with the cotton type, the fiber length, and the degree of twist contributing to the diversity of these fibers. Processing techniques and color applications also influence the value of cotton fiber identification. Other plant fibers used in the production of textile materials include flax (linen), ramie, sisal, jute, hemp, kapok, and coir. The identification of less common plant fibers at a crime scene or on the clothing of a suspect or victim would have increased significance.

25-19. The animal fiber most frequently used in the production of textile materials is wool, and the most common wool fibers originate from sheep. The end use of sheep's wool often dictates the fineness or coarseness of woolen fibers. Finer woolen fibers are used in the production of clothing; whereas, coarser fibers are found in carpet.

25-20. The diameter of fibers and the degree of scale protrusion of fibers are other important characteristics. Although sheep's wool is most common, woolen fibers from other animals may also be found. These include camel, alpaca, cashmere, mohair, and others. The identification of less common animal fibers at a crime scene or on the clothing of a suspect or victim would have increased significance.

MAN-MADE FIBERS

25-21. More than half of all fibers used in the production of textile materials are man-made. Some man-made fibers originate from natural materials, such as cotton or wood; others originate from synthetic materials. Polyester and nylon fibers are the most commonly encountered man-made fibers followed by acrylics, rayons, and acetates.

25-22. There are also many other less common man-made fibers. The amount of production of a particular man-made fiber and its end use influence the degree of rarity of a given fiber.

25-23. The shape of a man-made fiber can determine the value placed on that fiber. The cross section of a man-made fiber can be manufacturer-specific. Some cross sections are more common than others, and some shapes may only be produced for a short period of time. Unusual cross sections encountered during examination can add increased significance to a fiber association.

FIBER COLOR

25-24. Color influences the value given to a particular fiber identification. Several dyes are often used to give a fiber a desired color. Individual fibers can be colored before being spun into yarns. Yarns can be dyed, and fabrics made from them can be dyed. Color can also be applied to the surface of fabric, as found in printed fabrics. How color is applied and absorbed along the length of the fiber are important comparison characteristics. Color fading and discoloration can also lend increased value to a fiber association.

FIBER NUMBER

25-25. The number of fibers on the clothing of a victim identified as matching the clothing of a suspect is important in determining actual contact. If there is a great number of fibers, it is likely that contact actually occurred between these individuals.

FIBER LOCATION

25-26. Where the fibers are found also affects the value placed on a particular fiber association. The location of fibers on different areas of the body or on specific items at the crime scene influences the significance of the fiber association.

FIBER TRANSFER AND PERSISTENCE

25-27. Textile fibers are transferred to the surface of a fabric either by direct transfer (primary transfer) or indirect transfer (secondary transfer). The likelihood of transfer depends on the types of fabric involved in the contact and the nature and duration of the contact. Studies have shown that transferred fibers are lost rather quickly, depending on the types of fabrics involved and on the movement of the clothing after contact. For example, the clothing of a homicide victim would tend to retain transferred fibers for a longer period of time because the victim is not moving.

NATURE OF CONTACT

25-28. The type of physical contact between a suspect and a victim can determine the number of fibers transferred and the value placed on their discovery. Violent physical contact of an extended duration will often result in numerous fiber transfers.

MULTIPLE FIBER ASSOCIATIONS

25-29. Multiple fiber types found on different items of clothing or fabric from the suspect, victim, and crime scene greatly increase the likelihood that contact occurred between these individuals and the scene. Each associated fiber type is considered to be an independent event and multiple associations undermine a coincidence defense.

FIBERS AS EVIDENCE

25-30. Since fiber evidence is generally small in nature, care should be taken to prevent loss or contamination. The processing considerations given to fibers are the same as for hair evidence (*Chapter 24*). The following are several methods that could be used in the collecting of fiber evidence:

- Visual search.
- Alternate light source.
- Additional magnification search aid.
- Taping.

25-31. When obtaining samples of fabrics as possible fiber donors, the samples should be representative of all the types and colors in the fabric of the item. All items should be sealed and labeled for identification.

25-32. Recovery of evidence should be the most direct but least intrusive technique practical. This could include picking, scraping, or vacuuming.

SOIL

25-33. Soils and rock may vary throughout a localized area. The differences between two visibly different soils, such as sand or clay, are easily recognized. However, minor differences between similar appearing soils may only be revealed by a thorough examination of their mineral compositions. Therefore, standard samples of the soil collected from a crime scene should be collected for comparison with the soil recovered from a suspect (such as from his clothing, shoes, or vehicle).

SOIL COLLECTION METHOD

25-34. At scenes where there are distinct footwear and/or tire track impressions or areas of disturbed earth, soil samples should be collected from beside each impression. These standard soil samples should be collected from a depth equal to that of the impression in that area. For deep impressions, it may be necessary to collect several samples from throughout the depth of the impression. For scenes where there are no distinct footwear and/or tire track impressions or areas of disturbed earth and the soil within the potential crime scene area varies in color and texture, samples of each visually distinct soil should be obtained. A minimum of six standard soil samples should be collected.

25-35. If a scene is void of distinct impressions and appears to have soils of the same or nearly similar appearing soil, samples should be collected from throughout the scene in an attempt to obtain a collection of soils that are representative of any variety that may be present in the area. Under these circumstances, soil standards should be collected from the upper-most layer, not more than 1/4 to 1/2 inch, since the materials that are present at that depth are most representative of what may have been transferred. At a minimum, six standard soil samples should be collected. A garden trowel is a common tool that can be used to collect soil standards; however, the trowel should be thoroughly cleaned between each sample.

25-36. Plastic screw-top, urine specimen containers capable of holding at least 120 milliliters may be used for packaging standard soil samples. These containers should be available from the nearest hospital or clinic. Each container should be filled completely. Alternatively, if these types of sample containers are not available, collect approximately 1/2 to 1 cup from each area.

25-37. For situations where the suspect may have walked through wet or muddy soil at the crime scene and then used a vehicle to leave the scene, the floor mats and the brake, clutch, and gas pedals should be examined for the presence of soil similar to that at the scene. If the investigator finds similar soils, he must remove the mats and/or pedals from the vehicle and submit them to the laboratory for examination. Distinct clumps of soil that may be present should be collected separately. General debris that is present on most automotive floor mats or floorboards consists of material that has accumulated over a period of time and is, generally, of little or no value for comparison with soil samples from a crime scene. Similarly, the investigator should attempt to collect questioned soil as discrete clumps (packaged individually) from the tires, fenders, or wheel wells of the vehicle to ensure that each sample represents a single source. Soil from vehicle wheel wells may have been deposited in layers. Therefore, the soil should be sampled in such a way as to remove the full thickness of the soil and preserve the layers. Each of these samples should be packaged individually to ensure that each one represents a single source.

25-38. Suspects should be questioned as to the origin of any soil that may be present, such as on their shoes or in or on a vehicle. Soil samples from the locations indicated by the suspect should be collected. These are the suspect's alibi standards. Should the soil on a suspect's shoes match that from the crime scene, but not that from the location where the suspect said it came, the implications are obvious. But should it match the standards from where the suspect said it originated, it provides corroboration of the suspect's statement.

LABORATORY TESTING

25-39. Given a sufficient number of standard samples for comparison, laboratory testing can show if a questioned soil removed from a suspect's clothes, shoes, or vehicle could have come from the crime scene. In rare instances, sufficient, unique, and inclusive materials (for example, vegetable matter [such as seeds] and materials [such as paint chips or glass]) are present in both the standard and questioned soils. The questioned soil can be directly associated with a crime scene sample. Finally, laboratory comparisons of standard and questioned soils may demonstrate that the questioned soil could not have originated from the same source as the standard samples.

BUILDING MATERIALS, SAFE INSULATIONS, AND SIMILAR EVIDENCE

25-40. When a building is broken into, the suspects may damage or break through a variety of building materials such as glass, paint, plaster, fiberglass, insulating materials, sheetrock, cinder block, mortar, brick, and caulking and sealing materials. It is possible that these materials may have been transferred to the clothing of the suspect or to the tools used during the break-in. Standards of samples should be obtained from each of these materials.

25-41. Penetration of the walls of a safe may cause its insulation to be broken. Pieces of insulation and insulation dust may be scattered about the scene and become deposited in or on the suspect's clothing and on any tools used. Thus, known standards of the safe insulation should be collected as comparison standards. The suspect's clothing, shoes, and tools should be collected and submitted to the laboratory for examination. A close examination of the scene may also reveal footprints in the dust that should be preserved for comparison purposes.

GLASS FRACTURES AND FRAGMENTS

25-42. Even though glass is usually considered to be class characteristic evidence, variations in its composition and properties make it a potentially valuable type of physical evidence. When a piece of glass that has been broken is reconstructed (such as from a headlight in a hit and run accident), it may assume an individuality when the fractured pieces fit together. It is for these reasons that an investigator must be particular when collecting glass.

25-43. When a suspect breaks a window, glass particles rebound up to ten feet or more toward the direction from which the force is applied, thus allowing for the suspect to be showered in glass fragments. This process is known as backward fragmentation. Glass fragments can easily be embedded in the shoes and clothing of any individual who is within range of the breakage. Additionally, glass fragments can get into hair and wounds and on or in the skin of the suspect or individuals who are near the scene. The object or projectile used to break a window may also have glass fragments in it.

DETERMINING THE DIRECTION OF FORCE

25-44. It is essential that a crime scene technician and investigator understand the manners in which glass reacts to force. This knowledge is often critical in determining whether a crime has been committed in the manner presented by the parties at the scene.

25-45. Broken glass shows two kinds of fractures: primary (first-made fractures) and secondary (subsequent fractures). Primary fractures are radial. They look like the spokes of a wheel as they radiate away from the point of pressure, such as the point in which a rock is thrown or a projectile is shot through a window. Radial fractures start on the opposite side of the force. Secondary fractures are concentric and are the result of continuing pressure. They form a series of broken circles or arcs around the point of impact and between the radial lines. Concentric fractures start on the same side of the glass as the original force. (See *Figure 25-1.*)

25-46. On radial fractures, the direction of force used to break the window is on the same side as the almost parallel parts of the rib marks and opposite that of the perpendicular parts of the rib marks. The relationship of the force to the rib marks reverses for concentric fractures; the side of the

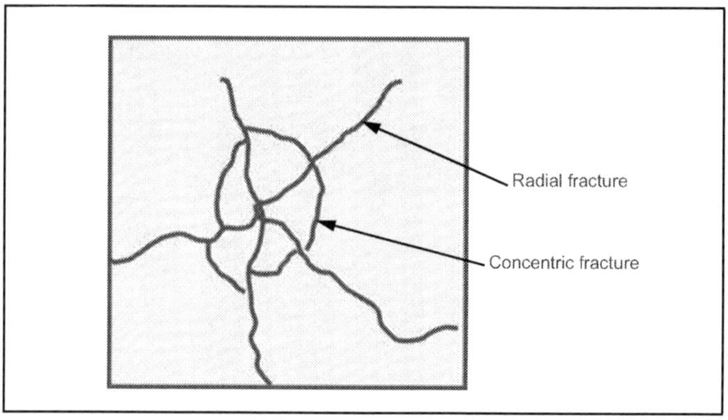

Figure 25-1. Example of a Radial and Concentric Fracture

perpendicular part of the rib mark is the side from which the breaking force came. (See *Figure 25-2.*) These observations should always be made on radial fractures of glass fragments nearest the point of impact (within the first concentric fracture).

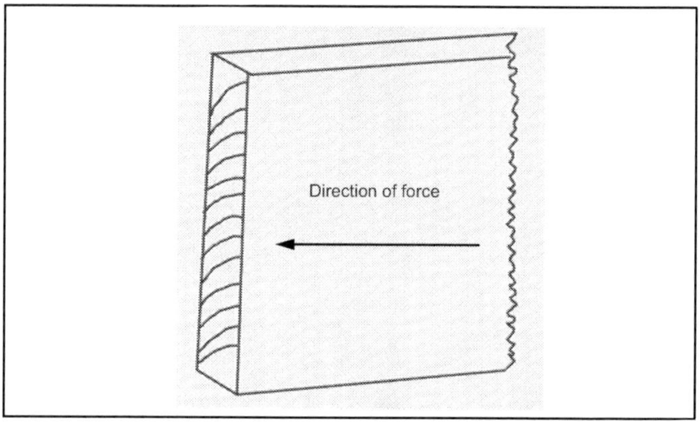

Figure 25-2. Example of a Concentric Fracture

COLLECTING THE GLASS

25-47. When glass from a window or doorframe is broken, pieces of glass often remain in the frame. The investigator should mark the original inside or outside surface and the orientation (such as top or bottom) of each remaining piece before its removal. If necessary, the entire frame may be removed if toolmarks or other evidence is relevant. Standards of wood, putty, paint, or other materials should be collected at this time. An investigator should do the following:

- Remove all larger pieces of glass first because they may contain latent prints. Handle the glass carefully because the prints could be another source of evidence.
- Photograph all glass fragments exactly as they are found and document their location on the crime scene sketch before collecting them.
- Consider what occurred at the crime scene from the time of the incident to the time of his arrival, realizing that moving glass pieces and extracting the remaining glass pieces in the window or doorframe may extend fractures, thereby, creating an opportunity for confusion or reversed findings by the lab technician. He must also determine whether the suspect caused the characteristics in the broken glass or whether other individuals, such as medical personnel or witnesses, introduced them.
- Wear rubber or cloth gloves while collecting glass.
- Pick up larger pieces of glass by their edges avoiding the flat surfaces as much as possible.
- Preserve the edges of glass fragments as the fragments are collected. The edges can help lab technicians determine the manner in which the glass was broken.
- Use rubber-tipped tweezers or similar type tools to handle smaller fragments. This will help prevent further damage to the fragments.
- Wrap each piece of glass separately and securely to avoid shifting and breakage during movement.
- Collect samples of known glass from each broken window or source.
- Package questioned pieces of glass separately from known pieces of glass.
- Submit all of the pieces of glass collected from the scene to help determine the point of impact after they are fitted together and to improve the likelihood of a match during forensics testing of recovered questioned pieces.
- Submit the victim's and suspect's clothing, with each clothing item packaged separately in a paper bag, to the lab. The lab can determine if glass fragments from the scene are on the suspect's clothing and can establish a link between the suspect and the scene.

PRESERVING GLASS COLLECTIONS

25-48. To preserve glass collections, wrap the larger glass or glass fragments in soft paper, cotton, or other similar material to prevent breakage. Never package glass fragments in glass containers. An investigator must be careful to avoid damage to prints or other substances on glass that are sent to the laboratory or saved as evidence. Smaller fragments may be wrapped in bond paper using the pharmacy fold. (See *Appendix I*.) The wrapped glass is placed in containers and stabilized so that the glass will not shift during movement. The wrappings and containers should be marked "Fragile." Evidence that will be examined by the laboratory must be packed carefully. Friction, shifting, or contact with other items can destroy or contaminate the evidence.

MARKING THE FRAGMENTS

25-49. The investigator should mark glass fragments with an indelible marker, a scriber, or a diamond point pencil. A piece of properly marked adhesive tape may also be used. The investigator initials, dates, and annotates the time on each piece of glass. Marks are annotated where there is no likelihood of latent prints or where material of evidentiary value may be present. The investigator should place marks on the side of the glass that was facing up when found (or on the inside if the glass was taken from a window frame or door). This helps in the reassembly of fragments and in the reconstruction of the incident. The investigation should include a sequence number that, when keyed, matches the investigator's notes, photographs, and sketches and will identify where the fragments were found. Fragments that are too small for markings should be placed in containers. Both the container and lid should be marked.

RECONSTRUCTING FRAGMENTS

25-50. Only the laboratory can truly reconstruct a piece of fractured glass. An investigator may want to place pieces of glass in relation to one another to get a better look at the fractures but should not do this on a continuous basis. How the reconstruction is done depends on the size and shape of the object. The fragments are not sent to the laboratory in a reconstructed form as damage may result during shipping and handling.

25-51. Investigators must take care not to rub the fractured edges of the glass against each other. This may cause more fracturing and destroy parts of the ridgeline marking. One way to avoid this is to keep the edges at least a pencil point's width apart. When as many pieces of the glass as possible are in place, the outlines may then be traced onto paper. The investigator should make notes on the paper and fit the markings on the pieces for future reference. If a more permanent reconstruction is needed later, the fragments may be fixed to a piece of plywood or heavy cardboard with tape or glue.

25-52. It is more difficult to reconstruct a curved or irregular-shaped piece of glass, such as a bottle or jar. Both the size and the shape of the object must be determined. Some pieces (such as automobile headlight lenses) may have patterns cast or cut into them. These can be compared and matched more easily than smooth glass surfaces. In many cases, the pattern may be matched independent of the fractured edges. But the exact matching of edges is still the most conclusive evidence of common source.

25-53. The laboratory may be able to complete a partial reconstruction on curved glass that originated from a bottle, lenses, or a similar curved item. This may allow an approximation of the size, such as the curvature. The investigator should keep in mind that these are only rough approximations. Lenses are made in round, oval, and other shapes with their spherical surfaces not always being completely regular in contour.

FIELD EXAMINATION OF FRACTURES

25-54. In the field, investigators must be able to distinguish between fractures caused by heat and those caused by blunt force. They must also be able to distinguish these kinds of fractures from those created by a high-speed impact like that of a bullet. A basic knowledge of the characteristics of safety glass is also helpful. There are two types of safety glass: laminated and tempered. Laminated glass will generally remain intact when fractured; tempered glass is made to produce many small fragments and will not remain intact when fractured.

BULLET HOLE FRACTURES

25-55. Checking glass for bullet holes may provide useful information. It may be possible to determine the direction from which a bullet was fired. Sometimes the sequence of a series of bullet holes can be learned. Sometimes the type of ammunition used and the distance from which the bullet was fired may also be learned.

25-56. Often, the direction from which a bullet enters a piece of glass is easily discernible. A bullet makes a somewhat clean-cut hole in the side of entrance. As it penetrates, it pushes glass fragments ahead of it. This causes a saucer-shaped or coning depression on the exit side with a greater diameter than the entrance hole. Determining direction becomes more difficult when several bullets enter safety glass closely together. When the last bullets enter the glass surface there are already a number of cracks. As a result, small pieces are knocked out around the holes on both sides.

25-57. Sometimes it is important to know which of two or more bullet holes in a pane of glass was made first. This may be determined from the fractures. When a fracture traveling across glass meets a fracture that is already present, the newer fracture will be stopped. If fractures from one bullet hole are stopped by those of another, it may be concluded that the blocking fracture was made first.

25-58. The angle from which a bullet enters a piece of glass may be found by the amount of chipping at the exit crater. If a bullet strikes glass straight on, chipping around the exit hole will be fairly even. If a bullet enters from the right of the glass, very little chipping will be found on the right side of the exit hole. Instead, there will be a lot of chipping around the left side of the exit hole. The entrance hole will show straight and short radial fractures on the right, while one or two long radial fractures should appear on the left. If a bullet enters from the left of the glass, these fractures will be reversed.

25-59. To learn the angle from which a bullet was fired, the bullet hole should be compared with test shots fired from varied known angles. The test shots should be fired through the same type of glass and under the same conditions with the same type of weapon and ammunition as the original bullet hole. The lab is equipped to conduct test fires of weapons and ammunition.

25-60. An ammunition type can sometimes be learned from the size and features of the bullet hole. Bullet holes in safety glass offer more evidence than those in window glass because safety glass fragments do not fall. When a bullet goes through a pane of glass in a sideways fashion, it is often hard to show the caliber of the bullet. Investigators should coordinate with the lab to learn the best way to submit evidence for this test. The lab can usually estimate the caliber and type of weapon used.

BLUNT-OBJECT FRACTURES

25-61. Glass fractures caused by a blunt object will show a pattern of fractures like, but not as regular as, those from a bullet. This difference is mainly due to the impacting force being dispersed over a greater area. It may be more difficult to tell which side the impact came from, but it can still be determined by the ridgelines on the edges of the radial fractures.

25-62. Anomalous fracture patterns can occur if the impact is close to a frame that is rigidly holding the glass pane. It is not recommended that the investigator attempt to partly reconstruct the object to find the radial and concentric fractures. The 3 R rule (in *radial* fractures, the stress lines are at *right* angles to the *rear*) applies most of the time; however, anomalous patterns do occur under some conditions.

HEAT FRACTURES

25-63. Recognizing heat fractures in glass can help you eliminate areas of concern in your investigation. Fractures due to heat are wave-shaped. They do not show a regular pattern of radial and concentric lines like fractures caused by impact. Heat fractures also show little, if any, curve patterns (stress lines) along the edges. Expansion of the glass (stretching action) occurs first on the side exposed to the heat. Glass splinters will often fall toward that side. Reconstruction of a glass object fractured by heat will show the wave-shaped fracture pattern.

25-64. If the stress lines are smooth, or almost smooth, and no point of impact is found, the investigator must consider other factors like the circumstances under which the fragments were found and their location. He may conclude that the fracture was due to excessive heat.

LABORATORY EXAMINATION OF FRAGMENTS AND FRACTURES

25-65. Glass fragments and fractures may yield important leads when examined by trained technicians. The lab can analyze glass fragments and fractures by a variety of means. In the case of comparing questioned glass fragments with a possible source, such as a broken window, a scientific examination of the physical and chemical properties of the glass could establish that the questioned glass could have originated from the window glass or eliminate that possibility altogether. Larger fragments of questioned glass can sometimes be physically matched to glass from the broken item. In this case, it is possible to conclude that the questioned and the suspected source were, at one time, a single unit. Examination of fractured glass may yield information as to the type of glass from which it originated. The manufacturer's name or logo may be imprinted or molded in the glass and could provide additional information. Examination of reconstructed glass items such as windowpanes may show the direction of a blow and the direction and angle of impact. It may also show the sequence of breaks and, in the case of multiple bullet holes, the sequence of the holes and the side from which they originated.

PAINT

25-66. Paint evidence commonly results from breaking and entering incidents, automobile accidents, and other crimes in which forcible contact has been made with painted surfaces. The type of paint evidence that results is usually in the form of chips and smears. When a suspect attempts to break into a safe, vault, window (sill), or door, paint can be transferred to and from the tools that are being used to commit the offense. Paint chips are frequently produced in automobile accidents in which two vehicles collide. These situations can also produce paint smear evidence, such as when two vehicles sideswipe each other.

COLLECTING PAINT EVIDENCE

25-67. Paint evidence should be carefully collected for examination by the USACIL. When the collection process is a result of an automobile accident, the investigator should—

- Search the accident scene for paint fragments and pieces of vehicle body parts.
- Pick up paint chips or smears by using tweezers or by placing a piece of clean paper under them and lifting them up.
- Place paint chips or smears in a box with a soft lining (to prevent altering the chips or smears).
- Pack multiple paint evidence separately.
- Submit entire vehicle components, such as a fender or bumper, when paint transfer is minimal and the severity of the crime warrants.
- Collect samples of questioned paint transfers from each vehicle area. In addition, collect control samples of paint from undamaged areas on the same body panel and directly adjacent to each area where a questioned transfer has been collected.
- Mark the evidence container with the exact location from where the paint evidence was collected, the date, the investigator's initials, and the case number.

25-68. When collecting paint evidence involving tools, the investigator should send the tools to the lab examiner. If the tools are suspected of having left impression marks, such as on a windowsill, he should send sample impressions with the tools. When collecting the paint samples from a surface, such as around the windowsill, the investigator should chip the paint from the surface to the foundation rather than scrape it. Chipping the paint allows its layered structure to remain intact. Each layer can then be a point of comparison.

COLLECTING CLOTHING

25-69. If clothing is collected as evidence, the investigator should not examine the items for paint fragments. To do so risks the loss of paint chips, especially those too small to be seen with the naked eye. These items should be marked for identification, packaged, and sent to the USACIL for examination.

PART FIVE

Acquisition of Police Information

Part Five describes the often dangerous task of obtaining police information and CRIMINT by operating undercover or employing surveillance techniques. Only the most experienced law enforcement personnel should employ these techniques and procedures. The chapters of Part Five provide the reader with information on the methods and motives of criminals and their associates and those individuals who provide valuable information about criminal activity to police.

Chapter 26

Surveillance Operations

In law enforcement, surveillance refers to the covert observation of individuals, places, or objects for the purpose of gathering police information or CRIMINT. In both tactical and nontactical environments, military police and CID personnel employ surveillance techniques in support of law enforcement and security operations. Surveillance techniques are often used to identify criminal activity associated with terrorism, organized crime, drug and contraband trafficking, and serious crimes against individuals.

OVERVIEW

26-1. A vital function of surveillance is the protection of military police and CID personnel performing undercover operations, such as drug suppression. Only experienced investigators should be assigned surveillance operations designed to protect undercover personnel. Failure to maintain observation can result in unnecessary risks to the undercover investigator.

26-2. The surveillance techniques described here are most often employed in drug suppression operations. However, these techniques can be easily modified for other police and security activities. Other surveillance objectives include—

- Supporting PIO.
- Obtaining evidence of a crime.
- Locating individuals by watching their places of activity and their associates.
- Checking the reliability of informants.
- Locating hidden property or contraband.
- Obtaining probable cause for requesting search warrants.

- Preventing the commission of an act of crime or apprehending a suspect in the commission of a crime.
- Obtaining information for later use during interrogations.
- Developing leads and information received from other sources.

QUALIFICATIONS

26-3. Police personnel are selected for surveillance operations based on their skill, experience, and resourcefulness. They must have well-developed observation and description skills. Surveillants must have patience and be able to endure long tedious hours of observation. Usually, police personnel who are selected are of average height and weight and devoid of unique physical features. The ability to blend in with their surroundings is extremely important. Sometimes they are chosen for ethnic or language qualifications. The type of surveillance and the area where it will take place are important factors to consider when selecting a surveillance team.

26-4. Every member of the surveillance team must know the elements of proof of various crimes to know when the suspect has committed an offense that warrants apprehension. Surveillance teams should not be too quick to apprehend suspects. They should keep the suspect under surveillance until the crime is completed, unless it would cause bodily harm to a victim. Continuing the surveillance, even after all elements of a crime have been committed, can also lead to other criminal information. The decision to continue the surveillance after a crime has been committed must be approved by the senior member of the surveillance team.

PLANNING

26-5. Surveillance operations are planned and coordinated. One of the initial tasks in planning and conducting surveillance is the designation of an investigator in charge. When a number of investigators are involved, a tactical plan is developed, outlining the duties of each investigator. Conducting a reconnaissance of the area that is to be surveilled and coordinating with other agencies should be planned. If the surveillance is likely to be lengthy, arrangements should be made for suitable relief. A prearranged, secure system of communicating with headquarters or superiors and central coordination must be established. Suitable signals for communicating information between surveillance investigators should be developed and thoroughly understood by all participants. The type of information that might be communicated through such signals would include the following:

- Take the lead.
- The suspect has stopped.
- The suspect has made a contact or a drop.
- There is countersurveillance.
- The suspect has spotted the surveillance.

Explanations for being at a particular place at a particular time should also be discussed in the eventuality that a suspect approaches and accuses the investigator of following him.

26-6. The plan may include a cover story for each person, specialized surveillance equipment, and communication needs. Surveillance operations should always be planned with a primary and alternate means of communications. Two-way radio contact is vital with vehicle surveillance. Cell phones can be used effectively for all types of surveillance operations; however, there are some precautions to consider. They include the following:

- Cell phone conversations can be intercepted.
- The ring volume must be set to a low setting or to vibrate mode to avoid drawing attention.
- The investigator should speak only loud enough to be heard on the cell phone but not by others around him.

26-7. Investigators should rehearse communication techniques to become familiar with what methods work best. What works best in one situation may not work in another.

26-8. Before initiating surveillance, team members should review all available information relating to the suspect to include the following:

- The names and aliases used by the suspect.
- The suspect's characteristics and mannerisms.
- Possible weapons involved.
- The suspect's habits and normal daily routine.
- The suspect's known criminal activity.
- The suspect's work and neighborhood environment.
- A full description of vehicles that the suspect may use.
- The description, names, and aliases of the suspect's known contacts or associates and their history with weapons, if known.
- Feedback resulting from the reconnaissance of the area where the surveillance will take place.
- Information resulting from the coordinated efforts of other agencies.

26-9. Each member of the surveillance team should know the scope and extent of crimes and activities in which the suspects are involved. For example, a drug dealer may also be involved in a fencing operation; he may be trading guns in exchange for drugs. Knowledge of all these activities will better prepare the team and greatly increase the success of the operation.

GATHERING INFORMATION

26-10. As part of the surveillance preparation, the investigators must consider their appearance. They must dress and adopt the demeanor of local inhabitants in order to blend into the setting. The type of clothing worn by investigators will determine if the concealment of weapons will be a problem. Investigators should also carry items, such as caps, jackets, and glasses to effect quick changes in appearance. They should carry sufficient money to pay for meals, transportation, or other expenses incurred during the surveillance. The investigator should have a reserve fund for use in emergencies.

26-11. A reconnaissance of the neighborhood should be performed to supplement file information. Someone who is familiar with the suspects should point out the suspects to the surveillance investigators. A physical reconnaissance should be made to study the areas where the surveillance will take place and to identify vantage points that are suitable for the investigators. Similarly, traffic conditions can be observed and the investigators can become familiar with the names and locations of streets in the area including locations of dead-end streets that may be used by the suspect to spot surveillance investigators. The reconnaissance will also yield information on the neighborhood and its inhabitants that would not be in police files.

MAINTAINING A SURVEILLANCE LOG

26-12. Whether it is a moving surveillance or a stationary surveillance, it is extremely important that the surveillance team maintain a log. Much like an investigator's notes, the surveillance log records the suspect's activities. The notes also document the significant activities of the surveillance teams. Some activity considered insignificant or unrelated to the surveillance at the time it is observed may become important later. It is then difficult, without notes, to recollect or reconstruct such events. Inadequate or improper notes can result in failure of prosecution of the violators due to insufficient evidence.

26-13. The log should specifically include the identity or detailed description of individuals and vehicles; the occurring activity; and the date, time, and place of occurrence. In addition, the investigators should include the weather conditions, the distance between the observation post and the site of the activity, and other factors affecting the surveillance. These notes should be so accurate and complete that the team can refer to them months later and recall in detail the activity observed.

SELECTING SURVEILLANCE EQUIPMENT

26-14. Technical surveillance devices can greatly enhance surveillance efforts. These devices range from digital cameras to more sophisticated satellite-tracking systems. Technical equipment, such as binoculars, night-vision devices, video cameras, and amplified listening equipment may be used for stationary surveillance.

26-15. Devices that are difficult to operate or hard to hide should not be selected as they may distract the surveillance team. Some electronic surveillance equipment may have certain legal restrictions. Investigators must always obtain advice from the office of the SJA before using electronic surveillance equipment and receive approval for its use from the appropriate headquarters.

SURVEILLANCE METHODS

26-16. There are three basic surveillance methods: loose, close, and a combination of the two. Loose surveillance can be used to spot-check a suspect. It can be used to compile long-term information on a subject. Loose surveillance is broken if the suspect seems to suspect that he is being observed. Close surveillance requires continued alertness on the part of the surveillance team. If the suspect is lost or is unfamiliar with the area, close surveillance must be continued under an alternate plan. Usually, a combination of the methods works best. Surveillance teams may need to move from a loose to a close surveillance because of an act or contact made by the suspect. If a place, such as a known crack house is under close surveillance, loose surveillance may be required at the same time on some of the individuals who frequent the place.

SURVEILLANCE TYPES

26-17. There are two general types of surveillance: stationary and mobile. Stationary surveillance is known as a stakeout. A stakeout is used when the suspect is stationary or when all of the important information can be learned at one place. However, in a stakeout, the surveillant may remain mobile moving from one vantage point to another for closer observation of the area or the suspect. A mobile surveillance is commonly known as tailing or shadowing. A mobile surveillance can be conducted on foot, in a vehicle, or a combination of the two. The choice depends on the suspect's movements.

STATIONARY SURVEILLANCE

26-18. A stationary or fixed surveillance is conducted to observe a home, building, or location to obtain evidence of criminal activity or to identify suspected offenders. Surveillance teams can use a parked van or a truck with a camper shell. However, the preferred method is to use another building that offers an unobstructed view of the area or person being observed. Unattended parked surveillance vehicles may create suspicion or become the target of criminal activity.

26-19. Primary consideration for beginning a stationary surveillance is whether the observation position affords the surveillance team a necessary vantage point to observe significant activity in detail without being observed. The position selected must give the team the opportunity to make close observations and allow them to see the details of the activity with the ability to identify the suspects involved.

26-20. The entrance and exit routes for the team must afford them the ability to come and go undetected. The suspects and their associates will be cautious and suspicious of any strangers in the surveillance area. In rural areas, investigators may have to walk a considerable distance in the dark with their supplies and equipment in order to avoid being detected.

26-21. Once inside the observation post, the team must not make any unnecessary noise or do anything that will draw the attention of anyone else in the area. An isolated observation post in a rural area requires that the team be very restricted in movement and have great patience and endurance.

Investigators perform stationary surveillance either by assuming a surface undercover role or by establishing an observation post.

- **Surface undercover role.** In urban areas, when an observation post is located in an apartment or home, it is often necessary to assume a surface undercover role in order to ensure the security of the surveillance team's identity. When an activity is in an area that does not readily lend itself to the use of a concealed observation position for a stationary surveillance, the team leader has to consider making observations from an open position. From an open position, the team's identity is kept hidden by assuming a role. Possible undercover roles include the following:
  - Power company lineman.
  - Telephone repairman.
  - Meter reader.
  - Road survey crew.
  - Road repair crew.
  - Door-to-door salesman.
- **Observation post.** An alternative to a surface undercover role is the installation of an observation post by the surveillance team. The most common observation post is a utility van or a pickup truck equipped with a camper shell referred to as a peek truck. This form of stationary surveillance has been used effectively in some areas, but the placing of the vehicle must be considered with respect to the type of activity in the area. There should be an obvious reason for where the vehicle is parked, such as because of a breakdown. The selection of a truck or van marked with slogans or advertisements work best.

MOBILE SURVEILLANCE

26-22. There are two methods of mobile surveillance: foot and vehicle. The means in which these methods are employed varies based on the circumstances presented, such as traffic congestion or rural area verses city area.

Foot Surveillance

26-23. There are three types of foot surveillance: the one-man foot surveillance, the two-man foot surveillance, and the ABC or three-man foot surveillance. Any of these methods may be used in a given situation, but the ABC foot surveillance is considered the most effective and is used more often.

- **One-man foot surveillance.** A successful one-man foot surveillance is extremely difficult to accomplish and should be avoided, if possible. There are situations that will suddenly require the initiation of a one-man foot surveillance, but the number of people in the area almost wholly dictates the distance between the surveillant and the suspect. For this reason, the distance between the suspect and the surveillant is usually either too great or too close. If the suspect and the surveillant are on the same side of the street, the suspect is kept in view at all times. It is necessary to be close enough to immediately observe the subject if he enters a building, turns a corner, or makes any other sudden similar moves. Generally, the efforts of a one-man foot surveillance are unsuccessful due to the limitations placed on the surveillant. He is in a position that does not allow him to be flexible, unlike the two-man or ABC foot surveillance. Flexibility is all-important during surveillance, and the one-man foot surveillance has few options open other than what the suspect, the circumstances, and the people in the area offer.

- **Two-man foot surveillance.** The use of at least two surveillants greatly increases the chances of success because the second investigator provides flexibility. With two surveillants, the position of the investigator directly behind the suspect can be changed frequently and allows for relatively close positioning behind the suspect. The use of two surveillants affords greater security against detection and reduces the risk of losing the suspect. Normally, both surveillants are on the same side of the street with the suspect, with the first investigator behind the suspect fairly close. The second surveillant is positioned behind the first with more distance between them. On streets that are not crowded, one surveillant may walk on the opposite side of the street.
- **ABC foot surveillance.** The ABC foot surveillance is considered the best and is the most frequently used. This method should be used whenever possible. The use of three investigators greatly reduces the risk of losing the suspect. A third team member provides greater variation in the position of the surveillants and allows a team member, who thinks that he has been identified or "burned," to discontinue his surveillance without affecting or having to stop the surveillance.

26-24. During a foot surveillance, the suspect may be either tail conscious or so suspicious of a surveillance that he takes action to prevent the surveillants from following him. He may take such action during the entire time that he moves from one place to another, or he may attempt to detect the surveillance only periodically when he feels his actions have become highly suspected. There is also the possibility that the suspect is following a normal route and that he is not particularly tail conscious or suspicious of surveillance.

26-25. The surveillant should be alert to the methods a suspect uses to detect and elude surveillance, but he must be careful not to interpret all actions by the suspect as an indication that the surveillance has been compromised. Of course, when the suspect takes such action, the surveillant is then alerted to be even more cautious to avoid being recognized by the suspect. Common methods used by a suspect to detect a foot surveillance include the following:

- Stopping abruptly and looking to the rear.
- Looking around casually.
- Reversing his course suddenly
- Stopping abruptly after turning a corner.
- Watching for reflections in shop windows.
- Entering a building and leaving immediately by another exit.
- Walking slowly and then rapidly at alternate intervals.
- Dropping a piece of paper to see if anyone retrieves it.

- Stopping to tie his shoestrings while looking around.
- Using an associate or friend in a business to watch for surveillance.
- Boarding a bus and riding a short distance or exiting just before it starts.
- Circling the block in a taxi.

26-26. Once the suspect has detected surveillance, he may use any of the following methods to elude the surveillance:

- Exiting a bus or subway just as the doors are about to close.
- Leaving a building through a rear or side exit.
- Entering a large crowd.
- Entering a theater and leaving immediately through an exit.
- Using a decoy.
- Using a side street or alley.

Vehicle Surveillance

26-27. Although foot surveillance is used frequently when critical moments of the investigation are at hand, vehicle surveillance is used more often as a means of gathering general intelligence. Investigators must feel as confident with the procedures of vehicle surveillance as they do with foot surveillance. Vehicles are used in the surveillance of premises, but their primary use and value is the surveillance of other vehicles. Foot surveillance can go with the suspect only so far. When the suspect becomes mobile, as he often does, then the surveillant must use vehicle surveillance. (See *Appendix J*.) The general principles that assist in conducting a successful vehicle surveillance are as follows:

- **Vehicle appearance.** The vehicle that is used on surveillance should be particularly suited for the purpose, both in appearance and speed. It should be inconspicuous in appearance with no noticeable features that would draw attention to the vehicle. It is preferable that the vehicle be a subdued, rather than bright, color. Experienced investigators have found that tail-conscious suspects are more apt to notice bright-colored surveillance vehicles. The color of the vehicle may often be further subdued and made indistinguishable at a distance by allowing road dirt and dust to accumulate on it. Official papers, manuals, handcuffs, and clipboards should be removed from the seats or the rear window shelf where they might be exposed to view. The inside dome light should be made inoperative by taping or disconnecting the doorpost switch to avoid accidentally illuminating the vehicle's interior at night. Communications equipment should not be visible.
- **Two surveillants per vehicle.** As effective as vehicle surveillance is in tailing a suspect from one location to another, normally some form of foot surveillance is required. When this need arises, an observer or second surveillant is an important adjunct to fill this void. Additionally, the observer allows the driver to give his full attention to driving while he operates the radio, makes notes of the suspect's activities, and observes the suspect. If the suspect leaves his vehicle to

complete some activity that the investigators need to observe, the observer should leave his vehicle in order to continue watching the suspect. His actions on foot must not be so unusual that he draws attention to himself from other individuals. The driver remains in the vehicle and maintains radio contact with the observer in order to give immediate mobility to the surveillance when the suspect's actions require it.

- **Driving skill.** An important factor contributing to a successful vehicle surveillance is the ability to drive skillfully under all types of traffic conditions. This is particularly true in heavy city traffic where a moment's hesitation could result in the loss of the subject. The driver must constantly anticipate traffic conditions ahead, especially for left turns, and keep his vehicle in the correct position behind the suspect. The surveillant must make quick and correct decisions to counter the suspect's tactics.
- **Police intervention.** During the course of any type of surveillance, particularly vehicle surveillance, the surveillants may be stopped or questioned by local police. The surveillants must identify themselves and briefly explain their presence. Normally, it is not necessary for the investigators to explain the nature of their investigation. Most interventions can be prevented through the coordination step of planning the surveillance.
- **Violation of traffic laws.** The safety of the surveillants, pedestrians, and other road users is the first priority in surveillance operations. Surveillants are required to operate vehicles in a safe and lawful manner. Deliberate violation of traffic rules and regulations is prohibited without prior consent of the installation commander. If approval is given, investigators must use good judgment and due caution. For example, when running a red light, the surveillants must first stop and then proceed only when it is safe to do so.
- **Nighttime surveillance.** Nighttime surveillance presents unusual problems to surveillance teams. One problem, because of darkness, is that the suspect is able to view any vehicle following him due to the presence of headlights. Another is that the suspect can use the darkness to elude surveillants by turning off his lights and driving onto a side road. Under medium traffic, the surveillants usually have the advantage. The suspect's field of vision is restricted to his rearview mirror, and he is unable to distinguish objects clearly because of the glare of the headlights. Surveillance vehicles that are equipped with cut-off switches for the headlights, taillights, and brake lights should be used because surveillants must be closer to the subject at night. This equipment enables the surveillance vehicle to alter its appearance by turning off or changing the brightness of the lights. If the suspect turns the corner, the tailing vehicle may alter one of its headlights before it turns the corner and appear to look like a different vehicle to the suspect.

SURVEILLANCE INITIATION

26-28. Surveillants must be aware that the tail conscious suspect routinely makes a check for surveillance at the beginning of his trip no matter where his start point is (home, work, or hangout). He may check for surveillance by driving his vehicle around several blocks in the immediate area of his departure point or by walking that same area. A surveillance team may have to be on foot in a hidden location several blocks away with binoculars or use a disguised vehicle such as a truck or van in order to initiate the surveillance.

26-29. When the suspect is near his destination, he may make routine checks again by observing vehicles behind him. He may make several approaches to his destination before he stops. With this in mind, surveillants must carefully position their vehicles to observe the suspect, deploy strategically, and avoid detection by the suspect.

SURVEILLANCE VEHICLES POSITIONS

26-30. Usually the surveillance will begin, if possible, in the area where the suspect's vehicle is parked. The team leader should be in a position where he can view the suspect's vehicle with the other surveillance vehicles deployed in the area covering the anticipated routes of travel. When the suspect leaves, the leader directs the appropriate car (possibly himself) to assume the lead position behind the suspect. The other cars will take up positions to the rear in a caravan style or on the paralleling streets.

DISTANCE FROM THE SUSPECT

26-31. The distance between the suspect and the lead surveillance vehicle is important and will vary with traffic conditions. It is easy to see that a correct distance between the subject and the surveillant is critical if observation is to continue for an extended period of time. This distance becomes greater as the suspect's speed increases, such as on highways, and is less when the suspect's speed decreases, such as in city traffic. Once behind the suspect, the lead surveillance team must use every means possible to avoid any prolonged viewing of their vehicle by the suspect in his rearview mirror. This can be accomplished by keeping one to two unrelated vehicles, referred to as cover vehicles, between the suspect and the lead surveillance vehicle. Cover vehicles operate differently on the highway than they do in city traffic.

- **Highway.** The decision when to have only one cover vehicle or at least two cover vehicles between the suspect and the surveillant in the lead vehicle is not easy. Some situations, such as an open highway with fast moving traffic, allow for two cover vehicles with few problems. Naturally, the more cover vehicles there are, the less chance there is of the suspect becoming aware of the surveillance. Too many cover vehicles, however, result in the loss of the suspect. When the number of cover vehicles is reduced to only two and the surveillance is on an open highway, there is usually little problem with the suspect being alerted or the surveillants losing the suspect. The use of only one vehicle as cover on the highway may not be enough if the suspect is tail conscious. If no other cover vehicles are available, allowing more distance between the surveillants and the suspect can alleviate this situation.

- **City traffic.** City traffic presents problems that are different from those found in highway traffic. The surveillants will frequently vary the number of cover vehicles used depending on how heavy traffic is and the type of streets. In city traffic, it is wise to have no more than two cover vehicles between the surveillant's vehicle and the suspect and one cover vehicle is often better. As traffic becomes more congested or traffic patterns are more difficult with extensive traffic lights, the use of two vehicles for cover can cause serious problems.

COMMUNICATIONS

26-32. Radio use during surveillance operations is kept to a minimum. The lead team has radio priority. It should give frequent locations of the suspect, the directions of his travel, the street or highway on which he is traveling, his approximate speed, and his lane of travel as the surveillance moves from one point to another. In areas unfamiliar to the surveillance team, the suspect's locations are given by reference to area landmarks, such as businesses, schools, manufacturing plants, and service stations.

26-33. The suspect's location must be given often enough so that the trailing surveillance vehicles feel comfortable about their positions. The frequent transmissions provided by the lead vehicle enable the other vehicles to determine how fast they should be traveling and whether they need to change their direction of travel. The following vehicles should not have to request the suspect's location. There is a tendency for team members, other than the lead vehicle, to periodically transmit their locations as the surveillance progresses. This is not necessary or helpful, unless a particular situation has developed where the lead vehicle needs the information. Generally, there should only be brief radio traffic, and it should be from the lead vehicle to the other surveillance vehicles as the suspect's location is given. Other radio traffic should be kept to a minimum.

Chapter 27

Undercover Operations

Going undercover is an investigative technique that is often used to gain police information when other efforts have proved impractical or have failed. An investigator goes undercover when he leaves his official identity and takes on a role to gain needed information. He associates with a person or becomes part of a group believed to have critical police information. The investigator must have supporting information, obtained by other means, such as police reports. The nature, habits, interests, and routines of a suspect must be studied. If an organization is the target, the purpose of the group and the names of members must be known. Undercover is an ideal technique used to investigate selling crimes involving drugs, stolen goods, frauds, contraband, and black marketing.

OVERVIEW

27-1. Undercover techniques may be used under the following circumstances:

- When information or evidence cannot be readily obtained in a traditional overt investigation.
- When an overt investigation will likely prove unsuccessful.
- When it is indicated that an undercover operation will reduce time and expenses involved in the completion of an investigation or produce evidence that cannot otherwise be amassed.

27-2. The improper use of an undercover operation could prove to be too costly in both resources and funding. Before making the decision to initiate this technique, the investigator should consider the following:

- The extent of the criminal activity and the results that will be achieved.
- The time factor.
- The complexity of the preparation involved. Complexity will likely increase the risk to personnel without a substantial benefit.
- The sophistication of the suspects as compared to the capabilities of investigative personnel.

PLANNING

27-3. This section provides general guidelines for planning an undercover operation. Each operation is different and must be planned in great detail.

27-4. Undercover operations are planned for a particular objective and with a certain end state in mind. The data or result of these operations must be specified before setting up the operation. The general objectives of an undercover operation are to obtain information, observe criminal activity, and collect evidence. These objectives must be clear and the degree of risk to the investigator must be assessed. Undercover operations are dangerous and should be used only when absolutely necessary and supported by a surveillance team. Investigators do not go undercover in a command without the knowledge of the installation PM and the appropriate USACIDC. The PM or USACIDC is responsible for advising appropriate leaders within the command of the operations. The advice of the SJA office is obtained for extensive or "heavy" undercover operations.

27-5. If the operation is to be conducted off the military installation, it is coordinated with civil authorities who may have an interest in the operation or who may have other covert personnel conducting similar operations. Coordination with local police is routine in all cases involving the civilian community. However, coordination should make as few people as possible aware of the operation. Only those individuals whose consent is needed and those who can distinctly add to the investigation should be informed.

27-6. Information useful to the investigation is assembled and reviewed by the undercover investigator and members of the surveillance team. Equipment and communication systems should be carefully selected. There should be a primary and alternate means to communicate. Normally, the undercover person does not work alone or independent of other investigators. A surveillance team should be planned. The primary focus of the surveillance team is to protect the undercover investigator and to record his actions. The surveillance team conducts its planning according to the procedures described in *Chapter 26.*

MISSION CHECKLIST AND OPERATION PLAN TEMPLATE

27-7. When planning a controlled-buy or buy-bust operation, investigative personnel should always complete an undercover mission checklist. (See *Appendix K.*)

27-8. Before any buy-walk, buy-bust, or raid operation, an operation plan should be completed. Investigators may use the fill-in-the-blank template at *Appendix L.*

RISK ASSESSMENT MATRIX

27-9. After completing the undercover mission checklist, the investigative team should complete a formal risk assessment matrix to ensure that all conceivable risks are identified. (See *Appendix M.*)

RISK MITIGATION WORK SHEET

27-10. A risk mitigation work sheet (see *Appendix N.*) is designed to aid investigative personnel in developing measures that will reduce the risk factors identified during the risk assessment process. These mitigating measures should seek to reduce or eliminate risks using countermeasures that will allow overall mission objectives to be achieved without unnecessary risk. Once the risk assessment matrix and the mitigation work sheet have been completed, a determination must be made as to whether or not identified risks are worth the potential gain. Local or command policies and procedures should identify the approval authority for missions designated as low, moderate, or high risk.

PRELIMINARY INVESTIGATION

27-11. The undercover investigator must have a thorough understanding of the overall objective to be reached as a result of his assignment. In essence, this is the time that he will be preparing for his role and planning to meet his overall objectives. In preparing for the assignment, the investigator should—

- Select his assumed identity.
- Determine the background story.
- Establish a means of communication.
- Gain knowledge of the suspect.
- Obtain information concerning local situations.
- Establish reporting procedures.

APPROACH SELECTION

27-12. The objective of the undercover investigation will generally dictate the type of approach the investigator will use. If the objective is an individual, the investigator will have to be accepted by the suspect. This can be accomplished through the use of a source or by the investigator becoming acquainted in the places the suspect frequents.

27-13. If the objective is a group of people, the investigator will have to determine how he can join them. Once again, the use of a source will probably be a necessity. However, the investigator may again consider frequenting establishments used by the suspects and establishing himself with them. A determination may also be made to gain the confidence of unwitting sources that are familiar with members of the group and, subsequently, obtaining an introduction.

27-14. Any one or a combination of these approaches may be used in most undercover situations. Whether the suspect accepts the investigator or not is depends on the investigator's ability in his role.

SUSPECT STUDY

27-15. Unless the undercover investigator has a thorough knowledge of the suspect, he will find himself frequently at a disadvantage. He must study the mannerisms, gestures, and speech of those that he deals with. Examining small details like tastes in food and music are also necessary. The investigator should, as a first step in his preparation, draw up a checklist of the details of the suspect's character and history. The following list is some of the information about a suspect that an investigator should be familiar with before entering an undercover role:

- **Name.** This should include the full name and any aliases or nicknames. This also includes any titles in relation to a job or public office, which the suspect may hold.
- **Addresses.** This should include past, present, residential, and business addresses.

- **Description.** This should include the basic description of the individual and any unusual traits and peculiarities. A photograph of the suspect should also be obtained.
- **Family and relatives.** This will assist the undercover investigator in his overall knowledge of the suspect.
- **Associates.** This knowledge is essential to an understanding of the suspect's activities.
- **Character and temperament.** The strength and weaknesses of the subject should be known. Likes, dislikes, and prejudices are particularly helpful.
- **Vices.** This should include vices, such as drug addiction, alcohol, and gambling.
- **Hobbies.** This knowledge could help the undercover investigator a in developing an acquaintance with the suspect. A common interest of this nature can cause a strong bond between the investigator and the suspect.
- **Occupation and specialty.** These may allow the investigator to establish a possible meeting ground with the suspect. These are also indicative of the character of the suspect.
- **Propensity for violence.** Information received from sources, background information from past arrest records, and personal observations made by the investigator will help identify a suspect's possible violent nature. The investigator should make every attempt to know a suspect's propensity for violence.

UNDERCOVER OBJECTIVES

27-16. When the decision to conduct undercover operations has been made, the investigator must be cognizant of the objectives of the operations. They include the following:

- Obtaining information and intelligence.
- Obtaining evidence for the prosecution.
- Determining if a crime is being planned or committed.
- Identifying individuals involved in criminal activity.
- Proving association between conspirators.
- Identifying witnesses and informers.
- Checking the reliability of informants.
- Locating contraband and stolen property.
- Determining the most advantageous time to make arrests or execute search warrants.

27-17. The undercover approach, through proper planning and execution, provides results as accurate and reliable as any other investigative technique. However, if careful preparation is disregarded, it is likely that the objectives of undercover operations will not be achieved.

LIMITATION OF UNDERCOVER ROLES

27-18. When assuming a fictitious identity as an undercover investigator, investigators may assume short-term or long-term roles as approved within the operational plan; however, there are three occupations that may not be assumed by investigative personnel. These three occupations are attorneys, members of clergy, and medical personnel. Consequently, if an investigation is initiated within a legal office, religious or chaplain's office, or medical facility, a role that will allow the covert team member access and credibility without violating the aforementioned prohibitions must be developed.

ASSIGNMENTS TYPES

27-19. The nature of undercover work can vary widely. An assignment may require the investigator to be in several different settings and situations that will test his resourcefulness, adaptability, and endurance.

27-20. An impromptu assignment places the investigator who is conducting an open investigation into an assumed role for the purpose of making inquiries. An example would be an investigator posing as a salesperson when calling on a third party or unwitting source.

27-21. A one-time assignment is one in which the investigator has received information that illegal activity may be taking place at a given location. An undercover visit to the suspect's location will assist in determining if the information is worthy of further investigation.

27-22. An extended assignment may last from a few days to several months. The majority of assignments received by an undercover investigator will fall within this category.

27-23. A penetration assignment is one that requires extensive planning and preparation as it involves the use of an undercover investigator over a period of several months to any extended length of time. This type of assignment is geared to the eventual penetration of the higher echelon criminal element.

27-24. A sting operation is designed to record criminal activities through direct observation and or participation. Most of these operations involve selling crimes, such as prostitution, narcotics, black marketing, or the sale of stolen property. In these operations, the investigator will generally act as a purchaser of the illicit service or product.

27-25. A reverse-sting operation is also designed to document criminal activities through direct observation and or participation. However, in these operations, the covert member assumes the role of the person providing the service or selling the illicit items. These operations require the approval of the USACIDC commanding general or other designated officials.

27-26. A storefront operation is generally used to purchase items that are likely stolen. Storefronts can take the form of a pawnshop or underground fencing operation. These operations require the approval of the USACIDC commanding general or other designated officials.

27-27. USACIDC and PM personnel are authorized to rent covert facilities and vehicles on a case-by-case basis. Authorization must be obtained from the designated authority before effecting such transactions. Normally, these operations are reevaluated for effectiveness every 30 days.

PERSONNEL SELECTION

27-28. Only the most experienced investigators should be selected for undercover operations. Undercover assignments test the investigator's adaptability, resourcefulness, and endurance. The investigator must be able to adjust his personality to the role he plays and have a clear understanding of the objective of the mission. He should be a skilled observer and a person of sound judgment. No one set of qualifications or attributes will enable an investigator to be successful in all types of undercover operations. Personnel should be selected based on the type and circumstances of the operation.

PRECAUTIONS AND POSSIBLE PITFALLS

27-29. An investigator cannot assume that he can conduct an undercover operation on his own, nor can he assume that all behavior is acceptable in order to "fit in." Undercover investigators should have a cover detail. They should also be aware of how to interact with the opposite gender and know when it is appropriate to consume alcoholic beverages. Failure to take note of these precautions and possible pitfalls could expose the investigation and cost the investigator his life.

USING A COVER DETAIL

27-30. Situations that require other investigators or law enforcement personnel to observe the activities of the undercover investigator may occur during the course of an undercover investigation. These actions would be primarily for the investigator's safety. They may also corroborate testimony regarding the undercover activity and transactions. Using other investigators in this manner is referred to as using a cover detail. The investigators comprising the cover detail must be supplied with all available information so that they may adequately perform their tasks and be of maximum help to the undercover investigator. If employed too long or indiscreetly, the cover detail may be discovered or suspected and seriously hinder the undercover operation.

BEING FRIENDLY WITH OPPOSITE GENDER ASSOCIATES

27-31. The undercover investigator must be aware of the possible consequences that could result from being overly friendly with opposite gender associates of the suspect. The investigator's relationship with opposite gender relatives and friends of the suspect should be such that he has their cooperation in obtaining the suspect's confidence, but should never extend to the point of provoking the suspect. The investigator should avoid situations, which would lead to accusations of improper or intimate relationships.

USING ALCOHOL OR SIMULATING THE USE OF DRUGS OR NARCOTICS

27-32. Generally, local policy prohibits an investigator from consuming alcoholic beverages while being armed and using or simulating the use of drugs unless he is in duress or as otherwise specified by the SOP or policy. The battalion commander or the author of the policy may make exceptions for undercover operations when prudent evidence exists to suggest that an operation could be jeopardized as a result of the policy. When an exception to the policy is approved, the investigator is held accountable for his actions and is expected to exercise mature, responsible judgment and professional behavior. The operational plan for undercover activity must address under what conditions the investigator may consume alcohol or simulate the use drugs.

CONTACT WITH THE SUSPECT

27-33. The manner in which an undercover investigator makes initial contact with a suspect is varied. It may be through a source, through a direct rapport developed by the investigator, through a deliberate means, or by chance. Regardless of how the contact is made, the undercover investigator must gain the suspect's confidence in order to develop information for the crime he is investigating.

APPROACHING THE SUSPECT

27-34. As the undercover assignment becomes operational, the investigator will use a source wittingly or unwittingly to reach the suspect, or he will be prepared to spend an extended period of time in establishing a relationship between himself and the suspect of the investigation. The approach used in the first contact with the suspect is, in most instances, the most critical point of the investigation. If this is accomplished smoothly and naturally, it should tend to offset any suspicion that the suspect may have and facilitate the establishment of a continuous association with him.

USING SOURCES

27-35. A witting source is a source that furnishes information to an investigator knowing full well his capacity as an investigator. The witting source may agree to accompany the undercover investigator in an initial contact with the suspect of an investigation using his association with the subject to effect an entrée for the investigator. This approach is the quickest, surest way to establish contact. The degree of success, however, is contingent on the amount of confidence the suspect has in the source (see *Chapter 28*) and the source's reliability.

MEETING THE SUSPECT (CHANCE ENCOUNTER)

27-36. A chance encounter between the undercover investigator and the suspect may occur on the spur of the moment or be a well-planned maneuver, either of which should appear to the suspect as a natural chain of events. Generally, this type of meeting causes little suspicion on the part of the suspect and provides the initial contact that the undercover investigator needs. The way the undercover investigator conducts himself during a chance meeting with a suspect will determine his acceptance or rejection by the suspect.

GAINING THE SUSPECT'S CONFIDENCE

27-37. After the first contact, the undercover investigator is immediately faced with the problem of avoiding suspicion. The investigator will have to appear to be the suspect's friend and may have to participate in some of the suspect's activities, in which case, the investigator must depend almost entirely upon his own judgment while ensuring that he does not violate the law or other ethical restrictions.

27-38. Usually the suspect's initial attitude will be suspicious and skeptical. The suspect may attempt to throw the investigator off balance by suddenly accusing him of being an investigator or belligerently calling him a source. This does not necessarily mean that the suspect is aware of the investigator's true identity, but only that he is trying to read the investigator's reaction or flush out a betrayer. A well-prepared undercover investigator anticipates such emergencies and immediately puts the suspect on the defensive, possibly by using counteraccusations.

27-39. To dispel the suspect's suspicions and gain his confidence, the undercover investigator may use the following techniques:

- Arrange to be arrested, questioned, or searched by other enforcement officers where the suspect can observe the action.
- Pretend disgust or anger with the suspect for questioning him.
- Appear as though he does not trust the suspect anymore than the suspect trusts him.

DEVELOPING INFORMATION FROM THE SUSPECT

27-40. The undercover investigator must listen. If he does all the talking or takes the lead in conversations, the suspect will not have the opportunity to talk. Occasionally, the investigator may have to initiate a conversation, especially about criminal activity. In this instance, the investigator should talk about activities other than those he is currently investigating, and eventually guide the conversation toward his real interest.

27-41. The undercover investigator, by listening, should learn everything possible about the suspect and his counterparts in the criminal activity with whom he does business and from whom he receives instructions or orders. If the suspect is in a position to have information about higher echelons in the criminal organization, the investigator should attempt to have the suspect introduce him to those individuals. This requires persistence and resourcefulness on the part of the investigator.

SPECIAL CONSIDERATIONS

27-42. Fronting of funds is defined as providing the suspect with money before the planned exchange of illicit items, such as narcotics or stolen property. As a general rule, investigators should not front funds; however, this may be done on a case-by-case basis. The fronting of funds should only be done when the risk of loss is minimal and there is an investigative gain from this action. Generally, a mid-level dealer may not have the money to obtain the drugs or other items and may request the funds to be fronted. This should only be done when the person to whom the funds will be fronted is fully identified and can be picked up at a later time should he simply steal the money. If funds are fronted, the investigator should impose a reasonable time limit that will require the suspect to go directly to his supplier to make the purchase, and investigative personnel must be set up to follow the suspect to his source. The gain in this situation would be identifying where the supply is coming from; consequently, failing to follow the suspect would negate the gain.

27-43. When making high-dollar purchases, the investigator may have to show the suspect that he actually has the money to make the deal (flashing funds). When it is determined that this is appropriate, there are several rules that should be followed, which are designed to safeguard both the money and the investigator. When a flash is warranted, the suspect should never know in advance that the money is to be at a given location, as this would provide him with the opportunity to try to steal the money. Such displays of good faith should be done by surprise without the knowledge of the suspect. The suspect should be taken away from other parties when showing him something (such as money) and with a third party should be present to display the money and then depart the area with the suspect watching. When the suspect sees the money depart the area, he will not be inclined to come back and steal it because he knows that it is no longer at that location.

27-44. An investigator must know the limits of his personal conduct when attempting to gain the suspect's trust. Sometimes an undercover investigator may act unruly or be disrespectful toward a law enforcement officer in an attempt to make the suspect more comfortable with them. However, if the undercover investigator is requested to commit a crime to gain the suspect's trust or prove that he is not a law enforcement officer, he must understand his limitations. In some cases, such criminal acts can be faked to make the suspect believe that the investigator committed a given crime without actually doing it. However, in other cases, there may be no conceivable means to be able to stage a crime. In these cases, the investigator may not commit acts for which he can be criminally culpable.

27-45. The most common defense used subsequent to an undercover operation is an entrapment defense. It is important for investigators to fully understand and appreciate what constitutes entrapment. Entrapment exists when an agent of the government entices someone to commit a crime that he was not predisposed to commit. For instance, merely soliciting drugs from a suspected drug dealer is not entrapment when he demonstrates that he could readily produce such items. An example of entrapment is when a government agent approaches an individual on the street who is not suspected of trafficking drugs, offers him an inordinate amount of money to make a buy, and tells him where to get it or offers sexual favors in exchange for the drugs and the individual produces the drugs even though he was not predisposed to commit such an offense.

Chapter 28

Sources

Anyone who provides information of an investigative nature to law enforcement personnel is a source. A source may provide an investigator with specific information about a particular case or background information that is useful in a number of investigations. Military law enforcement personnel should comply with all applicable laws and regulations governing the collection of information and/or intelligence on individuals to include common law, *Section 1385, Title 18, USC (The Posse Comitatus Act) (18 USC 1385 [The Posse Comitatus Act])*; law of war; DOD and DA directives and regulations; and SOFAs.

CATEGORIES

28-1. Sources generally fall into one of the following four groups:

- US and HN citizens.
- Police and security personnel.
- Mentally or emotionally challenged individuals.
- Criminals and their associates.

UNITED STATES AND HOST NATION CITIZENS

28-2. People within a community are often aware of criminal behavior. They may live in a high-crime area or have recently been a victim of crime. Others gain knowledge of criminal activity through their employment. A short list of some potential sources in the community include the following:

- Bartenders.
- Cab drivers.
- Barbers.
- Beauty shop operators.
- Hotel managers and employees.
- Insurance and other private investigators.
- Postmen.
- Public utility employees.
- Airport and railway personnel.
- Rental agency clerks.

28-3. Normally, citizens do not seek out the police to report criminal activity; however, they may confide in an officer they know and trust. Police officers should cultivate these contacts at every opportunity. The importance of citizens in fighting crime cannot be overstated.

POLICE AND SECURITY PERSONNEL

28-4. Police and security officers usually exchange information with fellow officers. Investigators should take every opportunity to meet as many federal, state, local, and HN officers as possible and establish a professional relationship with them. Sharing of police information and CRIMINT greatly enhances the ability of an organization to identify threats and interdict criminal activity.

MENTALLY OR EMOTIONALLY CHALLENGED INDIVIDUALS

28-5. Often, mentally and emotionally challenged individuals will try to provide information to law enforcement personnel. Sometimes criminals will say or do something incriminating in the presence of these individuals because they do not consider them a threat to their criminal activity. With experience, an investigator can identify these individuals whose information can often be attributed to newspapers, gossip, or hallucinations. Information from such individuals must be corroborated through an independent source.

CRIMINALS AND THEIR ASSOCIATES

28-6. The most valuable source is often the person who is a violator or has been associated with the criminal element. This person is usually in a better position to have substantive information regarding criminal activity.

28-7. A source that is closely associated with criminals is often difficult to control. Dependability and reliability are usually not characteristic of the criminal element. Therefore, it is critical that the investigator maintain firm control and management of the source. The legal, moral, and safety considerations of an investigation are the responsibility of the investigator, not the source. Sources who disregard the investigator's guidelines should be severed from further association with the investigation before major problems or violations occur.

SOURCE SELECTION

28-8. The most important consideration in selecting sources may be reliability. The investigator must evaluate both the source and his information to arrive at the facts. Source information must be tested for consistency by checking it against data from other sources. To test a source's reliability, an investigator should ask the person being evaluated about data known only to the law enforcement organization. This allows for a tentative degree of reliability to each source.

28-9. Investigators must maintain a good liaison and a friendly working relationship with investigators in their unit and in other law enforcement agencies. This will help them in developing quality sources. Sometimes, newer investigators will be referred sources by the more experienced investigators as a result of their being transferred or because they have more sources and investigations than they can effectively work.

28-10. Investigators should approach every individual that is arrested as a potential source. Some investigators may work an entire career by developing sources only from individuals arrested in cases that they have investigated. This technique is generally recommended for source development.

28-11. When considering an individual for selection, the investigator should review the prospective source's mental and physical health, age, education, and personality traits. An individual's experience, work record, financial status, and presence or lack of a criminal background should all be checked. Failing to look at the whole person can waste time and money.

28-12. A skillful investigator can develop a sense of gratitude in a potential source. In return for providing some ethical assistance, the source may show his thanks by giving information. Sometimes simple concern for the source's welfare may create a sense of gratitude.

SOURCE MOTIVES

28-13. Most investigators readily accept the tenet that sources are fundamental in law enforcement work. An investigator must understand what motivates a person to become a source, especially criminals or their associates.

28-14. A source's motivation is a key factor in determining his reliability. The investigator should attempt to find out what motive a person has for informing. A source gives information for any number of reasons.

CITIZENS AND LAW ENFORCEMENT PERSONNEL

28-15. Motivation for average citizens and law enforcement officers is usually easily recognized. The average citizen source ordinarily supplies information or performs a service because of his civic duty and a desire to see justice done. Law enforcement officers are motivated by the same basic factors as the average citizen plus their professional responsibility as sworn officers to enforce the law and investigate violations.

MENTALLY OR EMOTIONALLY CHALLENGED INDIVIDUALS

28-16. What motivates a mentally or emotionally challenged person to become a source is very complex and difficult to determine. In a majority of instances, these individuals give useless information and should not be used as sources. When mentally or emotionally challenged individuals possess valuable information or are able to provide a necessary service, they are usually inspired by one motive or a combination of motives associated with the criminal or associate source.

CRIMINALS AND THEIR ASSOCIATES

28-17. The investigator often uses criminals and their associates as sources. These sources present many legal, ethical, and moral problems for the investigator that are not usually associated with the other three classifications of sources. If an investigator is to properly instruct, control, protect, and effectively use sources and evaluate their information, it is critical, in every instance, that he knows the source's motivation. The following include the most common motivations of sources:

- **Fear.** It is said that self-preservation is the first law of nature. Therefore, many individuals turn to law enforcement with a desire to cooperate when they are afraid of something. Probably the most typical of this situation is the person who has been arrested for an offense and is afraid of going to jail. Since man is social by nature and does not like to live alone, he will sometimes cooperate with law enforcement in return for a consideration of leniency by the court regarding pending charges. Telling a potential registered source that his cooperation will be brought to the attention of the appropriate authority is acceptable. However, any speculation of promises by the investigator relative to the disposition of the charges pending against the source is beyond the authority of the investigator and must be avoided. Additional examples of individuals motivated by fear are victims of rackets or swindles and those afraid of criminal associates for a variety of reasons.
- **Revenge.** The person who is motivated by revenge is usually overwhelmed with a desire for retaliation and often has little concern about openly testifying in court or his identity being publicly exposed. The typical revenge-motivated source is a person who wishes to settle a grudge because someone else informed on him, took advantage of him or, in some manner, injured him. This type of source may exaggerate or make a report that is completely erroneous in an effort to accomplish his goal. A desire for revenge may arise from factors other than criminal activities, such as jealousy and quarrels over lovers, causing close friends to become bitter enemies.
- **Mercenariness.** Some individuals provide information or render a service to law enforcement strictly for a fee. They want to sell what they know for the highest price. Their information is usually good but may backfire when too much reliance is placed on the mercenary motive or the reward payments. Investigators must be cautious not to let financially motivated sources needlessly extend an investigation. Some, especially those on a continuous pay status during an investigation, will attempt this tactic by providing frivolous information knowing that payment will stop when the investigation is over.
- **Egotism or vanity.** Within the average citizen category of sources, these individuals usually subconsciously or consciously want to be law enforcement officers but for various reasons cannot qualify for law enforcement employment. The benefit of these individuals to law enforcement should not be too quickly discounted. Since a willingness to assist law enforcement is a positive characteristic of any source, these individuals can often be of substantial benefit if directed and motivated properly by a competent investigator.
- **Perverseness.** Characteristic of this type of motivation is the source that makes a disclosure hoping for some unusual advantage. This motivation is prevalent in the areas of vice and/or contraband crimes when the source is earning his living by questionable means and informs with the desire to eliminate his competition. He provides

trivial or worthless information while attempting to learn investigative techniques, the identities of undercover agents, or to direct attention away from himself. Law enforcement should not use perversely motivated sources unless the source's motivation is to eliminate his competition.

- **Repentance or reform.** Occasionally, a source will cooperate with law enforcement because he wants to repent his wrongdoing, has a desire to make restitution, or wants to break criminal alliances. This source may decide to cooperate for money, but subconsciously convinces himself that he is cooperating with law enforcement for altruistic reasons. When this type of source is properly managed, he can become an excellent continuing source of police information.

28-18. These motives are not necessarily all-inclusive. Many investigators have allied sources for life, even after a source's case has been adjudicated or the revenge motive situation no longer exists. By treating a source honestly and fairly, he will sometimes continue to provide information or service because of his appreciation or gratitude.

28-19. Investigators must be careful not to misinterpret a source's motivation. There are many instances where individuals have supplied information to law enforcement for purely unselfish reasons and then were completely "turned off" when offered money by an investigator.

28-20. For whatever reason sources volunteer information, they should never be cut short. Investigators should give them the chance to tell their story and then ensure that it is checked out. There is always that one chance that the information could be the missing link in an important case.

SOURCE INTERVIEW TECHNIQUES

28-21. One key to developing and gaining the cooperation of most sources is proper interviewing techniques. Most criminals or their associates do not readily offer information or agree to cooperate with law enforcement officials. However, an investigator who is a good listener and can communicate clearly and effectively can often gain their cooperation.

28-22. The criminal or his associate often has two principal concerns that must be addressed before he will seriously consider cooperating. First, he is concerned about his identity being registered. Secondly, he is concerned that the investigator or organization will address his specific motivation, such as monetary payments. Therefore, the investigator must be prepared to discuss these concerns with the potential source to gain his cooperation. If most of the following questions can be answered by a source or potential source, an investigator should be able to objectively evaluate the source and the information he has provided. The following are questions that an investigator should attempt to find the answers to when interviewing a source or potential source:

- What is his motivation?
- Has the source been reliable in the past?
- How intelligent is the source?
- How does he know about the violation?

- Does he have a personal interest?
- Does he have direct knowledge relative to the information?
- Does he have access to additional related information?
- Does he have reason to be vengeful toward the violator?
- Does he have enough experience to report the information accurately?
- Is he withholding some of the information?
- Has he fabricated information in the past?
- Is he willing to testify in court?

SOURCE IDENTITY PROTECTION

28-23. Every CID agent and MPI has a professional and ethical obligation to safeguard the identity of sources to the maximum extent provided by the law. Failure to fulfill this obligation may result in the death, injury, or intimidation of the source or his family. Reprisals against a source due to improper investigator techniques weigh heavily on the investigator's conscience and undermine the total law enforcement effort. As a general rule, an investigator should not place a source in a situation where the source's identity could be exposed unless the investigator has previously explained to the source that this could happen as a result of his cooperation.

28-24. The names of sources are privileged information as supported by the source privilege doctrine. This doctrine allows the government to withhold the identity of the source under most circumstances. The rationale for this privilege is twofold. It ensures a constant and continuing flow of information regarding illegal activities to law enforcement authorities. It protects the source of information from reprisals or revenge.

28-25. Confidentiality must be furnished to all sources that, regardless of motive, have provided law enforcement personnel with information concerning the criminal activities of others. Restricting the release of the source's identity, exercising appropriate security measures regarding communications between the source and the handler, and using caution in documenting the source's activities can normally provide confidentiality. Additional measures may include providing physical protection and transferring an individual to another location.

SOURCE CONTROL AND HANDLING

28-26. The investigator, not the source, directs and controls the investigation. This is not as basic as it may sound, especially for the inexperienced investigator who wants to do a good job and establish his reputation. He could fall prey to a perversely motivated source. There is nothing wrong in asking the source for suggestions or showing appreciation for quality information. In fact, these techniques are encouraged. However, the investigator must be the controlling and decision-making authority.

28-27. Investigators must avoid promising inflated monetary payments or making other promises to a source that cannot be kept or that are outside their authority. Source payments are not within the sole discretion of a case agent or investigator and are contingent upon factors, such as supervisory approval, budget, or unit and organization guidelines. Sources may be told that monetary payments are available and general payment guidelines. Inflated promises of payment or rewards may stimulate a source one time, but a source is not likely to assist in future investigations once a promise has been reneged upon. Advising a source of anticipated judicial disposition of a charge pending against him or indicating that a source will receive probation or a reduced sentence because of his cooperation are examples of commitments that are beyond the authority of an investigator.

28-28. In order to maintain proper control of the source and the direction of the investigation, the investigator must maintain frequent personal contact with the source. When personal contact is not possible, telephonic contact should be used. Personal meetings between the investigator and source are the best atmosphere for debriefing the source, developing a rapport, and issuing instructions. Prearranged, secure meeting places should be established. Using government buildings, police departments, and other official buildings to meet a source should be avoided to reduce the risk of exposing the source.

28-29. It is generally recommended that a second witnessing investigator be present at these meetings. The second investigator serves to corroborate anything said or done at the meeting. This could become important at a later time if the source makes incriminating accusations against the controlling investigator. Exposure of a second investigator to the source will make the transition of the source to another investigator easier, such as when the controlling investigator transfers or retires. Although some organizations do not make it mandatory for two investigators to meet with a source, they do recognize that meetings between investigators and sources of the opposite sex and meetings where sources are paid money are obvious situations that dictate the presence of two investigators.

28-30. A good rapport and mutual trust between an investigator and a source greatly enhance the likelihood of accomplishing the mission. Initially, creating this type of atmosphere may be difficult especially when dealing with a fear-motivated source that has recently been arrested. Generally, if the investigator is truthful and fair with the source, a solid professional relationship will eventually develop. An investigator should always conduct personal and criminal background checks on a source and attempt to corroborate the source's information through independent means.

SOURCE TESTIMONY

28-31. The confidentiality of a source's identity is a limited privilege recognized by military law. *Rule 507, MCM 2000,* provides that a source's identity is normally privileged against disclosure. Communications with these individuals are also privileged to the extent necessary to prevent disclosure of their identity. This privilege, however, is not absolute and is subject to various exceptions. Legal considerations in this area are both complex and subject to change.

28-32. At the trial, all disclosure orders made by a military judge will be complied with fully. However, in an appropriate case, such as where disclosure may jeopardize an important ongoing investigation, a recess to petition the convening authority to terminate the current judicial proceedings may be requested through the TC by USACIDC personnel.

ENTRAPMENT

28-33. The defense of entrapment exists when the design or suggestion to commit the offense originated in the government and the accused had no predisposition to commit the offense. The "government" includes agents of the government and individuals cooperating with the government, such as informants. The fact that individuals acting for the government afford opportunities or facilities for the commission of an offense does not constitute entrapment. Entrapment occurs only when the criminal conduct is the product of the creative activity of law enforcement officials.

28-34. Therefore, one of the first things discussed with a source is the law regarding entrapment. The investigator must ensure that the source understands the difference between providing an opportunity for a suspect to violate the law versus providing the inspiration.

Appendix A

Metric Conversion Chart

This appendix complies with current Army directives which state that the metric system will be incorporated into all new publications. *Table A-1* is a metric conversion chart.

Table A-1. Metric Conversion Chart

US Units	Multiplied By	Metric Units
Cup	0.2366	Liter
Degrees Fahrenheit	Subtract 32, multiply by 5/9	Degrees Celsius
Feet	0.3048	Meters
Gallon	3.7854	Liter
Inches	2.5400	Centimeters
Inches	0.0254	Meters
Inches	25.4001	Millimeters
Mil	25.4000	Micrometer
Ounces	28.3490	Grams
Ounces	28,400,000	Micrograms
Pounds	0.4536	Kilograms
Yards	0.9144	Meters
Metric Units	**Multiplied By**	**US Units**
Centimeter	0.3937	Inches
Degrees Celsius	Multiply by 9/5, add 32	Degrees Fahrenheit
Grams	0.0353	Ounces
Kilograms	2.2046	Pounds
Liter	4.2268	Cup
Liter	0.2641	Gallon
Meters	3.2808	Feet
Meters	39.3701	Inches
Meters	1.0936	Yards
Microgram	.0000000353	Ounces
Micrometer	0.0393	Mil
Millimeters	0.0394	Inches

Appendix B

Crime Scene Predeployment Equipment List

Army law enforcement managers plan for short-notice deployments by preparing crime scene predeployment kits (see *Table B-1*). This appendix provides an equipment list that is intended to be a guide, not a standard. In addition to this list, planners should also consider special tools and equipment used for death scene investigations, arson and/or explosive incidents, and electronic evidence collection (a list of those items is also included). Every deployment will present different challenges. Many of the items necessary for crime scene processing are not available in remote locations. Supervisors and planners must consider their transportation and lift capabilities. Predeployment kits are organized for a single lift. Refer to military police support on the battlefield consolidated TOE update (CTU) 2000-04 for details on mobility and lift capabilities.

Table B-1. Equipment List for a Crime Scene Deployment Kit

Photography Equipment	
Camera and case	Cleaning equipment
Film and floppy disks (if digital)	Batteries
Light source	Tripod
Paperwork and Supplies	
Crime scene sketch kits	Notebooks and folders
Envelopes (various sizes)	Tape
Necessary paperwork	Pens, pencils, paper, and erasers
Point of contact listing	Accordion files (preferably tabbed)
Fingerprint Kits	
Fingerprint cards	Ink and rollers
Automation	
Extension cords	Printers
Formatted floppy disks	Serial port cables and accessories
Laptop computers with backup software	
Evidence Collection and Holding	
Fingerprints	Portable fuming equipment
Fingerprint dusting kit	Hinge lifters
Palm print lifters	
Casting Impressions	
Casting kit	Dental stone
Shovels and trowels	

Table B-1. Equipment List for a Crime Scene Deployment Kit (Continued)

Firearms and Ammunition	
Primer residue collection kit	
Drugs	
Detection kits	*CID Form 36*
General	
Swabs	PPE
Questioned documents holders	Evidence collection bags (various sizes)
Paper bags	*DA Form 4002*
Shipping tape	Portable light source
Boxes	
Death Scene Investigations	
Gloves (universal precautions)	Writing implements (pens, pencils, and markers)
Body bags	Communication equipment (cell phone, pager, and radio)
Flashlight	Body identification tags
35-mm and digital cameras (with extra batteries and film)	Investigative notebook (for scene notes)
Measuring instruments (tape measure, ruler, and rolling measuring tape)	Official ID (for yourself)
Watch	Paper bags (for hands and feet)
Specimen containers (for evidence items and toxicology specimens)	Disinfectant (universal precautions)
Departmental scene forms	Inventory lists (clothes and drugs)
Blood collection tubes (syringes and needles)	Clean white linen sheet (stored in a plastic bag)
Paper envelopes	Business cards and office cards with phone numbers
Evidence tape	Medical equipment kit (scissors, forceps, tweezers, an exposure suit, scalpel handles, blades, disposable syringe, large-gauge needles, and cotton-tipped swabs)
Foul-weather gear (a raincoat and umbrella)	Tape or rubber bands
Phone listing (important phone numbers)	Evidence seal (use with body bags and locks)
Disposable (paper) jumpsuits, hair covers, and a face shield	Shoe covers
Pocketknife	Waterless hand wash
Trace evidence kit (tape)	Crime scene tape
Thermometer	Latent print kit
First aid kit	Plastic trash bags
Local maps	Photo placards (signage to identify the case in photos)
Gunshot residue analysis kits (SEM/EDS)	Hand lens (magnifying glass)
Boots (for wet conditions and construction sites)	Barrier sheeting (to shield body and area from public view)
Portable electric lighting	Reflective vest

Table B-1. Equipment List for a Crime Scene Deployment Kit (Continued)

Purification mask (disposable)	Basic hand tools (bolt cutter, screwdrivers, hammer, shovel, trowel, and paintbrushes)
Tape recorder	Video camera (with extra videotape and batteries)
Body bag locks (to secure body inside bag)	Presumptive blood test kit
Personal comfort supplies (insect spray, sunscreen, and a hat)	
Arson and Explosive Investigations	
Barrier tape	Tape measure (100 ft)
Clean, unused evidence containers (cans, glass jars, and nylon or polyester bags)	Compass
Decontamination equipment (buckets, pans, and detergent)	Evidence tags, labels, and tape
Gloves (disposable and work gloves)	Hand tools (hammers, screwdrivers, knives, and crowbars)
Lights (flashlights and spotlights)	Marker cones or flags
PPE	Photographic equipment
Rakes, brooms, and spades	
Electronic Evidence Collection	
Documentation tools	Stick-on labels
Cable tags	Indelible felt-tip markers
Disassembly and Removal Tools (in a Variety of Nonmagnetic Sizes and Types)	
Standard pliers	Flat-blade and cross-slotted screwdrivers
Hex-nut drivers	Needle-nose pliers
Secure-bit drivers	Small tweezers
Specialized computer screwdrivers (manufacturer specific)	Wire cutters
Star type nut drivers	
Package and Transport Supplies	
Antistatic bubble wrap	Cable ties
Evidence bags	Evidence tape
Packing materials (avoid materials that can produce static electricity, such as polystyrene peanuts)	Packing tape
Sturdy boxes of various sizes	Antistatic bags
Other Items	
Forms (keystroke/mouse click log, photo log, and *DA Form 4137*)	Hand truck
Seizure disk	List of contact telephone numbers for assistance

Appendix C

Evidence Room Inspection or Inventory Checklist

A checklist for evidence room inspection or inventory assists a supervisor in conducting an inspection of an evidence room. This inspection checklist (*Table C-1*) should be used as a guide and is not all encompassing.

Table C-1. Checklist for an Evidence Room Inspection or Inventory

REQUEST FOR EXCEPTION ***AR 195-5, paragraph 1-5***			
	Yes	**No**	**NA**
USACIDC activities. Were requests for exceptions to *AR 195-5* sent to Commander, USACIDC, ATTN: CIOP-PP-PO, 6010 6th Street, Fort Belvoir, VA 22060?			
Military police activities. Were requests for exceptions to *AR 195-5* sent to the installation commander with a copy furnished to the major Army commander concerned?			
Approved exceptions. Did the requesting agency keep a copy of the exception (if it was granted) until the deficiency was corrected or the exception expired? (Exceptions are usually for one year.)			
PRIMARY EVIDENCE CUSTODIAN ***AR 195-5, paragraph 1-6***			
Prerequisites for USACIDC. Is the custodian an accredited enlisted special agent or warrant officer approved by USACIDC group commanders (group commander)?			
Military police activities. Is the custodian a commissioned officer; an NCO, MOS 95B/C, grade E6 or above; or DA civilian employees with the grade determined by the local civilian personnel office (CPO)?			
Appointment.			
• USACIDC. Was the primary evidence custodian appointed in writing by one of the following: a field office, district office, group commander or commanding general (CG), or USACIDC?			
• Military police activities. Did the PM appoint the primary evidence custodian in writing?			
• Was a copy of the appointing document kept in the evidence room under file number 1e?			
• Was the appointing document brought forward for filing each year the custodian retained the position?			
NOTE: *AR 195-5, paragraph 1-4b*, is cited as the authority to appoint the primary evidence custodian.			

Table C-1. Checklist for an Evidence Room Inspection or Inventory (Continued)

ALTERNATE EVIDENCE CUSTODIAN ***AR 195-5, paragraph 1-7***			
	Yes	**No**	**NA**
Prerequisites for USACIDC. Is the alternate custodian an accredited enlisted special agent or warrant officer approved by the USACIDC group commander?			
Military police activities. Is the alternate evidence custodian a commissioned officer; NCO, MOS 95B/C grade E5 or above; or a DA civilian employee with the grade determined by the local CPO?			
Appointment.			
• USACIDC. Was the alternate evidence custodian appointed in writing by one of the following: a field office, district office, group commander or CG, or USACIDC?			
• Military police activities. Did the PM appoint the alternate evidence custodian in writing?			
• Was a copy of the appointing document kept in the evidence room under file number 1e?			
• Was the appointing document brought forward for filing each year that the alternate evidence custodian retained that position?			
NOTE: *AR 195-5, paragraph 1-4b,* is cited as the authority to appoint the alternate evidence custodian.			
Responsibilities. Did the alternate evidence custodian assume the duties of the primary custodian during his temporary absence?			
Temporary custodianship.			
• Did the alternate, on assuming duties of the primary evidence custodian, enter and sign the required statement in the evidence ledger immediately below the last entry?			
• Did the primary evidence custodian, on return from temporary absence, ensure that all entries on records pertaining to evidence taken in, released, and disposed of by the alternate custodian were correct and accurate?			
• Did the primary custodian enter and sign the required statement in the evidence ledger immediately below the last entry after ensuring that the records were correct and the evidence was accounted for and properly documented?			
• Did the primary evidence custodian, on return from temporary absence, find that the alternate had made an incorrect entry?			
▪ Did the primary custodian immediately inform the USACIDC district field office commander, SAC, or PM?			
▪ Did the primary custodian prepare a memorandum outlining the error and what was done to correct it?			
▪ Was the original memorandum filed with the proper *DA Form 4137* or in a file folder under number 195-5a if the error was not on a *DA Form 4137*?			
▪ Was a copy of the memorandum placed in the proper case file?			

Table C-1. Checklist for an Evidence Room Inspection or Inventory (Continued)

IDENTIFICATION OF EVIDENCE *AR 195-5, paragraph 2-1*			
	Yes	No	NA
Did the CID special agent or military policeman initially assuming custody of the evidence permanently mark the evidence for future identification by marking his initials and the date and time of acquisition directly on the item, to include major interchangeable parts?			
Was a *DA Form 4002* typed or printed legibly in blue or black ink with the document number, MPR and/or CID sequence number, item number, and remarks (the common name of the item)?			
• Did the *DA Form 4002* correspond with the *DA Form 4137?*			
• Was the *DA Form 4002* attached to each item of evidence to identify and control it?			
SEALING EVIDENCE ***AR 195-5, paragraph 2-2***			
Were the original *DA Form 4137* and the original and one copy of *DA Form 3655* put in a properly addressed, sealed envelope and attached to the outside (inner wrapper) container when sealed evidence was sent to USACIL?			
Were multiple items of fungible evidence sealed in separate containers and sent in one shipping package?			
NOTE: If nonfungible evidence is sent in the same package with fungible evidence from the same investigation, all evidence must be double wrapped.			
Fungible evidence.			
• Were wet or damp fungible items sealed in a container before sending them to USACIL after they were visually examined, field tested, and dried?			
• Was fungible evidence sealed in paper or manila envelopes, cardboard boxes, or wrapping paper?			
• Was fungible evidence sealed as early as possible in the chain of custody to reduce the number of persons having access to it?			
• Were all openings, joined surfaces, and edges sealed with paper packaging tape or the equivalent when fungible evidence was sealed in containers other than heat seal bags?			
NOTE: **No cellophane or masking tape will be used.**			
• Were the time and date of sealing, the initials or signature of the seizer, and the USACIDC control number or MPR number placed so that they appeared on both the tape and package?			
• Was a *DA Form 4002* properly completed and attached to the package to identify and control it?			
When heat seal bags were used to seal fungible evidence—			
• Were the procedures provided with the equipment used to heat and seal the bag?			
• Was a *DA Form 4002* properly completed and attached to the package to identify and control it?			

Table C-1. Checklist for an Evidence Room Inspection or Inventory (Continued)

	Yes	No	NA
Were items, such as small amounts of powders, hairs, fibers, small paint chips, or flakes (that will adhere to the inside because of electricity), placed in plastic bags?			
• Were the items put in paper wrappings or cardboard containers?			
• Were the containers marked for future identification with the time and date of sealing, the initials or signature of the seizer, and the USACIDC control number or the MPR number on the seal and container?			
• Was a *DA Form 4002* properly completed and attached to the containers to identify and control them?			
Was fungible evidence that was submitted to USACIL for serological testing sealed in any type of plastic container?			
Was fungible evidence packaged according to *FM 3-19.13*?			
Nonfungible evidence.			
NOTE: There is no requirement to seal nonfungible evidence; however, it may be sealed if desired.			
• Was nonfungible evidence handled the same as fungible evidence when sealed?			
• Was nonfungible evidence that may have other evidence on it or may contain other evidence (fibers, hairs, paint chips, and glass fragments) sealed in a proper container just as fungible evidence was? (This avoids loss or accidental contamination of evidence that could be present.)			
PREPARATION OF *DA FORM 4137* ***AR 195-5, paragraph 2-3***			
Was all physical evidence, regardless of how it was obtained, recorded on *DA Form 4137*?			
Did the CID special agent or military policeman who first acquired the evidence prepare the *DA Forms 4137* and make three copies?			
Were the initial entries in the "Chain-of-Custody" sections of all *DA Forms 4137* accurately completed by the CID special agent or military policeman who first acquired the evidence?			
Was the last copy of the *DA Form 4137* given to the person as a receipt when evidence was taken from him?			
Was any change in custody of fungible and nonfungible evidence, after USACIDC or a military policeman acquired it, recorded in the "Chain-of-Custody" section of the *DA Form 4137*?			
Did each person handling or processing the physical evidence preserve the integrity of the evidence while it was under his control and maintain the chain-of-custody entries on the original *DA Form 4137* and any other copies?			
Was "Sealed container received, contents not inventoried" or was the acronym "SCRCNI" noted in the "Purpose of Change of Custody" column of the *DA Form 4137* when sealed fungible evidence changed custody?			
Did the evidence custodian check the *DA Form 4137*?			
Did the evidence custodian submit the *DA Form 4137* with the evidence?			
Did the evidence custodian correct errors on the *DA Form 4137* when possible?			

Table C-1. Checklist for an Evidence Room Inspection or Inventory (Continued)

	Yes	No	NA
Did the CID special agent or military policeman who first received evidence from the non-Army law enforcement agency prepare a *DA Form 4137?*			
Did the CID special agent or military policeman mark the evidence for future identification if the releasing agency had not already done so and would not mark it at the time of release?			
Were receipts or chain-of-custody documents furnished by the releasing agency attached to the *DA Form 4137*?			
Did evidence custodians note in the "Purpose of Change of Custody" column "Sealed container received, contents not inventoried" or the acronym "SCRCNI?"			
Did evidence custodians breach or inventory the contents of sealed containers of evidence if they were marked "Sealed container received, contents not inventoried" or bore the acronym "SCRCNI?"			
PROCESSING OF *DA FORM 4137* BY THE EVIDENCE CUSTODIAN *AR 195-5, paragraph 2-4*			
Was physical evidence released to the evidence custodian the first working day after it was acquired?			
Was evidence acquired during nonduty hours properly secured in an approved temporary evidence container?			
Was access to the temporary evidence container controlled by the duty special agent or military policeman, as appropriate, pending release of the evidence to the evidence custodian?			
Did the evidence custodian and the CID special agent or military policeman properly complete the chain-of-custody entry on the copies of and the original *DA Form 4137* received with the evidence?			
Did the evidence custodian assign the correct document number to the original and the copies of *DA Forms 4002* and *4137?*			
Did the evidence custodian keep the original and the first copy of *DA Form 4137* and give the second copy to the CID special agent or military policeman to be placed in the appropriate case file after the chain of custody was completed and all copies were assigned the correct document number?			
Was the location of the evidence within the evidence room shown in pencil on the bottom margin of *DA Form 4137*?			
Were location changes kept current by the evidence custodian by erasing the previous entries on *DA Form 4137* and noting the new location?			
Did *DA Form 4137* files (kept by the evidence custodian) contain the original and the first copy of the form?			
Were *DA Forms 4137* put in number sequence in a folder no thicker than 3/4 inch or in a folder that contained no more than fifty documents per folder with the number and year of the documents in the folder shown on the outside (for example, 1-50, 1989)?			
Did the original *DA Form 4137* go with the evidence when it was temporarily released from the evidence room to USACIL, an *Article 32* investigating officer, or another agency for other official reasons?			

Table C-1. Checklist for an Evidence Room Inspection or Inventory (Continued)

	Yes	No	NA
Was the first copy of *DA Form 4137* detached from the original and placed in the proper suspense file?			
Were at least three suspense folders kept in the evidence room: one labeled "USACIL" for evidence sent to laboratories; one labeled "ADJUDICATION" for evidence on temporary release to *Article 32* investigation officers, courts, SJA officers, or other persons for legal proceedings; and one labeled "PENDING DISPOSITION APPROVAL" for the first copy of *DA Form 4137* when it was sent to the SJA for approval of disposition?			
Did the evidence custodian use other suspense folders as management tools as needed?			
Did the evidence custodian place the original *DA Form 4137*, to include all attached memorandum or associated documents (such as *DD Forms 281* and *1131* pertaining to .0015 funds) in a separate evidence voucher file under number 195-5a after all items of evidence listed on the form were *properly* disposed of?			
• Was a copy of the *DA Form 4137* sent to USACRC as an exhibit to the final report when the it pertained to an ROI or MPR?			
• Was a copy of the *DA Form 4137* sent as an exhibit to the supplemental report if the evidence was collected after the final report was submitted to USACRC?			
• Was a copy of the *DA Form 4137* attached to the office's file copy of the report?			
• Was a copy of *DA Form 4137* made from the suspense copy and placed in the evidence voucher file (noting the disposition of the original) if the original *DA Form 4137* was entered as a permanent part in the record of trial, accompanied evidence released to an external agency, or was not available for any other reason?			
EVIDENCE LEDGER ***AR 195-5, paragraph 2-5***			
Did the evidence ledger show accountability when cross-referenced with *DA Form 4137?*			
Did the evidence ledger account for document numbers assigned to *DA Form 4002* and *4137*?			
Is the evidence ledger a bound book and that follows a disposition schedule set by *AR 25-400-2* with a destruction date three years after final disposition is made of all items entered in it?			
Was the evidence ledger prepared with six columns that spanned two facing pages when the book was opened? (See *AR 195-5, Figure 2-2,* for column headings.)			
• Did the first page of the ledger and the first page beginning a new calendar year, as a minimum, show the proper column headings? (Each page does not need to show the column headings.)			
• Were both vertical and horizontal lines used to separate the entries?			
• Was blue or black ink used to post entries?			
NOTE: The lines separating the entries may be in a different color.			

Table C-1. Checklist for an Evidence Room Inspection or Inventory (Continued)

	Yes	No	NA
The columns in the evidence ledger should be completed as follows:			
• Document number and date received. Did this column contain the document number assigned to the evidence custody document and the date that the *DA Form 4137* was received in the evidence room?			
• CID control number or MPR number. Was the number assigned to the investigation that the evidence pertains to entered in this column?			
NOTE: CID units may use both the MPR and CID sequence number when the military police prepare an MPR on the same investigation.			
• Description of the evidence.			
▪ Was a brief description of the evidence entered in this column to include the item number from the *DA Form 4137*?			
▪ Was fungible evidence that was sealed in a container briefly described from data on the *DA Form 4137*?			
NOTE: This does not imply that the evidence custodian inventoried the items.			
• Date of final disposition.			
▪ Was the date the evidence was disposed of, as shown in the "Chain-of-Custody" section of the *DA Form 4137*, entered in this column?			
▪ Was the date of disposition for each item shown opposite its description when *DA Form 4137* contained several items and they were not disposed of on the same date?			
▪ Was only one date entered, followed by the words "all items" (for example, 25 Jan 89—all items), when all the items in an entry were disposed of on the same date?			
• Final disposition.			
▪ Was a brief note on the means of the final disposition entered in this column opposite the item description?			
▪ Was the means of disposal listed once, preceded by the words "all items" (for example, all items burned), when all items in the entry were disposed of in the same manner?			
▪ Did this column include a cross-reference to another *DA Form 4137* that contained evidence from the same investigator; names of owners, subjects, investigators; notations to show the presence of .0015 funds; or the results of laboratory examinations?			
▪ Was the notation SCRNI made when fungible evidence was received in a sealed container and was not inventoried?			
Was the ledger book completely filled before starting a new one?			
NOTE: If in large offices the number of entries made nearly fill a ledger, the remaining pages need not be used for the next year. A new ledger may be opened. Conversely, a small office may use only a few pages per year; therefore, the same ledger should be used for several years.			
Was a statement that the ledger pertains to *DA Forms 4137* (for example, 001-849 for calendar year 2002) entered after the last entry in the ledger for a calendar year?			

Table C-1. Checklist for an Evidence Room Inspection or Inventory (Continued)

	Yes	No	NA
Did the cover of the ledger book identify the organization or activity responsible for the evidence room and the dates spanned by the entries?			
MAINTENANCE OF EVIDENCE ***AR 195-5, paragraph 2-6***			
Was evidence stored so that the integrity and physical characteristics were maintained?			
Were items such as weapons kept in a clean, rust-free condition so their value as evidence was not destroyed?			
Were fingerprint cards that were obtained for comparison kept in the proper case file?			
Were fingerprint cards listed on *DA Form 4137* when obtained for comparison and sent to the laboratory with other evidence from the evidence room?			
NOTE: The description of the laboratory request is enough to connect the card with the evidence.			
Were fingerprint cards put back in the proper case file when they were returned from the laboratory?			
Were documents that could prove or disprove a point in question, such as insufficient fund checks, forged or altered documents, and other questioned documents (along with related standards or exemplars), kept in the evidence room?			
Were statements, records, and other documents routinely associated with an investigation kept in the evidence room?			
Were large items of evidence (for example, vehicles) kept in an impoundment lot, warehouse, or other reasonably secure place when it was necessary to retain them?			
TEMPORARY RELEASE OF EVIDENCE ***AR 195-5, paragraph 2-7***			
Was evidence removed from the evidence room only for permanent disposal or for temporary release for specific reasons? (The most common reasons for temporary removal are transmittal to a crime laboratory for forensic examination and presentation at a court-martial or hearing conducted under *Article 32, UCMJ.)*			
Did the person that evidence was released to, either temporarily or for permanent disposal, sign for it in the "Received By" column of the "Chain of Custody" section?			
Did the person sign on the original and first copy of the *DA Form 4137*? (The person receiving the evidence must safeguard it and maintain the chain of custody until it is returned to the evidence custodian.)			
Did the evidence custodian release the original *DA Form 4137* to the person who assumed temporary custody of the evidence or was it sent by first class registered mail or another means along with the evidence?			
Did the evidence custodian put the first copy of the *DA Form 4137* in the proper suspense folder after releasing the evidence?			
Was the original *DA Form 4137* properly annotated by the custodian and the person returning the evidence and put in the evidence document file when the evidence was returned?			

Table C-1. Checklist for an Evidence Room Inspection or Inventory (Continued)

	Yes	No	NA
Was the first (suspense) copy, with the chain of custody properly annotated, refiled with the original *DA Form 4137*?			
Were copies used and processed as stated above when items on the same *DA Form 4137* were temporarily released to more than one agency or person at the same time?			
Was a note made on the original *DA Form 4137* and the first copy stating that copies of the original form were made?			
Was the chain of custody for all evidence recorded on the first copy of the *DA Form 4137*?			
Were personnel receiving evidence, either on a temporary or permanent basis, required to present identification as necessary to ensure that evidence was only handled by authorized persons?			
Did packages contain evidence from only one investigation to ensure proper maintenance of the chain of custody?			
Was a laboratory report prepared in all cases and returned to the requestor with the evidence submitted, unless cancelled by the contributor? (USACIL will not make final disposition of any item submitted for examination.)			
Was the sealed container opened except for official purposes or when disposed of when fungible evidence was returned from the USACIL?			
Did the person who opened the container sign the chain-of-custody document?			
Did the person note in the "Purpose-of-Change" column that the seals were intact and the reason for opening?			
Did the person open the container by cutting it without damaging the seals?			
Was the evidence, with all prior containers of its sealed parts, put in a new container and resealed?			
Did any person who accidentally breached the sealed container note how or why it happened in the "Purpose of Change of Custody" column?			
Was the container then sealed in a new container as stated above?			
Were the remarks continued in the block immediately below the entry if the space in the "Purpose of Change of Custody" column was too small?			
Was a diagonal line drawn from the left side of the "Item Number" block through the "Received By" block if remarks were continued in the block immediately below the entry?			
Did the original *DA Form 3655* and one copy go with the evidence? The field office commander, SAC, or PM does not need to sign as the requestor. The evidence custodian, special agent, or military policeman may do this.			
DA Form 3655.			
• Was the address of the USACIL serving as the contributing office in the "TO" block?			
• Was the division that was making the examination shown on the "ATTENTION LINE" (for example, photography, firearms and toolmarks, questioned documents, latent prints, serological evidence, and trace evidence)?			

Table C-1. Checklist for an Evidence Room Inspection or Inventory (Continued)

	Yes	No	NA
• Was the contributor's return address in the "FROM" block?			
NOTE: If the evidence or report is to be returned to an office other than the one in the "FROM" block, the address that the report or evidence is to be returned to is noted in item 11.			
• Were items 1 through 9 filled out correctly?			
• Were the items being submitted annotated in items 10a and b?			
▪ Was the exhibit number the same as the item number shown on *DA Form 4137* where practical?			
▪ Was the description column detailed enough so that each item submitted could be identified and distinguished from any other item with the same request for examination?			
▪ Were the document numbers listed on *DA Form 4137* to ease identification when more than one *DA Form 4137* was involved with the same laboratory request?			
• Did item 11 describe the examinations desired by the contributor and furnish information required for processing and any other information that the requestor felt would help the examiner?			
NOTE: All serological, firearms, and trace evidence submissions must have background information in this section.			
Expeditious handling and examination of evidence is sometimes required.			
• Was expeditious handling necessary when—			
▪ The subject was being held in pretrial confinement?			
▪ A trial date was set?			
▪ Results were needed for an *Article 32* investigation?			
▪ Results were needed for an *Article 39a* session?			
▪ The analysis of covertly acquired controlled substances was needed for further investigation?			
▪ Other valid reasons existed?			
• Were most requirements made known to the USACIDC or military police activity by the SJA?			
• Did USACIDC and military police activities request expeditious handling by the USACIL by putting "EXPEDITE" on the *DA Form 3655* after the subject's name?			
• Was the reason for the request for expeditious handling included?			
• Was a preference for telephone or electronic mail notification of the laboratory results noted on *DA Form 3655*?			
• Did notification instructions include the name, telephone number, and e-mail address of the person to notify?			
• Was USACIL notified by the quickest means possible if such notice was required after the evidence had already been sent?			
• Did USACIL notify the requester via e-mail stating the reasons why it was not possible for it to receive such requests and immediately process them?			
Was a return receipt for registered mail used when evidence was mailed to the owner for final disposal?			

Table C-1. Checklist for an Evidence Room Inspection or Inventory (Continued)

	Yes	No	NA
Was the *DA Form 4137* sent with the evidence?			
Was the return receipt or registered mail permanently attached to the original *DA Form 4137* when it was returned?			
Was evidence properly handled to maintain the chain of custody if it was sent by means other than registered mail? (Evidence may be hand-carried or shipped by an installation transportation officer on a government bill of lading [GBL].)			
Was a copy of the shipping document obtained and attached to the suspense copy of *DA Form 4137* if evidence was shipped by GBL?			
Was a copy of the shipping document obtained and attached to the suspense copy of *DA Form 4137* if evidence was shipped by GBL?			
Was the shipping document kept with the suspense copy of *DA Form 4137* until—			
• Notice was received from the addressee indicating receipt of the evidence?			
• The evidence was returned to the evidence room?			
Were the original and duplicate custody documents properly annotated and sent with the evidence when the evidence was permanently transferred from one evidence room to another?			
Did the custodian who received this evidence enter the next document number of the receiving room on both copies?			
Was the previous document number lined through in such a manner that it remained legible?			
Was evidence for military judicial proceedings sent directly to the requesting SJA or other agency by first class registered mail?			
Did the SJA or agency return the evidence to the evidence custodian in the same manner?			
Were requests for USACIL analysis from non-CID or nonmilitary police Army activities (for example, SJA, unit commanders, alcohol and drug abuse program coordinators) coordinated with and made by the local CID elements?			
• Were items of material value or physical evidence in a criminal investigation mishandled, contaminated, or improperly accounted for?			
• Were USACIL elements made aware of incidents of suspected criminal activity as defined in *AR 195-2, paragraphs 3-2* through *3-3*?			
• Was help given when a CID element decided that it had no interest in the circumstances of direct referrals to a USACIL?			
• Did the CID provide proper laboratory request forms, addresses, and advice on packaging and transmittal?			
• Were the date, name, and unit of the CID representative coordinated with entered in item 11 of *DA Form 3655*?			
FINAL DISPOSITION ***AR 195-5, paragraph 2-8***			
Was evidence disposed of as soon as possible after it had served its purpose or was determined to be of no value as evidence?			
Was evidence released to trial counsel for judicial proceedings returned as soon as possible to the custodian for final disposition?			

Table C-1. Checklist for an Evidence Room Inspection or Inventory (Continued)

	Yes	No	NA
Did the trial counsel immediately notify the custodian so that the evidence custody document could be properly annotated if an item of evidence was made part of the trial record?			
Was the inclusion of evidence in the trial record considered the final disposition?			
Was the original evidence custody document sent to the SJA of the commander with general courts-martial jurisdiction over the subject when final action had been taken in known subject cases?			
Did the SJA complete the final disposal authority part of the *DA Form 4137* if the evidence was no longer needed? (When evidence must be retained, this part of the form is not completed; a brief statement giving the reason for keeping the evidence is furnished to the custodian on separate correspondence. In unusual cases, where there is a high risk of losing the original *DA Form 4137* [for example, isolated units that must mail the *DA Form 4137* to the servicing SJA for disposition approval], a letter or memorandum may be substituted for disposition approval.)			
Was enough information furnished to allow the SJA to make a decision when it was used?			
Was the return correspondence from the SJA giving disposition approval attached to the original *DA Form 4137* for file at USACRC?			
Was evidence in an investigation for which no subject was identified disposed of 3 months after the completion of the investigation without SJA approval?			
Was the evidence disposed of earlier with SJA approval?			
NOTE: In some situations, it may be advisable to keep the evidence longer than 3 months.			
Did the evidence custodian obtain the approval of the commander, SAC, PM, or representative, for disposal of evidence when the subject was not known?			
Was this approval given by completing the final disposal authority section of the original *DA Form 4137?*			
Were controlled substances received by the evidence custodian that did not apply to an investigation immediately disposed of under the authority outlined above? (Disposition may be made immediately after determining that the substance cannot be linked to a suspect.)			
Were items of possible evidence that were determined to have no value as evidence by the CID special agent or military police before they were sent to the evidence custodian disposed of by the special agent or military police?			
Was evidence disposed of only after consulting with the immediate supervisor and getting approval? (This does not include controlled substances that were found.)			
Was verbal permission obtained (followed by written permission) for the disposal of evidence in an investigation for which no subject had been identified?			
Was disposal of evidence done according to *AR 195-5*?			
Were items of evidence disposed of immediately when it was not practical or desirable to keep them (for example, automobiles, serial numbered items, items required for use by the owner, undelivered mail, large amounts of money, or perishable or unstable items)?			
NOTE: Do not enter evidence into the evidence room if it can be disposed of immediately.			
Was the disposal of evidence coordinated with the SJA?			
Did the SJA complete the final disposal authority portion of the *DA Form 4137*?			

Table C-1. Checklist for an Evidence Room Inspection or Inventory (Continued)

	Yes	No	NA
Was the final disposal authority part of the *DA Form 4137* completed (as in the case of evidence in an investigation for which no subject had been identified) when evidence was permanently released to an external agency?			
Did the SJA give legal advice when a legal question on methods of disposal arose?			
Was US government property released to the organization to which it was issued?			
Was personal property that was legal to own released to the rightful owner?			
Were US Government Treasury checks and money orders from an APO money order facility returned to the issuing APO?			
Were other types of negotiable instruments (for example, money orders, travelers checks, and checks owned by a business firm) released to the firm?			
Were negotiable instruments destroyed if the firm did not want them returned?			
Were US postal money orders that were not identified as the property of a specific person sent to the Military Money Order Division, Postal Data Center, P.O. Box 14965, St. Louis, MO 63182 (except for US Government Treasury checks and money orders from an APO money order facility)?			
• Was a letter of transmittal that had the investigative case number, date of final report of investigation, offense, and complete identification of both the subject and victim included?			
• Did the letter also state that information on the ROI or MPR could be obtained by contacting the USACRC, USACIDC, ATTN: (office symbol), 6010 6th Street, Fort Belvoir, VA 22060-5585?			
• Was a first-class letter sent immediately to the Military Money Order Division, Postal Data Center if it was foreseen that postal money orders would be kept as evidence for more than 120 days from the date received by the evidence custodian?			
• Did the first-class letter list the serial number and symbol number of each money order and the reasons for detention? (This may prevent the issue of duplicate money orders.)			
Were negotiable instruments and other documents obtained from a person returned?			
Were known document standards released to the agency, person from whom the standards were received, or the rightful owner, as appropriate?			
Were exemplars and other documents of no value to the person or agency from whom they were received placed in a case file or destroyed?			
Were items of personal property that no longer had evidentiary value and belonged to deceased or missing Army personnel released to the summary court officer appointed to dispose of the decedent's effects?			
Were controlled substances destroyed in the presence of a witness who was a CID special agent or in grade E6 or above? (The witness must not have been involved in the chain of custody.)			
Were controlled substances burned or was another method used that made it permanently useless?			
NOTE: For final disposal of marijuana used for training, see *paragraph 2-9*.			
Was counterfeit currency and coins and counterfeiting equipment released to the nearest office of the USSS, unless the USSS directed otherwise?			

Table C-1. Checklist for an Evidence Room Inspection or Inventory (Continued)

	Yes	No	NA
Was the disposal of counterfeiting equipment coordinated with the supporting USSS office before release?			
• Was evidence of this type that was seized in the Far East sent to the USSS office in Hawaii?			
• Was evidence of this type that was seized in Europe, Africa, or the Middle East sent to the USSS representative at the US Embassy in Paris France?			
• Was evidence of this type that was seized in the Caribbean Sea and Central and South America areas sent to the USSS office in Hato Rey, Puerto Rico?			
Firearms and ammunition.			
• Were US government firearms, ammunition, and explosives that were kept as evidence returned to the proper military unit?			
• Were US government firearms, ammunition, and explosives released to the installation accountable officer, per *AR 710-2, paragraph 2-5q*, if the unit could not be identified?			
• Were contraband firearms and ammunition that were kept as evidence disposed of after coordination was made with the USACIL firearms examiner?			
• Were legal personal weapons that were impounded for minor infractions (for example, failure to register per local laws) returned to the rightful owner when the legal requirements had been met and they were no longer needed as evidence?			
• Was nongovernment ammunition (live or inert) that was kept as evidence reported for turn-in to the supporting DRMO, per *DOD 4160.21-M*, for destruction using the procedures in *DA Pamphlet 710-2-1, Chapter 3*? (The evidence custodian may keep the items until the DRMO possesses the turn-in forms and provides specific disposal instructions.)			
Was evidence of obvious value turned in to the DRMO per *DOD 4160.21-M* when the owner was unknown or could not be located?			
Were items of evidence that were found at crime scenes, had no known owner, and were of no obvious value (for example, matchbooks, beer cans, bottles, glass fragments, and wooden sticks) destroyed by crushing, burning, or other methods used to render the items useless or harmless?			
Was money turned in to a US Army finance officer when the owner was not known or could not be located after reasonable attempts?			
Was a *DD Form 1131* filled out?			
Was a copy of *DD Form 1131* attached to the original custody document?			
Were post exchange items, commissary items, and items illegally introduced into a host country that were connected with black market, customs, and postal investigations disposed of per—			
• Local regulations?			
• Status of forces agreements?			
• Law or customs of the host country?			
Were limitation .0015 contingency funds (CID funds) that were held as evidence disposed of per *AR 195-4*?			

Table C-1. Checklist for an Evidence Room Inspection or Inventory (Continued)

	Yes	No	NA
Were CID funds that no longer had value as evidence promptly deposited with the local FAO on *DD Form 1131*?			
Did personnel who prepared the *DD Form 1131* ensure that the accounting classification cited was the same as the one on the voucher the CID funds were originally disbursed on before depositing funds with the FAO?			
Was a copy of *DD Form 1131* showing the return of funds given to the proper certifying and approving officer?			
Was the IRS office in the geographical area notified when $10,000 or more (US or foreign money) was imported outside the United States during a criminal investigation and was kept as evidence?			
Was a notice of levy sent by the IRS to the unit holding the funds if there was a tax liability?			
Were funds in the amount of the levy released to the IRS?			
Were the remaining funds released elsewhere (as required) per *AR 195-5, paragraph 2-7?*			
RELEASE OF MARIJUANA ***AR 195-5, paragraph 2-10***			
Was marijuana that was held by custodians and no longer had value as evidence released subject to the following?			
• Did the officer responsible for training marijuana detection dogs submit a written request to the evidence custodian?			
• Did the request list the approximate amount of marijuana needed for training?			
• Was a copy of this request attached to the custody document for the marijuana?			
• Was marijuana that was approved for final disposal, per *AR 195-5, paragraph 2-8*, released for training purposes?			
• Did the evidence custodian consider this a final disposal and annotate it on the original *DA Form 4137?*			
• Did the amount of marijuana released at one time exceed 200 grams per dog?			
• Was more released for the same dog until the current supply was exhausted or was no longer suitable for training purposes?			
• Did the person receiving the marijuana properly account for it?			
• Did the person dispose of any that remained if it was replaced or was no longer required for training?			
▪ Was the marijuana listed on a new *DA Form 4137*?			
▪ Was the first "Released By" section signed by the custodian releasing the marijuana, and was the first "Received By" section signed by the person receiving it?			
▪ Did dog handlers and other personnel handling the marijuana maintain the chain of custody?			

Table C-1. Checklist for an Evidence Room Inspection or Inventory (Continued)

	Yes	No	NA
▪ Did they secure the marijuana in an office safe or a similar container when it was not being used for training?			
▪ Was it noted on the *DA Form 4137* when the marijuana became too old for training, was lost, or was consumed by the dog?			
▪ Was the marijuana burned in the presence of a witness when all or part of it was determined to be of no further value for training and was this annotated on the *DA Form 4137?*			
▪ Was the *DA Form 4137* then filed in the reference or training files of the office responsible for the training?			
NOTE: Marijuana may be used for training CID and military police personnel in the techniques of identification and "field testing," subject to the provisions above.			
Was the unused marijuana, including residue, returned to the evidence custodian for final disposition?			
Did training personnel receive and return the marijuana on the same date? (Marijuana should not be kept any longer than needed to complete the training.)			
FIELD TESTING OF NONNARCOTIC CONTROL SUBSTANCES ***AR 195-5, paragraph 2-11***			
NOTE: Field testing of nonnarcotic controlled substances by CID agents and MPIs is authorized.			
Were the results of field tests verbally furnished to the commander concerned as soon as possible?			
Was evidence consumed through field testing deducted from the *DA Form 4137*?			
Was the chain-of-custody section marked to show the disposal of the amount consumed?			
Did the person whose name appears in the chain of custody on the *DA Form 4137* match the person who conducted the test as recorded on the *CID Form 36*?			
Was a copy of *CID Form 36* attached to both the original and first copies of the *DA Form 4137*?			
Was the evidence sent to USACIL for forensic analysis when the action commander indicated that a person would not be court-martialed for an offense that involved use or possession of marijuana?			
Was evidence always sent to the USACIL for analysis when court-martial action against the offender was considered?			
Were cases accepted that involved the simple use or possession of marijuana and for which the SJA or a representative decided that no analysis was required?			
Did the commander and SJA concerned coordinate closely to determine the commander's intentions and to ensure that proper USACIL support was provided when needed?			
Did the SJA or his representative promptly notify the proper CID or military police elements when the status of a case changed and there was no longer a need for laboratory analysis?			
INSPECTIONS ***AR 195-5, paragraph 3-1***			
Did the commander, SAC, or PM supervise the evidence custodian?			

Table C-1. Checklist for an Evidence Room Inspection or Inventory (Continued)

	Yes	No	NA
Was the evidence custodian directly responsible for the proper handling and processing of evidence?			
Did the evidence custodian inspect the evidence once each month?			
Was this inspection made in lieu of a physical security inspection per *AR 190-13*?			
Did the inspector determine that—			
• The evidence room was orderly and clean?			
• Structural and security requirements of *AR 195-5* were being met?			
• Evidence was being received, processed, safeguarded, and disposed of per *AR 195-5*?			
INVENTORIES ***AR 195-5, paragraph 3-2***			
Were inventories made—			
• Once during each calendar quarter?			
• When there was a change of the primary evidence custodian?			
• When there was a breach of security or the loss of evidence stored in the evidence room?			
Did the evidence custodian and a disinterested officer, appointed for this purpose, make a joint inventory of the evidence stored in the evidence room every quarter?			
Did the commander, SAC, or PM send a written request to the proper commander to appoint a disinterested commissioned or warrant officer?			
Did the appointing authority give the requesting activity a copy of the appointing document (memorandum)?			
Did the evidence custodian retain this until the inventory was complete and make a ledger entry per *AR 195-5, paragraph 3-2*?			
Was the disinterested officer a current member of USACIDC or a military police officer?			
Did the disinterested officer—			
• Become familiar with the provisions of *AR 195-5* before conducting the inventory?			
• Count the evidence in the evidence room and verify that it corresponded with that shown on *DA Form 4137*?			
• Cross-reference all *DA Forms 4137* (including those in suspense files) with entries in the evidence ledger to ensure the accountability of all evidence?			
• Ensure that copies of *DA Form 4137* in the suspense files were properly annotated with the—			
▪ Registered mail number if sent to USACIL or another agency?			
▪ Proper signature if released for court-martial, investigations under *Article 32, UCMJ*, or other official purposes?			
• Ask the evidence custodian to verify the weight of drug evidence or of controlled substances for inspection?			
• Provide the evidence custodian with a copy of the appointing document?			

Table C-1. Checklist for an Evidence Room Inspection or Inventory (Continued)

	Yes	No	NA
Did the incoming and the outgoing primary custodians make a joint physical inventory of all the evidence room when the evidence custodian was changed? (Joint inventories may be made along with quarterly inventories by disinterested officers; however, each type of inventory should be recorded separately.)			
Were all evidence records carefully checked to ensure proper documentation and accountability?			
Did the outgoing custodian resolve all discrepancies before the transfer of accountability?			
NOTE: A joint inventory does not need to be made when the alternate custodian replaces the primary custodian for 30 days or less.			
Was a joint inventory done before it was known whether the primary custodian would be gone 30 days or less?			
Was another inventory done when the primary custodian returned?			
Was an inventory done by the person assigned for the inquiry (see *AR 195-5, paragraph 3-3*) when evidence was lost or the security of the evidence room was breached?			
Was the inventory done in the presence of the evidence custodian?			
Were sealed containers of fungible evidence breached when making any type of inventory?			
Was the evidence sealed in a new container per *AR 195-5, paragraph 2-7d*, when it was breached?			
RECORDED INSPECTIONS ***AR 195-5, paragraph 3-2***			
Was the following statement entered in the evidence ledger immediately below the last entry and signed by the commander, SAC, or PM? I, (NAME), certify that on (DATE), per *AR 195-5*, inspected the evidence room. Evidence is being processed per *AR 195-5* (with no exceptions or the following exceptions). __________ __________ (Signature)__________			

Table C-1. Checklist for an Evidence Room Inspection or Inventory (Continued)

	Yes	No	NA
RECORDED INVENTORIES ***AR 195-5, paragraph 3-2***			
Were quarterly inventories recorded in the evidence ledger as follows? We the undersigned, certify that on (DATE), per *AR 195-5*, a joint inventory was made of the evidence room. All evidence was properly accounted for (with no exceptions or the following exceptions). ____________________ ____________________ (Signature of officer) (Signature of evidence custodian) (Printed name, grade, unit) (Memorandum, date, issuing HQ)			
Were change of custodian inventories entered in the evidence ledger immediately below the last entry and signed by both the incoming and the outgoing primary custodian? I, (NAME), assume the position of primary custodian and accept responsibility for all evidence shown on *DA Form 4137* in the evidence document files. A joint inventory was made on (DATE), with (NAME), the outgoing evidence custodian. Any discrepancies have been resolved to my satisfaction. (Signature of incoming primary evidence custodian) (Signature of outgoing primary evidence custodian or person appointed as primary evidence custodian per *AR 195-5, paragraph 3-b[3]*)			
Were all *DA Forms 4137* that were in the document file noted and signed to show the change of custody on satisfactory completion of the inventory?			
Did the custodian ensure that copies of *DA Form 4137* in the suspense file were properly annotated with the—			
• Registered mail receipt number if sent to a USACIL or another agency?			
• Proper signature if released for court-martial, investigations under *Article 32, UCMJ*, or other official purposes?			
Was a joint inventory done if the alternate custodian became the primary custodian due to death, extension of absence beyond 30 days, sudden illness, or emergency transfer of the primary custodian?			
Did the alternate custodian and a person appointed by the commander supervising the evidence room do the joint inventory?			
Was the "Released By" block of each *DA Form 4137* noted "NA-Custodian Unable to Sign" in this situation?			

Table C-1. Checklist for an Evidence Room Inspection or Inventory (Continued)

	Yes	No	NA
Did the alternate custodian complete the "Received By" block to accept custody for the evidence described on the *DA Form 4137*?			
Was the reason the primary custodian was unable to sign noted in the "Purpose of Change of Custody" block?			
Did the person appointed to make the inventory with the alternate custodian sign under the ledger entry that showed the inventory?			
INQUIRIES *AR 195-5, paragraph 3-3*			
Was an inquiry or investigation made per *AR 15-6* or *AR 195-3* if evidence was lost or the security of the evidence room was breached?			
Did the proper region, separate field office, or USACIL commander initiate inquiries or investigations?			
Were losses or breaches during evidence handling and the initiation of inquiries reported to Commander, USACIDC, ATTN: CIOP-ZA, 6010 6th Street, Fort Belvoir, VA 22060?			
Did the inquiry fail to account for or recover the evidence? (If this was the case, relief for accountability of the evidence must be granted. For military police activities, this will be done by the installation or activity commander with an information copy of the entire proceedings sent through the MACOM PM to HQDA (DAMO-ODL), Washington, DC 20310-0440. For CID activities, relief will be granted by the appropriate USACIDC group commander, with an information copy of the entire proceedings furnished to the commander, USACIDC, ATTN: CIOP-ZA, 6010 6th Street, Fort Belvoir, VA 22060-5585.)			
• Was relief from further accountability for evidence granted, allowing the custody document to be closed?			
• Did the relief from further accountability have any bearing on administrative or judicial actions taken against those responsible?			
SECURITY STANDARDS FOR EVIDENCE STORAGE *AR 195-5, paragraph 4-1*			
Was the evidence room set up using the security standards outlined in AR 195-5? (Evidence should be stored in a room designated as an evidence room; however, circumstances may necessitate use of some other structure or container. A structure, such as a vault, that exceeds or equals the requirements of *AR 195-5* is acceptable. Evidence storage facilities cannot be used to store property that is not evidence.)			
EVIDENCE ROOM *AR 195-5, paragraph 4-2*			
Is the evidence room in the same building as the operational or administrative staffs of the USACIDC unit or PM?			
Evidence room construction.			

Table C-1. Checklist for an Evidence Room Inspection or Inventory (Continued)

	Yes	No	NA
• Were the walls extended from the floor to the true ceiling? (Walls and ceilings may be made of masonry or wood. Walls or ceilings that are of wooden-stud construction must have a combined exterior and interior thickness of at least 1 inch. Permanently installed flooring [other than masonry] may be used if the floor cannot be breached without causing considerable damage to the building structure.)			
• Was number 6-gauge steel mesh with a 2-inch diamond grid, permanently attached to the interior wall or ceiling if walls and ceiling could not be constructed as above? (Walls or ceilings may also be lined with steel plates at least 1/8-inch thick.)			
• Was a prefabricated steel mesh cage installed within the room as an evidence facility due to the walls, flooring, or ceiling of the room not meeting structural standards? (The cage must be number 6-gauge steel, conform to OCE drawing 40-21-01, and be permanently attached to the floor. When a cage creates a space between the original walls of the room and the cage, the added space must not be used for anything except the processing of evidence.)			
Doorway construction.			
• Was there only one doorway that allowed access to and from the evidence room?			
• Did entrance into the evidence room require opening two successive doors?			
• Did the door to the cage serve as the second door if an interior steel mesh cage was used? (In this case the outer door was of solid core wood or metal.)			
• Were two doors hung one behind the other if a steel mesh cage was not used? (One door may be made of steel mesh welded to a steel frame. The second door may be of solid core wood with the exterior reinforced with a steel plate not less than 1/8 inch thick.)			
• Were vertical steel bars at least 1/3 inch thick and spaced no more than 4 inches apart if a barred door was used?			
• Were horizontal bars welded to the vertical bars and spaced so that openings did not exceed 32 square inches?			
• Were the doors hung so that the doorframe was not separated from the door casing? (Either door may be hung on the outside of the doorway.)			
• Were door hinges installed so that the doors could not be removed without seriously damaging the door or doorjamb?			
• Were all exposed hinge pins spot-welded or bradded to prevent removal? (This is not required when safety stud hinges are used or when the hinge pins are on the inside of the door. A safety hinge has a metal stud on the face of one hinge leaf and a hole in the face of the other leaf. As the door closes, the stud enters the hole and goes through the full thickness of the leaf. This creates a "bolting" effect.)			
• Was the outer door secured by one high-security, key-opened padlock? (These padlocks conformed to military specifications MIL-P43607 [GL] [High-Security Padlock], FSN 5430-799-8248. The changeable combination padlock for the inner door conforms to requirements of military specification FF-P-110, 1969.)			

Table C-1. Checklist for an Evidence Room Inspection or Inventory (Continued)

	Yes	No	NA
• Were all locks used with a heavy steel hasp and staple?			
• Were the hasp and staple attached with smooth-headed bolts or rivets that went through the entire thickness of the door or doorjamb?			
• Were the bolts or rivets spot-welded or bradded on the inside of the door?			
• Were heavy-duty hasps and staples attached so that they could not be removed when the doors are closed?			
• Was only one door used for evidence rooms under 24-hour surveillance? (If this was the case, the single door should be of solid wood or covered with metal to prevent anyone from looking into the evidence room. It should be secured with a high-security padlock as discussed above.)			
Window construction.			
• Were the number of windows kept to the essential minimum?			
• Were all windows covered with steel, iron bars, or steel mesh?			
• Were steel bars at least 3/8 inch thick and vertical bars spaced not more that 4 inches apart?			
• Were horizontal bars welded to the vertical bars and spaced so that the openings did not exceed 32 square inches?			
• Were the ends of the bars securely embedded in the wall or welded to a steel channel frame fastened securely to the window casing?			
• Was acceptable steel mesh that was made from high carbon manganese (no less than fifteen hundredths of an inch thick) with a grid of no more than 2 inches from center to center used?			
• Was number 6-gauge steel mesh with a 2-inch diamond grid used if high carbon manganese steel was not available?			
• Was the steel mesh welded or secured to a steel channel which was fastened to the building by smooth-headed bolts that went through the entire window casing?			
• Was the steel mesh spot-welded or bradded on the interior or cemented into the structure itself to prevent forced entry?			
• Were appropriate measure taken when air conditioners were installed in windows or outside walls?			
INTERNAL FIXTURES FOR EVIDENCE ROOMS ***AR 195-5, paragraph 4-3***			
Evidence rooms.			
• Were there containers for high-value items, narcotics, and weapons?			
▪ Was there at least one container for added security of high-value items (for example, watches and jewelry) and large quantities of narcotics evidence (for example, 1 ounce of heroin or 1 kilogram of marijuana)?			
▪ Were there bins or shelves to store small amounts of narcotic evidence (for example, a package of marijuana cigarettes or pillbox of capsules)?			
NOTE: Firearms and ammunition are always stored in containers.			

Table C-1. Checklist for an Evidence Room Inspection or Inventory (Continued)

	Yes	No	NA
▪ Were containers such as field safes, filing cabinets, lockers, or locally made containers able to be secured with at least one locking device?			
▪ Were containers that weighed less than 300 pounds secured to the structure so that removal would be as difficult as breaking into the container itself? (This may be done with a chain secured to the container and fastened to a part of the building; it could be fastened to a radiator, a water pipe, an eyelet installed for this specific purpose, or another similar object. When several containers are used, they may be fastened together without being fastened to the structure if the combined weight of all the containers fastened together is at least 300 pounds. The containers and any chains attached to them may be secured with either heavy-duty pin tumbler padlocks or combination padlocks.)			
• Were there shelves, bins, or cabinets in the evidence room to allow neat and orderly arrangement of evidence? (The use of adjustable shelves is recommended whenever possible. Uniform-sized envelopes arranged numerically by document numbers are recommended when storing small items [for example, controlled substances] on shelves.)			
• Was there a worktable in the evidence room large enough for the custodian to use while processing incoming and outgoing evidence?			
• Was there a refrigerator in the evidence room? (A refrigerator is not required to be a permanent fixture in the evidence room; however, one should be readily available to use when perishable or unstable items must be stored.)			
TEMPORARY EVIDENCE FACILITIES ***AR 195-5, paragraph 4-3***			
Was a temporary evidence facility needed due to the size, amount, and type of evidence collected; the physical location of the CID or military police element; or the time the evidence was acquired?			
Was a safe or a secure filing cabinet used for temporary storage of evidence during nonduty hours, pending release to the evidence custodian?			
Was access to the safe or filing cabinet restricted to the person who collected the evidence, the military police duty investigator, or the military police desk sergeant?			
Was a key-opened padlock used in lieu of a combination padlock?			
Was one key secured in a separate envelope in the safe that contained the combinations and extra keys to padlocks for the evidence room?			
Were there at least three temporary containers so that duty personnel had an individual container for use during the weekend?			
Were temporary containers secured to the structure or fastened together as explained above?			
Was a CONEX used as a temporary evidence room?			
Was a separate building or fenced enclosure used when there—			
• Were unusually large items of physical evidence (for example, motor vehicles)?			
• Was a large amount of recovered property that could not be placed in the evidence room?			

Table C-1. Checklist for an Evidence Room Inspection or Inventory (Continued)

	Yes	No	NA
Was a fenced enclosure used only when there was no suitable building? (Normally, evidence that requires a fenced enclosure can be processed, photographed, and released after consulting with the SJA. However, if an enclosure or separate building must be used for temporary storage, the responsible supervisor and the evidence custodian must take action to protect the evidence.)			
KEY AND COMBINATION CONTROL ***AR 195-5, paragraph 4-4***			
Did only the primary and alternate custodians know the combination for locks used in an evidence room?			
Were copies of all combinations recorded on *Standard Form (SF) 700*?			
Were these forms kept in sealed envelopes in the safe of the CID commander, SAC, or PM?			
Did each key-operated lock have two keys?			
• Was one key for each lock in the constant possession of the primary evidence custodian?			
• Was the duplicate key put in a separate sealed envelope and secured in the safe of the CID commander, SAC, or PM?			
Were lock combinations changed when the primary or alternate was changed?			
Were all combinations and key locks changed upon possible compromise?			
Were keys transferred from the primary custodian to the alternate only if the primary custodian was absent for more than 8 duty hours or 72 nonduty hours?			
Were master key padlocks or set locks used, in any capacity, in the evidence room?			

Appendix D

Electronic Devices

Electronic devices may contain data crucial to an investigation. Many electronic devices contain memory that requires continuous power to maintain the information, such as battery or alternating current (AC) power. Unplugging the power source or allowing the battery to discharge can result in lost data. After determining the mode of collection, collect and store the power supply adaptor or cable (if present) with the recovered device.

COMPUTER SYSTEMS

D-1. A computer system typically consists of a main base unit (sometimes called a central processing unit [CPU]), data storage devices, a monitor, a keyboard, and a mouse. It may stand alone or be connected to a network. There are many types of computer systems, such as laptops, desktops, tower systems, modular rack-mounted systems, minicomputers, and mainframe computers. Additional components include modems, printers, scanners, docking stations, and external data storage devices. For example, a desktop is a computer system consisting of a case, motherboard, CPU, and data storage device with an external keyboard and a mouse.

- **Primary uses.** Computer systems are used for all types of computing functions and information storage, including word processing, calculations, communications, and graphics.
- **Potential evidence.** The device itself may be evidence of component theft, counterfeiting, or remarking.

CENTRAL PROCESSING UNIT

D-2. Often called the chip, the CPU is a microprocessor located inside the computer. The microprocessor is located in the main computer box on a printed circuit board with other electronic components.

- **Primary uses.** The CPU performs all arithmetic and logical functions in the computer. It controls the operation of the computer.
- **Potential evidence.** The device itself may be evidence of component theft, counterfeiting, or remarking.

MEMORY

D-3. Memory is information stored on removable circuit boards inside the computer. It is usually not retained when the computer is powered down.

- **Primary uses.** Memory stores the user's programs and data while the computer is in operation.
- **Potential evidence.** The device itself may be evidence of component theft, counterfeiting, or remarking.

HARD DRIVES

D-4. A hard drive is a sealed box containing rigid platters (disks) coated with a substance capable of storing data magnetically. It can be encountered in the case of a PC as well as externally in a stand-alone case.

- **Primary uses.** A hard drive is used to store information, such as computer programs, text, pictures, video, and multimedia files.
- **Potential evidence.** The device itself may be evidence of component theft, counterfeiting, or remarking.

MEMORY CARDS

D-5. Memory cards are removable electronic storage devices that do not lose the information when power is removed from the card. It may even be possible to recover erased images from memory cards. Memory cards can store hundreds of images in a credit card-sized module. They can be used in a variety of devices, including computers, digital cameras, and PDAs. Examples are memory sticks, smart cards, flash memory, and flash cards.

- **Primary uses.** Memory cards provide additional, removable methods of storing and transporting information.
- **Potential evidence.** The device itself may be evidence of component theft, counterfeiting, or remarking.

MODEMS

D-6. Modems can be internal and external (analog, DSL, ISDN, and cable). A laptop computer may use a wireless modem or PC card.

- **Primary uses.** A modem is used to facilitate electronic communication by allowing the computer to access other computers and/or networks via a telephone line, wireless, or another communications medium.
- **Potential evidence.** Evidence is most commonly found in files that are stored on hard drives and storage devices and media. Files can be user-created, user-protected, computer-created, or other types of data files.

USER-CREATED FILES

D-7. User-created files may contain important evidence of criminal activity such as address books and database files that may prove criminal association, still or moving pictures that may be evidence of pedophile activity, and communications between criminals, such as email or letters. Also, drug deal lists may often be found in spreadsheets. The following are examples of user-created files:

- Audio/video files.
- Calendars.
- Documents or text files.
- Internet bookmarks/favorites.

USER-PROTECTED FILES

D-8. Users have the opportunity to hide evidence in a variety of forms. For example, they may encrypt or password protect data that are important to them. They may also hide files on a hard disk, within other files, or under an innocuous name. The following are examples of user-protected files:

- Compressed files.
- Encrypted files.
- Hidden files.
- Misnamed files.
- Password-protected files.
- Cryptography.

D-9. Evidence can also be found in files and other data areas created as a routine function of the computer's operating system. In many cases, the user is not aware that data are being written to these areas. Passwords, Internet activity, and temporary backup files are examples of data that can often be recovered and examined.

D-10. There are components of files that may have evidentiary value including the date and time of creation, modification, deletion, access, user name or identification, and file attributes. Even turning the system on can modify some of this information.

COMPUTER-CREATED FILES

D-11. The following are some examples of computer-created files:

- Backup files.
- Configuration files.
- Cookies.
- Hidden files.
- History files.
- Log files.
- Printer spool files.
- Swap files.
- System files.
- Temporary files.

OTHER DATA AREAS

D-12. The following are considered other data areas:

- Bad clusters.
- Computer date, time, and password.

- Deleted files.
- Free space.
- Hidden partitions.
- Lost clusters.
- Metadata.
- Other partitions.
- Reserved areas.
- Slack space.
- Software registration information.
- System areas.
- Unallocated space.

SMART CARDS, DONGLES, AND BIOMETRIC SCANNERS

D-13. A smart card is a small handheld device that contains a microprocessor capable of storing a monetary value, an encryption key or authentication information (password), a digital certificate, or other information. A dongle is a small device that plugs into a computer port that contains types of information similar to information on a smart card. A biometric scanner is a device connected to a computer system that recognizes physical characteristics of an individual (for example, fingerprint, voice, and retina).

- **Primary uses.** Smart cards provide access control to computers or programs or function as encryption keys.
- **Potential evidence.** Identification/authentication information on the card and the user, the level of access, configurations, permissions, and the device itself may be evidence.

NETWORK COMPONENTS

LOCAL AREA NETWORK CARD OR NETWORK INTERFACE CARD

D-14. LAN cards or network interface cards (NICs) are indicative of a computer network. Network components are network cards and associated cables. Network cards also can be wireless.

- **Primary uses.** A LAN/NIC card is used to connect computers. Cards allow for the exchange of information and resource sharing.
- **Potential evidence.** The device itself as well as the media access control (MAC) address may be evidence.

ROUTERS, HUBS, AND SWITCHES

D-15. Routers, hubs, and switches are used in networked computer systems. Routers, hubs, and switches provide a means of connecting different computers or networks. They can frequently be recognized by the presence of multiple cable connections.

- **Primary uses.** This equipment is used to distribute and facilitate the distribution of data through networks.
- **Potential evidence.** The devices themselves as well as configuration files for routers may be evidence.

SERVERS

D-16. A server is a computer that provides services for other computers connected to it via a network. Any computer, including a laptop, can be configured as a server.

- **Primary uses.** Servers provide shared resources such as email, file storage, Web page services, and print services for a network.
- **Potential evidence.** The device itself may be evidence of component theft, counterfeiting, or remarking.

NETWORK CABLES AND CONNECTORS

D-17. Network cables can be different colors, thicknesses, and shapes and have different connectors. Differences depend on the components they are connected to.

- **Primary uses.** Network cables and connectors connect components of a computer network.
- **Potential evidence.** The devices themselves may be evidence.

MISCELLANEOUS ELECTRONIC ITEMS

D-18. There are many additional types of electronic equipment (too numerous to be listed) that might be found at a crime scene. However, there are many nontraditional devices that can be an excellent source of investigative information and/or evidence. Some examples are credit card skimmers, cell phone cloning equipment, caller ID boxes, audio recorders, and Web TV. Fax machines, copiers, and multifunction machines may have internal storage devices and may contain information of evidentiary value. The search of this type of evidence may require a search warrant.

ANSWERING MACHINES

D-19. An answering machine is an electronic device that is part of a telephone or connected between a telephone and the landline connection. Some models use magnetic tape or tapes, while others use an electronic (digital) recording system.

- **Primary uses.** An answering machine records voice messages from callers when the called party is unavailable or chooses not to answer a telephone call. It usually plays a message from the called party before recording the message.
- **Special concerns.** Batteries have a limited life; data could be lost if they fail. Therefore, appropriate personnel (for example, an evidence custodian, lab chief, or forensic examiner) should be informed that a device powered by batteries is in need of immediate attention.
- **Potential evidence.** Answering machines can store voice messages and, in some cases, the time and date information about when the message was left. They may also contain other voice recordings. Answering machines may provide the following potential evidence:
  - Caller ID information.
  - Deleted messages.
  - The last number called.

- Memos.
- Telephone numbers and names.
- Tapes.

DIGITAL CAMERAS

D-20. A digital camera is a digital recording device for images and video. Related storage media and conversion hardware make it capable of transferring images and video to computer media.

- **Primary uses.** Digital cameras capture images and/or video in a digital format that is easily transferred to computer storage media for viewing and/or editing.
- **Potential evidence.** Digital cameras may provide the following potential evidence:
  - Images.
  - Removable cartridges.
  - Sound.
  - The time and date stamp.
  - Video.

HANDHELD DEVICES (PERSONAL DIGITAL ASSISTANTS AND ELECTRONIC ORGANIZERS)

D-21. **Description.** A personal digital assistant (PDA) is a small device that can be used for computing, as a telephone/fax, for paging, for networking, and for other functions. It is typically used as a personal organizer. A handheld computer approaches the full functionality of a desktop computer system. Some do not contain disk drives, but may contain PC card slots that can hold a modem, a hard drive, or another device. They usually include the ability to synchronize their data with other computer systems, most commonly by a connection in a cradle. If a cradle is present, attempt to locate the associated handheld device.

- **Primary uses.** Handheld devices are used for computing, storing information, and communicating.
- **Special concerns.** Batteries have a limited life; data could be lost if they fail. Therefore, appropriate personnel (for example, a evidence custodian, lab chief, or forensic examiner) should be informed that a device powered by batteries is in need of immediate attention.
- **Potential evidence.** Handheld devices may provide the following potential evidence:
  - An address book.
  - Appointment calendars/information.
  - Documents.
  - Emails.
  - Handwriting.
  - A password.

- A telephone book.
- Text messages.
- Voice messages.

PAGERS

D-22. A pager is a handheld, portable electronic device that can contain volatile evidence (such as, telephone numbers, voice mails, and emails). Cell phones and personal digital assistants also can be used as paging devices.

- **Primary uses.** Pagers are used for sending and receiving electronic messages, numeric (often telephone numbers) and alphanumeric (often including emails).
- **Special concerns.** Batteries have a limited life; data could be lost if they fail. Therefore, appropriate personnel (for example, an evidence custodian, lab chief, or forensic examiner) should be informed that a device powered by batteries is in need of immediate attention.
- **Potential evidence.** Pagers may provide the following potential evidence:
  - Address information.
  - Emails.
  - Telephone numbers.
  - Text messages.
  - Voice messages.

PRINTERS

D-23. A printer is one of a variety of printing systems (thermal, laser, inkjet, and impact) connected to the computer via a cable (serial, parallel, universal serial bus (USB), or firewire) or accessed via an infrared port. Some printers contain a memory buffer, allowing them to receive and store multiple page documents while they are printing. Some models may also contain a hard drive.

- **Primary uses.** Printers are used for printing text and images from the computer to paper.
- **Potential evidence.** Printers may maintain usage logs and time and date information, and they may store network identity information (if attached to a network). In addition, unique characteristics may allow for the identification of a printer. Printers may provide the following potential evidence:
  - Documents.
  - A hard drive.
  - Ink cartridges.
  - Network identity/information.
  - Superimposed images on the roller.
  - The time and date stamp.
  - The user's usage log.

REMOVABLE STORAGE DEVICES AND MEDIA

D-24. Removable storage devices and media are used to store electrical, magnetic, or digital information. Examples are floppy disks, CDs, digital versatile disks (DVDs), cartridges, and tapes.

- **Primary uses.** Removable storage devices and media are portable devices that can store computer programs, text, pictures, video, and multimedia files.
- **Potential evidence.** The device itself may be evidence of component theft, counterfeiting, or remarking. New types of storage devices and media come on the market frequently, so one must stay current on advances in technology to ensure that all devices are being checked for evidentiary value properly.

SCANNERS

D-25. A scanner is an optical device connected to a computer. It scans a document and sends it to the computer as a file.

- **Primary uses.** Scanners convert documents and pictures to electronic files, which can then be viewed, manipulated, or transmitted on a computer.
- **Potential evidence.** The device itself may be evidence. Having the capability to scan may help prove illegal activity (for example, child pornography, check fraud, counterfeiting, and identity theft). In addition, imperfections such as marks on the glass may allow for unique identification of a scanner used to process documents.

TELEPHONES

D-26. Telephones are a handset either by itself (as with cell phones), a remote base station (cordless), or connected directly to the landline system. They draw power from an internal battery, an electrical plug-in, or the telephone system.

- **Primary uses.** Telephones are used for two-way communication from one instrument to another using landlines, radio transmission, cellular systems, or a combination. Telephones are capable of storing information.
- **Special concerns.** Batteries have a limited life; data could be lost if they fail. Therefore, appropriate personnel (for example, an evidence custodian, lab chief, or forensic examiner) should be informed that a device powered by batteries is in need of immediate attention.
- **Potential evidence.** Many telephones can store names, telephone numbers, and caller identification information. Additionally, some cell telephones can store appointment information, receive electronic mail and pages, and may act as a voice recorder. Telephones may provide the following potential evidence:
  - Appointment calendars/information.
  - Caller identification information.
  - The electronic serial number.
  - Emails.

- Memos.
- Passwords.
- Telephone books.
- Text messages.
- Voice mail.
- Web browsers.

COPIERS

D-27. Some copiers maintain user access records and history of copies made. Copiers with the scan once/print many feature allow documents to be scanned once into memory, and then printed later.

D-28. Copiers may provide the following potential evidence:

- Documents.
- The time and date stamp.
- The user's usage log.

CREDIT CARD SKIMMERS

D-29. Credit card skimmers are used to read information contained on the magnetic stripe of plastic cards. Cardholder information contained on the tracks of the magnetic stripe may provide the following potential evidence:

- The card expiration date.
- Credit card numbers.
- The user's address.
- The user's name.

DIGITAL WATCHES

D-30. There are several types of digital watches available that can function as pagers that store digital messages. They may store additional information, such as address books, appointment calendars, emails, and notes. Some also have the capability of synchronizing information with computers.

FAX MACHINES

D-31. Fax machines can store preprogrammed telephone numbers and a history of transmitted and received documents. In addition, some contain memory that allows multiple-page faxes to be scanned in and sent at a later time and incoming faxes to be held in memory and printed later. Some may store hundreds of pages of incoming and/or outgoing faxes.

D-32. Fax machines may provide the following potential evidence:

- Documents.
- Film cartridges.
- Telephone numbers.
- Send and receive logs.

GLOBAL POSITIONING SYSTEMS

D-33. Global positioning systems (GPSs) can provide information on previous travel via destination information, waypoints, and routes. Some automatically store the previous destinations and include travel logs.

D-34. GPSs may provide the following potential evidence:

- Home location.
- Previous destinations.
- Travel logs.
- Waypoint coordinates.
- A waypoint name.

Appendix E

Affidavit/Authorization to Search and Seize or Apprehend Electronic Devices

Search authorizations issued by a US magistrate or civilian judge at the state or federal level or the consent of the property owner is required before searching privately owned electronic devices. However, in almost all cases, courts have held a relatively high standard with regard to the specificity of computer-related search authorizations. Investigative personnel seeking search authorization must be able to articulate specific and recent information pertaining to the individual items cited on the affidavit and authorization in order to establish probable cause. (See *Figures E-1, E-2,* and *E-3,* pages E-2, E-5, and E-6.) In many instances, information that is several months old cannot be used to generate probable cause, and more recent information gained through pretext phone calls or online undercover operations may be required to develop current and reliable information.

AFFIDAVIT SUPPORTING REQUEST FOR AUTHORIZATION TO SEARCH AND SEIZE OR APPREHEND

For use of this form, see AR 27-10; the proponent agency is OTJAG.

BEFORE COMPLETING THIS FORM, SEE INSTRUCTIONS ON PAGE 2

1. I, SA Art E. Fishal (Name), 1234 Camp Swamp Resident Agency, 10th MP Group (CID) (Organization or Address)

Camp Swamp, LA 00012

having been duly sworn, on oath depose and state that:

About 1500, March 1999, SA Cyber Stalker, FBI, Camp Swamp, Louisiana, notified this office that SSG Jim File, 123 Nowhere Avenue, Camp Swamp, Louisana, transmitted numerous digital photographs depicting children performing felletio and other sexual acts over the Internet. Further, SSG File established an F-Serve system on his home computer, which allowed other Internet users to access his database for the purpose of uploading and downloading pornographic photos of children.

In a preliminary investigation by the FBI around November 1998, SSG File transmitted numerous digital photographs depicting children performing felletio and other sexual acts over the Internet.

About 2040, 25 March 1999, SA Stalker, FBI telephoned (123) 555-1234 (the number listed by the Bell South Telephone Company as belonging to File). A male answered the call and identified himself as Jim File (continued, see attached).

2. The affiant further states that:
It is paramount to the resolution of this allegation that a search of File's quarters be conducted to locate and seize the computers and all related storage media. Further, it is important for the resolution of this investigation that SSG File be apprehended for the purpose of obtaining identification photographs and record fingerprints.

DA FORM 3744, SEP 2002 DA FORM 3744-R, MAR 85, IS OBSOLETE. USAPA V1.00ES

Figure E-1. Sample of DA Form 3744

3. In view of the foregoing, the affiant requests that an authorization be issued for a search of SSG Jim File and his quarters at
(the person) (and)

123 Nowhere Avenue (including all vehicles and storage facilities located at the premises).
(the quarters or billets) (and)

(the automobile) ( ) and (seizure) (apprehension) of Two computers described by the affaint,
(items/persons searched for)

all tapes, cassettes, cartridges, streaming tape, commercial software and hardware, computer disks, (continued, see attached).

TYPED NAME AND ORGANIZATION OF AFFIANT	SIGNATURE OF AFFIANT
SA Art E. Fishal 10th MP Group (CID)	Art E. Fishal

SWORN TO AND SUBSCRIBED BEFORE ME THIS 31st DAY OF March 1999 AT 1000

TYPED NAME, ORGANIZATION AND OFFICIAL CAPACITY OF AUTHORITY ADMINISTERING THE OATH	SIGNATURE OF AUTHORITY ADMINISTERING THE OATH
John A. Doe, Captain, JA, Military Magistrate Office of the Staff Judge Advocate Camp Swamp, LA 00012	John A. Doe

Sample

INSTRUCTIONS FOR

AFFIDAVIT SUPPORTING REQUEST FOR AUTHORIZATION TO SEARCH AND SEIZE OR APPREHEND

1. In paragraph 1, set forth a concise, factual statement of the offense that has been committed or the probable cause to believe that it has been committed. Use additional page if necessary.

2. In paragraph 2, set forth facts establishing probable cause for believing that the person, premises, or place to be searched and the property to be seized or the person(s) to be apprehended are connected with the offense mentioned in paragraph 1, plus facts establishing probable cause to believe that the property to be seized or the person(s) to be apprehended are presently located on the person, premises, or place to be searched. Before a person may conclude that probable cause to search exists, he or she must first have a reasonable belief that the person, property or evidence sought is located in the place or on the person to be searched. The facts stated in paragraphs 1 and 2 must be based on either the personal knowledge of the person signing the affidavit or on hearsay information which he/she has plus the underlying circumstances from which he/she has concluded that the hearsay information is trustworthy. If the information is based on personal knowledge, the affidavit should so indicate. If the information is based on hearsay information, paragraph 2 must set forth some of the underlying circumstances from which the person signing the affidavit has concluded that the informant (whose identity need not be disclosed) or his/her information was trustworthy. Use additional pages if necessary.

3. In paragraph 3, the person, premises, or place to be searched and the property to be seized or the person(s) to be apprehended should be described with particularity and in detail. Authorization for a search may issue with respect to a search for fruits or products of an offense, the instrumentality or means of committing the offense, contraband or other property the possession of which is an offense, the person who committed the offense, and under certain circumstances for evidentiary matters.

DA FORM 3744, SEP 2002 Page 2 of 3
USAPA V1.00

Figure E-1. Sample of DA Form 3744 (Continued)

DA Form 3744
Continuation Sheet

SA Art E. Fishal
31 March 1999

Paragraph 1 (continued):
SA Stalker conducted a pretext interview of SSG File, stating that she was a representative from a data software research corporation. SA Stalker requested that SSG File participate in a survey pertaining to computer usage, to which SSG File agreed. During the interview, SSG File stated that he had two computers in his residence (one IBM compatible Pentium processor, which he reported that he constructed himself, and the second an AST Advantage Pro 486 DX 33). SSG File stated that he built the first computer about 4 to 6 months ago and has been using it on the Internet ever since. He stated that the second computer had been in his possession for a significantly longer period of time, and that both computers were currently connected to the Internet. SSG File stated that he was extremely proficient in computer usage, construction, and on-line access (citing that he has been an on-line computer user for approximately 4 years and has been using World Net of Louisiana services for the past 1½ years). SSG File stated that he spent approximately 4 to 6 hours per day using his computers on-line for educational, business, and recreational needs. When asked how many of the occupants of his residence used on-line services, SSG File stated that he had three daughters (only one of which was old enough to use the computer; however, she could not do so without his access code, which she did not know). SSG File also stated that his wife occasionally used the computer, but did not do so frequently nor with any level of proficiency.

During the conversation, he stated that he had two computers at the time the pornographic photographs were transmitted, and he confirmed both computers were still located in his residence, which he verified to be 123 Nowhere Avenue, Camp Swamp, LA.

It is the experience of the affiant, that the USACIL has the capability to retrieve media that was deleted from storage, which in recent investigations has surpassed a 4-month period of time. Further, it is the training and experience of the affiant that child sex offenders (including child pornography) maintain trophies, such as photographs for many years, and it is likely the files containing child pornography are currently located on the requested storage media.

Paragraph 3 (continued):
disk drives, monitors, computer printers, modems, tape drives, disk application programs, data disks, operating systems disk, magnetic media, floppy disks, tape systems, hard drives, and other computer-related operation equipment. Computer photographs; graphic interchange formats; slides; or other visual depictions of such graphic interchange format equipment which may be or are used to visually depict child pornography or child erotica; information pertaining to the sexual interest in child pornography; sexual activity with children; or the distribution, possession, or receipt of child pornography or child erotica; or information pertaining to an interest in child pornography or child erotica.

Any and all correspondence pertaining to the possession, receipt, or distribution of visual depictions of minors engaged in sexually explicit conduct, as defined by Title 18, United States Code, Section 2256, whether transmitted or received using a computer, a facility or means of interstate commerce, common carrier, or mail.

Any and all correspondence pertaining to the travel in interstate commerce for the purpose of engaging in sexual activities with minors. The term minors as used, in this list of items to be seized, means persons under the age of 18 years.

Any and all books and magazines containing visual depictions of minors engaged in sexually explicit conduct, as defined in Title 18, United States Code, Section 2256.

Figure E-1. Sample of DA Form 3744 (Continued)

SEARCH AND SEIZURE AUTHORIZATION

For use of this form, see AR 27-10; the proponent agency is OTJAG

TO: *(Name and Organization of the person to whom authorization is given)*
SA Art E. Fishal, 10th MP Group (CID)
1234 Camp Swamp Resident Agency, Camp Swamp, LA 00012

(An affidavit) ~~*(a (sworn) or (unsworn) oral statement)*~~ having been made before me by SA Art E. Fishal *(Name of Affiant)*

10th MP Group (CID), 1234 Camp Swamp Resident Agency, Camp Swamp, LA 00012 *(Organization or Address of Affiant)*

(which affidavit is attached hereto and made a part of this authorization), and as I am satisfied that there is probable cause to believe that the matters mentioned in the affidavit are true and correct, that the offense set forth therein has been committed, and that the property to be seized is located ~~*(on the person)*~~ *(at the place)* to be searched, you are hereby ordered to search the ~~*(person)*~~ *(place)* known as

123 Nowhere Avenue (including all vehicles and storage facilities located at the premises.

for the property described as **(For an example of the required, detailed information needed for this see *Figure E-3*, page E-6.)**

bringing this order to the attention of the *(person searched)* ~~*(person in possession)* *(if any person be found at the place or on the premises searched)*~~. The search will be made in the *(daytime)* ~~*(night)*~~ and if the property is found there, you shall seize it, issue a receipt therefor to the person from whom the property is taken or in whose possession the property is found, deliver the property to:

Evidence Custodian, Camp Swamp Resident Agency, 10th MP Group (CID), Camp Swamp, LA *(Name and Organization of Authorized Custodian)*

and prepare a written inventory of the property. If there is no person at the searched place to whom the receipt may be delivered, the receipt will be left in a conspicuous location at the place or on the premises where the property is found.

Dated this 31st day of March, 1999.

TYPED NAME AND GRADE OF AUTHORIZING OFFICIAL	DUTY POSITION OF AUTHORIZING OFFICIAL
John A. Doe Captain, JA	Military Magistrate
ORGANIZATION OF AUTHORIZING OFFICIAL	**SIGNATURE OF AUTHORIZING OFFICIAL**
Office of the Staff Judge Advocate Camp Swamp, LA 00012	John A. Doe

DA FORM 3745, SEP 2002 DA FORM 3745-R, MAR 85, IS OBSOLETE USAPA V1.00ES

Sample

Figure E-2. Sample of DA Form 3745

The two computers described by the affiant; tapes; cassettes; cartridges; streaming tape; commercial software and hardware; computer disks; disk drives; monitors; printers; modems; tape drives; disk application programs; data disks; operating systems disk; magnetic media; floppy disks; tape systems; computer photographs; graphic interchange formats; photographs, slides, or other items used to visually depict child pornography or child erotica; information pertaining to sexual interest in child pornography or sexual activity with children; distribution, possession, or receipt of child pornography or child erotica; or information pertaining to an interest in child pornography or child erotica.

The term "minors" as used in this list of items to be seized, means persons under the age of 18 years. All correspondence pertaining to the possession, receipt, or distribution of visual depictions of minors engaged in sexually explicit conduct as defined in Title 18, USC, Section 2256,whether transmitted or received using a computer, a facility or means of interstate commerce, common carrier, or mail.

All books and magazines containing depictions of minors engaged in sexually explicit conduct as defined in Title 18, USC, Section 2256.

All originals, copies, and negatives of depictions of minors engaged in sexually explicit conduct as defined in Title 18, United States Code, Section 2256.

All envelopes, letters, and other correspondence offering to transmit, through interstate commerce (to include US mail or computer), any depictions of minors engaged in sexually explicit conduct as defined by Title 18, United States Code, Section 2256. All envelopes, letters, and other correspondence identifying persons transmitting any visual depiction of minors engaged in sexually explicit conduct through interstate commerce as defined by Title 18, USC, Section 2256.

All books, ledgers, and records bearing on the production, reproduction, receipt, shipment, orders, requests, trades, purchases, or transactions of any kind involving the transmission of any depiction of minors engaged in sexually explicit conduct as defined by Title 18, USC, Section 2256.

All address books, mailing lists, supplier lists, mailing address labels, and all documents and records pertaining to the preparation, purchase, and acquisition of names or lists of names to be used in connection with the purchase, sale, trade, or transmission of any depiction of minors engaged in sexually explicit conduct as defined by Title 18, USC, Section 2256.

All address books, names, and lists of names and addresses of minors depicted while engaged in sexually explicit conduct as defined by Title 18, USC, Section 2256.

All diaries, notebooks, notes, and any other records reflecting personal contact and any other activities with minors engaging in sexually explicit conduct as defined by Title 18, USC, Section 2256.

All materials and photographs depicting sexual conduct between adults, minors, and adults and minors.

Figure E-3. Sample of a Detailed Property Description

Appendix F

Violent Crime Scene Checklist

Conducting a felony criminal investigation can be complicated. There are a number of important and detail-oriented activities that need to be completed within the first 24 to 48 hours. Keeping accountability of these tasks can be a daunting mission, made more arduous when the felony involves a violent crime and an entire team is involved in completing them. The following checklist (*Figure F-1*, pages F-2 through F-20) is an exceptional investigative tool, that allows the case agent to maintain a written record of the activities and also ensure that all the essential tasks are completed in a timely and thorough manner.

ROI # ________ - ____ - CID - _____ - ________

US ARMY CRIMINAL INVESTIGATION COMMAND

VIOLENT CRIME SCENE CHECKLIST

HOMICIDE/DEATH/RAPE/SERIOUS ASSAULT

COMPLETED BY: ____

TIME AND DATE: __________________________

FOR OFFICIAL USE ONLY

Figure F-1. Sample of a Violent Crime Scene Checklist

ROI # ________ - ____ - **CID** - _____ - __________

1. **INITIAL NOTIFICATION:** Time: ____________ Date: ________________

   Notification Recorded: Y or N

2. **INFORMATION RECEIVED BY:** Military Police / 911 / CID / Civilian Police /
   Other: ______________________________

   Rank: ________ Name: ______________________________

   SSN: ________________ Position Title: ______________________________

   Unit/Address: ______________________________

   City: ________________ State: __________ Zip: __________

   Home Phone: ________________ Unit Phone: ________________

3. **PERSON MAKING REPORT:** Military Police / Civilian Police / Victim / Witness / Suspect /
   Other: ______________________________

   Rank: ________ Name: ______________________________

   SSN: ________________ Position Title: ______________________________

   Unit/Address: ______________________________

   City: ________________ State: __________ Zip: __________

   Home Phone: ________________ Unit Phone: ________________

   Relation to Victim: Spouse / Parents / Child / Friend / Family Member / Unit Representative (CQ/CDR/1SG) / Other: ________________

4. **DETAILS OF INITIAL REPORT:** ______________________________

   ____

   ____

FOR OFFICIAL USE ONLY

Figure F-1. Sample of a Violent Crime Scene Checklist (Continued)

ROI # ________ - ____ - CID - _____ - _________

5. CID NOTIFICATIONS (State Times):

SAC: ____________ EMT: ____________ Military Police: ____________ Fire: ____________

Medical Exam: ____________ Civilian PD: ____________ Other FED: ____________

FBI: ____________ Unit: ____________ PAO: ____________ Other: ____________

6. CID ARRIVAL AT CRIME SCENE: Time: ____________ Date: ____________

Outside Temperature: ______________ Inside Temperature: ______________

Weather Conditions: Drizzle / Rain / Snow / Sleet / Clear / Cloudy / Overcast / Fog

Natural Light Conditions: Dawn / Daylight / Dusk / Night

General Outside Visibility: ______________ Visual Factor: ______________

Weather Previous 24 Hours: ____

7. POLICE ARRIVAL AT SCENE: Military Police / MPI / Civilian Police / Detective / Other: ______________

Time Dispatched: __________ Time Arrived: __________

Organization/Department: ____

Summary of Information: ____

____

____

____

Rank/Name: ______________________________ SSN: ______________

Phone: ________________________

Rank/Name: ______________________________ SSN: ______________

Phone: ________________________

FOR OFFICIAL USE ONLY

Figure F-1. Sample of a Violent Crime Scene Checklist (Continued)

ROI # ________ - _____ - CID - _____ - ________

a. Was first aid attempted on victim? Y or N ________ If yes, describe: ____________________

____

b. Was there any person(s)/vehicle(s) observed departing the scene? Y or N ____________

If yes, describe: ____

____

c. Do the police patrol the area? Y or N ______ Last time in area: ________ am/pm

Patrolman Identification: ____________________________ Interviewed: Y or N ________

8. **MEDICAL PERSONNEL AT SCENE:** Dispatched: ____________ Arrived: ____________

Organization/Facility Name: ____

Address: ____

Rank/Name: ____ SSN: ______________

Phone: ______________________ Interviewed By: ____________________________

Rank/Name: ____ SSN: ______________

Phone: ______________________ Interviewed By: ____________________________

Rank/Name: ____ SSN: ______________

Phone: ______________________ Interviewed By: ____________________________

a. Describe position in which victim was found? ______________________________

____

b. Was a doctor at scene? Y or N ________ Name: ______________________________

FOR OFFICIAL USE ONLY

Figure F-1. Sample of a Violent Crime Scene Checklist (Continued)

ROI # ________ - ____ - CID - _____ - __________

c. Were vital signs of victim taken at the scene? Y or N _______ If yes, by whom: __________

Name: ____

Results: Respiration __________ Heart Rate __________

Blood Pressure _______/_______

d. Victim pronounced dead by: ____

at ____________ am / pm.

e. Describe medical treatment given to victim: ____

____

If no medical treatment was given to victim, explain: ____

____

f. Were there any statements made by victim? ____

____

g. Did victim appear to be under the influence of drugs or alcohol? ______________________

Describe: ____

h. Were there any drugs or alcohol present? ____

i. Were there any unusual odors? Y or N _________ Describe: ______________________

____

j. Were there any person(s) present upon arrival of medical personnel? Y or N ____________

Who: ____ __________

k. Did the medics alter or contaminate the crime scene? Y or N ______________________

Explain: ____

FOR OFFICIAL USE ONLY

Figure F-1. Sample of a Violent Crime Scene Checklist (Continued)

ROI # ________ - ____ - CID - _____ - __________

9. **OTHER PERSONS AT CRIME SCENE:**

Rank/Name: ______________________________ SSN: ____________

Unit/Address: ______________________________

Unit Phone: ____________ Home Phone: ______________

Actions at Scene: ______________________________

Rank/Name: ______________________________ SSN: ____________

Unit/Address: ______________________________

Unit Phone: ____________ Home Phone: ______________

Actions at Scene: ______________________________

10. **CRIME SCENE EXAMINATION:**

a. CRIME SCENE LOCATION: (Mark the appropriate area)

Main Post / Housing / Billets / Unit Area / Motor Pool / Training Area / Range / Open Field / Recreation Area / Waterfront / Dump / Landfill / Residential / Commercial / Private Property / Public Use Area / Other: ______________________________

Address: ______________________________

City: ______________ State: ________ Zip: __________

The crime scene was located on the ________ floor of a _____ story structure. The building was constructed of ____________ and ____________. _____ persons were occupying the house/room and their exact location was: ______________________________

____

____

b. OUTDOOR CRIME SCENE:

1) On-post Grid Coordinates: ______________________________

FOR OFFICIAL USE ONLY

Figure F-1. Sample of a Violent Crime Scene Checklist (Continued)

ROI # ________ - ____ - CID - _____ - ________

2) Off Post: City: ______________________ State: ______________________

County: ______________________ Country: ______________________

3) Odors Detected? Y or N______ Describe: ______________________

4) Type of Road: Black Top / Dirt / Rock / Trail / Footpath / Fire Break /
Other: ______________________

5) Road Conditions: Dry / Dusty / Oiled / Wet / Muddy / Snow / Ice / Other: __________

6) Common Name for Area: ______________________

	Nearest Building	Nearest Street	Distance	Direction	Common Name
On Post					
Off Post					

c. INDOOR CRIME SCENE:

1) Crime search authority obtained: Time: ________ Date: ______________

Who authorized? ______________________

2) Primary entrance/exit located: ______________________

Describe: ______________________

3) Are windows/doors equipped with locking devices? Y or N ________

Describe: ______________________

4) Are windows equipped with screens? Y or N ________ With shades? Y or N _____

5) Was there forced entry? Y or N ________

Describe: ______________________

FOR OFFICIAL USE ONLY

Figure F-1. Sample of a Violent Crime Scene Checklist (Continued)

ROI # ________ - ____ - CID - _____ - __________

6) Were there indications of efforts to disable alarm/telephone? Y or N ________

Describe: ____

7) How did the person making discovery enter? ____________________

8) Is there evidence of other crimes committed? Robbery / Rape / Burglary / Arson

____

9) Is there obvious property missing from scene? Y or N __________

Describe: ____

10) Describe negative evidence: ______________________________

____

11) Were there any weapons noted at the scene? Y or N ________

12) Were any weapons removed? Y or N ________

13) Were there signs of struggle? Y or N ________

Describe: ____

____

14) Did death/assault occur where victim was found? Y or N ________

15) Was victim removed? Y or N ________

Describe: ____

16) Were there indications that suspect remained at location following assault? Y or N _____

Describe: ____

17) Describe crime area (general condition/evidence noted/fragile evidence):

____

____

FOR OFFICIAL USE ONLY

Figure F-1. Sample of a Violent Crime Scene Checklist (Continued)

ROI # ________ - ____ - CID - ____ - ________

18) Was there dated material present (mail, letters)? Y or N ____________________

19) Was the answering machine on? Y or N ________

Messages: ____

20) Were there fingerprints developed? Y or N ________

Location: ____

21) Were there footprints developed? Y or N ________

Location: ____

22) Were there prints lifted using the electrostatic dust print lifter? Y or N ______

Location: ____

23) What were the UV light examination results? ____________________

____

24) What were the alternate light source examination results? ________________

____

25) Were the lights, water, television, radio, VCR, and stereo on or off? ____________

26) What station was the television and radio on? ____________________

27) Was the heater or air conditioner fan on or off? ____________________

28) Were there pets present? Y or N ________ Type: ____________________

29) Were there odors detected? Y or N ____________

Describe: ____

30) What was the telephone number? ______________________________

FOR OFFICIAL USE ONLY

Figure F-1. Sample of a Violent Crime Scene Checklist (Continued)

ROI # ________ - ____ - CID - _____ - __________

31) Was there food or drinks lying around? Y or N________________________________

Where: ____

32) List the POCs for scene documentation:

a) Photos taken by: ____

b) Crime scene sketch prepared by: ______________________________

c) Video taken by: ____

33) What were the results of the search beyond the crime scene?

____

____

____

34) Was crime scene security established? Y or N ______________________

Who: ____

Where: ____

____

35) Was the victim found at this/another location? Y or N ________________

Describe: ____

____

36) Were there indications of multiple vehicle(s) or person(s)? Y or N ____________

Describe: ____

____

FOR OFFICIAL USE ONLY

Figure F-1. Sample of a Violent Crime Scene Checklist (Continued)

ROI # ________ - ____ - CID - _____ - _________

37) Were there tire tracks in the area? Y or N ________

Width of Vehicle: _________ inches Length of Tracks: _________ feet

Depth of tracks is/is not consistent with vehicle found at scene? _______

Explain: ____

d. PHYSICAL EVIDENCE COLLECTED AT CRIME SCENE:

	Tire Casts Y or N	Foot Casts Y or N	Soil Samples Y or N	Insect Samples Y or N	Vegetation Samples Y or N
Location					
Number					
Type					
Describe					

e. EVIDENCE NOTED:

____

____

____

11. CANVASS INTERVIEWS CONDUCTED BY: ______________________________

12. WITNESS(ES) INFORMATION:

Rank: ________ Name: ______________________________

SSN: ____________________ Home/Work Phone: ______________________

Unit/Address: ______________________________________

City: ____________________ State: _______ Zip: _________

Synopsis: ____

____

____

FOR OFFICIAL USE ONLY

Figure F-1. Sample of a Violent Crime Scene Checklist (Continued)

ROI # ________ - ____ - CID - ____ - ________

Rank: ________ Name: ________________________

SSN: ________________ Home/Work Phone: ________________

Unit/Address: ________________________

City: ________________ State: ________ Zip: ________

Synopsis: ________________________

____

____

____

____

13. VICTIM DATA (DECEASED): Y or N VICTIM #: ________

Military / Family Member / Other U.S. Civilian: ________________
Foreign: Military / Civilian Nationality: ________________

a. PERSONAL INFORMATION:

Rank: ________ Name: ________________________

SSN: ________________ DOB: ____________ Age: ______

Unit/Address: ________________________

City: ________________ State: ________ Zip: ________

Height: ______ Weight: ______ Hair: ______ Eyes: ______ Race: ______

Sex: ______ Complexion: ________ Glasses: ______ Mustache/Beard: ______

1) Personal Identification Basis: Witness / Friend / Family Member / Unit

Rank: ______ Name: ________________________

SSN: ________________ Home/Work Phone: ________________

FOR OFFICIAL USE ONLY

Figure F-1. Sample of a Violent Crime Scene Checklist (Continued)

ROI # ________ - ____ - CID - _____ - _________

Unit/Address: ____________________

City: ____________ State: ________ Zip: __________

2) Document Identification Basis:

a) Military Identification: Active Duty / Reserve Service: ____________

b) Driver's License: Y or N State: __________ Photo: Y or N

Drivers License #: ______________ Restrictions: __________

3) Other Identification Basis (Describe):

a) Tattoos: ____________________

b) Wedding Ring: ____________________

c) Watch: ____________________

d) Other Jewelry: ____________________

b. POSTMORTEM DESCRIPTION:

1) Victim was found at: Time: ________ Date: __________

2) Examination was conducted at: Time: ________ Date: __________

a) Body temperature: ________ Taken by: ______________

b) Temperature was taken by what method? ______________

c) Outside temperature: __________

d) Inside temperature: __________

e) Was the victim inside or outside? ______________

f) Was victim above or below ground? ______________

FOR OFFICIAL USE ONLY

Figure F-1. Sample of a Violent Crime Scene Checklist (Continued)

ROI # _______ - ____ - CID - _____ - _________

g) Was there rigor mortis present? Y or N ____________________

Describe: ____________________

____

h) Was there livor mortis present? Y or N ____________________

Describe: ____________________

____

i) Was there decomposition or discoloration? Y or N ____________________

j) Was there insect/larva present? Y or N ________ Type: ____________________

Location: ____________________

k) Was there animal damage? Y or N ____________________

Describe: ____________________

l) What was the victim's activity at time of death? ____________________

____

m) Were there indications that victim was moved or transported? Y or N ____________________

Describe: ____________________

n) Were there indications of drug abuse/overdose? Y or N ____________________

Describe: ____________________

o) Were drugs found at scene? Y or N ____________________

p) Was there pornographic literature present? Y or N ____________________

Describe: ____________________

q) Were items protruding from/inserted into body? Y or N ____________________

Describe: ____________________

FOR OFFICIAL USE ONLY

Figure F-1. Sample of a Violent Crime Scene Checklist (Continued)

ROI # ________ - ____ - CID - _____ - _________

r) Was a GSR kit used on victim? Y or N ______________________________

s) Were victim's hands bagged? Y or N ______________________________

t) Were there broken nails/hairs/fibers/objects in hands? Y or N ______________

Describe: ____

u) Was there a suicide note? Y or N ______________________________

Location: ____

v) Was the victim bound? Y or N _________ (mouth / hands / feet)

Describe: ____

Type bond: Rope / Leather / Handcuff / Tape / Plastic / Other: ______________

Was there an escape mechanism in place? Y or N ______________________

Describe: ____

w) Was the clothing consistent to where it was found? Y or N ______________

x) Was the clothing soiled / damaged / mussed / stained / missing? Y or N _________

FOR OFFICIAL USE ONLY

Figure F-1. Sample of a Violent Crime Scene Checklist (Continued)

ROI # ________ - ____ - CID - ____ - ________

14. **INJURY WORKSHEET:**

Victim _____ of _____ victims. Body sketched prepared? Y or N __________

Circle Type Injury	Injury #1	Injury #2	Injury #3	Injury #4
Location of Injury				
Gunshot	Entrance Exit	Entrance Exit	Entrance Exit	Entrance Exit
Contact	Hard Close	Hard Close	Hard Close	Hard Close
Range	Intermediate Distance	Intermediate Distance	Intermediate Distance	Intermediate Distance
Type Gun				
Vehicle/ Fall/ Other				
Sharp Force Type Weapon				
Type Injury	Cut Stab Chop Defensive	Cut Stab Chop Defensive	Cut Stab Chop Defensive	Cut Stab Chop Defensive
Sexual Assault Mutilation				
Blunt Force Object Weapon				
Skin	Laceration Contusion Abrasion Defensive	Laceration Contusion Abrasion Defensive	Laceration Contusion Abrasion Defensive	Laceration Contusion Abrasion Defensive

FOR OFFICIAL USE ONLY

Figure F-1. Sample of a Violent Crime Scene Checklist (Continued)

ROI # ________ - ____ - CID - _____ - _________

Circle Type Injury	Injury #1	Injury #2	Injury #3	Injury #4
Broken	Bones Teeth	Bones Teeth	Bones Teeth	Bones Teeth
Pattern Injuries (Describe)				
Asphyxia	Choke Strangle Ligature Marks Other	Choke Strangle Ligature Marks Other	Choke Strangle Ligature Marks Other	Choke Strangle Ligature Marks Other
Self-Inflicted Wounds				
Hesitation Marks				

15. VEHICLE INFORMATION (additional vehicles on back):

a. Were there vehicles present? Y or N ____________________________

b. How many vehicles were there? ____________________________

c. Was NCIC conducted on the vehicles? Y or N ____________________________

	Make/Model	Year	Color	License	Odometer Reading	VIN #	Registered
1.							
2.							

FOR OFFICIAL USE ONLY

Figure F-1. Sample of a Violent Crime Scene Checklist (Continued)

ROI # ________ - ____ - CID - ____ - __________

16. **SUSPECT DATA:**

a. PERSONAL INFORMATION (based on initial information):

1) SUSPECT #1:

Rank: ________ Name: ______________________________

SSN: ______________ DOB: ______________ Age: ________

Unit/Address: ______________________________

City: ______________ State: ________ Zip: __________

Height: ______ Weight: ______ Hair: ______ Eyes: ______ Race: ________

Sex: ____ Complexion: __________ Glasses: ______ Mustache/Beard: ________

Build: __________ Injuries: ______________________________

2) SUSPECT #2:

Rank: ________ Name: ______________________________

SSN: ______________ DOB: ______________ Age: ________

Unit/Address: ______________________________

City: ______________ State: ________ Zip: __________

Height: ______ Weight: ______ Hair: ______ Eyes: ______ Race: ________

Sex: ____ Complexion: __________ Glasses: ______ Mustache/Beard: ________

Build: __________ Injuries: ______________________________

FOR OFFICIAL USE ONLY

Figure F-1. Sample of a Violent Crime Scene Checklist (Continued)

ROI # ________ - ____ - CID - ____ - _________

3) SUSPECT #3:

Rank: ________ Name: ____

SSN: ________________ DOB: _________________ Age: ________

Unit/Address: ____

City: ___________________ State: _______ Zip: _________

Height: _______ Weight: ______ Hair: _______ Eyes: _______ Race: _________

Sex: ____ Complexion: __________ Glasses: ______ Mustache/Beard: _________

Build: ____________ Injuries: ____

b. SUSPECT VEHICLE INFORMATION: Time: __________ Date: _________________

1) Information received from: ____

a) Year: _________ Make: __________________ Model: _____________________

Color: _______________ License Plate # : _________________ State: ________

Damage: ____

Owner: ____

b) Year: _________ Make: __________________ Model: _____________________

Color: _______________ License Plate # : _________________ State: ________

Damage: ____

Owner: ____

2) BOLO: Y or N NCIC: Y or N Time: ____________ Date: ______________

Agency: ______________________________ Results: ________________________

____

FOR OFFICIAL USE ONLY

Figure F-1. Sample of a Violent Crime Scene Checklist (Continued)

Appendix G

Fingerprinting Procedures

Fingerprints provide the most consistent and unique evidence characteristics available at a crime scene; however, latent prints discovered at a scene are only as valuable as the record prints obtained for comparison. When collecting record prints, keep in mind that the ridge detail is the most valuable part of the print and an excessive amount of ink will obliterate and/or obstruct detail. Keep the ink to a minimum, be mindful of ridge details, and collect as many surfaces as possible. *Figure G-1*, page G-2, holds examples of the various surfaces that are obtained during record print collection and are likely to be deposited at a crime scene.

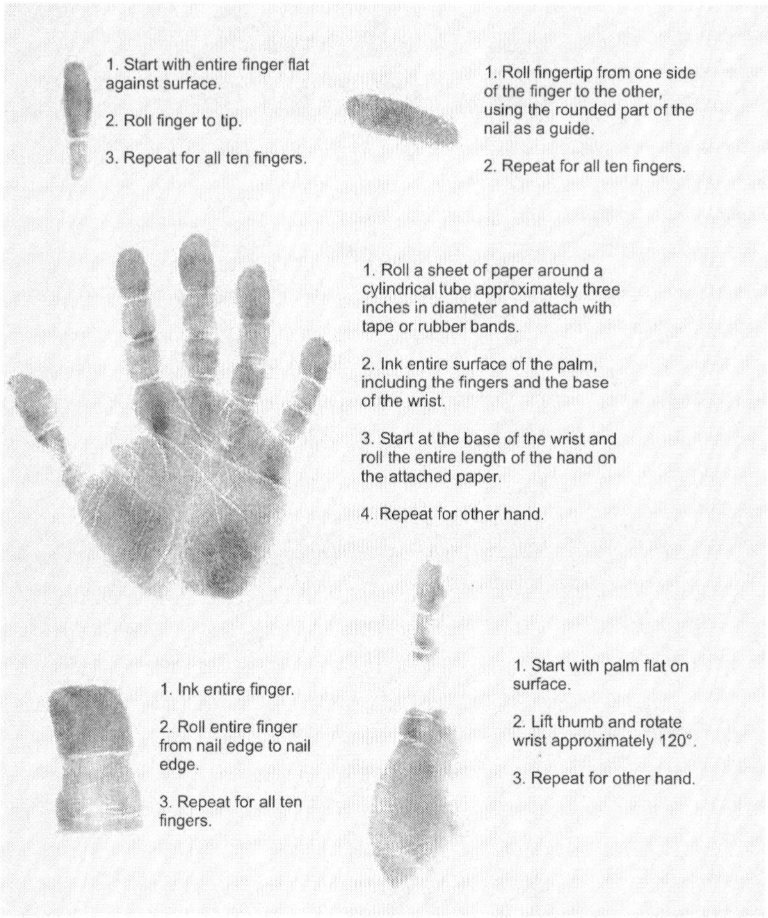

Figure G-1. Procedures for Taking Record Finger and Palm Prints

Appendix H

Tire Chart

The information in *Figure H-1* describes how to read a tire sidewall. It is very useful for investigators who are required to cast tire impressions and identify tire characteristics.

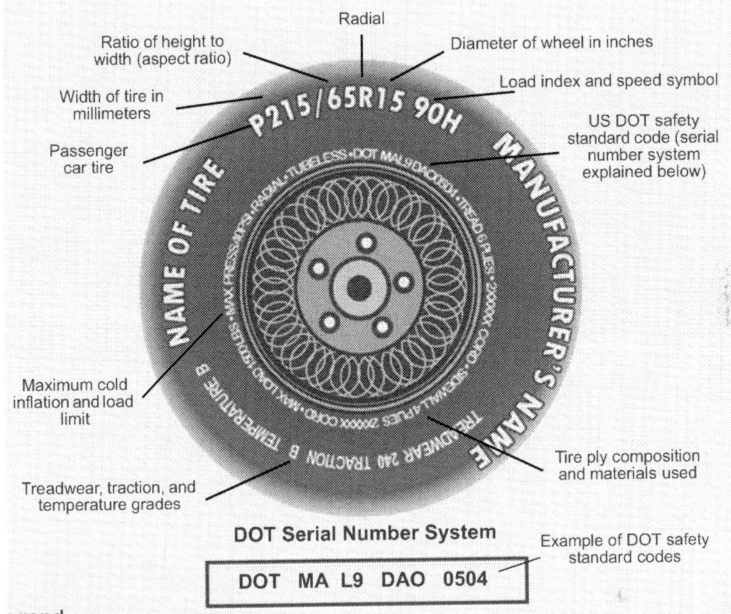

Legend

DOT = the tires meet or exceed the Department of Transportation's safety standards.
MA = the code number assigned by the DOT to the manufacturing plant.
L9 = the tire size (in this case, P215/65R15).
DAO = a group of optional symbols that the manufacturer can use to identify the brand or other significant characteristics of the tire.
0504 = the date the tire was made (fifth week of 2004). (The first two numbers designate the week the tire was manufactured and the last two numbers indicate the year.)

Figure H-1. Tire Chart

Appendix I

Pharmacy Fold

A pharmacy fold is used as a quick, reliable evidence collection tool by investigators. It is ideal for collecting powders and small particles of evidence. The steps for the pharmacy fold are as follows (see *Figure I-1*):

Step 1. Obtain or cut a square piece of paper the approximate size needed to hold the material in question.

Step 2. Make a diagonal fold (as shown in A). Fold B and C together, then fold at D.

Step 3. Open and place the sample in the center of the square and refold in the same manner.

Step 4. Tuck triangle E into the slot formed by the fold of D and initial and date the formed packet.

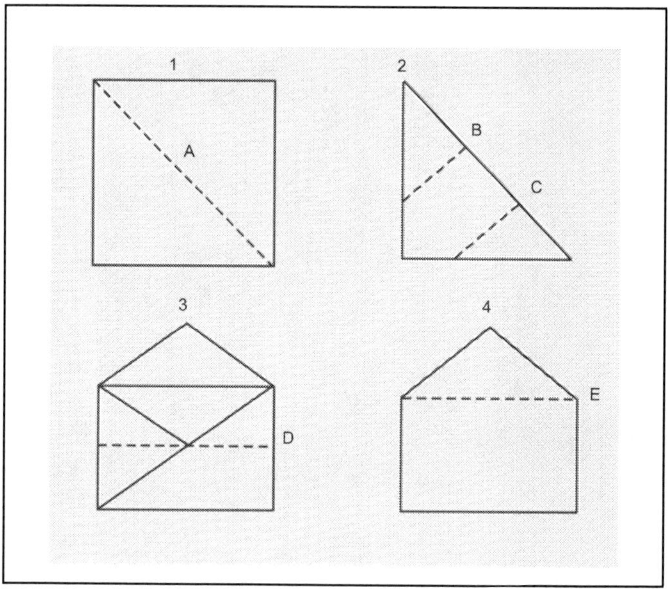

Figure I-1. Steps for the Pharmacy Fold

Appendix J

Types of Surveillance

This appendix describes the surveillance types in detail and is accompanied by graphics. It is intended to clarify/compliment the surveillance types described in *Chapter 26* and is not meant to be a stand-alone appendix.

BASIC ABC PROCEDURE WITH NORMAL PEDESTRIAN TRAFFIC

J-1. Besides tailing the subject from the rear on foot, one agent observes him from the opposite side of the street. The subject is thus bracketed, and the agents are provided with strategic positions from which to observe him, especially when he turns a corner or reverses his direction. (See *Figure J-1*, page J-2.)

J-2. The following are basic positions for foot surveillance:

- Surveillant A is located to the rear of the subject with a reasonable distance between them. A reasonable distance is determined by the number of people on the street between surveillant A and the subject. The fewer the people, the greater the distance should be. A common mistake of many investigators is to have too much distance between himself and the subject.
- Surveillant B follows surveillant A and his responsibility is to keep surveillant A in sight and to detect associates of the subject. The distance between surveillant A and surveillant B is slightly more than the distance between surveillant A and the subject.
- Surveillant C is on the opposite side of the street and slightly to the rear of the subject. Surveillant C's responsibility is to keep both the subject and surveillant A in sight.

VARIATIONS ON ABC PROCEDURE WITH LITTLE PEDESTRIAN TRAFFIC

J-3. In a situation where there is a noticeable lack of people on the street, the basic ABC positions may be varied.

J-4. The following are variations of basic positions:

- Two surveillants may be on the opposite side of the street, usually designated as surveillant B and surveillant C.
- Surveillant A remains to the rear of the suspect, but surveillant B remains to the rear of surveillant C. Surveillant B concentrates on keeping surveillant A and surveillant C in view. (See *Figure J-2*, page J-3.)

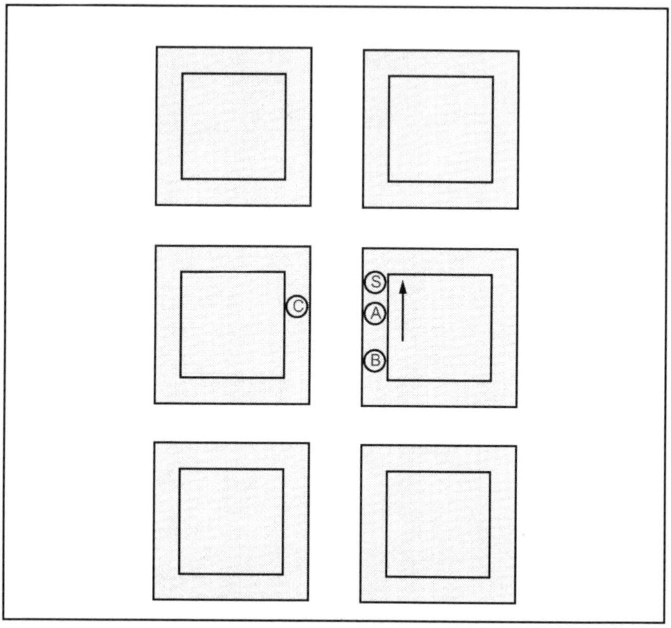

Figure J-1. Foot Surveillance Example 1

ABC PROCEDURE ON A VERY CROWDED STREET

J-5. On very crowded streets, surveillant C's position is on the same side of the street as surveillant A and surveillant B and all surveillants are close to the subject. (See *Figure J-3*, page J-4.)

J-6. In this variation, surveillant A should follow the subject very closely. Surveillant B concentrates on keeping surveillant A in view. Surveillant C concentrates on keeping surveillant B in view.

J-7. Three to five surveillants are usually used with the ABC method. Six to eight individuals can also be used, but more than this leads to confusion and causes the surveillance team to become unwieldy.

ON-FOOT LEADING SURVEILLANCE

J-8. In situations where the surveillants are confident that the subject is likely to follow a particular route for a period of time, then the leading surveillance method may be used. The basic ABC method is employed except that one surveillant walks in front of the subject and leading him along the expected route that the subject is going to take. This can be very effective in

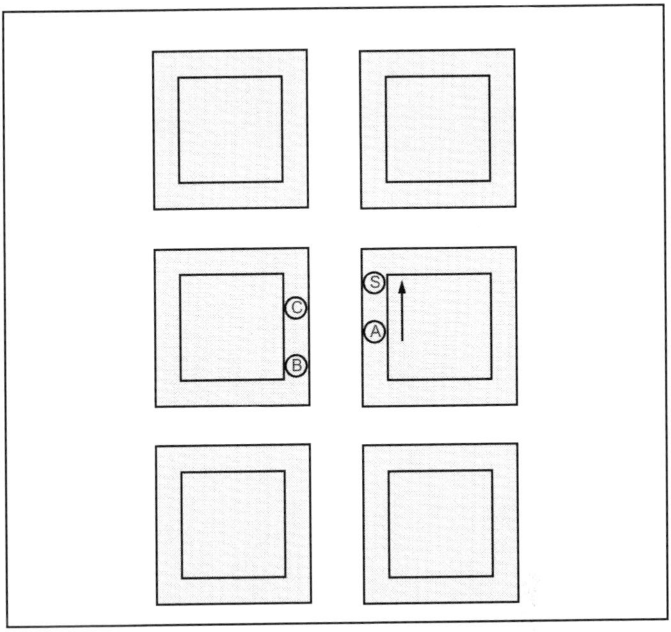

Figure J-2. Foot Surveillance Example 2

that the subject (in most cases) is not suspicious of individuals in front of him and walking in the same direction. (See *Figure J-4*, page J-5.)

POSITION CHANGES

J-9. Except for varying the positions in the basic ABC method, the positions are not changed—only the surveillants occupying the positions are changed. Position changes are usually made at intersections. (See *Figures J-5, J-6, J-7*, and *J-8*, pages J-6, J-7, J-8, and J-9.)

J-10. A frequent change in the ABC positions must be made if surveillance is to be carried out successfully. These changes should be made, in most cases, when the subject turns a corner at an intersection. The following is the recommended procedures to rotate the positions when the subject turns a corner:

- As the subject approaches the intersection, surveillant C should lead the subject and reach the intersection first. By pausing at the intersection and turning to face the subject, surveillant C can watch and signal to surveillant A and surveillant B concerning the subject's actions. (See *Figure J-5*, page J-6.)

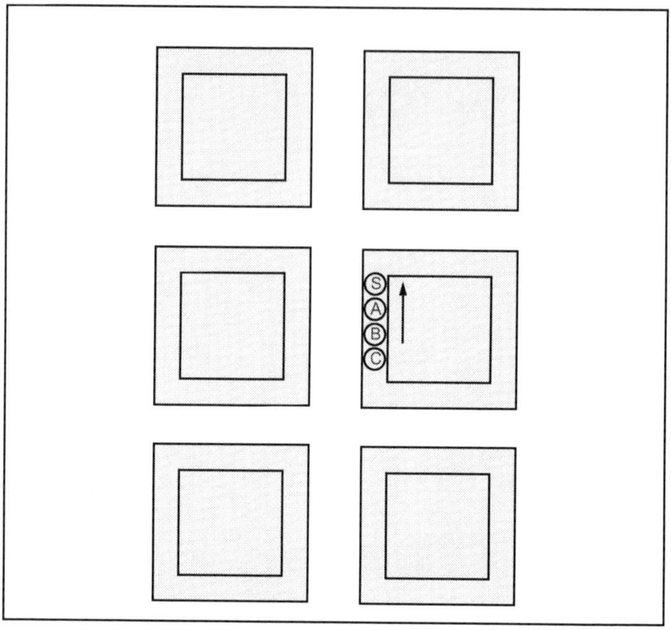

Figure J-3. Foot Surveillance Example 3

- When the subject turns right, he may suddenly be out of sight of surveillant A and surveillant B. They then must rely on surveillant C to keep them informed of the subject's movements. (See *Figure J-6*, page J-7.)
- Assuming that the subject continues to walk after turning right, surveillant A would cross the intersection to the opposite side of the street where the subject is now walking. Surveillant C would cross the intersection and walk in the same direction as the subject and on the same side of the street. Surveillant B would turn right (as the subject did) and walk in the same direction and the same side of the street as the subject.
- A (now across the street from the subject) will assume surveillant C's position and responsibilities. Surveillant B and surveillant C (now on the same side of the street and walking behind the subject) will become surveillant A and surveillant B and assume the responsibilities of those positions. Whether surveillant B or surveillant C assumes surveillant A's position will depend on which one has been seen last or not at all by the subject. (See *Figure J-7*, page J-8.)

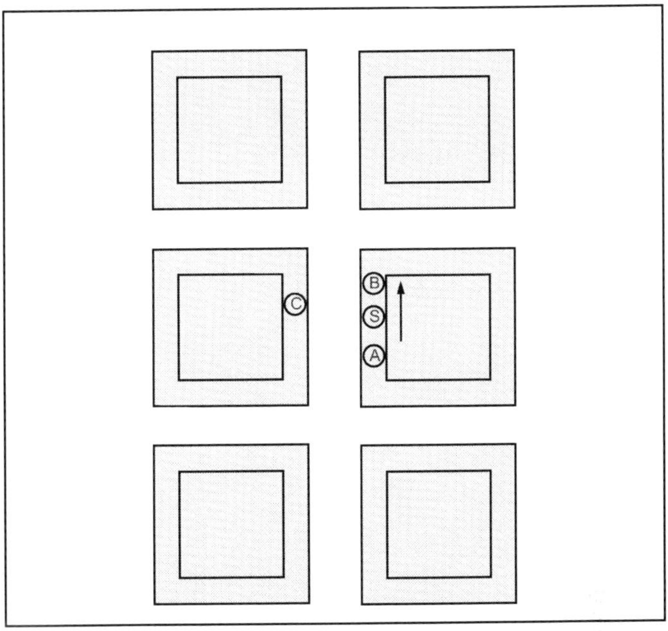

Figure J-4. Foot Surveillance Example 4

J-11. This is one recommended procedure for changing the ABC positions at an intersection. Circumstances might suggest another change. Whatever position rotation is made, it must be done frequently, usually at intersections. (See *Figure J-8*, page J-9.)

ONE-VEHICLE SURVEILLANCE

J-12. The use of only one surveillance vehicle should be avoided if at all possible. It is difficult to achieve a successful surveillance with only one vehicle. The subject must be kept in view and followed constantly by the same vehicle. This greatly increases the chances of detection by the subject. The lone surveillance vehicle must use all available traffic cover in remaining out of the subject's view. In heavy traffic, the surveillance vehicle must remain very close to the subject or risk losing him. On rural roads and highways greater distance must be allowed, even to the extent of losing sight of the subject at times. The surveillants should make every effort to alter the appearance of their vehicle so that it does not present the same picture to the subject each time it is in view. This may be accomplished by doing the following:

- Changing seating arrangements within the surveillance vehicle.
- Donning and removing hats, coats, and sunglasses.

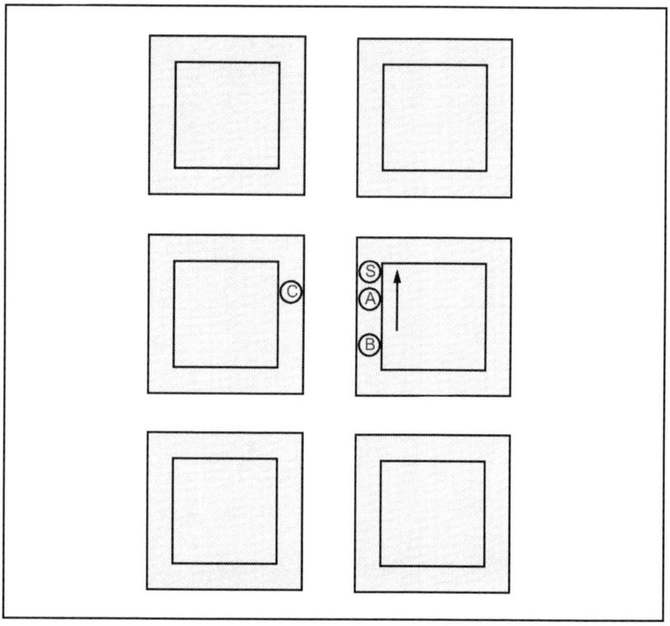

Figure J-5. Foot Surveillance Example 5

- Changing vehicle license plates.
- Turning into side streets or roads, and then returning back into traffic to resume the tail.

J-13. Changing the appearance of the surveillance vehicle is especially important in one-vehicle surveillance. The same principles can be applied during a prolonged multiple-vehicle surveillance.

J-14. In one-vehicle surveillance, the observer from the vehicle will travel on foot many times to take up a position of observation after the subject has turned a corner or parked. From his position, the observer can give further directional signals (by hand or radio) to the driver as to the subject's activity.

TWO-VEHICLE OR MULTIPLE-VEHICLE SURVEILLANCE

J-15. Frequent change of the surveillants' position in the lead or in the A position is an important factor in determining the success of the surveillance. Because this position is immediately behind the subject, it must be changed often to avoid detection of the lead surveillance team. The lead position is often referred to as the eyeball position.

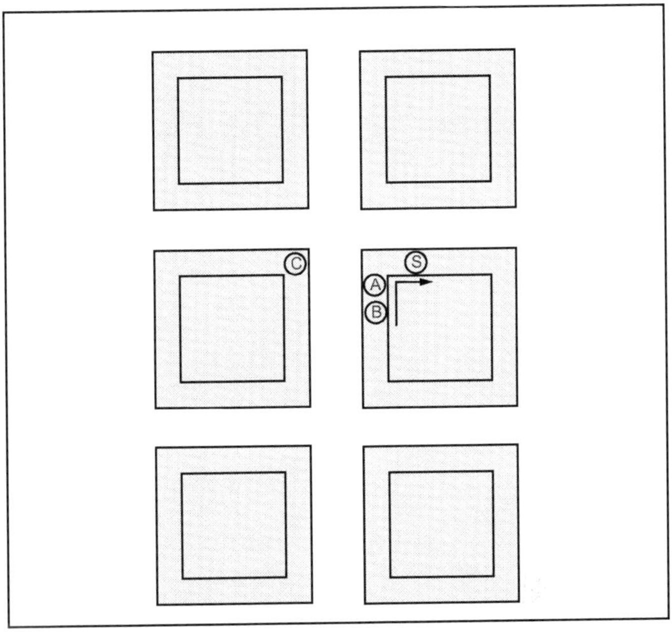

Figure J-6. Foot Surveillance Example 6

J-16. Once surveillance begins, the surveillance vehicles are positioned behind the subject as vehicle #1 (lead vehicle) and vehicles #2, #3, and #4. The surveillants have to be continually aware of the need to switch the position of the lead vehicle. The initial positions may be in caravan style with all vehicles on the street or highway behind the subject, or one or more vehicles may be situated on parallel streets.

J-17. The length of time that a vehicle should be in the lead vehicle position is a matter of judgment for the surveillant. It is based on traffic and street or highway conditions. Intersections or open highways offer the best opportunities for changing vehicle positions. On highways it is a simple matter for vehicle #1 (lead vehicle) to reduce its speed slightly, while vehicle #2 assumes the lead position. Vehicle #1 either drops back and gradually becomes the last vehicle, or it may turn off, allowing the subject to see him turn. This action can be very effective in dispelling fears or suspicions of the subject. Parking areas or intersections can be very effective to convince a subject that he is not being tailed by allowing him to see the lead vehicle turn off. (See *Figure J-9*, page J-10.)

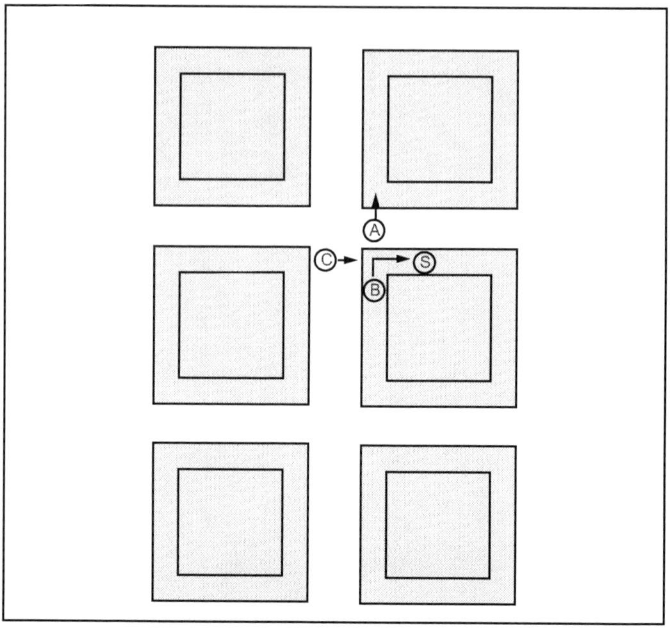

Figure J-7. Foot Surveillance Example 7

LEFT TURNS

J-18. Intersections can present problems that may result in the loss of the subject if handled incorrectly. This is particularly true with left turns (see *Figure J-10*, page J-11). If a subject never made a turn at an intersection, the chances of losing him would be greatly reduced. When the subject does make turns at intersections, serious problems can arise for surveillants, even when the subject is making no effort to lose them. Usually, surveillants handle right turns without any problems. However, left turns present problems.

J-19. The actions of the surveillants during left turns are dictated by traffic conditions and traffic lights for a few seconds. Left turns are made from a variety of situations and present numerous possibilities for the subject. They may be executed from highways or streets wide enough for two lanes of traffic or from a divided four-lane highway or street. They may also be made from a left-turn-only lane or from any turn lanes (against oncoming traffic). Left turns are made with or without traffic lights.

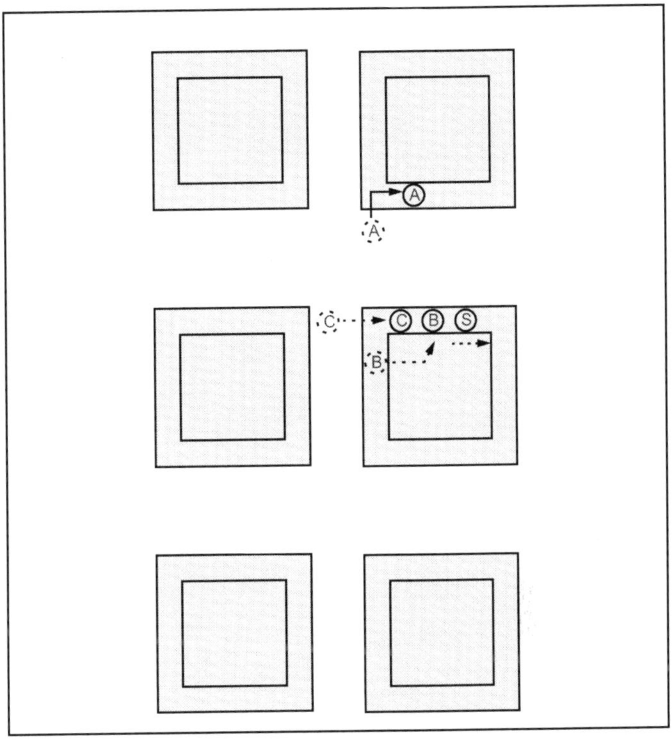

Figure J-8. Foot Surveillance Example 8

J-20. These traffic situations present potential problems in maintaining surveillance, losing the subject, and compromising surveillance. Vehicle #1 (lead vehicle) surveillants must be especially alert and use their best judgment in making the decision of whether to relinquish their position to vehicle 2 at left-turn situations.

J-21. The key to success in handling left turns is for vehicle #1 (lead vehicle) to maintain its position until it is sure that vehicle #2 is able to assume the lead position and follow the subject through the left turn. If it is apparent to the surveillants in vehicle #1 that vehicle #2 would be unable to follow the subject through the left turn because of traffic conditions, vehicle #1 should hold its position and delay making the change until after the turn is made. (See *Figures J-10* and *J-11*, page J-11 and J-12.)

LEFT-TURN-ONLY LANE WITHOUT TRAFFIC SIGNAL

J-22. A left-turn-only lane presents a problem when the subject is delayed at the intersection because of oncoming traffic. The recommended procedure is for vehicle #1 (lead vehicle) to continue through the intersection after it has

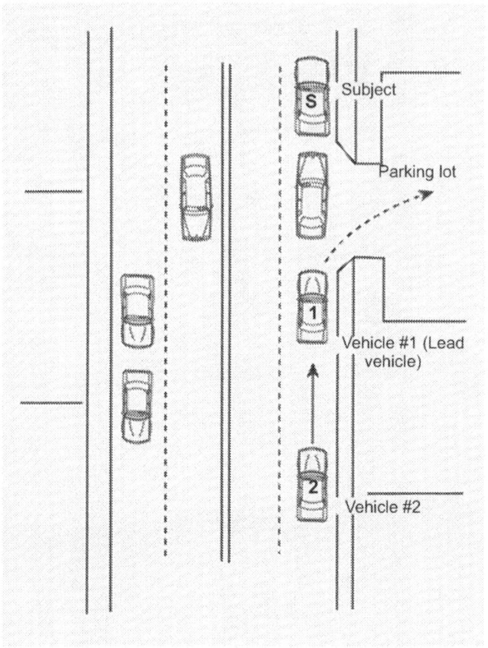

Figure J-9. Two-Vehicle Surveillance

determined that vehicle# 2 is close enough to assume the lead vehicle position behind the subject. Vehicle #2 assumes the lead vehicle position in the left turn lane and makes the turn behind the subject as quickly as possible. Vehicle #1 moves through the intersection and performs a U-turn as soon as it can safely do so and gets back to the intersection to assume a position behind the subject. If for some reason vehicle #2 is not close enough to assume the lead vehicle position at the intersection, vehicle #1 should not hesitate to get into the left-turn lane behind the subject and make the turn with him. (See *Figure J-12*, page J-13.)

GENERAL RULE OF THUMB FOR TURNS

J-23. During the surveillance of a vehicle, the subject may turn left or right or make a U-turn. The turns may be legal or prohibited by law. Some turns will be normal for the route that the subject is traveling, others will be for the sole purpose of detecting or losing surveillance. When the subject makes a turn as a detection maneuver, he will be looking behind him to see if a particular

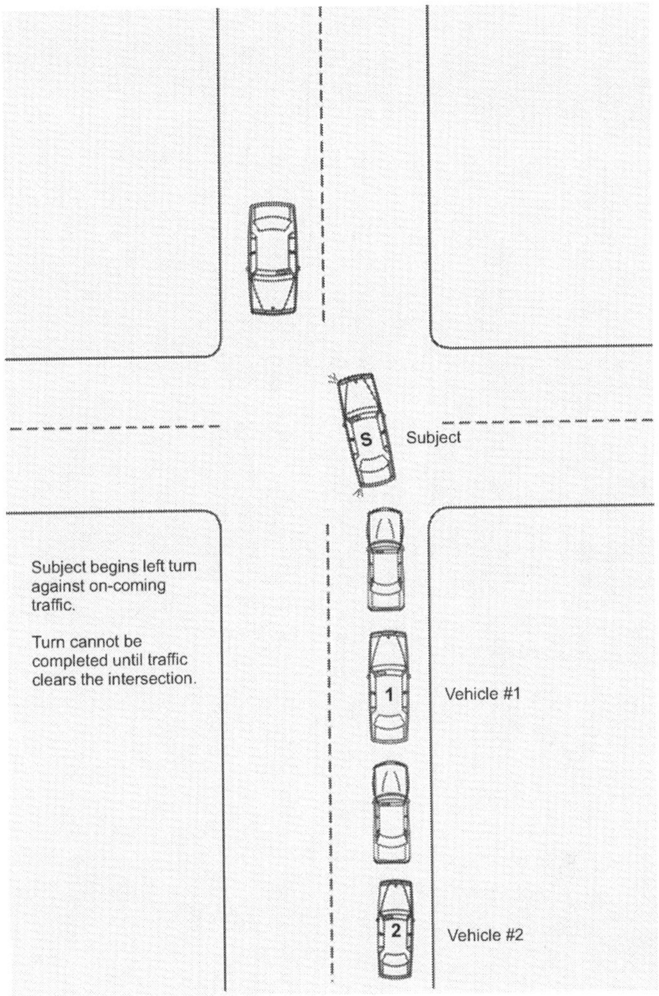

Figure J-10. Two-Vehicle Surveillance—Left Turn (Two-Lane Highway) Example 1

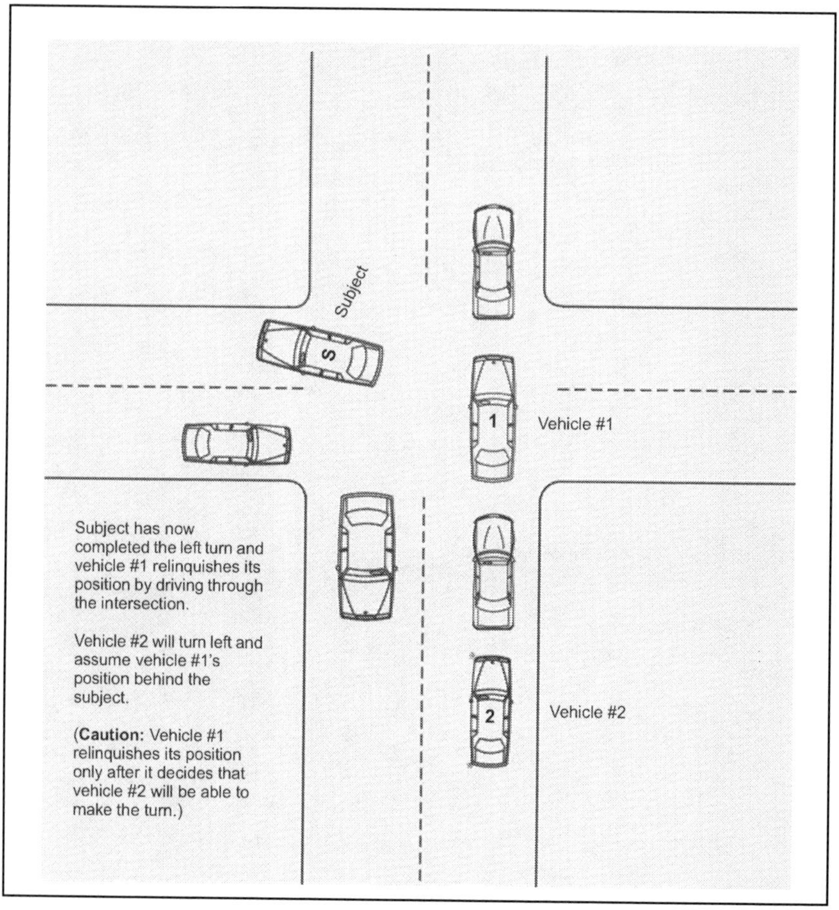

Figure J-11. Two-Vehicle Surveillance—Left Turn (Two-Lane Highway) Example 2

vehicle turned also. If he sees the same vehicle on several successive turns, he is convinced that he is under surveillance.

J-24. Always remember that when the subject turns, vehicle #1 (lead vehicle) should continue straight. This procedure cannot be followed with every turn but should be used whenever it is practical to do so.

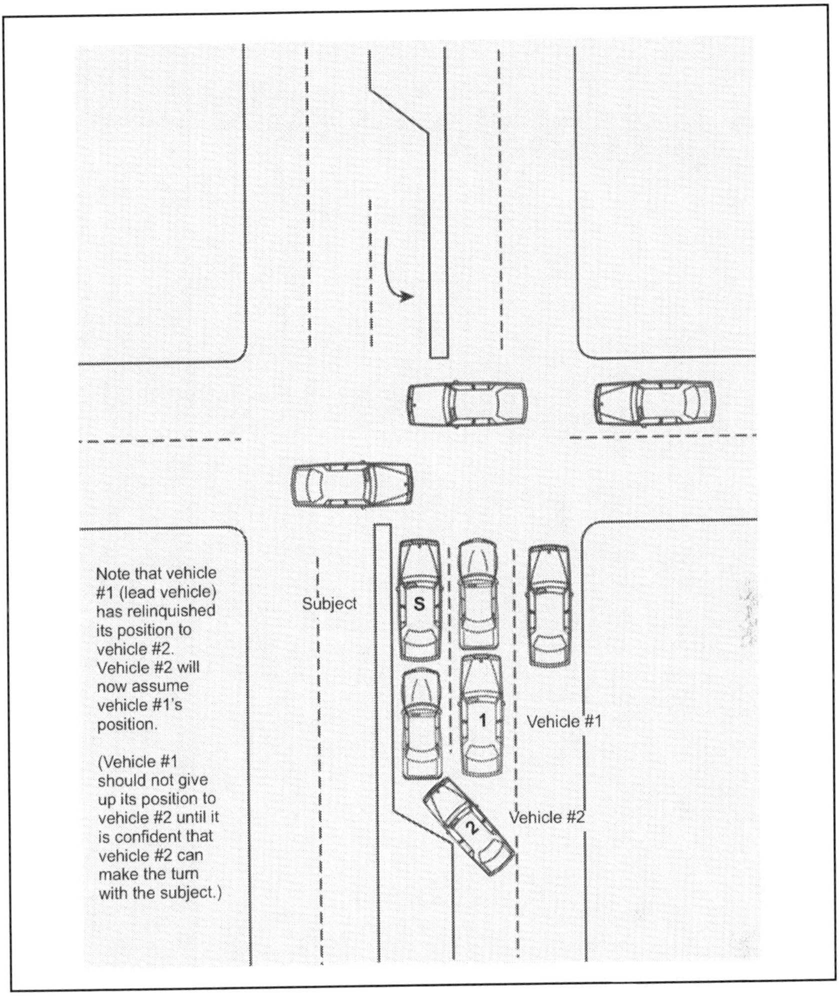

Figure J-12. Left-Turn-Only Lane Without a Traffic Signal

VEHICLE LEADING SURVEILLANCE

J-25. A variation of the standard ABC method of foot surveillance is to use one surveillant in front of the subject, and the subject is lead for a period of time as surveillance proceeds. When this method can be used, it is very effective. It is called leading surveillance. The same technique can be used during vehicle surveillance and is just as effective. Its effectiveness can be traced to the subject's natural reaction about being concerned with who is behind him and not who is in front of him.

J-26. This technique cannot be used in all surveillance situations. It be used only when the surveillant can anticipate the subject's travel route because of previously obtained reliable information or the subject's likely route as indicated by his actions. In some situations, such as a subject driving at a very slow speed, the leading surveillance method is the most effective technique. When the subject is led in this manner, the surveillants in the vehicle behind the subject should not be as close as they usually would. The surveillants leading the subject should have at least one cover vehicle between themselves and the subject.

PARALLEL SURVEILLANCE

J-27. When the subject's route of travel is fairly well established and can be anticipated for short periods of time or if the surveillance is being conducted in the city or an urban area, parallel surveillance may be used. As the name implies, surveillance vehicles travel on streets that are parallel to that of the subject. It is useful in removing surveillance vehicles from the subject's view on streets with little or no traffic and gives flexibility to the surveillants in the paralleling vehicles when the subject turns.

J-28. Paralleling vehicles arrive at cross streets at about the same time as the subject, or shortly before he does. This method is most effective in residential areas where the traffic is light. If the subject stops, slows down, or increases his speed, the timing of the paralleling vehicles is disrupted and the subject may be lost. When a lead vehicle is positioned behind the subject, this problem can be eliminated. This vehicle should be farther behind the subject than it usually would be, and from this position, it can advise the paralleling vehicles of the subject's location if this becomes necessary. Once the subject turns at a cross street, each surveillance vehicle will adjust according to his position at the time the turn is made and assume a new position. (See *Figure J-13.*)

PROGRESSIVE SURVEILLANCE

J-29. Some criminal activities do not lend themselves to being penetrated by routine surveillance methods. If the subjects involved are wary or extremely tail conscious or if their activities are conducted very late at night or early in the morning and involve traveling considerable distances, then any routine surveillance is likely to be compromised. It is best to use experienced investigators who are familiar with surveillance techniques and the area.

J-30. Several surveillance methods may be considered in these situations, such as using some type of electronic surveillance device that allows tailing the vehicle from a great distance (or using aircraft when possible). However, the best method is probably using the progressive surveillance technique.

J-31. The progressive surveillance technique is particularly effective when the subject's route of travel takes him away from the city to more rural areas and when part of his route of travel is unknown. This method involves

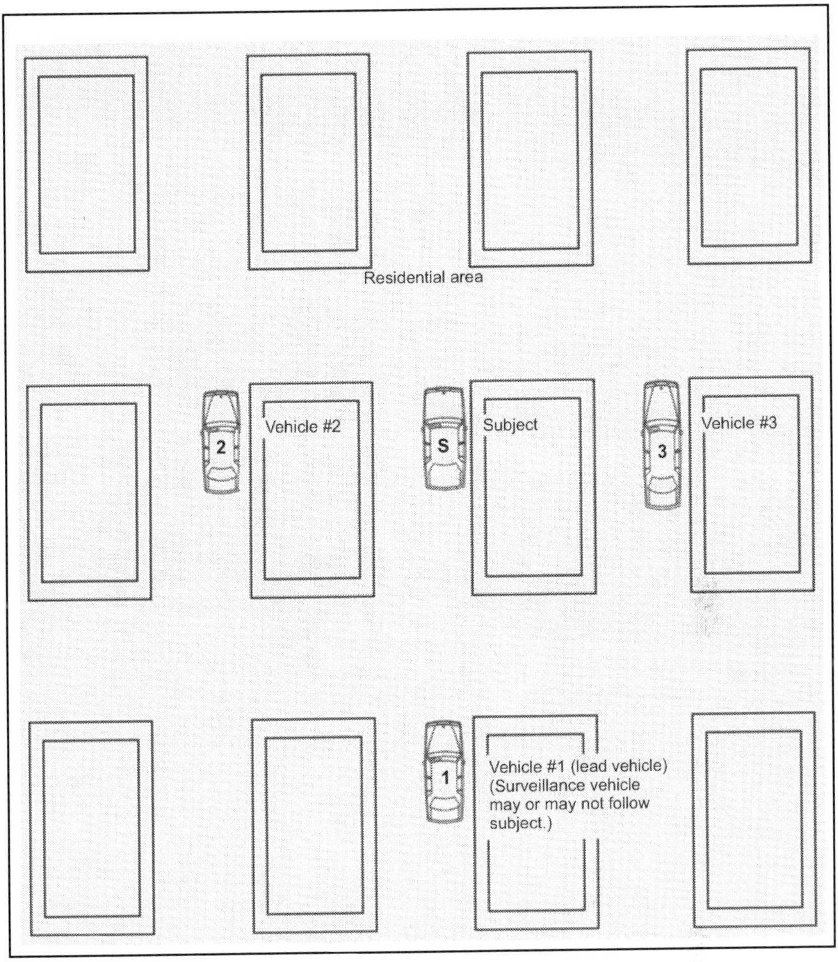

Figure J-13. Parallel Surveillance

stationing surveillance vehicles at various points, preferably at intersections along the suspected or known route. The vehicles, or a surveillant on foot, are hidden at the intersections. When the subject fails to reach a particular intersection after passing an earlier observation position, it is possible to determine where he turned off. The surveillance is then resumed from that point.

J-32. Due to the time-consuming nature of this type of surveillance, its use is limited to those occasions where it is known that the subject makes regular trips to the same destination. In another version of this method, the surveillance vehicles may tail the subject for a short distance after he passes their position. The surveillance must be done with caution, or its purpose will be defeated.

FOOT AND VEHICLE SURVEILLANCE

J-33. In some situations, it may be possible and desirable to use a vehicle in conjunction with foot surveillance. This method is used most often in city or urban areas where traffic is light and the vehicle can be quickly moved from one street to another.

J-34. One or two investigators in a vehicle provide support to surveillants conducting the foot surveillance. If two investigators are in the vehicle, one can act as a reserve foot surveillant to assume a position as a member of the foot surveillance team when a position change is required. The driver stays alert to the subject's location and acts as additional eyes for the foot surveillants. A vehicle in this type of surveillance can give the foot surveillant an important advantage—the ability to cover several blocks very quickly. This is particularly important when observation of the subject is lost. The driver must be careful to stay out of view of the subject.

J-35. Another important advantage of having a vehicle available during foot surveillance is that the surveillants will be assured of transportation if the subject should board a bus, taxicab, or any other form of transportation.

SURVEILLANT'S EVASIVE TACTICS RESPONSES

J-36. A subject will often use many tactics to detect surveillance. Surveillants must be alert and prepared to counter such tactics. These tactics are used when the subject suspects that he is under surveillance. They may also be used as part of his routine driving patterns to make any possible surveillance difficult. Some of the tactics commonly used by the subject are—

- **Performing U-turns in the street.** Vehicle #1 (lead vehicle) notifies the other vehicles of the subject's action and continues through the intersection. Vehicle #2 responds by driving into a parking lot or an off-street area and assuming the lead vehicle position after the subject has completed the U-turn and has passed by. Vehicle #1 performs a U-turn when it is safe to do so or makes successive turns to get back into a surveillance position, either as vehicle #2, #3, or #4. (See *Figure J-14.*).
- **Making left or right turns.** The lead vehicle notifies the other vehicles of the subject's action and continues through the intersection. Vehicle #2 turns at the intersection behind the subject and assumes the lead vehicle position. Vehicle #1 does a U-turn or makes successive turns to get back into a surveillance position. If more than two vehicles are involved in the surveillance, the other vehicles have an

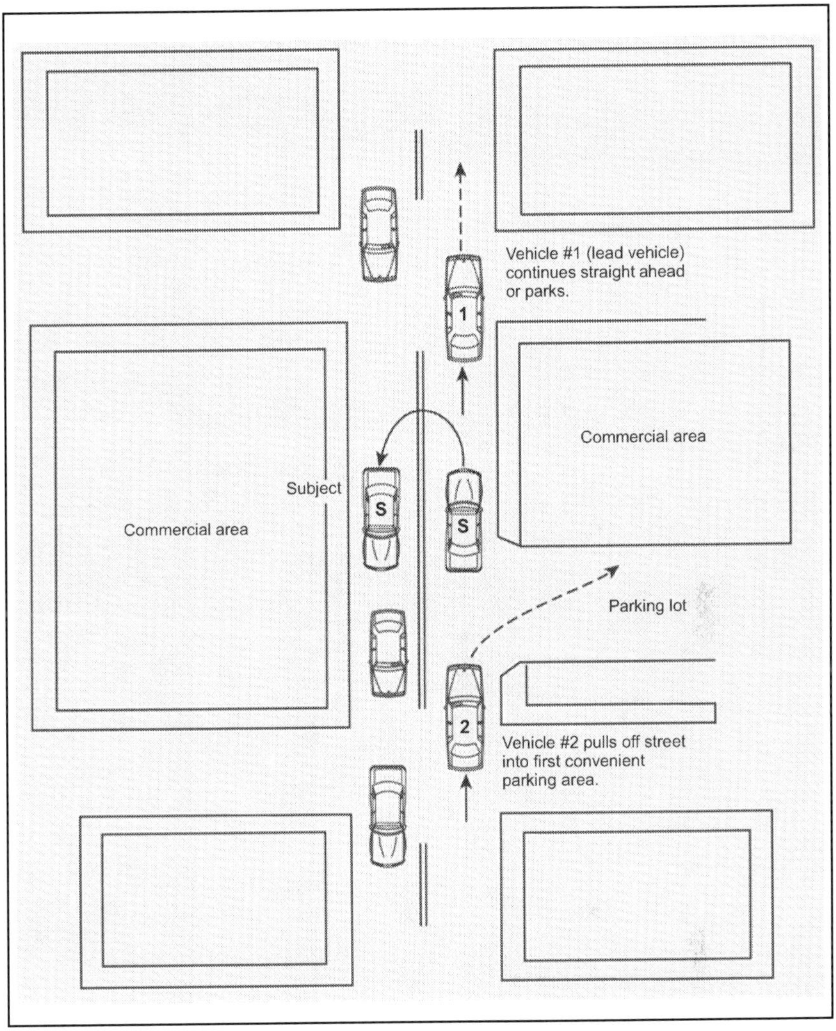

Figure J-14. U-Turns in the Street

option of turning at the intersection where the subject turned or turning in the same direction on paralleling streets, depending on traffic conditions and their distance from the subject. (See *Figures J-15* and *J-16*.)

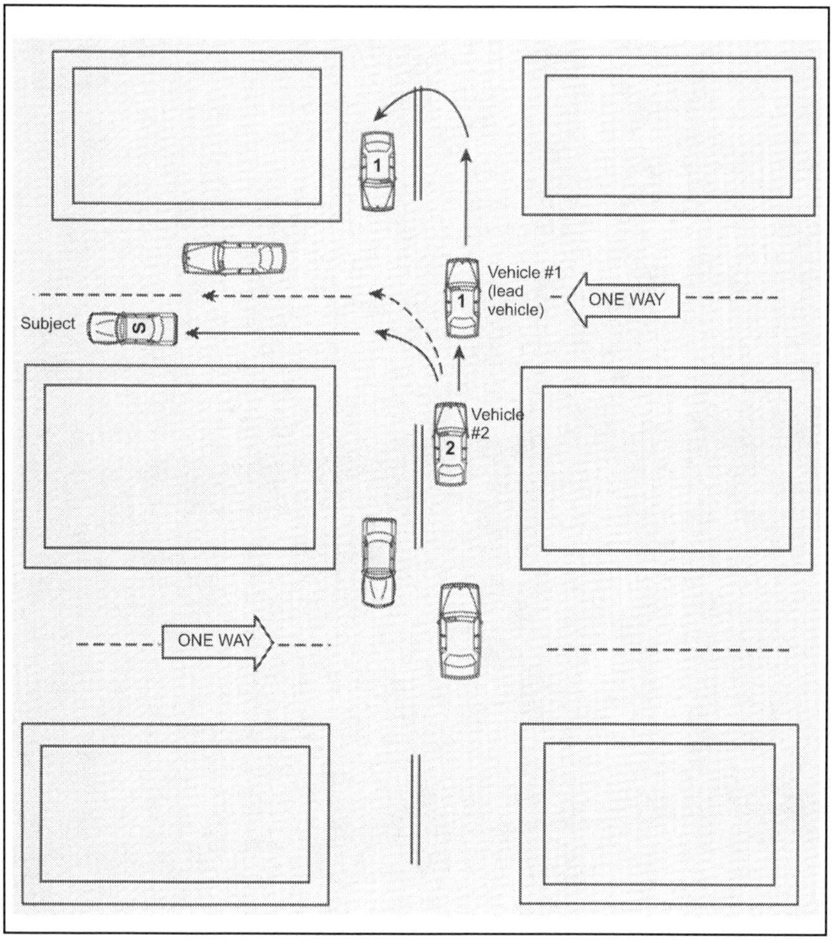

Figure J-15. Left Turns

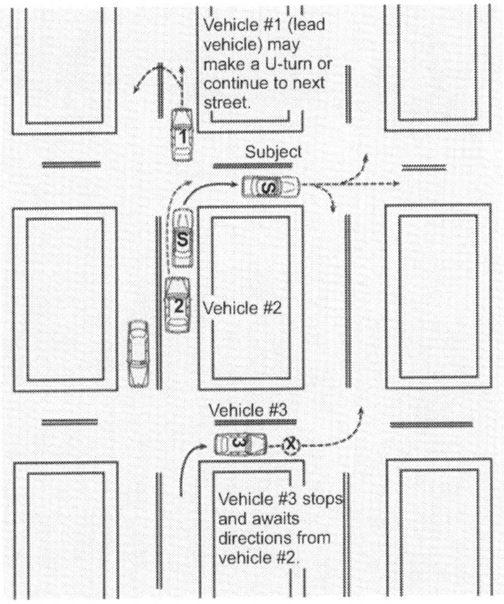

Figure J-16. Right Turns

- **Turning at an intersection and stopping.** Vehicle #1 (lead vehicle) observes the subject as he drives through the intersection and notifies the other vehicles of the subject's action. Instead of vehicle #2 turning at the intersection behind the subject, it stops at the intersection and an observer goes on foot to continue observing. Vehicle #1 has the option of making a U-turn and coming back to the intersection or making successive turns and getting into position for the subject's expected route of travel when he continues. When the subject does continue, vehicle #2 assumes the lead vehicle position. Other vehicles involved in the surveillance use this opportunity to take up positions to either parallel the subject or change to a closer surveillance position in anticipation of assuming the lead position at the next stop.

J-37. In one-vehicle surveillances, (see *Figure J-17*, page J-20) the subject should not be followed around the corner if traffic is sparse. The observer should dismount and travel on foot to the corner to observe the subject's actions. This procedure should also be followed in a multiple-vehicle surveillance when a drive-by is not possible or practical, such as in rural or isolated areas when the subject makes an abrupt turn and is suddenly out of sight. In rural or isolated areas, subjects may use the following tactics:

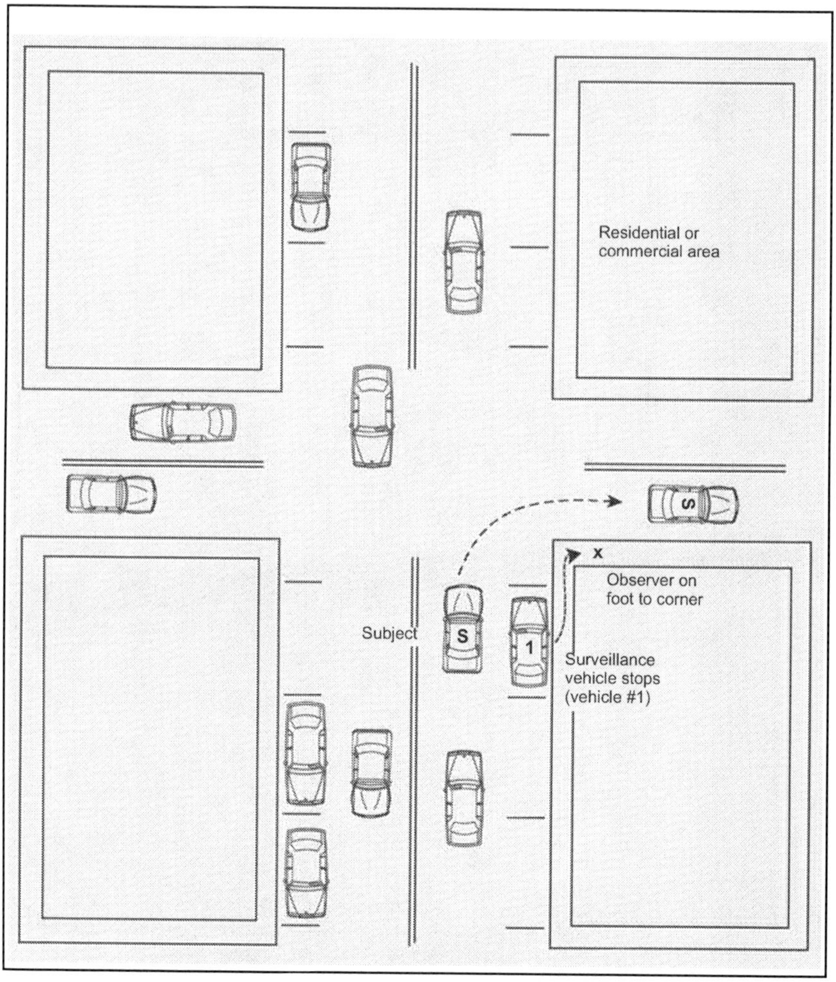

Figure J-17. Subject Turns Corner—One-Vehicle Surveillance

- **Stopping just beyond a curve or hill crest.** Vehicle #1 (lead vehicle) notifies vehicle #2 of the subject's actions and drives on by. Vehicle #2 stops and pulls off the road or onto a side road. An observer from vehicle #2 goes on foot to observe the subject's activities. Vehicle

#1 stops out of sight of the subject and attempts to have an observer on foot observe the subject from the side opposite of vehicle #2. (See *Figure J-18.*)

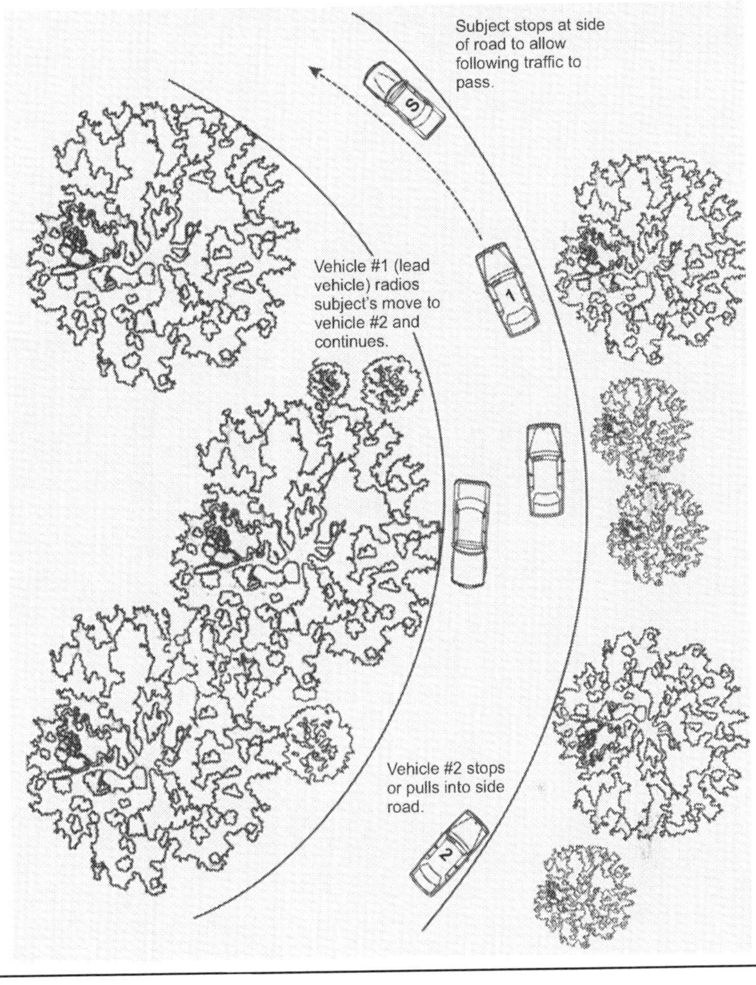

Figure J-18. Subject Stops Beyond a Curve

- **Turning into an alley.** This tactic appears at first to be only a repeat of a right turn, but it requires a different approach. Vehicle #1 (lead vehicle) must stop before reaching the alley and street and allow an observer to get out of the vehicle to approach on foot. Vehicle #2 should hold its position until the surveillant in vehicle #1 can determine what the subject's next move will be. A drive-by will be ineffective in establishing the subject's intentions in this situation because of the limited or restricted visibility. (See *Figure J-19.*)

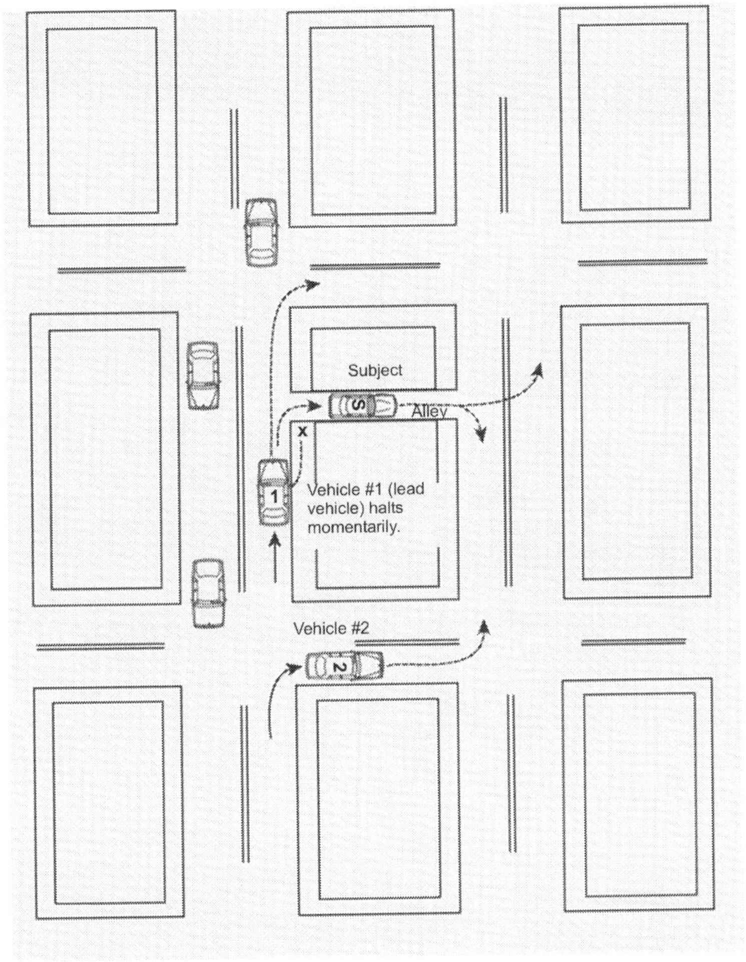

Figure J-19. Subject Turns Into an Alley

Appendix K

Undercover Mission Checklist

The undercover mission checklist, *Figure K-1*, should be used during the initial planning phases and postoperation task organization. The checklist will assist the leader in developing the overall operation plan.

STAGE 1: TRANSACTION POINT.

1. Have you surveilled the transaction point in depth?

   Yes__________ No__________

2. Have you checked the number of entrances and exits? If so, give the specific number of each.

   Entrances__________ Exits__________

3. Summarize where the transaction point is located in detail. Be specific.

   ____

   ____

4. Have you found specific surveillance points?

   Yes__________ No__________

5. Draw a map of exits and entrances to the transaction point.

STAGE 2: BRIEFING OF ASSISTING AGENTS.

Date:________________

Start time:______________

Location of the briefing:____

1. How many police officers are going to be used in support of this operation?__________________

2. List specific responsibilities by duty, name, and team number:

   Special Duties:

   a. Command post:______________________________ Team number ____________

   b. Surveillance point:____________________________ Team number ____________

Figure K-1. Checklist for Undercover Missions

c. Takedown team____________________ Team number __________

d. Countersurveillance:____________________ Team number __________

3. List the vehicles/operators that will be used in the operation:

	Vehicle	Driver	Call sign
a.	__________	__________	__________
b.	__________	__________	__________
c.	__________	__________	__________

4. Have you explained the location of the transaction point to all assisting police officers in detail? Provide officers with a map of the transaction point (if needed).

Yes__________ No__________

5. Have you instructed the assisting officers where you want them to set up for surveillance on the operation?

Yes__________ No__________

6. Is there any specific equipment that will be used in support of this operation? If yes, specify what type of equipment and who will be using it.

Yes__________ No__________

	Type of equipment	Used by
a.	__________	__________
b.	__________	__________
c.	__________	__________

7. Have the assisting officers been told the route the CI/UCA will travel when moving to and from the transaction site?

Yes__________ No__________

8. Have the assisting officers met the CI/UCA, or have they been provided with a photograph as well as a complete description of the type of clothing the CI/UCA will be wearing?

Yes__________ No__________

9. Have all of the assisting officers been briefed on the arrest signals/trouble signals that will be used in support of this operation?

Yes__________ No__________

10. Have all assisting officers been briefed on physical force and arrest procedures?

Yes__________ No__________

11. Are all the police officers armed, and has all equipment been checked and accounted for before initiating this surveillance to ensure that the equipment is on hand and functional?

Yes__________ No__________

Figure K-1. Checklist for Undercover Missions (Continued)

STAGE 3: PAPERWORK.

Has all paperwork in support of this mission been prepared?

Yes__________ No__________

STAGE 4: PERSONNEL/COMMAND OFFICIALS BRIEFED.

1. Have the following individuals been briefed about this operation?

   a. Chief of police.

   Yes__________ No__________

   b. Task force commander.

   Yes__________ No__________

   c. Prosecutor (if appropriate).

   Yes__________ No__________

2. Has coordination been effected with all law enforcement agencies in the area where the operation is to take place to ensure that they are aware that a police operation will be ongoing?

   Yes__________ No__________

STAGE 5: PREBUY PROCEDURES.

1. Which police officer will be controlling the confidential informant (if appropriate)?

____

2. List the time, date, and location of the meet.

   Time:________________

   Date:________________

   Location:____

3. Has the informant been strip-searched?

   Yes__________ No__________

   By whom:______________________

   Time began:______________________

   Time ended:______________________

4. List all items the informant is wearing in detail (color, type, and condition).

   a. Shirt:____

   b. Pants:____

   c. Socks:____

   d. Jacket:____

Figure K-1. Checklist for Undercover Missions (Continued)

e. Underwear:____

f. Billfold:____

g. Shoes:____

h. Money:____

5. Has the informant been given confidential funds to purchase the illegal drugs?

Yes__________ No__________

By whom:____________________

Time:____________________

Amount provided:____________________

Amount returned:____________________

6. Has the informant been briefed on which route to follow from the start to finish?

Yes__________ No__________

7. Has the informant been told what to do if he gets into trouble?

Yes__________ No__________

8. Has the informant been told what to do with the drugs once he has them?

Yes__________ No__________

STAGE 6: PREBUY SURVEILLANCE.

1. The prebuy surveillance team should be deployed at least one hour before the scheduled transaction time. This team should accomplish the following:

- Survey the transaction site to ensure that it is safe to allow the CI/UCA into the area.
- Attempt to detect countersurveillance that the subject may deploy to the transaction site before his arrival.
- Determine if the subject is already at the transaction site. If so, did he arrive alone or did associates accompany him?

2. The prebuy surveillance team will contact the group supervisor and give him the all-clear signal when they are ready to allow the CI/UCA to move to the transaction site. The CI/UCA should not move to the transaction site until he is told to do so.

NOTE: Remember that prebuy surveillance could save the life of the CI/UCA.

STAGE 7: SURVEILLANCE OF THE ACTUAL OPERATION.

All police officers supporting the surveillance operation should keep personal notes of what they observe during the course of the surveillance. On completion of the operation, the team should prepare a surveillance report.

Figure K-1. Checklist for Undercover Missions (Continued)

STAGE 8: OUTBRIEFING OF CONFIDENTIAL INFORMATION.

1. Has the informant been strip-searched:

   Yes__________ No__________

   By whom:____________________________

   Time began:__________________________

   Time ended:__________________________

2. Have you obtained evidence from the informant?

   Yes__________ No__________

3. Has the evidence been field-tested?

   Yes__________ No__________

4. Has the evidence been turned over to the evidence custodian?

   Yes__________ No__________

   Time:_______________

   Date:_______________

5. Has the informant given a statement detailing the activities he conducted in support of this operation?

   Yes__________ No__________

6. Has the informant been given back any items that were taken from him before the initiation of this operation?

   Yes__________ No__________

7. Have all police officer's notes been collected and an investigative entry made in the case file concerning the overall operation?

   Yes__________ No__________

Figure K-1. Checklist for Undercover Missions (Continued)

Appendix L

Operation Plan Template for Undercover Operations

The standard five-paragraph operation order (OPORD) format must be used for undercover operation plans. Mission leaders can use the operation plan (OPLAN) format in *Figure L-1* to include specific information pertaining to undercover operations.

Team ______________ **Case Number** ________ **Agent** ______________ **Date** ______________

I. SITUATION.

A. Type of operation (buy-bust, preliminary bust, search warrant, and so forth).

____

B. Type of drug and amount anticipated.

____

C. Locations (address and/or description).

____

(Include any codes used to describe the location.)

D. Suspect(s) name, description, and prior criminal background:

1. ____
2. ____

E. Possible weapons possessed or available to suspect(s).

1. ____
2. ____

F. Countersurveillance suspected (if possible, description of vehicle).

____

____

____

____

Figure L-1. Template for Undercover Operation Plans

G. Background information (obtained from informant, surveillance, investigations, record checks, and so forth).

II. MISSION.

Instructions.

III. EXECUTION.

A. Overall concept of operation.

B. Specific duties.

NAME	UNIT NO	VEHICLE	ASSIGNMENT

C. Coordinating Instructions.

IV. SERVICE SUPPORT.

A. Clothing and equipment.

B. Meals, breaks, and relief —if necessary.

C. Bust signals (primary, secondary, and rip-off).

Figure L-1. Template for Undercover Operation Plans (Continued)

V. COMMAND AND SIGNAL.

A. Command.

____

____

B. Communications.

____

____

NOTE: For security reasons, plans should be collected at the end of each operation and destroyed; file copies should be retained for inclusion in case file.

VI. DIAGRAM OF LOCATION, AND SO FORTH (OPTIONAL).

Figure L-1. Template for Undercover Operation Plans (Continued)

Appendix M

Risk Assessment Matrix

The risk assessment matrix, *Figure M-1*, is a management tool that assists leaders in the identification of hazards or hotspots involved with any mission. The matrix provides a method of identifying and quantifying risks. Leaders must continuously conduct the risk assessment process throughout the mission.

Planning				
Guidance	In-Depth	Adequate	Minimal	Risk
Vague	3	4	5	
Implied	2	3	4	
Specific	1	2	3	
Supervision				
Command/Control	Jurisdiction	Joint Jurisdiction	No Jurisdiction	Risk
OPCON	3	4	5	
Attached	2	3	4	
Organic	1	2	3	
Intelligence				
Source	Reliable	Unknown	Unreliable	Risk
Vague	3	4	5	
Moderate	2	3	4	
Specific	1	2	3	
Target Analysis				
Violence Potential	Specifics Known	Specifics Vague	Specifics Unknown	Risk
High	3	4	5	
Medium	2	3	4	
Low	1	2	3	

Figure M-1. Risk Assessment Matrix

Key Personnel				
Task	Highly Qualified	Qualified	Minimally Qualified	Risk
Complex	3	4	5	
Routine	2	3	4	
Simple	1	2	3	
Personnel/Equipment (Including Vehicles)				
Suitability	Availability High	Availability Medium	Availability Low	Risk
Low	3	4	5	
Medium	2	3	4	
High	1	2	3	
Weather				
Temperature (F)	Clear/Dry Day/Night	Drizzle/Rain Day/Night	Fog/Snow/Ice Day/Night	Risk
0° to 31°	2/3	3/4	4/5	
32° to 59°	1/2	2/3	3/4	
60° to 85°	1/2	2/3	3/4	
85°+	2/3	3/4	4/5	
			Total:	
Quick Reference Guide				
Low Risk		Caution		High Risk
0	12		24	35

Figure M-1. Risk Assessment Matrix (Continued)

Appendix N

Risk Mitigation Work Sheet

The risk mitigation work sheet, *Figure N-1*, aids mission leaders and their supervisors with the development of risk management methods to meet mission requirements. This work sheet helps the leader conduct operations without unnecessary risks.

NOTE: Identify "hot spots" from the initial assessment and use the empty block to mitigate the risk down to an acceptable risk level.		
Initial Assessment	Planning	Mitigated Assessment
	Supervision	
	Intelligence	
	Target Analysis	
	Key Personnel	
	Personnel/Equipment (Including Vehicles)	
	Weather	
	TOTALS	

Figure N-1. Risk Migration Work Sheet

Glossary

1SG	first sergeant
AAS	agent activity summary
AC	alternating current
AD	active duty
AFIP	Armed Forces Institute of Pathology
AFIS	Automated Fingerprint Identification System
AIR	agent investigation report
AL	Alabama
am	ante meridiem
amp	amputation
APO	Army Post Office
AR	Army regulation
ARIMS	Army Records Information Management System
ARNG	Army National Guard
ASCLD	American Society of Crime Laboratory Directors
attn	attention
Aug	August
ave	avenue
BATF	Bureau of Alcohol, Tobacco, and Firearms
bde	brigade
BIOS	basic input-output system
bldg	building
bn	battalion
BOLO	be on the lookout
C4	composition 4
CA	California
cal	caliber
CB	composition B
CBP	US Customs and Border Protection
CCIR	commander's critical information requirement
CCIU	computer crime investigation unit
CD	compact disk
cdr	commander
CDR	compact disk recordable
CD-ROM	compact disk–read only memory
CFR	Code of Federal Regulations

CG	commanding general
CI	confidential informant
CID	Criminal Investigation Division
civ	civilian
cmd	command
co	company
CODIS	Combined DNA Index System
CONEX	container express
CONUS	continental United States
CPO	civilian personnel office
CPT	captain
CPU	central processing unit
CQ	charge of quarters
CRIMINT	criminal intelligence
CSA	Controlled Substances Act
CTU	consolidated TOE update
CWA	Clean Water Act
CW3	Chief Warrant Officer, W3
D	day
DA	Department of the Army
DAIG	Department of the Army Inspector General
DC	defense counsel
DC	District of Columbia
DD	Department of Defense
DEA	Drug Enforcement Administration
Dec	December
DFAS	Defense Finance and Accounting Service
div	division
DL	driver's license
DNA	deoxyribonucleic acid
DOB	date of birth
DOD	Department of Defense
DODD	Department of Defense Directive
DODI	Department of Defense Instruction
DOJ	Department of Justice
DOT	Department of Transportation
DRMO	defense reutilization and marketing office
DSL	digital subscriber line
DSN	Defense Switched Network
DVD	digital versatile disk

DVD-R	digital versatile disk–recordable
E3	private first class
E4	specialist
E5	sergeant
E7	sergeant first class
E/E	entry and exitway
EEFI	essential elements of friendly information
EMS	emergency medical services
EMT	emergency medical technician
EO	executive order
EOD	explosive ordnance disposal
EPA	Environmental Protection Agency
etc	et cetera
F	Fahrenheit
FA	field artillery
FAO	finance and accounting officer
FBI	Federal Bureau of Investigation
FD	federal document
Feb	February
fed	federal
FEDEX	Federal Express
FFCA	Federal Facilities Compliance Act
FFIR	friendly force information requirement
FinCEN	Financial Crimes Enforcement Network
FM	field manual
FO	field operation
FOIA	Freedom of Information Act
FPO	Fleet Post Office
FSN	federal stock number
ft	fort
FTC	Federal Trade Commission
fwd	forward
GA	Georgia
GBL	government bill of lading
GCM	general courts-martial
GHB	gammahydroxybutyrate or gamma hydroxybutyric acid
GPS	Global Positioning System
GS	general schedule
GSA	General Services Administration
GSR	gunshot residue

HAZMAT	hazardous material
HHC	headquarters and headquarters company
HIV	human immunodeficiency virus
HN	host nation
HQ	headquarters
HQDA	Headquarters, Department of the Army
HUMINT	human intelligence
I&I	interviews and interrogations
IC	incident commander
IMINT	image intelligence
INTSUM	intelligence summary
inv	investigator
IPC	interpersonal communication
IQ	intelligence quotient
IRS	Internal Revenue Service
ISDN	integrated services digital network
ISO	International Organization for Standardization
ISP	internet service provider
JA	judge advocate
JAG	judge advocate general
JAGC	judge advocate general's corps
Jan	January
JP	joint publication
LA	Louisiana
LAB	laboratory accreditation board
LAN	local area network
lb	pound
lic	license
LOD	line of duty
LSD	lysergic acid diethylamide
M	month
MAC	media access control
MACOM	major Army command
MAJ	major
MANSCEN	United States Army Maneuver Support Center
MARKS	Modern Army Record-Keeping System
MASINT	measurement and signature intelligence
max	maximum
MCM	Manual for Courts-Martial
MDMA	3,4-methylenedioxymethamphetamine

med	medical
MEVA	mission-essential vulnerable areas
MFR	memorandum for record
MICR	magnetic ink character recognition
MILVAN	military-owned demountable container
mm	millimeter
MO	Missouri
MO	modus operandi
MOS	military occupational specialty
MOU	memorandum of understanding
MP	military police
MPI	military police investigator
MPR	military police report
MSDS	material safety data sheet
N	no
N	north
NA	not applicable
NAFI	National Association of Fire Investigators
NCIC	National Crime Information Center
NCO	noncommissioned officer
NFI	not further identified
NFIRS	National Fire Incident Reporting System
NFPA	National Fire Protection Association
NIBIN	National Integrated Ballistics Information Network
NIBRS	National Incident-Based Reporting System
NIC	network interface card
no	number
Nov	November
NSN	national stock number
O3	captain
OCONUS	outside continental United States
OCE	Office of the Chief of Engineers
ODCSOPS	Office of the Deputy Chief of Staff for Operations and Plans (Army)
OPCON	operational control
OPLAN	operation plan
OPORD	operation order
opt	optional
OSHA	Occupational Safety and Health Administration
OTJAG	Office of the Judge Advocate General
P2P	phenyl-2-propanone

PAO	public affairs office
PC	personal computer
PCP	phencyclidine
PD	police department
PDA	personal digital assistant
PDR	Physician's Desk Reference
PETN	pentaerythrite tetranitrate
PFC	private first class
PIO	police intelligence operations
PIR	priority intelligence requirements
PL	public law
PLGR	precision lightweight GPS receiver
pm	post meridiem
PM	provost marshal
PO	post office
POL	petroleum, oils, and lubricants
POW	prisoner of war
PPE	personal protective equipment
PX	post exchange
R&S	reconnaissance and surveillance
RA	regular army
RA	resident agency
RCRA	Resource Conservation and Recovery Act
rec'd	received
reg	regular
rel	release
ret	return
ROI	report of investigation
ROS	report of survey
S2	Intelligence Officer (US Army)
SA	special agent
SAC	special agent in charge
SARA	Superfund Amendments and Reauthorization Act
SCRCNI	sealed container received, contents not inventoried
sec	second
SEM/EDS	scanning electron microscopy/energy dispersive (X ray) spectroscopy
Sep	September
SF	standard form
SGT	sergeant
SIDS	sudden infant death syndrome

SIGINT	signal intelligence
SIPRNET	Secret Internet Protocol Router Network
SITREP	situation report
SJA	staff judge advocate
SLR	single-lens reflex
SN	serial number
SOFA	Status of Forces Agreement
SOP	standing operating procedure
SP	standard play
SPOTREP	spot report
SPR	small particle reagent
SSA	Social Security Administration
SSG	staff sergeant
SSN	social security number
st	street
TC	trial counsel
TCN	third country nationals
TDA	table of distribution and allowances
THC	tetrahydrocannabinol
tng	training
TNT	trinitrololuene
TO	theater of operations
TOE	table of organization and equipment
TRADOC	United States Army Training and Doctrine Command
TTP	tactics, techniques, and procedures
TX	Texas
UCA	undercover agent
UCMJ	Uniform Code of Military Justice
UN	United Nations
UPS	United Parcel Service
US	United States
USACIDC	United States Army Criminal Investigation Command
USACIL	United States Army Criminal Investigation Laboratory
USACRC	United States Army Crime Records Center
USAMPS	United States Army Military Police School
USAPA	United States Army Publishing Agency
USAR	United States Army Reserve
USB	universal serial bus
USC	United States Code
USCIS	United States Citizenship and Immigration Services

USFA	United States Fire Administration
USNG	United States National Guard
USPS	Unites States Postal Service
USSS	United States Secret Service
UV	ultraviolet
VA	Virginia
VCR	video cassette recorder
VIN	vehicle identification number
WA	Washington
WAN	wide area network
WMD	weapons of mass destruction
Y	year
Y	yes

Bibliography

AFIP Form 1323. *AFIP/Division of Forensic Toxicology—Toxicological Request Form.*

AR 15-6. *Procedures for Investigating Officers and Boards of Officers.* 30 September 1996.

AR 190-13. *The Army Physical Security Program.* 30 September 1993.

AR 190-30. *Military Police Investigations.* 1 June 1978.

AR 190-45. *Law Enforcement Reporting.* 20 October 2000.

AR 195-2. *Criminal Investigation Activities.* 30 October 1985.

AR 195-3. *Acceptance, Accreditation, and Release of US Army Criminal Investigation Command Personnel.* 22 April 1987.

AR 195-4. *Use of Contingency Limitation .0015 Funds for Criminal Investigative Activities.* 15 April 1983.

AR 195-5. *Evidence Procedures.* 28 August 1992.

AR 195-6. *Department of the Army Polygraph Activities.* 29 September 1995.

AR 200-1. *Environmental Protection and Enhancement.* 21 February 1997.

AR 25-400-2. *The Army Records Information Management System (ARIMS).* 18 March 2003.

AR 27-10. *Military Justice.* 6 September 2002.

AR 525-13. *Antiterrorism.* 4 January 2002.

AR 600-20. *Army Command Policy.* 13 May 2002.

AR 670-1. *Wear and Appearance of Army Uniforms and Insignia.* 5 September 2003.

AR 70-12. *Fuels and Lubricants Standardization Policy for Equipment Design, Operation, and Logistic Support.* 1 May 1997.

AR 710-2. *Inventory Management Supply Policy Below the National Level.* 25 February 2004.

AR 735-5. *Policies and Procedures for Property Accountability.* 10 June 2002.

CFR, Title 29, Labor; Part 1910, *Occupational Safety and Health Standards*; Section 120, *Hazardous Waste Operations and Emergency Response.*

CFR, Title 40, *Protection of Environment;* Part 311, *Worker Protection*; Section 1, *Scope and Application.*

CFR, Title 48, *Federal Acquisition Regulations System*; Part 1, *Federal Acquisition Regulation.*

CID Form 28. *Agent Activity Summary (AAS).*

CID Form 36. *Field Test Analysis on Non-Narcotic Substances.*

CID Form 88. *Wanted Poster.*

CID Form 94. *Agents Investigation Report (AIR).*

CID Pamphlet 195-5. *Criminal Individual Performance Evaluation Instructions.* 13 March 1998.

CID Regulation 195-1. *Criminal Investigation Operational Procedures.* 1 July 2003.

Constitution of the United States.

Controlled Substance Act (PL 95-633). 10 November 1978.

Clean Water Act, Federal Water Pollution Control Act Amendment (as amended through public law [PL] 107-303). 27 November 2002. 1977.

DA Form 2028. *Recommended Changes to Publications and Blank Forms.*

DA Form 2823. *Sworn Statement.*

DA Form 3643. *Daily Issues of Petroleum Products.*

DA Form 3655. *Crime Lab Examination Request.*

DA Form 3744. *Affidavit Supporting Request for Authorization to Search and Seize or Apprehend.*

DA Form 3745. *Search and Seizure Authorization.*

DA Form 3881. *Rights Warning Procedure/Waiver Certificate.*

DA Form 4002. *Evidence/Property Tag.*

DA Form 4137. *Evidence/Property Custody Document.*

DA Pamphlet 27-1. *Treaties Governing Land Warfare.* 7 December 1956.

DA Pamphlet 710-2-1. *Using Unit Supply System.* 31 December 1997.

DD Form 281. *Voucher of Emergency or Extraordinary Expense Expenditures.*

DD Form 1131. *Cash Collection Voucher.*

DD Form 2701. *Initial Information for Victims and Witnesses of Crime.*

DD Form 2702. *Court-Martial Information for Victims and Witnesses of Crime.*

DD Form 2703. *Post-Trial Information for Victims and Witnesses of Crime.*

DD Form 2704. *Victim/Witness Certification and Election Concerning Inmate Status.*

DD Form 2705. *Victim/Witness Notification of Inmate Status.*

DD Form 2706. *Annual Report Victim and Witness Assistance.*

DOD 4160.21-M. *Defense Materiel Disposition Manual.* August 1997.

DODD 1030.1. *Victim and Witness Assistance.* 13 April 2004.

DODD 5500.7. *Standards of Conduct.* 30 August 1993.

DODD 5525.7. *Implementation of the Memorandum of Understanding Between the Department of Justice and the Department of Defense Relating to the Investigation and Prosecution of Certain Crimes.* 22 January 1985.

DODI 1030.2. *Victim and Witness Assistance Procedures.* 4 June 2004.

DODI 5505.8. *Investigations of Sexual Misconduct by the Defense Criminal Investigative Organizations and Other DoD Law Enforcement Organizations.* 6 June 2000.

EO 9397. *Numbering System for Federal Accounts Relating to Individual Persons.* 22 November 1943.

FBI Form FD 249. *Arrest and Institutional Fingerprint Card.*

Federal Facilities Compliance Act (PL 102-386). 6 October 1992.

Federal Rules of Evidence. 1 December 2002.

FM 10-67-1. *Concepts and Equipment of Petroleum Operations.* 2 April 1998.

FM 101-5. *Staff Organization and Operations.* 31 May 1997.

FM 27-10. *The Law of Land Warfare.* 18 July 1956.

FM 3-0 (100-5). *Operations.* 14 June 2001.

FM 3-25.26. *Map Reading and Land Navigation.* 20 July 2001.

Geneva Conventions.

Hague Conventions.

JP 1-02. *Department of Defense Dictionary of Military and Associated Terms.* 12 April 2001.

JP 1-06. *Joint Tactics, Techniques, and Procedures for Financial Management During Joint Operations.* 22 December 1999.

MCM. 2000.

MCM, Part III, *Military Rules of Evidence.* 2000.

MCM, Part III, *Military Rules of Evidence;* Rule 507, *Identity of Informant.* 2000.

MCM, Part III, *Military Rules of Evidence;* Rule 901, *Requirement of authentication or identification.* 2000.

MCM, Part III, *Military Rules of Evidence;* Rule 902, *Self-authentication.* 2000.

MCM, Part III, *Military Rules of Evidence;* Rule 903, *Subscribing witness' testimony unnecessary.* 2000.

Memorandum of Understanding Between the Department of Justice and Defense Relating to the Investigation and Prosecution of Certain Crimes. August 1984.

Physician's Desk Reference. 57th ed., Thompson PDR, Montvale, NJ, 2003.

Resource Conservation and Recovery Act (PL 94-580). 21 October 1976.

SF 700. Security Container Information.

Superfund Amendments and Reauthorization Act (PL 99-499). 17 October 1986.

UCMJ. 2000.

UCMJ; Article 31(b), Compulsory self-incrimination prohibited. 2000.

UCMJ; Article 32, Investigation. 2000.

UCMJ; Article 39a, Sessions. 2000.

UCMJ; Article 80, Attempts. 2000.

UCMJ; Article 112a, Wrongful use, possession, etc., of controlled substances. 2000.

UCMJ; Article 118, Murder. 2000.

UCMJ; Article 119, Manslaughter. 2000.

UCMJ; Article 120, Rape and carnal knowledge. 2000.

UCMJ; Article 121, Larceny and wrongful appropriation. 2000.

UCMJ; Article 122, Robbery. 2000.

UCMJ; Article 123, Forgery. 2000.

UCMJ; Article 123a, Making, drawing, or uttering check, draft, or order without sufficient funds. 2000.

UCMJ; Article 124, Maiming. 2000.

UCMJ; Article 125, Sodomy. 2000.

UCMJ; Article 126, Arson. 2000.

UCMJ; Article 127, Extortion. 2000.

UCMJ; Article 128, Assault. 2000.

UCMJ; Article 129, Burglary. 2000.

UCMJ; Article 130, Sodomy. 2000.

UCMJ; Article 134, General Article. 2000.

UCMJ; Article 136(b)4, Authority to administer oaths and to act as a notary. 2000.

USC, Title 5, Government Organization and Employees; Part I, The Agencies Generally; Section 303, Oaths to Witnesses.

USC, Title 5, Government Organization and Employees; Part III, Employees; Section 2951, Reports to the Office of Personnel Management.

USC, Title 10, Armed Forces.

USC, Title 10, Armed Forces; Section 301, Definitions.

USC, Title 10, Armed Forces; Subtitle B, Army; Part I, Organization; Section 3012 Department of the Army: seal.

USC, Title 18, Crimes and Criminal Procedure; Part I, Crimes; Section 2256, Sexual Exploitation and Other Abuse of Children.

USC, Title 18, Crimes and Criminal Procedure; Part IV, Correction of Youthful Offenders; Chapter 401, General Provisions.

USC, Title 18, Crimes and Criminal Procedure; Part IV, Correction of Youthful Offenders; Chapter 403, Juvenile Delinquency.

USC, Title 18, Crimes and Criminal Procedure; Section 1028, Identity Theft and Assumption Deterrence Act.

USC, Title 18, Crimes and Criminal Procedure; Part I, Crimes; Section 1385, Use of Army and Air Force as posse comitatus.

USC, Title 21, Food and Drugs; Chapter 13, Drug Abuse Prevention and Control; Subchapter I, Control and Enforcement; Part A, Introductory Provisions; Section 802(16), Definitions.

USC, Title 21, Food and Drugs; Chapter 13, Drug Abuse Prevention and Control; Subchapter I, Control and Enforcement; Part B, Authority to Control; Standards and Schedules; Section 812, Schedules of Controlled Substances.

USC, Title 32, National Guard.

USPS Publication 52. Hazardous, Restricted, and Perishable Mail. July 1999.

Index

V

W

Made in United States
Orlando, FL
05 November 2026

90229676R10350